7th RILEM International Conference on Cracking in Pavements

RILEM Bookseries

Volume 4

RILEM, The International Union of Laboratories and Experts in Construction Materials, Systems and Structures, founded in 1947, is a non-governmental scientific association whose goal is to contribute to progress in the construction sciences, techniques and industries, essentially by means of the communication it fosters between research and practice. RILEM's focus is on construction materials and their use in building and civil engineering structures, covering all phases of the building process from manufacture to use and recycling of materials. More information on RILEM and its previous publications can be found on www.RILEM.net.

For further volumes:
http://www.springer.com/series/8781

A. Scarpas · N. Kringos · I. Al-Qadi · A. Loizos
Editors

7th RILEM International Conference on Cracking in Pavements

Mechanisms, Modeling, Testing, Detection, Prevention and Case Histories
Volume 1

Editors

A.(Tom) Scarpas
Delft University of Technology
Delft
The Netherlands

Niki Kringos
KTH Royal Institute of Technology
Stockholm
Sweden

Imad L. Al-Qadi
University of Illinois at Urbana-Champaign
Urbana-Champaign
USA

Andreas Loizos
National Technical University of Athens
Athens
Greece

ISSN 2211-0844
ISBN 978-94-007-4565-0
Printed in 2 Volumes
DOI 10.1007/978-94-007-4566-7
Springer Dordrecht Heidelberg New York London

e-ISSN 2211-0852
e-ISBN 978-94-007-4566-7

Library of Congress Control Number: 2012937232

© RILEM 2012

This work is subject to copyright. All rights are reserved by the Publisher, whether the whole or part of the material is concerned, specifically the rights of translation, reprinting, reuse of illustrations, recitation, broadcasting, reproduction on microfilms or in any other physical way, and transmission or information storage and retrieval, electronic adaptation, computer software, or by similar or dissimilar methodology now known or hereafter developed. Exempted from this legal reservation are brief excerpts in connection with reviews or scholarly analysis or material supplied specifically for the purpose of being entered and executed on a computer system, for exclusive use by the purchaser of the work. Duplication of this publication or parts thereof is permitted only under the provisions of the Copyright Law of the Publisher's location, in its current version, and permission for use must always be obtained from Springer. Permissions for use may be obtained through RightsLink at the Copyright Clearance Center. Violations are liable to prosecution under the respective Copyright Law.

The use of general descriptive names, registered names, trademarks, service marks, etc. in this publication does not imply, even in the absence of a specific statement, that such names are exempt from the relevant protective laws and regulations and therefore free for general use.

While the advice and information in this book are believed to be true and accurate at the date of publication, neither the authors nor the editors nor the publisher can accept any legal responsibility for any errors or omissions that may be made. The publisher makes no warranty, express or implied, with respect to the material contained herein.

Printed on acid-free paper

Springer is part of Springer Science+Business Media (www.springer.com)

Preface

Because of vehicular and environmental loading, pavement systems have been deteriorating at a rapid rate. A series of six earlier RILEM Conferences on Cracking in Pavements in Liege (1989) (1993), Maastricht (1996), Ottawa (2000), Limoges (2004) and Chicago (2008) have clearly demonstrated that cracking constitutes one of the most detrimental, frequent and costly pavement deterioration modes.

Unfortunately, despite intense international efforts, there is still a strong need for methodologies that enable the construction and rehabilitation of crack resisting and/or tolerant pavements which at the same time are smooth, quiet, efficient, cost effective and environmentally friendly.

In the resent past, new materials, laboratory and in-situ testing methods and construction techniques have been introduced. In addition, modem computational techniques such as the finite element method enable the utilization of sophisticated constitutive models for realistic model-based predictions of the response of pavements. The **7th RILEM International Conference on Cracking in Pavements** aims to provide an international forum for the exchange of ideas, information and knowledge amongst experts involved in computational analysis, material production, experimental characterization, design and construction of pavements.

All submitted contributions were subjected to an exhaustive refereed peer review procedure by the Scientific Committee, the Editors and a large group of international experts on the topic. On the basis of their recommendations, 129 contributions which best suited the goals and the objectives of the Conference were chosen for presentation and inclusion in the Proceedings.

The strong message that emanates from the accepted contributions is that, by accounting for the idiosyncrasies of the response of pavement engineering materials, modern sophisticated constitutive models in combination with new experimental material characterization and construction techniques provide a powerful arsenal for understanding and designing against the mechanisms and the processes causing cracking and pavement response deterioration. As such they enable the adoption of truly "mechanistic" design methodologies.

The Editors would like to thank the Scientific Committee and the pavement engineering research community who took the responsibility of reviewing the manuscripts and ensuring the excellent quality of the accepted papers and the members of the Organizing Committee for their contribution to the management of the Conference affairs.

We hope that the Conference will contribute to the establishment of a new generation of asphalt and concrete pavement engineering design methodologies based on rational mechanics principles and in which computational techniques, advanced constitutive models and material characterisation techniques shall constitute the backbone of the design process.

Delft, March 2012

The Editors

A.(Tom) Scarpas
Delft University of Technology, The Netherlands

Niki Kringos
KTH Royal Institute of Technology, Sweden

Imad L. Al-Qadi
University of Illinois at Urbana-Champaign, USA

Andreas Loizos
National Technical University of Athens, Greece

Contents

Volume 1

Laboratory Evaluation of Asphalt Concrete Cracking Potential

Characterization of Asphalt Mixture's Fracture Resistance Using the Semi-Circular Bending (SCB) Test 1
L.N. Mohammad, M. Kim, M. Elseifi

The Flexural Strength of Asphalt Mixtures Using the Bending Beam Rheometer 11
M.I. Turos, A.C. Falchetto, G. Tebaldi, M.O. Marasteanu

Experimental Study of the Precracking 21
R. Mitiche_Kettab, A. Boulanouar, A. Bali

Comparison between 2PB and 4PB Methodologies Based on the Dissipated Energy Approach 31
M. Pettinari, C. Sangiorgi, F. Petretto, F. Picariello

Evaluation of Thermal Stresses in Asphalt Layers Incomparison with TSRST Test Results 41
M. Pszczoła, J. Judycki

A Four-Point Bending Test for the Bonding Evaluation of Composite Pavement 51
M. Hun, A. Chabot, F. Hammoum

Assessment of Cracking Resistance of Bituminous Mixtures by Means of Fenix Test 61
R. Miró, A. Martínez, F. Pérez-Jiménez, R. Botella, G. Valdés

Development of Dynamic Asphalt Stripping Machine for Better
Prediction of Moisture Damage on Porous Asphalt in the Field 71
M.O. Hamzah, M.R.M. Hasan, M.F.C. van de Ven, J.L.M. Voskuilen

Effect of Wheel Track Sample Geometry on Results 83
P.M. Muraya, C. Thodesen

Performance of 'SAMI'S in Simulative Testing 93
O.M. Ogundipe, N.H. Thom, A.C. Collop, J. Richardson

Towards a New Experimental and Numerical Protocol for
Determining Mastic Viscosity 103
E. Hesami, D. Jelagin, B. Birgisson, N. Kringos

Interference Factors on Tests of Asphalt Biding Agents Destinated to
Paving Works Using a Statistic Study 115
E.F. Amorim, A.C. de Lara Fortes, L.F.M. Ribeiro

Development of an Accelerated Weathering and Reflective Crack
Propagation Test Methodology 125
K. Grzybowski, G.M. Rowe, S. Prince

Pavement Cracking Detection

The Use of Ground Penetrating Radar, Thermal Camera and Laser
Scanner Technology in Asphalt Crack Detection and Diagnostics....... 137
T. Saarenketo, A. Matintupa, P. Varin

Asphalt Thermal Cracking Analyser (ATCA) 147
H. Bahia, H. Tabatabaee, R. Velasquez

Using 3D Laser Profiling Sensors for the Automated Measurement of
Road Surface Conditions 157
J. Laurent, J.F. Hébert, D. Lefebvre, Y. Savard

Pavement Crack Detection Using High-Resolution 3D Line Laser
Imaging Technology 169
Y. (James) Tsai, C. Jiang, Z. Wang

Detecting Unbounded Interface with Non Destructive Techniques 179
*J.-M. Simonin, C. Fauchard, P. Hornych, V. Guilbert, J.-P. Kerzého,
S. Trichet*

New Field Testing Procedure to Measure Surface Stresses in Plain
Concrete Pavements and Structures 191
D.I. Castaneda, D.A. Lange

**Strain Measurement in Pavements with a Fibre Optics Sensor
Enabled Geotextile** 201
O. Artières, M. Bacchi, P. Bianchini, P. Hornych, G. Dortland

Field Investigation of Pavement Cracking

**Evaluating the Low Temperature Resistance of the Asphalt Pavement
under the Climatic Conditions of Kazakhstan** 211
B. Teltayev, E. Kaganovich

**Millau Viaduct Response under Static and Moving Loads Considering
Viscous Bituminous Wearing Course Materials** 223
S. Pouget, C. Sauzéat, H. Di Benedetto, F. Olard

**Material Property Testing of Asphalt Binders Related to Thermal
Cracking in a Comparative Site Pavement Performance Study** 233
A.T. Pauli, M.J. Farrar, P.M. Harnsberger

**Influence of Differential Displacements of Airport Pavements on
Aircraft Fuelling Systems** 245
A.L. Rolim, L.A.C.M. Veloso, H.N.C. Souza, P.L. de O. Filho,
L.V. de A. Monteiro

**Rehabilitation of Cracking in Epoxy Asphalt Pavement on Steel
Bridge Decks** 255
L. Chen, Z. Qian

Long-Term Pavement Performance Evaluation 267
L. Petho, C. Toth

Structural Assessment of Cracked Flexible Pavement 277
L.W. Cheung, P.K. Kong, G.L.M. Leung, W.G. Wong

**Comparison between Optimum Tack Coat Application Rates as
Obtained from Tension- and Torsional Shear-Type Tests** 287
S. Hakimzadeh, N.A. Kebede, W.G. Buttlar

**Using Life Cycle Assessment to Optimize Pavement
Crack-Mitigation** 299
A.A. Butt, D. Jelagin, B. Birgisson, N. Kringos

**Preliminary Analysis of Quality-Related Specification Approach for
Cracking on Low Volume Hot Mix Asphalt Roads** 307
D.J. Mensching, L.M. McCarthy, J.R. Albert

**Evaluating Root Resistance of Asphaltic Pavement Focusing on
Woody Plants' Root Growth** 317
S. Ishihara, K. Tanaka, Y. Shinohara

20 Years of Research on Asphalt Reinforcement – Achievements and
Future Needs 327
A.H. De Bondt

Concrete Pavement Strength Investigations at the FAA National
Airport Pavement Test Facility 337
E.H. Guo, D.R. Brill, H. Yin

Pavement Cracking Modeling Response, Crack Analysis and Damage Prediction

The Effects Non-uniform Contact Pressure Distribution Has on
Surface Distress of Flexible Pavements Using a Finite Element
Method 347
D.B. Casey, A.C. Collop, G.D. Airey, J.R. Grenfell

Finite Element Analysis of a New Test Specimen for Investigating
Mixed Mode Cracks in Asphalt Overlays............................. 359
M.R.M. Aliha, M. Ameri, A. Mansourian, M.R. Ayatollahi

Modelling of the Initiation and Development of Transverse Cracks in
Jointed Plain Concrete Pavements for Dutch Conditions 369
M. Pradena, L. Houben

Pavement Response Excited by Road Unevennesses Using the
Boundary Element Method 379
A. Almeida, L.P. Santos

Discrete Particle Element Analysis of Aggregate Interaction in
Granular Mixes for Asphalt: Combined DEM and Experimental
Study 389
G. Dondi, A. Simone, V. Vignali, G. Manganelli

Recent Developments and Applications of Pavement Analysis Using
Nonlinear Damage (PANDA) Model 399
E. Masad, R.A. Al-Rub, D.N. Little

Laboratory and Computational Evaluation of Compact Tension
Fracture Test and Texas Overlay Tester for Asphalt Concrete 409
E.V. Dave, S. Ahmed, W.G. Buttlar

Crack Fundamental Element (CFE) for Multi-scale Crack
Classification 419
Y. Huang, Y. (James) Tsai

Cracking Models for Use in Pavement Maintenance Management 429
A. Ferreira, R. Micaelo, R. Souza

Multi-cracks Modeling in Reflective Cracking 441
J. Pais, M. Minhoto, S. Shatnawi

Using Black Space Diagrams to Predict Age-Induced Cracking 453
G. King, M. Anderson, D. Hanson, P. Blankenship

Top-Down Cracking Prediction Tool for Hot Mix Asphalt Pavements ... 465
C. Baek, S. Thirunavukkarasu, B.S. Underwood, M.N. Guddati, Y.R. Kim

A Theoretical Investigation into the 4 Point Bending Test 475
M. Huurman, R. Gelpke, M.M.J. Jacobs

Multiscale Micromechanical Lattice Modeling of Cracking in Asphalt Concrete 487
A.D. Banadaki, M.N. Guddati, Y.R. Kim, D.N. Little

Accelerated Pavement Performance Modeling Using Layered Viscoelastic Analysis 497
M. Eslaminia, S. Thirunavukkarasu, M.N. Guddati, Y.R. Kim

Numerical Investigations on the Deformation Behavior of Concrete Pavements 507
V. Malárics, H.S. Müller

Fatigue Behaviour Modelling in the Mechanistic Empirical Pavement Design 517
M.F. Saleh

Theoretical Analysis of Overlay Resisting Crack Propagation in Old Cement Concrete Pavement 527
Y. Zhong, Y. Gao, M. Li

Calibration of Asphalt Concrete Cracking Models for California Mechanistic-Empirical Design (CalME) 537
R. Wu, J. Harvey

Performance of Concrete Pavements and White Toppings

Shear Failure in Plain Concrete as Applied to Concrete Pavement Overlays 549
Y. Xu, J.N. Karadelis

Influence of Residual Stress on PCC Pavement Potential Cracking 561
X. Li, D. Feng, J. Chen

Plain Concrete Cyclic Crack Resistance Curves under Constant and Variable Amplitude Loading 571
N.A. Brake, K. Chatti

Influence of External Alkali Supply on Cracking in Concrete
Pavements 581
C. Sievering, R. Breitenbücher

Plastic Shrinkage Cracking Risk of Concrete – Evaluation of Test
Methods 591
P. Fontana, S. Pirskawetz, P. Lura

Compatibility between Base Concrete Made with Different Chemical
Admixtures and Surface Hardener 601
M.T. Pinheiro-Alves, A.R. Sequeira, M.J. Marques, A.B. Ribeiro

Compatibility between a Quartz Surface Hardener and Different Base
Concrete Mixtures... 607
M.T. Pinheiro-Alves, A. Fernandes, M.J. Marques, A.B. Ribeiro

Suitable Restrained Shrinkage Test for Fibre Reinforced Concrete:
A Critical Discussion... 615
A. Reggia, F. Minelli, G.A. Plizzari

Influence of Chemical Admixtures and Environmental Conditions on
Initial Hydration of Concrete 625
A.B. Ribeiro, V.A. Medina, A.M. Gomes

Application of Different Fibers to Reduce Plastic Shrinkage Cracking
of Concrete... 635
T. Rahmani, B. Kiani, M. Bakhshi, M. Shekarchizadeh

Volume 2

Fatigue Cracking and Damage Characterization of Asphalt Concrete

Evaluation of Fatigue Life Using Dissipated Energy Methods 643
C. Maggiore, J. Grenfell, G. Airey, A.C. Collop

Measurement and Prediction Model of the Fatigue Behavior of Glass
Fiber Reinforced Bituminous Mixture 653
I.M. Arsenie, C. Chazallon, A. Themeli, J.L. Duchez, D. Doligez

Fatigue Cracking in Bituminous Mixture Using Four Point Bending
Test 665
Q.T. Nguyen, H. Di Benedetto, C. Sauzéat

Top-Down and Bottom-Up Fatigue Cracking of Bituminous
Pavements Subjected to Tangential Moving Loads 675
Z. Ambassa, F. Allou, C. Petit, R.M. Eko

Fatigue Performance of Highly Modified Asphalt Mixtures in Laboratory and Field Environment 687
R. Kluttz, J.R. Willis, A.A.A. Molenaar, T. Scarpas, E. Scholten

A Multi-linear Fatigue Life Model of Flexible Pavements under Multiple Axle Loadings 697
F. Homsi, D. Bodin, D. Breysse, S. Yotte, J.M. Balay

Aggregate Base/Granular Subbase Quality Affecting Fatigue Cracking of Conventional Flexible Pavements in Minnesota 707
Y. Xiao, E. Tutumluer, J. Siekmeier

Fatigue Performance of Asphalt Concretes with RAP Aggregates and Steel Slags 719
M. Pasetto, N. Baldo

Fatigue Characterization of Asphalt Rubber Mixtures with Steel Slags 729
M. Pasetto, N. Baldo

Fatigue Cracking of Gravel Asphalt Concrete: Cumulative Damage Determination 739
F.P. Pramesti, A.A.A. Molenaar, M.F.C. van de Ven

Fatigue Resistance and Crack Propagation Evaluation of a Rubber-Modified Gap Graded Mixture in Sweden 751
W. Zeiada, M. Souliman, J. Stempihar, K.P. Biligiri, K. Kaloush, S. Said, H. Hakim

On the Fatigue Criterion for Calculating the Thickness of Asphalt Layers 761
M. Livneh

Acoustic Techniques for Fatigue Cracking Mechanisms Characterization in Hot Mix Asphalt (HMA) 771
M. Diakhaté, N. Larcher, M. Takarli, N. Angellier, C. Petit

Fatigue Characteristics of Sulphur Modified Asphalt Mixtures 783
A. Cocurullo, J. Grenfell, N.I.M. Yusoff, G. Airey

Effect of Moisture Conditioning on Fatigue Properties of Sulphur Modified Asphalt Mixtures 793
A. Cocurullo, J. Grenfell, N.I.M. Yusoff, G. Airey

Fatigue Investigation of Mastics and Bitumens Using Annular Shear Rheometer Prototype Equipped with Wave Propagation System 805
M. Buannic, H. Di Benedetto, C. Ruot, T. Gallet, C. Sauzéat

Effect of Steel Fibre Content on the Fatigue Behaviour of Steel Fibre
Reinforced Concrete 815
M.F. Saleh, T. Yeow, G. MacRae, A. Scott

Effect of Specimen Size on Fatigue Behavior of Asphalt Mixture in
Laboratory Fatigue Tests 827
N. Li, A.A.A. Molenaar, A.C. Pronk, M.F.C. van de Ven, S. Wu

Evaluation of the Effectiveness of Asphalt Concrete Modification

Long-Life Overlays by Use of Highly Modified Bituminous Mixtures ... 837
D. Simard, F. Olard

Investigation into Tensile Properties of Polymer Modified Bitumen
(PMB) and Mixture Performance 849
E.T. Hagos, M.F.C. van de Ven, G.M. Merine

Effect of Polymer Dispersion on the Rheology and Morphology of
Polymer Modified Bituminous Blend 859
I. Kamaruddin, N.Z. Habib, I.M. Tan, M. Komiyama, M. Napiah

Effect of Organoclay Modified Binders on Fatigue Performance 869
N. Tabatabaee, M.H. Shafiee

Effects of Polymer Modified Asphalt Emulsion (PMAE) on Pavement
Reflective Cracking Performance 879
Y. Chen, G. Tebaldi, R. Roque, G. Lopp

Characterization of Long Term Field Aging of Polymer Modified
Bitumen in Porous Asphalt 889
D. van Vliet, S. Erkens, G.A. Leegwater

Bending Beam Rheological Evaluation of Wax Modified Asphalt
Binders 901
G.L. Baumgardner, G.M. Rowe, G.H. Reinke

Reducing Asphalt's Low Temperature Cracking by Disturbing Its
Crystallization 911
E.H. Fini, M.J. Buehler

Mechanistic Evaluation of Lime-Modified Asphalt Concrete
Mixtures 921
A.H. Albayati

Crack Growth Parameters and Mechanisms

Determination of Crack Growth Parameters of Asphalt Mixtures 941
M.M.J. Jacobs, A.H. De Bondt, P.C. Hopman, R. Khedoe

Contents

Differential Thermal Contraction of Asphalt Components 953
I. Artamendi, B. Allen, C. Ward, P. Phillips

Mechanistic Pavement Design Considering Bottom-Up and Top-Down-Cracking 963
A. Walther, M. Wistuba

Strength and Fracture Properties of Aggregates 975
I. Artamendi, C. Ward, B. Allen, P. Phillips

Cracks Characteristics and Damage Mechanism of Asphalt Pavement with Semi-rigid Base 985
A. Sha, S. Tu

Comparing the Slope of Load/Displacement Fracture Curves of Asphalt Concrete 997
A.F. Braham, C.J. Mudford

Cracking Behaviour of Bitumen Stabilised Materials (BSMs): Is There Such a Thing? 1007
K. Jenkins

Experimental and Theoretical Investigation of Three Dimensional Strain Occurring Near the Surface in Asphalt Concrete Layers 1017
D. Grellet, G. Doré, J.P. Kerzrého, J.M. Piau, A. Chabot, P. Hornych

Reasons of Premature Cracking Pavement Deterioration – A Case Study 1029
D. Sybilski, W. Bańkowski, J. Sudyka, L. Krysiński

Effect of Thickness of a Sandwiched Layer of Bitumen between Two Aggregates on the Bond Strength: An Experimental Study 1039
S. Mondal, A. Das, A. Ghatak

Hypothesis of Existence Semicircular Shaped Cracks on Asphalt Pavements 1049
D. Hribar

Quantifying the Relationship between Mechanisms of Failure and the Deterioration of CRCP under APT: Cointegration of Non-stationary Time Series 1059
E.R. de Vos

Influence of Horizontal Traction on Top-Down Cracking in Asphalt Pavements 1069
C.S. Gideon, J.M. Krishnan

Evaluation, Quantification and Modeling of Asphalt Healing Properties

Predicting the Performance of the Induction Healing Porous Asphalt Test Section 1081
Q. Liu, E. Schlangen, M.F.C. van de Ven, G. van Bochove, J. van Montfort

Determining the Healing Potential of Asphalt Concrete Mixtures – A Pragmatic Approach 1091
S. Erkens, D. van Vliet, A. van Dommelen, G.A. Leegwater

Asphalt Durability and Self-healing Modelling with Discrete Particles Approach 1103
V. Magnanimo, H.L. ter Huerne, S. Luding

Quantifying Healing Based on Viscoelastic Continuum Damage Theory in Fine Aggregate Asphalt Specimen 1115
S. Palvadi, A. Bhasin, A. Motamed, D.N. Little

Evaluation of WMA Healing Properties Using Atomic Force Microscopy 1125
M. Nazzal, S. Kaya, L. Abu-Qtaish

Cracking and Healing Modelling of Asphalt Mixtures 1135
J. Qiu, M.F.C. van de Ven, E. Schlangen, S. Wu, A.A.A. Molenaar

Reinforcement and Interlayer Systems for Crack Mitigation

Effects of Glass Fiber/Grid Reinforcement on the Crack Growth Rate of an Asphalt Mix 1145
C.C. Zheng, A. Najd

Asphalt Rubber Interlayer Benefits in Minimizing Reflective Cracking of Overlays over Rigid Pavements 1157
S. Shatnawi, J. Pais, M. Minhoto

Performance of Anti-cracking Interface Systems on Overlaid Cement Concrete Slabs – Development of Laboratory Test to Simulate Slab Rocking 1169
K. Denolf, J. De Visscher, A. Vanelstraete

The Use of Bituminous Membranes and Geosynthetics in the Pavement Construction 1181
P. Hyzl, M. Varaus, D. Stehlik

Contents XVII

Stress Relief Asphalt Layer and Reinforcing Polyester Grid as
Anti-Reflective Cracking Composite Interlayer System in Pavement
Rehabilitation 1189
G. Montestruque, L. Bernucci, M. Fritzen, L.G. da Motta

Characterizing the Effects of Geosynthetics in Asphalt Pavements 1199
S. Vismara, A.A.A. Molenaar, M. Crispino, M.R. Poot

Geogrid Interlayer Performance in Pavements: Tensile-Bending Test
for Crack Propagation 1209
A. Millien, M.L. Dragomir, L. Wendling, C. Petit, M. Iliescu

Theoretical and Computational Analysis of Airport Flexible
Pavements Reinforced with Geogrids 1219
M. Buonsanti, G. Leonardi, F. Scopelliti

Optimization of Geocomposites for Double-Layered Bituminous
Systems... 1229
F. Canestrari, E. Pasquini, L. Belogi

Sand Mix Interlayer Retarding Reflective Cracking in Asphalt
Concrete Overlay 1241
J. Baek, I.L. Al-Qadi

Full Scale Tests on Grid Reinforced Flexible Pavements on the French
Fatigue Carrousel 1251
P. Hornych, J.P. Kerzrého, J. Sohm, A. Chabot, S. Trichet, J.L. Joutang,
N. Bastard

Thermal and Low Temperature Cracking of Pavements

Low-Temperature Cracking of Recycled Asphalt Mixtures 1261
N. Tapsoba, C. Sauzéat, H. Di Benedetto, H. Baaj, M. Ech

Thermal Cracking Potential in Asphalt Mixtures with High RAP
Contents 1271
Q. Aurangzeb, I.L. Al-Qadi, W.J. Pine, J.S. Trepanier, I.M. Abuawad

Micro-mechanical Investigation of Low Temperature Fatigue
Cracking Behaviour of Bitumen 1281
P.K. Das, D. Jelagin, B. Birgisson, N. Kringos

The Study on Evaluation Methods of Asphalt Mixture Low
Temperature Performance 1291
T. Yiqiu, Z. Lei, S. Liyan, J. Lun

Cracking Propensity of WMA and Recycled Asphalts

Permanent Deformations of WMAs Related to the Bituminous Binder Temperature Susceptibility 1301
F. Petretto, M. Pettinari, C. Sangiorgi, A. Simone

Cracking Resistance of Recycled Asphalt Mixtures in Relation to Blending of RA and Virgin Binder 1311
M. Mohajeri, A.A.A. Molenaar, M.F.C. Van de Ven

Warm Mix Asphalt Performance Modeling Using the Mechanistic-Empirical Pavement Design Guide 1323
A. Buss, R.C. Williams

Shrinkage and Creep Performance of Recycled Aggregate Concrete 1333
J. Henschen, A. Teramoto, D.A. Lange

Effect of Reheating Plant Warm SMA on Its Fracture Potential 1341
Z. Leng, I.L. Al-Qadi, J. Baek, M. Doyen, H. Wang, S. Gillen

Fatigue Cracking Characteristics of Cold In-Place Recycled Pavements 1351
A. Loizos, V. Papavasiliou, C. Plati

Author Index 1361

RILEM Publications 1367

RILEM Publications Published by Springer 1377

Characterization of Asphalt Mixture's Fracture Resistance Using the Semi-Circular Bending (SCB) Test

Louay N. Mohammad[1,*], Minkyum Kim[2], and Mostafa Elseifi[3]

[1] Professor, Dept. of Civil and Environmental Engineering and Louisiana Transportation Research Center, Louisiana State University, 4101 Gourrier Ave, Baton Rouge, LA 70808, U.S.A
Louaym@Lsu.edu
[2] Research Associate, Louisiana Transportation Research Center, 4101 Gourrier Ave, Baton Rouge, LA 70808, USA
[3] Assistant Professor, Dept. of Civil and Environmental Engineering, Louisiana State University, 4101 Gourrier Ave, Baton Rouge, LA 70808, U.S.A.

Abstract. Pavement cracking is a major distress mode in asphalt pavements. The fracture resistance is an important factor that relates to pavement cracking. This paper describes the evaluation of the fracture resistance of asphalt mixtures using the semi-circular bending (SCB) test. The mechanism of the SCB test is based on the elastic-plastic fracture mechanics concept that leads to the laboratory determination of the critical strain energy release rate, also called the critical value of J-integral (J_c). Asphalt mixtures from nine rehabilitation field projects throughout the state of Louisiana were evaluated in this study. The critical strain energy release rate of plant produced-laboratory compacted (PL) asphalt mixtures was evaluated using a three-point SCB test. In addition, field cracking measurements for those projects that have been trafficked for approximately ten years were performed. Four types of asphalt binders and two nominal maximum aggregate sizes were included in those mixtures. Analysis of the results indicated that there is a good correlation between the critical values of J-integral (J_c) and the field cracking rate. The J_c value increased as the cracking rate decreased. Results of this studysupport to use the semi-circular bend test to evaluate the fracture resistance of asphalt mixtures.

1 Introduction

Fatigue cracking is a major fracture failure mode in flexible pavements, which is caused by repeated traffic loadings over time. The occurrence of fatigue cracking is dependent upon the structural capacity of the pavement, quality of construction, asphalt mixtures' fracture resistance, etc. With respect to materials, the fracture

---

[*] Corresponding author.

resistance of asphalt mixture can be experimentally evaluated using several laboratory fracture tests. To name few, these test methods include conventional repeated loading beam fatigue tests and surrogate monotonic fracture tests such as the indirect tensile (IDT) strength and the semi-circular bending (SCB) test. While the conventional repeated loading beam fatigue tests measure either the number of load repetitions to a certain level of damage accumulation [1, 2] or the rate of energy dissipation over a long testing time [3] as the index of fatigue resistance, monotonic fracture tests measure either toughness [4] or the critical strain energy release rate [5] during a single displacement controlled loading until failure as the index of fracture resistance. In recent years, the SCB test method has been investigated vigorously by many researchers [6-12] due to its simplicity in specimen preparations and less time-consuming test procedure.

Researchers at the Louisiana Transportation Research Center (LTRC) investigated the SCB test device as a candidate test method for characterizing the fracture resistance of asphalt mixtures [5]. In the study, 13 plant mixed-laboratory compacted (PL) Superpave asphalt mixtures were measured for their critical strain energy release rate, i.e., critical value of J-integral (J_c). These mixtures were collected from field projects across the state of Louisiana based upon a test factorial of four different grades of asphalt binders (AC-30, PAC-40, PG70-22m, and PG76-22m), two nominal maximum aggregate sizes (NMAS, 19- and 25-mm), and fourgyratory compaction levels (75, 96, 109, and 125 gyrations). It was found that the J_c values of those various asphalt mixtures were sensitive to the changes in the mixture design parameters such as the asphalt binder grades and the NMAS. Based upon the observed reasonable sensitivity of J_c values, the study concluded that the SCB test can be a valuable tool to evaluate the fracture resistance of asphalt mixtures as an indicator of fatigue cracking performance in asphalt pavements.

In the past 10 years, actual in-field cracking performances of these Superpave mixtures have been regularly monitored by the Louisiana Department of Transportation and Development (LADOTD) through its Pavement Management System (PMS). LADOTD collects roughness, rutting, cracking, patching, and faulting data every two years to analyze the deterioration rates of its pavement network [13]. A comparison analysis between the laboratory fracture resistance of the aforementioned asphalt mixtures measured by the SCB test and the field cracking performance of the corresponding pavements obtained from the PMS databaseis conducted in this study.

2 Objective and Scope

The objective of this study was to evaluate the effectiveness of the SCB test method for predicting fatigue cracking performance of asphalt pavements by analyzing the relationship between SCB measured J_c values of hot-mix asphalt (HMA) and the field performance of asphalt pavements. In this study, nine field projects, which were originally included in the earlier LTRC investigation [5], were revisited. The J_c values of plant mixed-laboratory compacted (PL) asphalt mixtures from these nine field projects, at the time of constructions, were measured by the SCB device at an intermediate temperature of 25°C. Corresponding cracking data of these field projects were retrieved from the Louisiana PMS database. A regression analysis was performed to evaluate relationship between the SCB and field performance against cracking.

2.1 Field Projects and Asphalt Mixtures

Table 1 presents a description of the nine rehabilitation projects as well as the characteristics of the asphalt mixtures installed in these projects. As shown in this table, four different grades of asphalt binders (AC-30, PAC-40, PG70-22m, and PG76-22m) and two nominal maximum aggregate sizes (NMAS, 19- and 25-mm) were used in these projects.

Table 1. Field Projects and Asphalt Mixtures Studied

Field Project	Construction Year	Design Traffic Level[1]	Asphalt Grade	NMAS (mm)	Mineral Composition
LA874	1999	Level 1	PG70-22m	19	Limestone
LA361	2000	Level 1	PG70-22m	19	Granite
LA121	1999	Level 1	AC-30	19	Limestone
LA22	1998	Level 1	AC-30	19	Limestone
LA4	1998	Level 1	AC-30	25	Limestone
US90	1999	Level 2	PAC-40	25	Limestone
US61	1999	Level 2	PAC-40	25	Limestone
I12	2000	Level 3	PG76-22m	19	Limestone
I49	2001	Level 3	PG76-22m	19	Limestone

[1] Level 1: <3 million ESAL, Level 2: 3-30 million ESAL, Level 3: >30 million ESAL.

2.2 Experimental Method: Semi-Circular Bending (SCB) Test

The SCB test is performed to characterize the fracture resistance of asphalt mixtures in terms of the critical strain energy release rate or the critical value of J-integral (J_c). The J_c is a function of the rate of strain energy change per notch depths (dU/da), as shown in Equation (1), and represents the consumed strain energy while a unit area of fractured surface is formed in a mixture. Therefore, the higher J_c values indicate that a material is tougher to resist cracking and crack propagation.

To determine J_c, specimens with three different notch depths were tested. *Figure 1* shows the 3-point bending test configuration and typical specimen dimensions of the SCB test. The SCB test is conducted according to the test procedure adopted by Mohammad et al. [6]. Three replicate specimens are tested for each notch depth at 25°C. The three notch depths typically selected are 25.4-mm, 31.8-mm, and 38.0-mm based on an 'a/r_d' ratio (the ratio of notch depth to the radius of the specimen, as shown in *Figure 1*), which is desirable to range from 0.5 to 0.75. Although the rate of strain energy change per notch depths (dU/da) can be calculated with only two different notch depths, having three notch depths increases the accuracy of J_c calculation. Applying a constant cross-head deformation rate of 0.5 mm/min, the SCB specimens are loaded monotonically on an MTS machine until fracture failure occurs. The load and deformation data are recorded continuously and used to generate a series of load versus deformation curves (*Figure 2*), from which the critical value of J_c is determined using Eqn. (1):

$$J_c = -\left(\frac{1}{b}\right)\frac{dU}{da} \qquad (1)$$

where,
  J_c = critical strain energy release rate (kJ/m^2);
  b = sample thickness (m);
  a = notch depth (m); and
  U = strain energy to failure (kN-m or kJ).

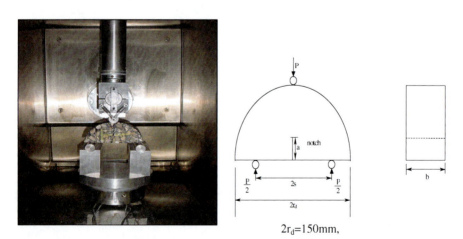

$2r_d = 150mm$,

Fig. 1. SCB Test Setup and Specimen Configuration

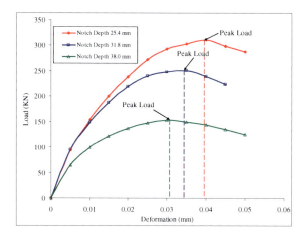

Fig. 2. Typical Load Deformation Curves from the SCB Test

Figure 2 presents a typical load-deformation plot obtained from the SCB test. Three areas under the load-deformation curves, until peak loads reached, are calculated to represent the strain energy values to failure (U) of specimens with three different notch depths. The average values of U (calculated from three replicate results) are then plotted against the different notch depths to compute a slope of a linear regression line, which is dU/da in Equation (1). The critical value of facture resistance, J_c, is then computed by dividing dU/da by the width, b, of the specimens.

2.3 Field Measurement of Cracking

Table 1 summarizes the cracking survey data of the nine field projects obtained from the Louisiana PMS database. LADOTD pavement distress data collection system currently uses the Automated Road Analyzer (ARAN) system, and the pavement network is surveyed once every 2 years [13]. Field cracking datacollected during the 2009 surveycycle were used in this study. Four types of cracking patterns are collected separately by the ARAN system, which are 'Transverse,' 'Longitudinal,' 'Alligator,' and 'Random' crackings at three severity levels of low, medium, and high. Longitudinal and random cracking patterns were not considered in this analysis as these distressesmay not be closely related to HMA fatigue resistance. For example, longitudinal cracking patterns typically form at the construction joints and many discontinuities on the pavement surface such as patching are counted as random cracking. Therefore, these two crack patterns were excluded from further analysis and only the 'Transverse' and 'Alligator' crackings were considered.

In the ARAN system, transverse cracking is reported in linear feet per 0.1 mile segment (ft/0.1-mile) and the alligator cracking is reported in square feet per 0.1

mile segment (ft²/0.1-mile). Therefore, each project consisted of different numbers of unit segmentswhere three different levels of transverse and alligator cracking were reported. For example, PMS data of LA874 pavement were collected for the 3-mile long stretch of the project, which included cracking counts in 30 individual segments. Each individual segment contained low, medium, and high severity transverse cracking counts and alligator cracking counts as well. These raw data were first averaged throughout the entire project extent using Eqn. (2) and were multiplied by 10 to represent cracking counts in 1-mile unit length:

$$C_{\_avg} = \frac{1}{n}\left[\sum_{i=1}^{n}\{(C_L)_i + (C_M)_i + (C_H)_i\}\right] \times 10 \qquad (2)$$

where,
C_avg = average crack count in a project for either transverse (TC) or alligator cracking (AC)
C_L, C_M, C_H = Crackings in Low, Medium, and High level of severity
n = number of 0.1-mile unit segments in a project

No weight factors for different severity of cracking were considered when calculating the average cracking counts of both transverse and alligator cracks. To combine transverse and alligator cracking, which have different reporting units, into a single composite cracking index (C_{comp}), it was necessary to convert the unit of the alligator cracking (ft²/mile) into the same linear-feet scale by taking the square root. The average transverse cracking and the square root of the average alligator cracking were then summed up as shown in Eqn. (3):

$$C_{comp} = TC_{\_avg} + \sqrt{AC_{\_avg}} \qquad (3)$$

where, C_{comp} = Composite cracking index, ft/mile

Hence, the composite cracking index (C_{comp}) defined in Eqn. (3) represents a simple indicator of cracking performance in thefield without considering crack severity, total pavement structural, climatic conditions, and applied traffic loading.

2.4 Discussions of Results

Figure 3 presents the SCB measured critical strain energy release rate (J_c) for the nine asphalt mixtures. The value ranged from 0.74 to 1.57 kJ/m² with the average of 1.05 and the median of 0.96 kJ/m². The wearing course asphalt mixture in LA4 project showed the highest J_c value, while the wearing course mixture in LA874 showed the lowest J_c value. Therefore, based solely on the measured J_c values, it was expected that the field cracking performance of the LA4 pavement would be better than that of the LA874 pavement. Likewise, I12 pavement is expected to have superior cracking resistance than LA361, US61, or I49 pavements based on SCB test results.

Table 2. PMS Field Cracking Data

Project	Survey Year	Project Length (miles)	Average Cracking		
			TC_avg (ft/mile)	AC_avg (ft²/mile)	C_{comp} (ft/mile)
LA874	2008	3.00	377.7	225.3	392.7
LA361	2008	6.93	4071.2	3309.6	4128.7
LA121	2008	9.90	358.9	1631.4	399.3
LA22	2008	7.14	1374.0	985.5	1405.4
LA4	2009	6.03	4.8	2.5	6.4
US90	2008	1.30	1473.8	0.0	1473.8
US61	2008	7.51	802.8	12.7	806.3
I12	2008	1.88	276.3	0.0	276.3
I49	2008	12.33	620.1	68.7	628.4

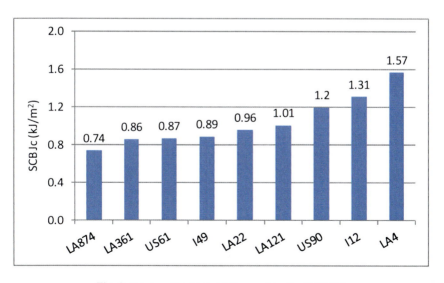

Fig. 3. Fracture Resistance Measured by the SCB (J_c)

Table 2 summarizes the comparison between field crack counts and J_c values. When comparing the J_c values of asphalt mixtures and the cracking resistance of pavements, it is desirable to take into account other influencing factors such as the entire pavement structure, climatic effects, and applied traffic loadings for more accurate comparisons. In Figure 3 the composite cracking index (C_{comp}) was normalized by the applied Equivalent Single Axle Load (ESAL), so that the cracking performance of pavements can be expressed per every million ESALs applied. The 'applied ESALs' in these pavements were assumed as the fraction of the pavement ages at the time of the last field performance survey to the 20-year design life multiplied by the 'design ESAL.' Structural aspects of each and every field projects were not considered due to the lack of accurate information regarding the underlying existing pavement structures. It is noted, however, that the thickness of the asphalt overlays including both wearing and binder course lifts of eight field projects were similar around 89 ~ 102 mm (3.5 ~ 4.0 inches). The only exception was I12 pavement, which had 102 mm wearing and 138 mm binder course lifts. The climatic effects were not considered since the nine projects were not geographically far apart from one another.

Table 3. Comparison between Cracking Rate and J_c

Project	Design ESAL (x10^6)	Years in Service	Applied ESAL (x10^6)	C_{comp} (ft/mile)	Cracking Rate per Million ESAL (ft/mile/million)	SCB Jc (kJ/m^2)
LA874	3	9	1.35	425.1	314.9	0.74
LA361	3	8	1.2	4253.1	3,544.3	0.86
LA121	6	9	2.7	486.6	180.2	1.01
LA22	6	10	3	1473.3	491.1	0.96
LA4	6	11	3.3	9.8	3.0	1.57
US90	10	9	4.5	1473.8	327.5	1.2
US61	10	9	4.5	814.0	180.9	0.87
I12	30	8	12	276.3	23.0	1.31
I49	30	7	10.5	646.4	61.6	0.89

As shown in Table 3, the LA4 project, which showed the highest J_c value, developed the least amount of cracking in the field. Moreover, the cracking rate per million ESALs was considerably lower than the other eight pavements. The I12 project, which had the thickest asphalt overlay, was the second best pavement in terms of the amount of crack and cracking rate. The LA874 project that showed the lowest J_c value, on the other hand, appeared to have much more crack counts and higher cracking rate than the two best performing pavements. Nevertheless, the LA874 project maintained decent field cracking performance compared to other pavements (e.g., US90, LA22, and LA361), while the second worst pavement in terms of J_c value (LA361) showed the worst field cracking performance among the nine projects.

Figure 4 depicts the correlation between the cracking rates and the J_c values. The solid line through the measured data points is an exponential regression model on a semi-log plot of cracking rate versus the SCB J_c. A strong downward trend of the regression line was observed as the J_c values increase, which indicates that the cracking rate of pavements decreases as the fracture resistance of asphalt mixtures (J_c values) increases. The coefficient of determination (R^2) of the regression was 0.58, which means that approximately 58% of the variability in the cracking rates observed from those nine field projects can be explained by the sole independent variable, SCB J_c value. It is noted that other influencing factors on the cracking performance of asphalt pavements, such as the structural and environmental aspects, were not taken into account. These results support the suitability of the SCB test method for estimating the fatigue cracking performance of asphalt pavements in the field.

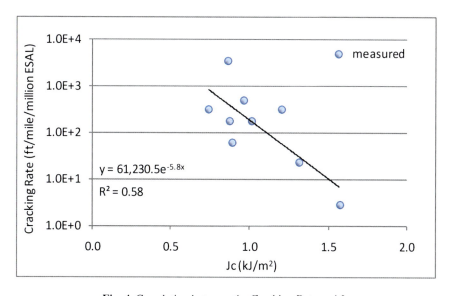

Fig. 4. Correlation between the Cracking Rate and J_c

3   Summary and Conclusions

In this study, laboratory measured fracture resistance of nine asphalt mixtures used in Louisiana was compared with field cracking performance data obtained from the PMS database. The semi-circular bending (SCB) test method was utilized in the laboratory to measure the critical fracture energy release rate (J_c) of asphalt mixtures at intermediate service temperature of 25°C. Field cracking performance data were measured by the automated road analyzer (ARAN) system. Transverse and alligator crack counts in the field were combined into a single cracking index that was normalized with respect to million ESALs.It was found that the SCB measured J_c values demonstrated a good correlation with field cracking performance data. This observation demonstrates that the SCB test is a suitable test method for estimating the fatigue cracking performance of asphalt pavements at intermediate service temperatures.

Further research is underway to extend the scope of comparison between laboratory-measured intermediate fracture properties and actual field cracking performance with the ultimate objective of developing performance-based specification criteria for fatigue cracking.

Acknowledgements. This work was supported by the Louisiana Transportation Research Center (LTRC) in cooperation with the Louisiana Department of Transportation and Development (LADOTD). The contribution of staffs in the asphalt laboratory and the Engineering Material Characterization and Research Facility (EMCRF) to this project is acknowledged.

References

1. Tayebali, A.A., Deacon, J.A., Monismith, C.L.: Trans. Res. Rec. 1545, 89 (1996)
2. Tsai, B.-W., Harvey, J.T., Monismith, C.L.: J. Assoc. Asph. Pav. Technol. 73, 623 (2004)
3. Carpenter, S.H., Ghuzlan, K.A., Shen, S.: Trans. Res. Rec. 1832, 131 (2002)
4. Zhang, Z., Roque, R., Birgisson, B., Sangpetngam, B.: J. Assoc. Asph. Pav. Technol. 70, 206 (2001)
5. Wu, Z., Mohammad, L.N., Wang, L.B., Mull, M.A.: J. ASTM Int. 2(3), 127 (2005)
6. Mohammad, L.N., Wu, Z., Aglan, M.: In: Fifth International RILEM Conference on Reflective Cracking in Pavements, Limoges, France, p. 375 (2004)
7. Mull, M.A., Stuart, K., Yehia, A.: J. Mater. Sci. 37, 537 (2002)
8. Molenaar, A.A.A., Scarpas, A., Liu, X., Erkens, S.M.J.G.: J. Assoc. Asph. Pav. Technol. 71, 794 (2002)
9. Mohammad, L.N., Kabir, M.D., Saadeh, S.: In: Proceedings of the 6th RILEM International Conference on Cracking in Pavements, Chicago, IL, USA, p. 427 (2008)
10. Mull, A.M., Othman, A., Mohammad, L.: In: Proceedings CD-Rom of TRB 85th Annual Meeting, Washington, DC, USA (2006)
11. Birgisson, B., Montepara, A., Napier, J., Romeo, E., Roncella, R., Tebaldi, G.: Trans. Res. Rec. 1970, 186 (2006)
12. Shu, X., Huang, B., Vukosavljevic, D.: Constr. Building Mater. 22, 1323 (2008)
13. Khattak, M.J., Baladi, G.Y., Zhang, Z., Ismail, S.: Trans. Res. Rec. 2084, 18 (2008)
14. Louisiana Standard Specifications for Roads and Bridges, State of Louisiana Department of Transportation and Development (2000)

The Flexural Strength of Asphalt Mixtures Using the Bending Beam Rheometer

Mugurel I. Turos[1], Augusto Cannone Falchetto[2], Gabriele Tebaldi[3], and Mihai O. Marasteanu[4]

[1] Scientist, University of Minnesota, USA
[2] Ph.D. Candidate, University of Minnesota, USA
[3] Assistant Professor, University of Parma, Italy
[4] Associate Professor, University of Minnesota, USA

Abstract. Asphalt mixture creep stiffness and strength are needed in the low temperature algorithm of the AASHTO Mechanistic Empirical Pavement Design Guide to predict low temperature performance. A procedure for obtaining creep stiffness by testing thin mixture beams with a Bending Beam Rheometer was previously developed at University of Minnesota. Preliminary work investigating the possibility of also obtaining bending strength by testing thin mixtures beams is presented in this paper.

Indirect Tensile (IDT), and Direct Tension (DT) strength tests are performed on eleven mixtures. The same eleven mixtures are used to perform three sets of tests using the proposed method named Bending Beam Strength (BBS). First set is performed to investigate the reliability and reproducibility of BBS testing method, and the validity of the measuring concept. Weibull modulus is calculated as part of the analysis. Second set of tests is done to investigate the effect of temperature, conditioning time and loading rate on the measured strength of three mixtures. Third set consists of tests performed at three different temperatures on eight mixtures. IDT, DT and BBS experimental determined strengths are first compared without using any transformation, and results are found to be statistically different. The results are then transformed to take into account the size of the samples and the testing configuration. The statistical analysis indicates that BBS strength values are similar to the values obtained with IDT test method.

1 Introduction

Thermal cracking is the main type of distress for asphalt pavements built in cold climates. This type of failure is the result of large tensile stresses caused by severe temperature drops combined with embrittlement of the asphalt mixture at low temperatures. Current specifications used to characterize and select asphalt binder and mixture for pavement applications at low temperatures are based on binder Bending Beam Rheometer (BBR) [1] and mixture Indirect Tensile Test (IDT) [2]. Recently, a simpler test method was proposed to determine the low temperature creep compliance of asphalt mixtures with BBR [3]. In the same research effort it was also shown, through microstructural analysis, that BBR specimens were representative of the entire asphalt mixture matrix for the materials analysed [3].

This paper presents the exploratory research conducted to investigate the possibility of expanding the previous test method [3] to determine the strength properties of asphalt mixtures. An extensive experimental work is first conducted. IDT [2], DT [4] and the proposed test, named Bending Beam Strength (BBS), are performed on eleven asphalt mixtures. Critical issues, such as statistical distribution, temperature effect, loading rate, are first considered. Then size effect, geometry and testing configuration differences among IDT, DT and BBS are investigated through weakest link theory [5].

2 Materials

Eleven asphalt mixtures (Table 1) produced using different asphalt binders, aggregates, and containing various amounts of reclaimed asphalt pavement (RAP) and roofing shingles were used in the experimental work. These materials were used in the 2008 reconstruction of test cells at MnROAD facility [6]. Loose mix sampled during construction was used to prepare all mixture specimens tested in the laboratory. An additional asphalt mixture, labelled W, prepared with a PG 58-28 asphalt binder and granite aggregates was used to investigate the statistical failure distribution of the BBS test. All mixtures were Superpave gyratory compacted to 7% air voids. The nominal maximum aggregate size (NMAS) was between 4.75mm and 12.5mm.

Table 1. Asphalt mixtures

ID	Description	Binder PG	Content	RAP
A	Novachip	70-28	5.1%	none
B	level 4 Superpave	64-34	5.4%	none
D	4.75 taconite Superpave	64-34	7.4%	none
E	WMA* wear course	58-34	5.2%	up to 20%
F	WMA* non wear	58-34	5.5%	up to 20%
G	non wear	58-28	5.2%	30% non fractioned
H	non wear	58-28	5.5%	30% fractioned
I	non wear	58-34	5.5%	30% fractioned
K	shoulder mix	58-28	4.8%	none, 5% manufactured waste shingles
L	control for WMA*	58-34	5.2%	up to 20 %
M	shoulder mix	58-28	5.0%	none, 5% manufactured tear off shingles

WMA= Warm Mix Asphalt.

3 Strength Test Methods

3.1 Indirect Tensile (IDT) Test

IDT strength test [2] is performed on asphalt mixture cylindrical specimens that are 40mm thick, 150mm in diameter, and have a volume of 706cm^3. Loading is diametrically applied at a constant displacement rate of 12.5 mm/min until failure occurs. Tensile strength σ_{IDT} is calculated according to:

$$\sigma_{IDT} = (2P/\pi \cdot bD) \tag{1}$$

where P is the failure load at which the difference between vertical and horizontal deformation is maximum and b and D are the thickness and the diameter of the specimen respectively. To avoid damage to the LVDT's, P is usually assumed as the peak load; for this reason a correction formula was proposed in the past [7] to evaluate the tensile nominal strength σ_{IDT}^{U} obtained with this method:

$$\sigma_{IDT}^{U} = (0.78\sigma_{IDT}) + 38 \tag{2}$$

IDT strength test was performed on all eleven mixtures at three different temperatures related to the asphalt binder grade used to prepare the mixtures: PG low limit + 22°C, PG low limit + 10°C, and PG low limit - 2°C. Specimens were conditioned for 2 hours at the testing temperature and three replicates were tested at each temperature.

3.2 Direct Tension Test (DT)

DT tests were performed on asphalt mixture cylindrical specimens 150mm long, with a 50.8 mm diameter, and a volume of 304cm^3. A modified Thermal Stress Restrained Specimen Test (TSRST) [8] device was used for testing DT specimens in order to have an upper plate displacement rate comparable with the strain rate in the BBS. Due to limited availability of material only two replicates of mixtures D, E and K were tested at a temperature corresponding to PG low limit - 2°C after 2 hours conditioning in the climatic chamber

3.3 Bending Beam Strength (BBS)

The flexural beam test presents similarities to the real field loading of pavements. In this test, a simple supported beam is loaded with either a concentrated force applied at the midpoint or with two equal concentrated forces applied at the two third points of the beam. The testing method has been used to predict the fatigue resistance of asphalt mixtures [9] and to determine the flexural creep stiffness of asphalt binders [1] and asphalt mixtures [3]. For a concentrated load applied at the midpoint of a beam the nominal flexural strength σ_{3PB} is calculated as:

$$\sigma_{3PB} = (3PL/2bh^2) \tag{3}$$

where P is the peak load, L is the beam span, b is the width of the beam and h is the thickness. Previously, BBR was used to obtain asphalt mixture creep stiffness

[3]. BBR was slightly modified to also perform three point bending strength tests on thin asphalt mixture beams with $L=101.6$mm, $b=12.5$mm, $h=6.25$mm and volume of 9.7cm^3; details on specimen preparation can be found elsewhere [3]. An upgraded 45N load cell and an in-house system consisting of a small centrifugal pump that is used to fill with water, at a constant rate, a plastic box attached to the loading shelf of the BBR, where used to load the mixture beams at constant loading rates. This prototype device (Figure 1) is referred to as Bending Beam Strength (BBS) in this paper.

Fig. 1. Bending Beam Strength (BBS) apparatus

Three sets of tests were performed. First, the statistical distribution of BBS test was investigated through histogram strength testing on asphalt mixture. Then, the temperature, conditioning time, and loading rate effects on the measured strength of three of the eleven mixtures were evaluated. Finally, BBS tests were performed on the other eight mixtures at the same three temperatures used for IDT tests.

3.3.1 BBS Strength Statistical Distribution

A preliminary evaluation of the BBS results was done through histogram testing. Twenty six BBS beam replicates cut from a gyratory compacted cylinder made of mix W (PG58-28) were tested at -24°C and using a loading rate of 3N/min. Loading rate was selected in such a way that BBS test had a duration comparable to BBR creep test to take advantage of the current BBR software capabilities. Conditioning time was set to 1h because of the reduced dimension of the BBS specimen [3]. Strength results did not follow a normal distribution and lognormal distribution was disregarded as suggested by other authors [10]. Weibull statistical distribution was then selected and strength data was plotted in the Weibull plane, where the failure probability P_f is plotted against the natural logarithm of the nominal flexural strength (Figure 2).

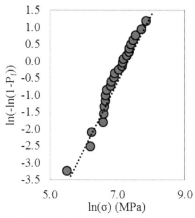

Fig. 2. Bending Beam Strength (BBS) histogram, mixture W

The experimental results form an almost straight line and appear to follow a Weibull statistical distribution. From the histogram, the Weibull modulus m, representing the slope of the line fitting the experimental data was calculated and a value of 11.59 was obtained. Since m has to be an integer number [10], it was rounded to 12. This value is similar with previous results obtained by other authors for cement concrete [11]. More recent studies in concrete found $m = 24$ [12].

3.3.2 Effect of Temperature, Conditioning Time, and Loading Rate

The effect of temperature, conditioning time and loading rate, on the measured strength, was investigated on mixtures D, E and K (Table 1). A 2^3 factorial design analysis was run to study the individual and the joint effect of these factors on BBS fracture strength. Table 2 presents the factorial design with statistical factors and levels.

Table 2. Experimental factorial design for BBS

ID	Temperature (°C)	Conditioning time (hour)	Loading rate (N/min)	Number of replicates
D	-24; -36	1; 24	3; 13	3 per factor/level
E	-24; -36	1; 24	3; 13	3 per factor/level
K	-18; -30	1; 24	3; 13	3 per factor/level

Testing temperatures were selected based on binder PG grade: PG+10°C and PG-2°C [3]. A higher loading rate of 13N/min was selected based on BBS device and centrifugal pump capabilities. The mean nominal flexural strength for mixture D, E and K are shown in Table 3 for all factor level combinations. The p-values for the statistical analysis of the 2^3-factorial design are summarized in Table 4; a p-value below the set significance level (0.05) provides evidence of the influence of the specific factor on the response. The analysis was conducted separately for each mixture, since differences due to mix design were expected.

Table 3. BBS nominal strength results for mixtures D, E and K

Mix D PG 64-34								
Conditioning Time (h)	1	24	1	24	1	24	1	24
Loading Rate (N/min)	3	3	13	13	3	3	13	13
Temperature (°C)	-24	-24	-24	-24	-36	-36	-36	-36
Mean Nominal Strength (MPa)	10.0	9.6	10.3	9.6	7.2	9.0	10.4	9.4
Mix E PG 58-34								
Conditioning Time (h)	1	24	1	24	1	24	1	24
Loading Rate (N/min)	3	3	13	13	3	3	13	13
Temperature (°C)	-24	-24	-24	-24	-36	-36	-36	-36
Mean Nominal Strength (MPa)	7.5	7.1	7.0	6.6	4.7	5.0	5.2	5.8
Mix K PG 58-28								
Conditioning Time (h)	1	24	1	24	1	24	1	24
Loading Rate (N/min)	3	3	13	13	3	3	13	13
Temperature (°C)	-18	-18	-18	-18	-30	-30	-30	-30
Mean Nominal Strength (MPa)	6.3	5.7	5.9	6.0	4.8	5.9	6.2	5.3

Table 4. p-value for the factorial analysis

ID	Conditioning Time	Loading Rate	Temperature
D	0.85	**0.01**	**0.01**
E	0.93	0.85	**0.00**
K	0.84	0.56	0.15

Results show that only the main factors are statistically significant. Temperature affects the strength of mixtures D and E (the lower the temperature, the lower the strength) and loading rate affects the strength of mixture D (the higher the rate, the higher the strength). Conditioning time does not affect the mean flexural strength of the material.

4 Comparison between BBS, IDT, and DT Mixture Strength Tests

4.1 Size Effect Approach

In order to compare equivalent strength values, obtained through different types of tests and for samples having different volumes, the statistical size effect caused by the randomness of material strength and the equivalency between the three point bending test and the tension test have to be considered. . The statistical size effect theory was initially introduced by Weibull in 1939 and is based on the model of a chain [5]. The failure load of a chain is determined by its weakest link. The longer the chain is, the smaller the strength value that is likely to be observed in the chain is. Weibull described this strength behavior using a special form of extreme value distribution, later named Weibull distribution in his honor [5]. This results in a definite relationship between mechanical load and the failure probability of the part if the distribution parameters are known. Using a level of strength at which the failure probability becomes 63.2% (σ_0), the Weibull modulus (m) becomes a measure of the distribution of strengths. The failure probability P_f is calculated as:

$$P_f = 1 - \exp[-(\sigma/\sigma_0)^m] \qquad (4)$$

where σ_0 is a scale parameter, σ is the structure stress, and m is the Weibull modulus (shape parameter).

Since the number of possible material defects is dependent on the volume of the part, the volume under load must be taken into account. The strength of large parts is thus smaller than what is measured on test samples. Using Weibull statistics, a relationship between strength obtained from the same material on specimen with different volume can be obtained:

$$\sigma_{structure} = \sigma_{test-specimen} \cdot (V_{test-specimen}/V_{structure})^{1/m} \qquad (5)$$

When specimens with different geometry are used, an equivalence formulation can be determined based on equation (4) and on an adimensionalized expression of imposed stress $s(x)$, where x refers to the coordinate system:

$$P_f = 1 - \exp\left\{-\left\{\int_V [s(x)]^m dV(x)\right\}(\sigma_N/\sigma_0)^m\right\} \qquad (6)$$

σ_N is the nominal strength and V is the effective volume of the specimen.

In case of a beam subject to three-point bending, equation (3) can be substituted into equation (6) and the integral can be solved over the volume of the beam subject to tensile stress (effective volume). For a uniform tension case, such as direct tension tests equation (6) can be further simplified as:

$$P_f = 1 - \exp\left[-\frac{V}{V_0}\left(\frac{\sigma}{\sigma_0}\right)^m\right] \qquad (7)$$

V_0 is a reference volume and V is the volume of the tested specimen. By equating the arguments inside the exponential function of equation (6) and (7) and subsequently introducing the volume correction given by equation (5), it is possible to relate the mean nominal flexural strength σ_N^B obtained from a three-point bending configuration with a given volume V_B to the mean nominal tensile strength σ_N^U measured from direct tension test with a different volume V_U:

$$\frac{\sigma_N^B}{\sigma_N^U} = \left[4 \cdot (1+m)^2\right]^{(1/m)} \left(\frac{V_U}{V_B}\right)^{1/m} \qquad (8)$$

4.2 BBS Strength Tests for Test Comparison

The remaining eight mixtures of Table 1 (A, B, F, G, H, I, L and M) were tested with BBS at three different temperatures. As a result of the factorial analysis performed on mixture D, E, and K, only one loading rate of 3N/min, and one conditioning time of 1 hour were considered. Three replicates were tested at each temperature. Analysis of variance (ANOVA) and Tukey HDS comparison showed that temperature is a significant factor affecting the mean strength value, while, overall, the difference in strength between the eight different mixture was not significant except for mixture A and F.

BBS, IDT, and DT strength test results were first directly compared. Then, the size and geometry corrections were considered and the IDT, DT, and BBS specimens were converted to a unitary volume subjected to uniaxial force based on equations (5) and (8) respectively. Figure 3a and 3b show the mean strength values before and after the conversion.

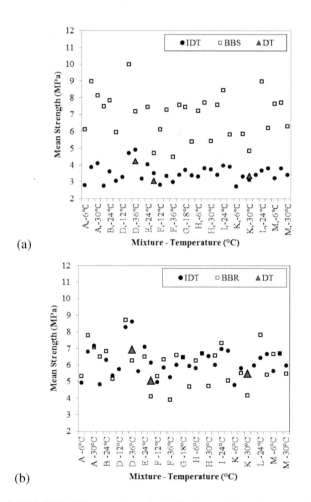

Fig. 3. IDT, BBS, DT strength results (a) before and (b) after corrections

Statistical analysis of the pre and post transformed data was performed using ANOVA. BBS uncorrected strength values are much higher than the IDT and DT strength results. However, after corrections were applied, no significant difference was found between BBS and IDT strength values at a 0.05 significance level. BBS and DT strength values are still statistically different; however, a p-value close to 0.05 was obtained, implying that DT results should be further investigated since only a limited number of replicates and mixtures were used in this study.

5 Summary and Conclusions

This paper presents the preliminary work in the development of a modified Bending Beam Rheometer capable of performing strength tests on thin asphalt mixture beams. Eleven mixtures were tested to evaluate the statistical distribution and parameters affecting the BBS strength measurements. Geometry, specimen volume, and testing configuration were considered to compare BBS strength to IDT strength and DT strength using statistics and weakest link theory.

BBS strength results follow a Weibull distribution, typical of quasi-brittle materials. BBS uncorrected strength values were much higher than the IDT and DT strength results. However, after size and geometry corrections were applied, no significant differences were found between BBS and IDT strength from a statistical view point.

A follow up comprehensive analysis that takes into account Representative Volume Element, multiple histogram testing and energetic statistical effect on strength is currently in progress to further validate the concept and extend its application to both laboratory and field samples.

References

[1] American Association of State Highway and Transportation Officials, Standard Method of Test for Determining the Flexural Creep Stiffness of Asphalt Binder Using the Bending Beam Rheometer (BBR). AASHTO T313-10-UL (2010)

[2] American Association of State Highway and Transportation Officials, Standard Method of Test for Determining the Creep Compliance and Strength of Hot Mix Asphalt (HMA) Using the Indirect Tensile Test Device. AASHTO T322-07-UL (2007)

[3] Marasteanu, M.O., Velasquez, R., Zofka, A., Cannone Falchetto, A.: Development of a simple performance test to determine the low temperature creep compliance of asphalt mixture, IDEA Program Final Report NCHRP-133 – Transportation Research Board of the National Academies (2009)

[4] Boltzman, P., Huber, G.: In: Strategic Highway Research Program SHRP-A-641, Washington DC (1993)

[5] Bazant, P.Z., Planas, J.: Fracture and Size Effect in Concrete. CRC Press, London (1998)

[6] Johnson, A., et al.: In: 2008 MnROAD Phase II Constrction Report, Minnesota Department of Transportation, Maplewood, MN (2009)

[7] Christensen, D.W., Bonaquist, R.F.: In: NCHRP Report 530, Washington DC (2004)
[8] Standard Test Method for Thermal Stress Restrained Specimen Tensile Strength, former TP10-93
[9] American Association of State Highway and Transportation Officials, Standard Method of Test for Determining the Fatigue Life of Compacted Hot-Mix Asphalt (HMA) Subjected to Repeated Flexural Bending. AASHTO T321-07-UL (2007)
[10] Bazant, Z.P., Pang, S.-D.: Journal of Mechanics and Physics of Solids 55, 91–131 (2007)
[11] Zech, B., Wittmann, F.H.: In: 4th International Conference on Structural Mechanics in reactor Technology, vol. H, pp. 1–14 (1977)
[12] Rocco, C.G.: Doctoral Thesis, Universidad Politecnica de Madrid, Spain (1995)

Experimental Study of the Precracking

Ratiba Mitiche_Kettab, Azzouzi Boulanouar, and Abderrahim Bali

National Polytechnic School Algiers

Abstract. Pre-cracking technique allows obtaining a controlled cracking, finer and less evolutive. The idea in reconsidering the design of precracked foundations has grown up with this technique. Studies are being developed, mainly in the field of structural analysis based on finite element methods.

Up today, a reduction in thickness compared to pre-cracked foundation thicknesses obtained without pre-cracking is not allowed and must be considered as experimental.

Tests have been conducted on precracked test boards with and without impregnation layer of cement treated base and construction on site. The execution of the microcrack has been performed three days after the start of implementation (from 48 to 72 hours after the execution of the cement treated base). It was conducted by applying 2 passes with *vibration* and without vibration; after that the state of microcracking has been confirmed.

Microcracking did not occur without vibration and six samples were collected each time (24 and 25 days after execution) for each of the three modes 0, 2 and 4 passes of vibrating compactors.

To confirm the reduction in strength due to microcracking observed on these samples, compression and indirect tensile tests were carried out 28 days after execution.

The results obtained show that, after inclusion of the microcrack, there is no much influence on the compressive strength of cemented treated base. Accordingly, we may consider that the inclusion of the microcrack does not affect the performance of the cemented gravel.

1 Introduction

Cement-bound aggregate, regardless of the nature of the binder, lead, during material setting, to shrinkage cracking. As a result of temperature variations and trucks traffic, these cracks cross wearing course layers and can thus cause degradations to surface and pavement structure respectively. The purpose of pre-cracking, during hydraulic material placing, is to develop transversal discontinuities in order, in the manner of concrete pavements, to introduce a discontinuity in the cement- bounded aggregate to be subsequently transformed into fine cracks. This reveals on the material surface a preferential crack path to prevent uncontrolled spread. The objective is to reduce the severity of the cracks rising up to the surface of the coated material so that it will require little or no maintenance.

2 Evolution of Cracks Up to the Surface Layers

The propagation of a crack within the wearing course of a pavement whose base layer is cracked may result basically either from:

- interface peeling off (horizontal propagation)
- or vertical propagation as an extension to the existing crack

This alternative is governed by the ratio of the effort, causing the crack propagation in a given direction, to the resistance to this propagation, opposed by the material.

3 Pre-cracking

Pre-cracking consists of causing shrinkage crack within the base at a desired place. It can be seen that, by causing the crack every two or three meters, the cracks rising up to the surface of the road pavement is straight and that can obviously facilitate a possible maintenance.

The second observation which can be, involves the evolution of cracks risen up to the surface. The always correspond to a pre-cracking of the base, and on the other hand, they are fine and their evolution is much less damaging than a natural crack.

Three pre-cracking techniques have been developed by French road companies. Basically these are the pre-cracking CRAFT process (Automatic Creation of transverse cracks) and , the pre- cracking JOINTS ASSESTS process and the pre-cracking OLIVIA process.

With these three pre-cracking techniques that permit obtaining a controlled cracking, finer and less progressive, the idea arose in order to reconsider the design of pre-cracked road foundations.

4 Micro Cracking Results

4.1 Introduction

To confirm the effects of micro-cracking, tests on drilled cores were made at different locations with and without emulsion impregnation.

Micro-cracking is performed three days after starting the implementation (from 48 to 72 hours after the achievement of cement treated base 3 (GC3). Micro cracking was carried out by applying 2 passes of roller with and without vibration respectively and following this, the state of micro-cracking has been reached and confirmed.

Experimental Study of the Precracking

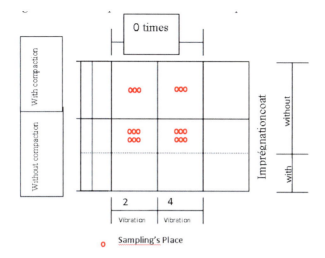

Fig. 1. Position of drilled core samples

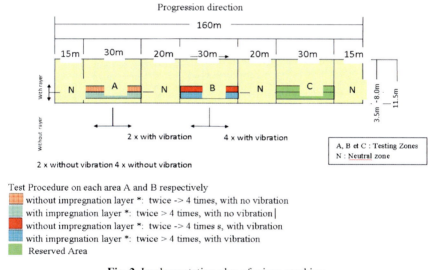

Fig. 2. Implementation plan of micro-cracking

It has been noted that no micro crack occurred without vibration. Six samples were collected each time (24 and 25 days after of test section) for each of the three modes 0, 2 and 4 passages of vibrating rollers. Compressive and indirect tensile tests were carried out at 28 din strength due to the observed micro cracking on the considered samples. Figure 1 shows the position of the drilled samples. Figure 2 shows the overall plan for the achievement of micro cracking. It highlights the areas with or without impregnation in order to compare them later.

4.2 Test Sections Results

4.2.1 State of the Surface Cracks after Applying Compaction with Vibration

Following compaction and vibration, micro-cracks appear on the surface. Figures 3 and 4 show their location and length according to the number of vibrations roller passages (2 or 4).

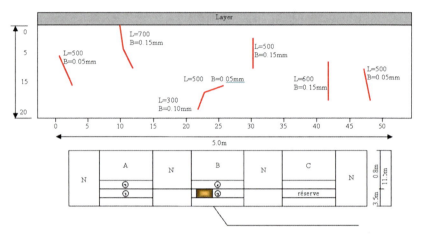

Fig. 3. State of cracks on the surface after compaction with vibration

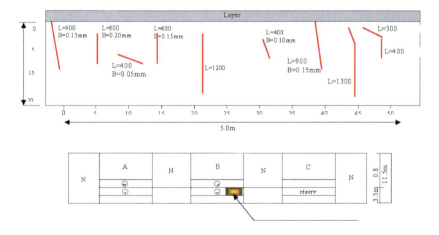

Fig. 4. State of cracks on the surface after compaction twice vibration (4 times)

Cores have been drilled as indicated and situated on Figure 5. Compressive and tensile test were carried out on these specimens.

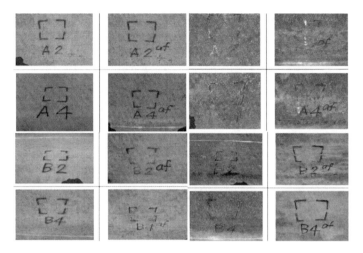

Fig. 5. Location of cores in the test section

4.2.2 Compression Results of the Cement Treated Base 3 (GC3)

The compressive test results obtained according to the implementation of the test section are presented in Table 1 and Figure 6.

Table 1. Compression test results

Standard reference: EN 14227-1 Paragraph 6.5

	Compared to the weight of the mixture			3,75
Weight percentage of cement (%)	Based on the aggregates weight			(3,90)
Frequency of rolling compaction	Vibration	0	twice	4 times
Compressive strength test σ_{28}	number.1	13,65	14,70	16,27
	number.2	12,77	14,04	14,15
	number.3	13,86	13,29	12,08
	Mean of three	13,76	14,00	14,18
	Mean+20%	16,11	16,81	17,00
	Mean-20%	10,74	11,21	11,33
	Mean	13,43	14,01	14,17

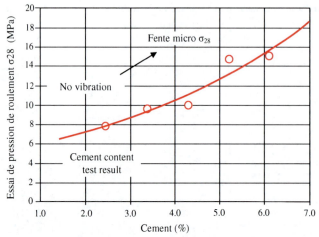

Fig. 6. Diagram of the direct compression test (3.75% Cement)

These results show that the introduction of micro-crack does not have much influence on the compressive strength of cement bound aggregate. Indeed, one can say that the strength is almost the same. Accordingly, we may consider that the inclusion of the micro-crack does not affect the performance of the GC3.

4.2.3 Results of Indirect Tension

The features provided by this test are:

1. Tensile force: $R_t 360$ = more than 1.15 MPa
2. Tensile force: $R_t 28 = 0.69$ MPa over ($R_t 360 = 1.15 \times 0.6$)
3. Indirect tensile force: $R_{it} 28$ = more of 0.86 MPa ($R_{t28} = 0.69/0.8$).

The results of the indirect tensile stress by varying the frequency of vibrations are given in Table 2 and Figure 7.

Table 2. Indirect tensile test results

Weight percentage of cement (%)	Compared to the weight of the mixture			3,75
	Based on the aggregates weight			(3,90)
Frequency of rolling compaction	Vibration	0 time	2 time	4 time
Indirect tension (MPa) σ_{28}	number.1	1,17	0,78	0,94
	number.2	1,34	1,13	1,12
	number.3	1,51	1,04	1,24
	Mean of three	1,34	1,09	1,09
	Mean+20%	1,61	1,18	1,32
	Mean -20%	1,07	0,79	0,88
	Mean	1,34	0,98	1,10

Experimental Study of the Precracking

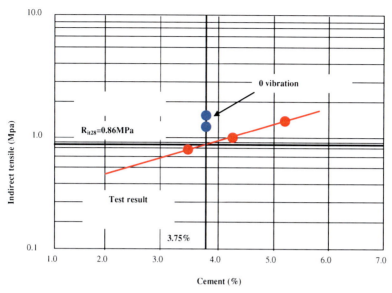

Fig. 7. Diagram of indirect tensile test (MPa) (3.75% Cement)

4.2.4 Results of Mechanical Testing of Micro Cracked Samples

Four passages of mechanical vibrating compacter (20 tones with a low vibration) seemed appropriate according to the observation made when performing the micro-cracking test. Laboratory testing will allow defining with better precision the number of roller passages.

It has been checked that the elastic modulus as well as the indirect tensile strength of the micro- cracked samples met the required value whatever the number of passages. It can be therefore suggested for the obtaining of micro cracks, 4 passes with a vibrating roller 20t with low vibration. GC3 formulation Tests were performed according to the roadway LCPC-SETRA and AFNOR Standards. However, it appears that test results obtained on drilled cores are much higher that required by these Standards, suggesting that in general, the values required have been prescribed with a wide safety margin compared to those obtained on road construction site. It has been checked that the strength obtained on the drilled cores samples is higher than that required for design. It seems therefore convenient that only the in-situ-density control will be performed though quality control.

4.3 In Situ Test Results

Direct compressive and tensile tests were performed as well as for the test section in order to validate them. Figure 8 shows the location of cores.

It seems therefore convenient that only the in-situ density control will be performed trough quality control.

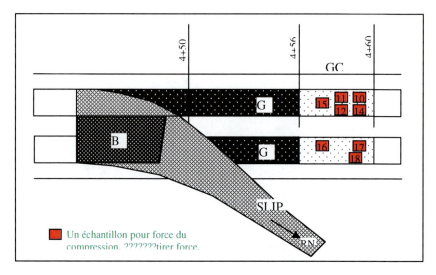

Fig. 8. Location of core samples

Tensile test results (stress and elastic modulus) are given in Table 3.

Table 3. Direct and indirect test results

Sample number	Test	Rit(MPa)	E(MPa)
C-10	Direct		25177
C-12	Indirect	1.20	
C-11	Indirect	0.92	
C-14	Indirect	1.24	
C-15	Direct		19893
C-16	Direct		16832
C-17	Direct		24819
C-18	Indirect	1.19	
Mean		1.14	21680.25
Mean+ 20%		1.37	26016.30
Mean – 20%		0.91	17344.20

Compressive and tensile tests results obtained on GC3 cores are summarized in Table 4.

Table 4. Compressive and tensile tests Results on the GC3 cores

NO	Number core cement	Diameter of core			thickness of core					Weight before test	type of test	Final result
		1	2	Average	1	2	3	4	Average			
1	C-10	147,0	147,1	147,1	125,0	124,5	125,2	128,5	125,3	4916	Direct Comp.	242,90
2	C-12	145,2	145,8	145,5	127,2	126,0	125,9	125,2	126,1	5011	Indirect Tens.	34,44
3	C-11	146,0	146,3	148,2	131,7	130,8	130,2	130,6	130,8	5138	Indirect Tens.	27,49
4	C-14	148,0	147,0	147,5	126,8	126,0	126,0	126,0	126,2	5095	Indirect Tens.	36,25
5	C-15	146,0	145,9	146,0	125,2	125,0	125,0	125,0	125,1	4933	Direct Comp.	265,56
6	C-16	148,0	145,9	146,0	128,0	128,3	128,2	127,9	128,1	5044	Direct Comp.	272,81
7	C-17	146,8	146,8	146,8	126,5	125,2	126,0	126,8	126,1	4936	Direct Comp.	214,80
8	C-18	148,2	148,0	146,1	128,2	128,2	129,0	128,0	128,4	5142	Indirect Tens.	35,05

Cement content: 3.75%
Water content: 5.10%.

The results obtained on site are compliant to the standards and validate the test section results. It can be noted that even in site conditions which generally are not the same as those of test sections (because all the precautions to reach better performance are taken with test sections whereas on site there are always anomalies related to the quantity of treated base material used), the cement treated base always shows an improved strength even with the application of the technique of micro-cracking.

5 Conclusion

Micro-cracks are fine and their evolution is much less damaging than a natural crack and does not affect the strength of the cement treated base.

The results obtained with pre-cracking technique which allows obtaining a controlled cracking, finer and less progressive, especially after the protection of cement treated base with an emulsion impregnation layer, show that it permits retaining the resistance of the cement treated base and increases the lifetime and durability of the road ensuring a high load bearing applied mainly by the heavy weight trucks.

References

[1] Azzouzi, B.: Behavior of stabilized base course according to the specific climate in Algeria. Magister Thesis EcolePolytechnique Algiers (2009)
[2] Belattaf:Socio-Economic Impacts and Environmental East-West Highway in Algeria. 3 Days of Development GRES, Université Montesquieu-Bordeaux IV (2009)
[3] Lefort, M., Sicard, D.l., Merrien, P.: Technical anti-cracking guide job pavements. The Department of Civil Engineering and tracks STBA in partnership with the Western Regional Laboratory, Paris (2009)
[4] STANDARD FRENCH, Mixed treated and untreated mixed with hydraulic binders. EN 13286-2, AFNOR, Paris (2004)
[5] Technical advice SETRA (1998): "CRAFT CBC" (July 1998)
[6] Technical advice SETRA: Joint Assets (SACER), (July 1997/July 2002)
[7] Notes for SETRA: "Limits and interest in clogging of pavement shrinkage cracks semi-rigid" (1990)

Comparison between 2PB and 4PB Methodologies Based on the Dissipated Energy Approach

M. Pettinari[1], C. Sangiorgi[2], F. Petretto[1], and F. Picariello[3]

[1] PhD Student, DICAM, University of Bologna, Bologna
[2] Researcher, DICAM, University of Bologna, Bologna
[3] Head of Pavement Laboratory, Elletipi srl, Ferrara

Abstract. Two and four point bending tests are among the most common methodologies adopted in Europe and United States for the fatigue characterization of asphalt mixes. Both tests tend to simulate the flexural stresses generated by traffic applying uniaxial rather than triaxial loading. The main differences between these procedures are: the direction of load application, the constrains and the volume of material subjected to fatigue.

In this study, based on Fatigue data, the Elletipi horizontal two point bending (2PB) and a traditional four point bending (4PB) results are compared. The peculiarity of the 2PB flexural device is the horizontal position of the trapezoidal specimen during the test.

All tests were performed in strain controlled conditions at different temperatures. The research focuses on the influence of the loading waveform by comparing the effects of sinusoidal and haversine loads of equivalent strain amplitudes. The Ratio of Dissipated Energy Change (RDEC) approach, based on the energy balance classical theory, was the application of choice for the analysis of results. Finally, the influence of specimen volume on fatigue resistance was assessed performing horizontal 2PB tests on trapezoidal specimens of different thickness.

1 Introduction

Beam fatigue (4PB) and classical trapezoidal fatigue (2PB) simulate the flexural stress pattern found in situ, but apply uniaxial rather than triaxial stresses. Loading can be in either the controlled-stress or the controlled-strain mode to better simulate the range of conditions encountered in real pavements. Testing equipment measures the stiffness, phase angle, cycles to failure and dissipated energy.

In terms of differences, beam fatigue under third-point loading over trapezoidal fatigue has a larger portion of the specimen subjected to a uniform maximum stress level. Thus the likelihood is greater in beam testing that test results will reflect the weaknesses that naturally occur in asphalt-aggregate mixes [1]. Differently from a traditional 2PB equipment, the horizontal device here presented is vertically loaded and a 25 kN UTM loading frame can be used to run it. The 2PB specimen is therefore bended in the same direction as the 4PB one (Figure 1).

The main purpose of this research step is to compare the 4PB and the Elletipi 2PB test results using the Ratio of Dissipated Energy Change approach. Included variables, other than applied strain and temperature, are the load waveform and the specimen thickness.

The Elletipi horizontal 2PB was designed by the Elletipi pavement laboratory. The negligible influence of the test configuration on the stress and strain conditions was verified by FEM analysis.

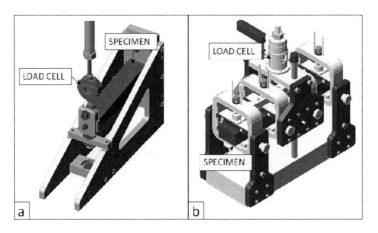

Fig. 1. Elletipi 2PB device (a), 4PB device (b)

2 The Ratio of Dissipated Energy Approach

The fatigue behaviour of asphalt mixes could be studied applying different methods and failure criteria. Above the first, the dissipated energy approaches provide a rational explanation of this critical phenomenon. In general, the energy dissipated in a testing cycle depends on the energy dissipated in the previous cycles and, consequently, on the material physical conditions. The relative amount of energy dissipation created by each additional load cycle will produce crack extension, plastic deformation and thermal energy [2].

The Ratio of Dissipated Energy Change (RDEC) approach sets forth the fatigue damage eliminating the dissipated energy that does not produce crack extension [3]. During cyclic fatigue loading, the viscoelastic bituminous mixes traces different paths for the unloading and loading cycles and creates an hysteresis loop. In a strain controlled test, the area inside the loop, the Dissipated Energy (DE), is calculated using equation (1):

$$DE_i = \pi \sigma_i \varepsilon_0 \sin(\delta_i) \quad (1)$$

where ε_0 is the strain amplitude, σ_i is the stress amplitude and δ_i is the phase angle. The DE for each load cycle, i, is the sum of the viscoelastic energy dissipation, E_i^η, and the energy dissipation due to the damage propagation, E_i^ξ. Ghuzlan and Carpenter [2, 3] defined that the material failure is imminent when the magnitude of the dissipated energy, between consecutive cycles, undergoes a significant increase. To quantify the change in dissipated energy they introduced the Ratio of Dissipated Energy Change (RDEC) calculated as follows (2) after Bhasin [4]:

$$RDEC_a = \frac{|DE_a - DE_b|}{DE_a \times (b-a)} = \frac{|E_a^\xi + E^\eta - E_b^\xi - E^\eta|}{(E_a^\xi + E^\eta) \times (b-a)} \approx \frac{|E_a^\xi - E_b^\xi|}{(b-a)} \frac{1}{E^\eta} = \frac{\Delta E^\xi}{E^\eta} \quad (2)$$

where ΔE^ξ is the energy dissipated per cycle due to damage.

The relationship between the RDEC and the number of Load Cycles (LC) is represented by a damage curve which can be divided into three stages. The RDEC value at the 50% stiffness reduction cycle (Nf_{50}) is defined as the Plateau Value (PV). The PV is the percent of dissipated energy in a load cycle that is causing actual damage relative to the following load cycle. It can be seen as a fundamental energy parameter to represent the asphalt mixes fatigue behavior because it carries the effects of aggregate, binder properties and loading conditions. The PV can be also calculated with the following equation (3):

$$PV = \frac{1 - \left(1 + \frac{100}{Nf_{50}}\right)^f}{100} = RDEC_{Nf_{50}} \quad (3)$$

where f is the slope of the regression curve DE-LC (Figure 4) until a 0.5 Stiffness Ratio (SR) value is attained. According to Shen and Carpenter [3, 5], a unique relationship (4) exists between PV and the number of load cycles at failure point (Nf_{50}), for different mixes, loading and testing conditions:

$$PV = c(Nf_{50})^d \quad (4)$$

where c and d are regression constants.

The uniqueness of the PV-Nf_{50} relationship provides a way to study both Fatigue Endurance Limit (FEL) and healing [2, 6, 7].

3 Research Program and Initial Testing

The study here presented is carried out in three phases and involves:

- Asphalt mix design, asphalt compaction and specimen preparation;
- Elletipi 2PB and 4PB fatigue testing;
- Data analysis with the Energetic Approach.

The asphalt mix object of this research, a Warm Mix Asphalt with 6.4% of binder by mass of aggregates, is Field Mixed and Field Compacted (FMFC). The asphalt mix compaction was completed using a set of field molds and the obtained slabs were cut in to specimens of the following dimensions:

- *Beam* (PR) - height 51 mm, width 65 mm, length 380 mm;
- *Trapezoidal* (TR) - major base 56 mm, minor base 25 mm, length 250 mm and height 25 mm.
- *Trapezoidal Big* (TR Big) - major base 56 mm, minor base 25 mm, length 250 mm and height 50 mm.

Fig. 2. Field compaction with specific mold

A total of 51 specimens was produced each with 6 cut faces. Table 1 summarizes the air-voids distribution categorized by specimen type, test temperature, and tensile strain level for the fatigue test.

In terms of fatigue test, the *ASTM D7460 Flexural Controlled-Deformation Fatigue Test* method was followed. The trapezoidal specimens (TR) were subjected to 2PB using a sinusoidal waveform at a loading frequency of 10 Hz.

With regard to the prismatic specimens (PR), 18 were tested with an haversine waveform and 9 with a sinusoidal one. As stated by Pronk [8], using an haversine deformation with a peak-peak value of 200 µε should render a fatigue life comparable to that obtained applying a sine deformation with amplitude of 100 µε. Therefore, in order to better compare all the fatigue results obtained with both waveforms, the strain level, in the 4th column (Table 1) and after, represents the semi- amplitude of the peak-peak applied strain.

Testing was performed in dry conditions at different strain levels and different temperatures. Failure point (Nf_{50}) was assumed as the number of loading cycles corresponding to the 50% Initial Stiffness reduction. The Nf_{50} for each specimen not reaching the failure within 5 million of loading was extrapolated using the Weibull Survivor function, as suggested by ASTM D7460 [9].

Table 1. Testing configurations and air voids content for the study mix

Specimen type	Number of specimens for sample	Load Waveform	strain level (με)	Temperature (°C)	Av. air voids Content (%)
TR	3	sinusoidal (sin)	100	10	6.2 ± 0.3
TR	3	sinusoidal (sin)	200	10	7.1 ± 0.2
TR	3	sinusoidal (sin)	400	10	7.5 ± 0.1
TR	3	sinusoidal (sin)	100	20	6.9 ± 0.3
TR	3	sinusoidal (sin)	200	20	6.3 ± 0.4
TR	3	sinusoidal (sin)	400	20	6.4 ± 0.1
TR Big	3	sinusoidal (sin)	200	10	8.0 ± 0.3
TR Big	3	sinusoidal (sin)	400	10	7.7 ± 0.4
PR	3	haversine (has)	100	10	6.6 ± 0.2
PR	3	haversine (has)	200	10	6.0 ± 0.1
PR	3	haversine (has)	100	20	7.3 ± 0.5
PR	3	haversine (has)	200	20	7.5 ± 0.3
PR	3	haversine (has)	100	30	6.8 ± 0.1
PR	3	haversine (has)	200	30	6.9 ± 0.2
PR	3	sinusoidal (sin)	200	10	7.7 ± 0.4
PR	3	sinusoidal (sin)	300	10	7.7 ± 0.3
PR	3	sinusoidal (sin)	400	10	6.5 ± 0.1

Figure 3 illustrates the Initial (after 100 cycles) Stiffness and Phase Angle comparison at various strain levels, temperatures, and fatigue test types. The following observations were made:

- Initial Stiffness and Phase Angle appear to be generally independent from the flexural test used; besides, the air voids seem to affect these initial characteristics (Table 1);
- the influence of testing temperature is similar on both test types;
- the variability of results is of the same magnitude for both test types.

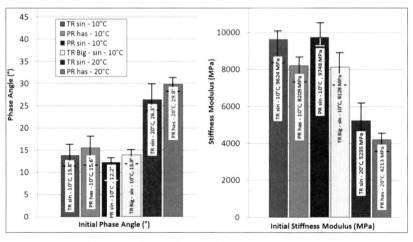

Fig. 3. Summary boxplots of Initial Phase Angle and Stiffness

4 Dissipated Energy Evaluation

4.1 Comparison of Dissipated Energy Curves

The DE is connected with the test conditions, materials properties and damage level. In particular, the phase angle and the stress amplitude are the basic variables.

With regard to the study mix and testing methods, Figures 4 and 5 summarize the DE curves respectively of 4PB and Elletipi 2PB fatigue tests.

The mixture, in the same test conditions, exhibits similar DE curves and fatigue behavior with both flexural tests. For example, the 4PB tests at 400 µɛ and 10°C, show an Initial DE and a Nf_{50} similar to those obtained with the corresponding 2PB tests. All test results show that fatigue resistance is highly related to the Dissipated Energy curves. Independently from the test conditions, at lower Initial DEs correspond longer fatigue lives; this should be taken into account when large testing programs are to be completed. Furthermore, temperature affects the Initial DE and curve shape as well as the failure point.

Figures 4 and 5 denote the existence of an Initial DE value for which the specimens do not seem to be affected by fatigue damage. In fact with both fatigue methods and at any temperature, specimen with Initial DE below 0.08 kPa do not reduce their level of dissipated energy per cycle for all the duration of the test. This asphalt mixture behavior could be highly related with the Fatigue Endurance Limit concept [3, 5, 7] and with the existence of an energy level that the mix is capable to dissipate without being affected by damage and below which the material shows an extraordinary long fatigue life. 4PB and Elletipi 2PB tests describe this characteristics with comparable straight plots.

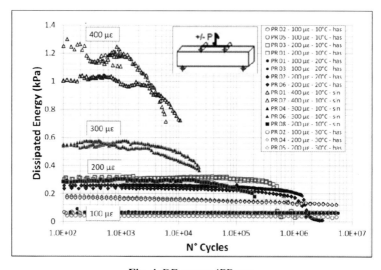

Fig. 4. DE curves 4PB test

On the basis of previous considerations, the Initial DE is strongly related with the number of load applications at 50% Stiffness reduction (Figure 6). In particular, all the measured Nf_{50} describe a specific trend at each temperature: higher temperatures, tests last longer.

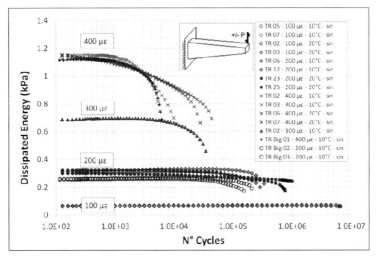

Fig. 5. DE curves Elletipi 2PB test

The thicker Trapezoidal specimen (TR Big) performed like the normal one (TR) for both tested strain levels. Thus, at 10°C, the specimen thickness is not affecting the Elletipi 2PB fatigue behavior. Comparing PR tests at 10°C, even the differences, in terms of fatigue results, between haversine and sinusoidal load waveform are not so relevant. This is confirming what stated by Pronk [10].

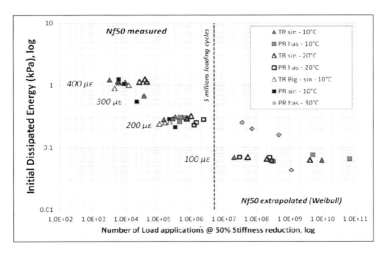

Fig. 6. Initial DE vs Number of Load applications at 50% Stiffness reduction

4.2 Ratio of Dissipated Energy Approach Application

The PV for each test was calculated by means of equation (3). The Nf_{50} for the incomplete fatigue tests and the slope f of the DE curve were calculated respectively with the Weibull Survivor Function and the Power Model [2, 7]. The results are plotted in Figure 7.

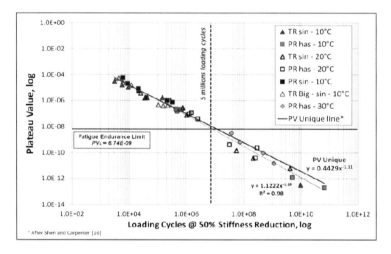

Fig. 7. PV-Nf_{50} for the study mix in all testing conditions

Upon analyzing the results, it is possible to infer the following:

- using the PV-Nf_{50} relation suggested by Carpenter and Shen [2, 3, 7], the PVs obtained from the two flexural fatigue tests are on the same trend line. In addition the difference between the Unique PV line and the trend obtained from the tests is higher where the Nf_{50} were estimated with the Weibull function (FEL conditions);
- all the PVs obtained at 100 µε are below the PV_L line defined by Carpenter [2] and the corresponding Initial DE is lower than 0.08 kPa and approximately constant during the tests.
- 100 µε could be considered a condition of Endurance Limit for the study mix.

5 Conclusions and Recommendations

The Elletipi 2PB device and its loading configuration can be considered a reliable fatigue testing equipment and it provides results comparable to those obtained by a traditional 4PB system. In 1998, Pronk [10] has shown the existence of a difference in lifetime between traditional 2PB and 4PB: the first test recording the shortest life. With the Elletipi horizontal 2PB, this distinction is not so remarkable.

The effect of the specimen thickness, comparing the TR with the TR Big at 10°C, is not considerable. In fact the failure points reached during the test with both strain levels (200 and 400 µε) are similar and the spread of the failure points seems more connected with the variability in air voids content than with the increase in thickness.

As for the Dissipated Energy approach, the Nf_{50} is evidently related with the Initial DE of an asphalt mixture and this correlation does not sensibly change from Elletipi 2PB to 4PB. In addition the tested asphalt mixture exhibits a *DE limit* below which no significant damage occurs. When the Initial DE is above the *DE limit*, the fatigue resistance of the mix can be modeled by a traditional fatigue law: this permits to estimate the failure points even within the very first testing cycles. When it is below, the failure is more related to the mix anisotropic characteristics and healing capabilities, hence difficult to preview. This *DE limit* could be considered a material property related to the Fatigue Endurance Limit of the mix itself. With the RDEC approach those two different performances can be explained for both tests with an individual fatigue curve.

This work confirms the validity of the Energetic Approach application for studying the fatigue resistance of asphalt mixes.

References

[1] SHRP A-404, Fatigue Response of Asphalt-Aggregate Mixes. National Research Council, Washington, DC (1994)
[2] Carpenter, S., Shen, S.: Dissipated Energy Approach to study Hot-Mix Asphalt Healing in Fatigue. Transportation Research Record: Journal of Transportation Research Board, No. 1970, pp. 178–185 (2006)
[3] Shen, S., Airey, G.D., Carpenter, S.: A dissipated Energy Approach to fatigue evaluation. Road Materials and Pavement Design 7(1) (2006)
[4] Bhasin, A., Castelo Branco, V.T.F., Masad, E., Little, D.N.: Quantitative Comparison of Energy Methods to Characterize Fatigue in Asphalt Materials. Journal of Materials in Civil Engineering © ASCE (February 2009)
[5] Shen, S., Carpenter, S.H.: Application of the Dissipated Energy Concept in Fatigue Endurance Limit Testing, Transportation Research Record: Journal of the Transportation Research Board, No. 1929, 165–173 (2005)
[6] Pettinari, M.: Performance evaluation of low environmental impact asphalt concretes using the Mechanistic Empirical design method based on laboratory fatigue and permanent deformation models, Ph.D thesis. University of Bologna (2011)
[7] NCHRP report 646, Validating the Fatigue Endurance Limit for Hot Mix Asphalt, Washington, DC (2010)
[8] Pronk, A.C., Poot, M.R., Jacobs, M.M.J., Gelpke, R.F.: Haversine Fatigue Testing in Controlled Deflection Mode: Is It Possible? In: Transportation Research Board Annual Meeting, No 10-0485, Washington, DC (2010)
[9] ASTM D7460, Standard Test Method for Determining Fatigue Failure of Compacted Asphalt Concrete Subjected to Repeated Flexural Bending, ASTM International
[10] Molenaar, A.A.A.: Predicting of fatigue cracking in Asphalt Pavements. Do we follow the right approach? Transportation Research Record: Journal of the Transportation Research Board, No. 2001, 155–162 (2007)

Evaluation of Thermal Stresses in Asphalt Layers Incomparison with TSRST Test Results

M. Pszczoła and J. Judycki

Gdansk University of Technology, Poland

Abstract. The paper presents the results of calculations and laboratory determination of thermal stressesat low temperatures. The modified Hills and Brien's method was used to calculate the thermal stresses in asphalt layers of pavements and the results were compared againstthe values obtained at a laboratory with the Thermal Stress Restrained Specimen Test (TSRST) method. The laboratory investigations were conducted using plain grade bitumen, modified bitumen with SBS elastomer modification and multigrade type bitumen. It was found that the type of bitumen binder in asphalt concrete is of significant importance to the value of the calculated thermal stresses. For thecooling rate of 10°C/h the lowest value was obtained for asphalt concrete produced with the use of multigrade type bitumen. This fact can be an indication of a better resistance to low temperature cracking. The thermal stresses were had the highest values for asphalt concrete produced with plain bitumen. A good correlation was obtained betweenthe thermal stresses calculated withthe Hills and Brien's procedure and the values of thermal stresses determinedwith the TSRST method.

1 Introduction

Drops of temperature induce thermal stresses in asphalt layers of pavements.These stresses are one of the main problems in behaviour of pavement during winter, especially in countries located in the zone of relatively cold continental climate.As a result, transverse cracks form in asphalt pavement.Thermal stresses are a consequence of changes in the temperature of asphalt layers of pavements and are particularly high when the drop is sudden, resulting in loss of elasticity and increase of stresses in excess of their relaxation capacity.The level of stresses depends on a number of factors related to the properties of asphalt layer and the rate of temperature drop.The purpose of the present research is to determine thermal stresses which may assist in predicting low-temperature cracking.The procedure developed by Hills and Brien [1] was used and the calculated values were compared with the thermal stresses obtained in laboratory conditions withthe TSRST method (Thermal Stress Restrained Specimen Test)defined by the AASHTO procedure TP10.

2 Calculations of Thermal Stresses

Thermal stresses were calculated with the Hills and Brien's method for the same asphalt mixes for which TSRST fracture temperatures and stresses were determined under laboratory conditions. The calculations were carried out for 0/16 mm asphalt concrete mixes produced with the use of the following bitumens:

- 50/70plainbitumem,
- DE 80B modified bitumen,
- 50/70multigrade bitumen.

2.1 The Method of Calculation

The advantage of the Hills and Brien's method, which is based on quasi-elastic solution is the simplicity of thermal stress calculations.However, it has also some weaknesses.These include, without limitation, ignored relaxation of stresses and arbitrarily assumed loading time and temperature gradient.In this method the thermal stresses are calculated with the following relationship:

$$\alpha \Sigma S(t,T)\Delta T < \sigma_x < \frac{1}{1-\mu}\alpha \Sigma S(t,T)\Delta T \tag{1}$$

where:
σ_x - accumulatedthermal stresses for the pre-defined cooling rate V_T,
α - coefficient of thermal contraction, assumed to be independent of temperature variations,
$S(t,T)$ - stiffness modulusdepending on the loading time t and temperature T,
ΔT - temperature increment- for calculations assumed $\Delta T=2^{\circ}C$,
μ - Poisson's ratio.

The term of the left-hand side of the relationship (1) describes the stresses in an infinite viscoelastic bar and the right-hand term - in infinite viscoelastic layer.Finalny, taking into account the road surface geometry the thermal stresses were calculated as an arithmetic average of the left-hand and right-hand terms of the formula (1).

According to Yoder and Witczak[2] the Poission's ratio μ for asphalt concrete varies between 0.25 and 0.5 depending on the temperature.At low temperatures it is closer to 0.25 and grows with the increase of temperature.For analysing the thermal stresses at temperatures lower than +6°C the Poisson's ratio has been taken at a constant value of 0.25.

Other input assumptions:

1. At +6°C the asphalt layers are free from thermal stresses.
2. The drop of temperature below +6°C is linear in time (as in TSRST).
3. Cooling rate has been taken at $V_T=10°C/h$, which corresponds to the cooling rate during laboratory testing with the TSRST method.
4. The thermal contraction coefficient α for asphalt concrete has been taken at: $\alpha=2.2\times10^{-5}$ 1/°C.

Modified method for determination of the stiffness modulus of asphalt concrete

In the original Hills and Brien's [1] method of calculating thermal stresses the stiffness modulus of asphalt concretewas determined on the basis of the Van der Poelnormographand the relationships developed by Heukelom and Klomp [3].However, the Hills and Brien's method was modified by the authors in order to obtain more accurate values of the stiffness modulus of asphalt concrete.Its values have been adopted for the respective bitumen grades on the basis of creep curves determined in bending of 50x50x300 mm specimens under constant load.Bending of specimens was carried out under a constant load at the following temperatures:

- 0°C,
- -5°C,
- -10°C,
- -15°C.

Besides the temperature also the loading time is relevant to the stiffness modulus value.In the analysis of thermal stresses the loading time was calculated with the following equation:

$$t = \frac{\Delta T}{V_T} \tag{2}$$

where:
  t - loading time, s
  ΔT - as in equation(1) – temperature range $\Delta T=2°C$
  V_T - cooling rate, °C/h

For cooling rate of V_T=10 °C/h loading time calculated with equation (2) is t=720s.

The stiffness modulus S(t,T) was calculated with the following equation:

$$S_{(t,T)} = \frac{\sigma}{\varepsilon_{(t,T)}} \tag{3}$$

where:
$S_{(t,T)}$ - stiffness modulus depending on the loading time and temperature, MPa
σ - stress determined for each specimen in bending under constant load, MPa
$\varepsilon_{(t,T)}$ - strain of specimen bent under constant load at a given test temperature T derived for t=720s loading time.

The values of stiffness modulus determined in creep test for loading time t=720s corresponding to the cooling rate of V_T=10°C/h and at different testing temperatures are given in Table 1.

Table 1. The values of stiffness modulus of asphalt concrete for different testing temperatures and bitumens

Temperature during creep test [°C]	Stiffness modulus of asphalt concrete depending on bitumen at $t=720s$ [MPa]:		
	50/70 plain bitumen	DE 80B SBS-modified bitumen	50/70 multigrade bitumen
0	1509	825	484
-5	2101	1325	1015
-10	4161	2781	1524
-15	8263	2437	2126
-20	13725	4474	3776

The values of the stiffness modulus at intermediate temperatures and temperatures lower than -15°C (reaching down to -20°C) and higher than 0°C (up to max. +4°C) were determined by interpolation and extrapolation of laboratory results obtained for the temperature range between 0°C and -15°C. This procedure was used to determine the values of stiffness modulus for each type of bitumen at 2°C increments within the temperature range between +4°C and -20°C.

2.2 Results of Calculations

The calculated thermal stresses are presented in Table 2 and in Fig. 1.

Table 2. Calculated thermal stresses for cooling rate of $V_T=10°C/h$

Temperature [°C]	Calculated thermal stresses: $\sigma=\Sigma\Delta\sigma$ [MPa]:		
	50/70 plain bitumen	DE 80B SBS modified bitumen	50/70 multigrade bitumen
+6	0,000	0,000	0,000
+4	0,044	0,034	0,019
+2	0,099	0,074	0,042
0	0,169	0,120	0,070
-2	0,257	0,175	0,104
-4	0,367	0,239	0,145
-6	0,507	0,314	0,194
-8	0,682	0,402	0,255
-10	0,904	0,505	0,328
-12	1,183	0,627	0,418
-14	1,535	0,769	0,526
-16	1,979	0,936	0,657
-18	2,538	1,132	0,817
-20	3,242	1,361	1,011

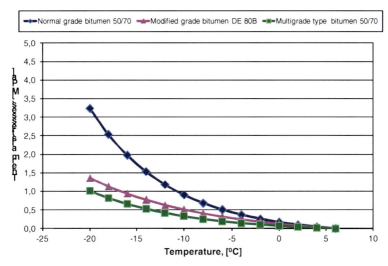

Fig. 1. The effect of bitumen on the calculated thermal stresses for $V_T=10°C/h$

The results of calculations show that the type of bitumen is highly relevant to the level of thermal stresses which develop in the asphalt layer. The highest values of thermal stresses were obtained for asphalt layer with plain bitumen of 50/70 grade.Higher values of thermal stresses may indicate greater risk of low-temperature cracking.The lowest values were obtained for asphalt layer with 50/70 multigrade bitumen.The values of thermal stresses for asphalt layer with DE 80B modified bitumen were slightly higher than for 50/70 multigrade bitumen and significantly lower than obtained for asphalt layer with 50/70 plain bitumen.

3 Laboratory Method for Determination of Thermal Stresses

3.1 TSRST Test Method

The testing method named Thermal Stress Restrained Specimen Test or TSRST in short [4] is used for determining the resistance to thermal cracking on specimens restrained from contracting and subjected to cooling at a constant rate of 10°C/h.

The first concepts of the TSRST method were developed by Monismith et al. [5] and applied on wider scale by Arand [6].The test methodology was developed on the basis of AASHTO TP 10-93 procedure.As a standard solution the tests are carried out with MTS apparatus.The specimens for TSRST test were 50x50x250 mm rectangular beams.Circular steel platens were glued to the specimens to enable securingthem in the loading frame.Extensometers were attached to three sides of specimens to measure the specimen displacements. The temperature sensor was attached to the fourth side.The above described stand was placed in the temperature chamber and secured in the strength tester frame.The specimen prepared for testing and the whole MTS stand are presented in Fig. 2.

Fig. 2. Specimen ready for testing with TSRST method

3.2 Laboratory Test Results

The relationships between the thermal tensile stress and the temperature in TSRST test for asphalt concrete mixes produced with three bitumen types are presented in Figure 3. Each chart presents the average results obtained in testing of two samples.

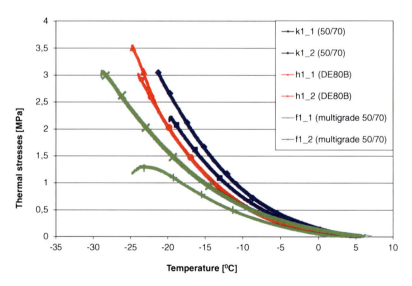

Fig. 3. Thermal tensile stress vs. temperature in TSRST testing of asphalt concrete mixes tested with three bitumen types

4 Comparison of Calculated vs. Tested Thermal Stresses

Figures 4 to 6 show thermal stresses calculated with the Hills i Brien's method for cooling rate of $V_T=10°C/h$ compared against the results of laboratory tests with TSRST method at temperature reaching down to -20°C.

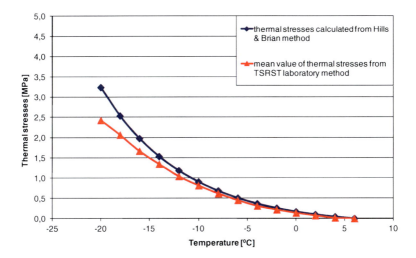

Fig. 4. Calculated vs. laboratory determined values of thermal stresses in asphalt concrete containing 50/70 plain bitumen

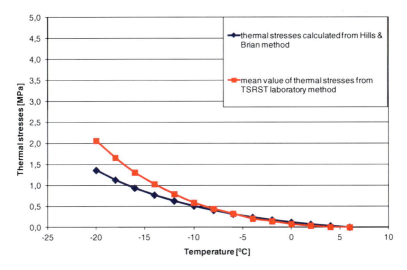

Fig. 5. Calculated vs. laboratory determined values of thermal stresses in asphalt concrete containing DE 80B modified bitumen

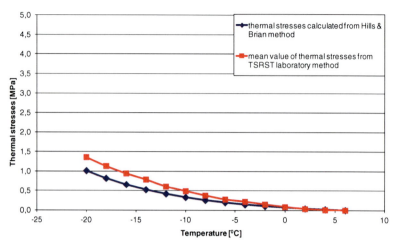

Fig. 6. Calculated vs. laboratory determined values of thermal stresses in asphalt concrete containing 50/70 multigrade bitumen

The comparison of thermal stresses obtained in TSRST testing were compared against the values calculated with Hills and Brien's method for cooling rate of $V_T=10°C/h$ showed satisfactory level of correspondence between calculated and laboratorymeasured values, especially in the temperature range between+6°C and+10°C.

5 Conclusssions

It can be concluded that the Hills and Brien's method is to some degree inaccurate as it ignores the effect of stress relaxation. On the other hand, simple application make it a suitable tool for quick estimation of thermal stresses in asphalt layers of pavements and thus it may be used as one of the tools applied in prediction of low-temperature cracking.The conclusions were more accurate than in the original Hills and Brien'a method because data of stiffness modulus of asphalt concretes were taken from laboratory test of creep test at different temperatures. This is supported by the TSRST results which showed a quite satisfactory degree of correspondence between the tested and calculated thermal stresses.The type of bitumen used for mix production had a strong effect on the values of analysed thermal stresses.The lowest level of thermal stresses at fracture of specimens cooled at a rate of 10°C/h were obtained for asphalt concrete produced with the use of 50/70 multigrade bitumen.

References

[1] Hills, J.F., Brien, D.: The fracture of bitumens and asphalt mixes by temperature induced stresses. In: Proceedings of the Association of Asphalt Paving Technologists, vol. 35, pp. 292–309 (1966)

[2] Yoder, E.J., Witczak, M.W.: Principles of pavement design, 2nd edn., pp. 280–282. A Wiley Interscience Publication (1975)
[3] Heukelom, W., Klomp, A.J.G.: Road design and dynamic loading. In: Proceedings of the Association of Asphalt Paving Technologists, vol. 33, pp. 92–125 (1964)
[4] AASHTO TP10 – Standard Test Method for Thermal Stress Restrained Specimen Tensile Strength
[5] Monismith, C., Secor, G., Secor, K.: Temperature induced stresses and deformations in asphalt concrete. In: Proceedings Association of Asphalt Paving Technologists, vol. 34 (1965)
[6] Arand, W.: Behaviour of asphalt aggregate mixes at low temperatures. In: IV International RILEM Symposium, Budapest (1990)
[7] Pszczoła, M.: Low temperature cracking of asphalt layers of pavements, Ph.D. thesis, Gdansk University of Technology, Gdansk, Poland (2006)

A Four-Point Bending Test for the Bonding Evaluation of Composite Pavement

M. Hun, Armelle Chabot, and F. Hammoum

LUNAM Université, IFSTTAR, Route de Bouaye, CS4, F-44344 Bouguenais Cedex, France

Abstract. The aim of this paper is to present a specific four-point bending test with a specific model to help investigate the crack initiation and propagation at the interface between layers of composite pavements. The influence of the geometry on the delamination phenomenon in specimens is analyzed. Considering the deflection behavior of specimens, both experimental and analytical results are compared. Two different types of interface (concrete / asphalt and asphalt / concrete) are tested in static conditions. Different failure mechanisms whose mainly delamination is observed. The crack mouth opening displacement is monitoring by means of linear variable differential transducer (LVDT). The strain energy release rate is provided and compared successfully to the literature.

1 Introduction

Due to shrinkage phenomenon occurred in cement materials, the existing vertical crack through the cement concrete layer combined to environmental and traffic loadings affects the durability of composite pavements made with asphalt and cement materials. Two main problems have to be investigated: i) debonding mechanisms at the interface between two layers; ii) reflective cracking phenomenon through asphalt overlay or corner cracks in concrete overlay. This paper deals with the study of debonding. Previous research works have proposed some experimental devices to characterize the bond strength of asphalt-concrete interface in mode I [1]. But the combined normal and shear stresses near the edge of the layer as the vertical crack usually initiates and propagates the delamination [1]. The optimum design incorporating these variables has not been done yet. Mixed mode test to evaluate the delamination resistance is needed. On site, only few devices [4-5] allow testing the bond strength in mixed-mode. The literature review offers interesting ideas especially those on reinforced concrete beams and on concrete beams strengthened with composite materials [6].

In this paper, we propose to adapt existing four-point bending test (4PB) to bi-material specimens made with asphalt and cement material layers as illustrated in Figure 1. By using a specific elastic model, the influence of the specimen geometry and the material characteristics on internal stresses is presented. Then, experimental program is described and a discussion on static results is given.

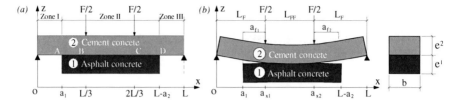

Fig. 1. (a) Schematic of test configuration, (b) Schematic adapted for calculating strain energy release rate calculation.

2 Quasi-analytical Investigation

The Multi-particle Model of Multi-layer Materials with 5 equilibrium equations per layer (M4-5n, n: total number of layers) [2] used to calculate stress and strain energy release rate on the 4PB test (Figure 1.b) is briefly presented. Considering homogenous, elastic and isotropic material assumptions, the specimen design is studied in order to optimize stresses to cause delamination between layers.

2.1 Introduction to the M4-5n

The M4-5n has five kinematic fields per layer i ($i \in \{1,...,n\}$): the average plane displacement $U_\alpha^i(x,y)$, the average out of plane $U_3^i(x,y)$ and the average rotations $\Phi_\alpha^i(x,y)$ ($\alpha \in \{1,2\}$). Stress field is assumed to be written with polynomial approximation in z (vertical direction) per layer i (characterized by e^i, E^i, υ^i, its thickness, Young modulus and Poisson ratio parameters). Its coefficients are expressed with the use of the classical Reissner generalized stress fields in (x,y) per layer i. These polynomial approximations have the advantage to define the normal stresses $\nu^{i,i+1}(x,y)$ and the shear stresses $\tau_\alpha^{i,i+1}(x,y)$ at the interface between i and i+1 layers. Theses stress fields are responsible for the delamination between layers at the edge or cracking location points. Hellinger-Reissner's formulation reduces the real 3D problem to the determination of regular plane fields (x,y) per layer i and interface i, i+1 (and i-1, i). This model can be viewed as superposition of n Reissner's plates, connected by means of an elastic energy that depends on the interlaminar stress fields [2]. The M4-5n advantage is to give finite value of stresses near the edge or crack permitted to identify easily delamination criteria [3].

In order to simplify the analysis, the 4PB test presented in Figure 1.a is simulated under the assumption of plane strain. Then, the mechanical fields depend only on the variable x. The problem is divided in three zones (see Figure 1.b). By mean of shear forces $Q_1^i(x)$ of layers 1 and 2, linking conditions of displacements, forces and moments between zones, the first and last single layer zone ($x \in [0, a_1]$ and $x \in [L - a_2, L]$) allow to pass on the support conditions of the beam at the bilayer zone ($x \in [a_1, L - a_2]$). On this central zone (where $n = 2$), different manipulations of M4-5n equations let to put finally into a system of second order differential equations in function of x only with the form Eqn. (1)

$$AX''(x) + BX(x) = C \quad \text{with} \quad X(x) = \begin{pmatrix} U_1^1(x) \\ \Phi_1^1(x) \\ Q_1^1(x) \\ U_1^2(x) \\ \Phi_1^2(x) \end{pmatrix} \quad (1)$$

where A, B, and C are the analytical matrices functions of geometric parameters, elastic characteristics of material behaviors and loading conditions specified (Figure 1.a). The expression of A, B, and C are given in Eqn. (2-4):

$$A = \begin{pmatrix} -\frac{e^{1^2}E^1}{2(1-v^{1^2})} & & \frac{e^{1^2}E^1}{12(1-v^{1^2})} & 0 & 0 & 0 \\ \frac{4}{15}\left(\frac{e^{1^2}}{(1+v^1)} + \frac{e^1 e^2 E^1 (1+v^2)}{E^2(1-v^{1^2})}\right) & & 0 & 0 & 0 & 0 \\ \frac{e^1}{5(1+v^1)} - \frac{e^1 E^1(1+v^2)}{5E^2(1-v^{1^2})} & & 0 & -\frac{13}{35}\left(\frac{e^1}{E^1} + \frac{e^2}{E^2}\right) & 0 & 0 \\ -\frac{e^1 e^2 E^1}{2(1-v^{1^2})} & & 0 & 0 & 0 & \frac{e^{2^2}E^2}{12(1-v^{2^2})} \\ \frac{e^1 E^1}{1-v^{1^2}} & & 0 & 0 & \frac{e^2 E^2}{1-v^{2^2}} & 0 \end{pmatrix} \quad (2)$$

$$B = \begin{pmatrix} 0 & 0 & -1 & 0 & 0 \\ -1 & -\frac{e^1}{2} & \left(\frac{1+v^1}{5E^1} - \frac{1+v^2}{5E^2}\right) & 1 & -\frac{e^2}{2} \\ 0 & -1 & \frac{12(1+v^1)}{5e^1 E^1} + \frac{12(1+v^2)}{5e^2 E^2} & 0 & 1 \\ 0 & 0 & 1 & 0 & 0 \\ 0 & 0 & 0 & 0 & 0 \end{pmatrix}; \quad C = \begin{pmatrix} 0 \\ -\frac{1+v^2}{5E^2} \cdot \frac{F}{2\times 1000} \\ \frac{12(1+v^2)}{5e^2 E^2} \cdot \frac{F}{2\times 1000} \\ -\frac{F}{2\times 1000} \\ 0 \end{pmatrix}, \text{if } x \in \left[a_1, \frac{L}{3}\right[\quad (3)$$

$$C = \begin{pmatrix} 0 \\ 0 \\ 0 \\ 0 \\ 0 \end{pmatrix}, \text{if } x \in \left[\frac{L}{3}, \frac{2L}{3}\right[; \quad C = \begin{pmatrix} 0 \\ \frac{1+v^2}{5E^2} \cdot \frac{F}{2\times 1000} \\ -\frac{12(1+v^2)}{5e^2 E^2} \cdot \frac{F}{2\times 1000} \\ -\frac{F}{2\times 1000} \\ 0 \end{pmatrix}, \text{if } x \in \left[\frac{2L}{3}, L - a_2\right] \quad (4)$$

The shear stresses $\tau_1^{1,2}(x)$ and normal stresses $v^{1,2}(x)$ of M4-5n at the interface between layer 1 and 2, are obtained analytically in function, respectively, of the unknowns of the system of Eqn. (1) and their derivative by the Eqn. (5) of interface behavior, and the equilibrium equation of shear forces of Eqn. (6). The sum of shear force of layers has to verify the condition as indicating in Eqn. (7).

$$\tau_1^{1,2}(x) = 15E^1 E^2 \frac{\left(U_1^2(x) - U_1^1(x) - \frac{e^1}{2}\Phi_1^1(x) - \frac{e^2}{2}\Phi_1^2(x) + \frac{1+\upsilon^1}{5E^1}Q_1^1(x) + \frac{1+\upsilon^2}{5E^2}Q_1^2(x)\right)}{4(e^1 E^2(1+\upsilon^1) + e^2 E^1(1+\upsilon^2))} \quad (5)$$

$$v^{1,2}(x) = -Q_1^{1'}(x) \quad (6)$$

$$Q_1^1(x) + Q_1^2(x) = \frac{F}{2 \times 1000} \text{ if } x \in \left[a_1, \frac{L}{3}\right[; \; 0 \text{ if } x \in \left[\frac{L}{3}, \frac{2L}{3}\right[; \; -\frac{F}{2 \times 1000} \text{ if } x \in \left[\frac{2L}{3}, L - a_2\right] \quad (7)$$

Eqn. (8) gives the M4-5n elastic energy W_e. According to linear elasticity theory for a system under constant applied load, the energy release rate can be expressed as in Eqn. (9) in case of the crack propagation along the interface (Figure 1.b).

$$W_e = \begin{bmatrix} \frac{(1-\upsilon^{2^2})}{2e^{2^3} E^2}\left(\frac{F}{1000}\right)^2 a_{x1}^3 + \frac{3(1+\upsilon^2)a_{x1}}{10e^2 E^2}\left(\frac{F}{1000}\right)^2 + \frac{e^1 E^1}{2(1-\upsilon^{1^2})}\int_{a_{x1}}^{a_{x2}}[U_1^{1'}]^2 dx \\ + \frac{e^{1^3} E^1}{12(1-\upsilon^{1^2})^2}\int_{a_{x1}}^{a_{x2}}[\Phi_1^1]^2 dx + \frac{e^2 E^2}{2(1-\upsilon^{2^2})}\int_{a_{x1}}^{a_{x2}}[U_1^{2'}]^2 dx + \frac{e^{2^3} E^2}{12(1-\upsilon^{2^2})^2}\int_{a_{x1}}^{a_{x2}}[\Phi_1^{2'}]^2 dx \\ + \frac{13e^1}{70E^1}\int_{a_{x1}}^{a_{x2}}[Q_1^{1'}]^2 dx + \frac{e^2}{2E^2}\int_{a_{x1}}^{a_{x2}}\left(\frac{(2Q_1^{1'}+Q_1^{2'})^2}{4} + \frac{17}{140}(Q_1^{2'})^2\right) dx \\ + \frac{6(1+\upsilon^1)}{5e^1 E^1}\int_{a_{x1}}^{a_{x2}}[Q_1^1]^2 dx + \frac{6(1+\upsilon^2)}{5e^2 E^2}\int_{a_{x1}}^{a_{x2}}[Q_1^2]^2 dx + \frac{1}{5}\int_{a_{x1}}^{a_{x2}}\left(\frac{(1+\upsilon^1)}{E^1}Q_1^1 + \frac{(1+\upsilon^2)}{E^2}Q_1^2\right)\frac{e^1 E^1}{1-\upsilon^{1^2}}U_1^{1'} dx \\ + \frac{2}{15}\left(\frac{e^1(1+\upsilon^1)}{E^1} + \frac{e^2(1+\upsilon^2)}{E^2}\right)\int_{a_{x1}}^{a_{x2}}\left[\frac{e^1 E^1}{1-\upsilon^{1^2}}U_1^{1'}\right]^2 dx \\ + \frac{(1-\upsilon^{2^2})}{2e^{2^3} E^2}\left(\frac{F}{1000}\right)^2 (L-a_{x2})^3 + \frac{3(1+\upsilon^2)(L-a_{x2})}{10e^2 E^2}\left(\frac{F}{1000}\right)^2 \end{bmatrix} \quad (8)$$

$$G = \frac{\partial W_e}{\partial A} \quad (9)$$

Both the methods of adimentionalisation and numerical resolution of equations by the Newmark finite difference scheme used by Pouteau [4] and Le Corvec [7] are adapted to this test. This method is programmed under the free software Scilab. For a symmetrical case, the excellent convergence of normal and shear stresses at the interface between layers at $x = a_1$ and $x = a_2$ is obtained in [8]. It has shown that the discretization of the x variable into 1200 elementary segments is sufficient. One simulation takes few seconds (CPU time). Interface ruptures are expected in mixed mode (mode I and II). The results have been compared successfully with finite element calculations and different static tests on Alu/PVC structure [8].

2.2 Effect of the Specimen Geometry and Material Characteristics on Stress Field

In the following, M4-5n simulations are done on material characteristics presented in Table 1. The total load of 4kN is chosen. The specimen geometry takes into account the space constraints of the test and heterogeneity of used material (span length 420mm, width 120mm, each layer thickness 60mm). Half of total load is applied at each third of span length.

The equivalent elastic modulus of the asphalt material depends of the temperature and the loading speed conditions. Simulations are performed for $1 < E_2/E_1 < 60$ in a symmetric case $a_1 = a_2 = 70$mm. Figure 2 shows that the more the Young modulus ratio between asphalt material (layer 1) and concrete material (layer 2) decreases, the more the tensile stress intensity at the bottom of layer 2 is maximal at points A and D relative to point B and C, and the more the intensities of normal and shear interface stresses are raised in absolute value at these points. This M4-5n parametric analysis indicates that the tensile stress at the bottom of the concrete layer 2 is in competition with interface stresses depending on the modulus of the asphalt. This variation influences the specimen rupture mode during the test.

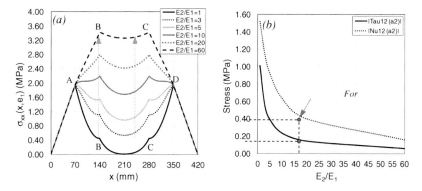

Fig. 2. Effect of Young modulus ratio between layers: (a) on the tensile stress at the bottom of layer 2 ($a_1=a_2=70$mm), (b) on the interface normal ($v^{1,2}$) and shear ($\tau^{1,2}$) stresses at $x=a_2=70$mm

Due to the specimen symmetry of preliminary results presented in [8], delamination can occur first or simultaneously with failure in concrete on either side of the specimen. To reduce the experimental cost for measuring the crack propagation and to get the maximum areas of damage towards one edge only, asymmetric specimens are explored numerically in the following. The length a_1 is fixed to 40mm with respect to the allowable distance from support to the edge of layer 1. For a low asphalt modulus condition, Figure 3 shows M4-5n simulations for a variable a_2 length. In Figure 3.a, the more the length a_2 increases, the more the tensile stress intensity at the bottom of concrete layer 2 is increasing under the loading point C and the more interface normal and shear stresses increase at the edge ($x = a_2$) (Figure 3.b). The parametric analysis confirms that the intensity of interface stresses at the edge ($x = a_2$) is increasing from 20% to 60% compared to those on the other side when the length a_2 is increasing. A compromise is still to be found between the tensile stress at the base of cement concrete layer and the shear stresses as well as the normal stresses at the edge of the interface.

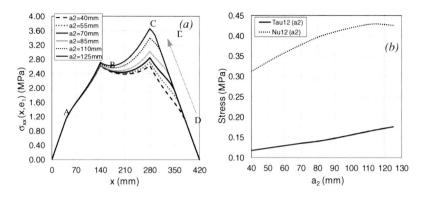

Fig. 3. Effect of variation of length a_2 (E_1 = 1600MPa): (a) on the tensile stress at the base of concrete layer, (b) on the normal ($v^{1,2}$) and shear ($\tau^{1,2}$) stress at $x=a_2$

3 Experimental Program

In this study, two types of interface were tested; (a) type I – concrete over asphalt known as Ultra Thin Whitetopping (UTW), (b) type II – asphalt bonded with concrete by a tack coat layer. The crack monitoring technique is investigated.

3.1 Test Specimens

In order to allow evaluation of bonding behavior, only one type of asphalt and cement concrete were used for all samples (see Table 1). A semi-coarse bituminous mix with aggregate size 0/10 and bitumen grade 35/50 is used. The cement CEM I 52.5R and the aggregate size 0/11 are used for cement concrete layer. Two types of specimen were made; (a) type I – concrete over asphalt known as Ultra Thin Whitetopping (UTW), (b) type II – asphalt overlay concrete with an intermediate tack coat layer. For type I, the cement concrete layer was cast directly on the prefabricated asphalt slab. For type II, the surface of concrete layer was cleaned by water blasting before tack coat placement. Then, the tack coat was placed on the concrete layer. A classical emulsion (C69 B 4) used for tack coat was kept at 45°C in autoclave. After placing the emulsion with the dosage of 0.4kg/m² of residual binder on concrete layer and leaving for 24 hours, the asphalt layer was placed and compacted by means of the plate compactor developed by LCPC. The composite slabs were sawed into a required dimension (see Table 2).

Table 1. Material characteristics

Material	E (MPa)	v	% air void	R_t (MPa)	R_c (MPa)
Cement concrete	34878MPa	0.25	2.57	3.46	47.67
Asphalt concrete	11258 (15°C, 10Hz)	0.35	9.59	-	-

Table 2. Dimensions of bilayer specimens and static test conditions (0.7mm/min)

Specimen name	L/e/b/a$_1$/a$_2$ (mm)	L$_{total}$ (mm)	Temperature (°C)	Test duration (s)
Type I-PT-3-1	420/60/120/70/70	480	20.0	20.0
Type I-PT-3-2	420/60/120/70/70	480	21.0	18.0
Type I-PT-1-3	420/60/100/70/70	480	20.0	28.0
Type I-PT-3-3	420/60/100/70/70	480	22.0	25.0
Type I-PT-1-1	420/60/120/70/70	480	21.0	17.5
Type I-PT-1-2	420/60/100/70/70	480	4.0	10.5
Type II-PT-1-1	420/60/120/70/70	480	6.0	13.4
Type II-PT-1-3	420/60/100/70/70	480	20.5	12.5
Type II-PT-2-1	420/60/120/40/70	480	22.0	11.0
Type II-PT-2-3	420/60/100/40/70	480	20.5	8.3

3.2 Test Setup and Conditions

To avoid any problems with the viscoelasticity and the thermo-susceptibility of asphalt material, the loading points and supports are placed on the concrete layer (Figure 1). The specimen geometry is designed to simulate the maximum stress intensity towards the edges of interface. Testing was performed by a hydraulic press. A linear variable differential transducer (LVDT) placed in the middle height of specimen section at the midspan was employed for measuring the deflection and controlling the imposed displacement test. The 4PB tests were conducted for bilayer specimens for various environmental conditions. During the test, the specimen was placed in a climatic chamber. The test temperatures and loading rates for each specimen are shown in Table 2.

3.3 Crack Propagation Monitoring

An ideal way of measuring the crack growth should give the possibility of continuous crack length determination without influencing the specimen or the delamination process itself. LVDT technique was chosen for this study. It consisted on using two LVDT per specimen edge fixed on asphalt layer and its respective ends supported on aluminum sheets attached to concrete layer (Figure 4). The two LVDT were placed at $d_0=10$ and $d_0+d_1=40$ mm distances from the edge. Figure 4.b represents the crack mouth opening displacement (CMOD) measured by the LVDT sensors in function of load and time. The crack length of delamination l_f is determined by knowing w_1 and w_2 values measured by LVDT during the test. Its expression is given in Eqn. (10) (Figure 4.b).

$$l_f = d_0 + d_1 + \frac{w_2 d_1}{w_1 - w_2} \tag{10}$$

4 Results and Discussion

4.1 Identification of Failure Phenomenon and Influence of Interface between Layers

Various kinds of failure mode were exhibited by the bilayer specimens under 4PB test around 20°C. Typical specimens after failure are depicted in Figure 4.a.

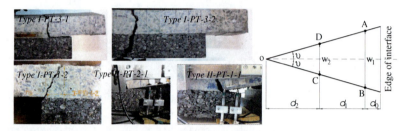

Fig. 4. (a) Typical failures of specimens, (b) Schematization of CMOD

Both types I and II specimens were delaminated by this proposed test. At 20°C, most (80%) of type I specimens were delaminated at the interface between layers. Only for one specimen (Type I-PT-3-2), a failure was observed in the central zone between the loading location points. The crack is propagated vertically from the bottom of the asphalt layer to the top of the concrete layer. Figure 2.a shows that a maximum tensile stress exists in this central part when the modulus ratio is high and if any defect in the material exists the crack can occurred. At low temperature (4°C), one specimen failure was located at the bottom of the concrete layer between central and edge zones (Type I-PT-1-2) which confirms the previous elastic modeling (point A and D of Figure 2.a). For the type II specimen, all specimens were delaminated at the interface between layers not only at low temperature (6°C) but also at high temperature (20°C).

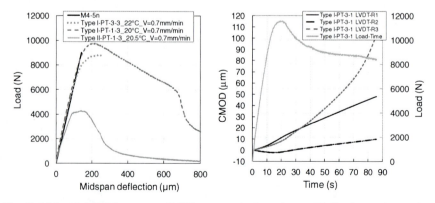

Fig. 5. (a) Load-deflection curve of different types of specimen; (b) Crack mouth opening displacement measured by LVDT

In the modeling, the asphalt modulus value was taken from its master curve at the test temperature and by converting the static test duration (T) into the frequency (f=1/T). In Figure 5.a, it is shown that the maximum load of type I specimen is about 50% more higher than the maximum load of type II specimen. The dissymmetric specimens were successfully delaminated as explained previously in the M4-5n analytical analysis. The combined approach with the 4PB test and M4-5n can evaluate the bonding between layers.

4.2 Stress Intensity at Edge of Interface and Energy Release Rate

According to the experimental results, the delamination is usually dissymmetric. Knowing the failure load (experimentally determined) for a specimen pre-crack length a_0, the energy $W(a_0)$ stored in the specimen for this load was calculated. It performs the same calculation for a pre-crack length a_0+da, and the energy release rate was calculated by the relation presented in Eqn. (9).

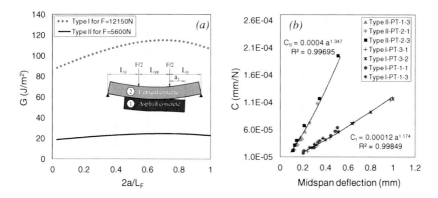

Fig. 6. (a) Evolution of G for different types of specimen; (b) Compliance curves $C = R.a^n$ for different types of specimen (at 20°C)

The evolution of the energy release rate is given in function of the normalized crack length $2a/L_F$ (Figure 6.a). Based on the derivative of the energy release with respect to the crack length, the crack growth is stable at the crack length of 50mm. In the other way, the energy release rate can also be determined experimentally by using the compliance method. From the load-deflection curve, the relation of compliance is determined. The compliance C is, in general, expressed by C = u/P where u is the midspan deflection related to the load P. Figure 6.b represents the compliance curve for different type of specimens. The compliance versus respective crack length a can be plotted and fit with the expression $C = Ra^n$. Therefore, the energy release rate can be found as $G = \dfrac{P^2}{2b} nRa^{n-1}$. Table 3 shows a summary of the interface stress intensity and the energy release rate which are comparable to the values found in literature [1]. The results show that the interface normal and shear stress of type I specimen are approximately 50% higher than those of type II specimen. But due to the self weight effect of asphalt layer on failure, the type II specimen test needs to be improved.

Table 3. Stress intensity at the interface and energy release rate

Specimen type	Failure load (N)	τ (MPa)	ν (MPa)	$G (J/m^2)$ for a crack length of 2mm	
				Model	Experiment
Type I	9760 - 12150	0.36 – 0.41	0.99 – 1.14	88 - 98	82 - 106
Type II	4300 - 5600	0.17 – 0.21	0.49 – 0.59	19 - 28	64 - 83

5 Conclusions

Experimental results on bilayer specimens, in accordance with quasi-analytical analysis given by the M4-5n, have demonstrated that the proposed 4PB test can determine the interface behavior of bilayer materials, asphalt-concrete and concrete-asphalt. The crack growth was monitoring by means of a LVDT technique. An approximate crack length was obtained. For better measuring the crack length and understanding the failure phenomenon, the Digital Image Correlation technique will be used for the next experimental campaign. For the geometry chosen, the specific test has shown mixed mode failure at the interface between layers. Comparisons with experimental results and analysis of failure modes given above demonstrate that the M4-5n can be used effectively for designing the specimen and as well as for analyzing the test.

References

[1] Tschegg, E.K., Macht, J., Jamek, M., Steigenberger, J.: ACI Materials Journal. Title no. 104-M52, 474–480 (2007)
[2] Chabot, A.: Analyse des efforts à l'interface entre les couches des matériaux composites à l'aide de Modélisations Multiparticulaires des Matériaux Multicouches (M4), ENPC - PhD thesis (June 1997)
[3] Caron, J.F., Diaz, A.D., Carreira, R.P., Chabot, A., Ehrlacher, A.: Comp. Sc. and Technology 66(6), 755–765 (2006)
[4] Pouteau, B.: Durabilité mécanique du collage blanc sur noir dans les chaussées, PhD thesis, Ecole Centrale de Nantes (December 2004)
[5] Chabot, A., Pouteau, B., Balay, J.-M., De Larrard, F.: In: Al-Qadi, Scarpas, Loizos (eds.) Proc. of the 6th Int. RILEM Conf. Pavement Cracking. CRC Press (2008)
[6] Achintha, M., Burgoyne, C.J.: Construction and Building Materials 25, 2961–2971 (2011)
[7] Le Corvec, G.: Simulations des effets du retrait du béton de ciment sur la flexion de matériaux de chaussées fissurées, Master thesis, Univ. Nantes (2008)
[8] Hun, M., Chabot, A., Hammoum, F.: In: Proc. of the $20^{\text{ème}}$ Congrès Français de Mécanique, paper n. 569, Besançon, France (2011)

Assessment of Cracking Resistance of Bituminous Mixtures by Means of Fenix Test

R. Miró[1], A. Martínez[1], F. Pérez-Jiménez[1], R. Botella[1], and G. Valdés[2]

[1] Technical University of Catalonia, BarcelonaTech, Barcelona, Spain
[2] Universidad de la Frontera, Temuco, Chile

This paper shows the application of a new direct tensile test developed at the Road Research Laboratory of the Department of Transport and Regional Planning of the Technical University of Catalonia. The test is called Fénix test and is aimed at the assessment of cracking resistance of different types of bituminous mixtures at different temperatures. Fénix test calculates the dissipated energy during the cracking process of the material and the softening phase in the load-displacement curve of the test. The test procedure consists of subjecting one half of a 63.5 mm thick cylindrical specimen of a 101.6 mm diameter prepared by Marshall or gyratory compaction to a tensile stress at a constant displacement velocity (1 mm/min) and specific temperature. A 6 mm-deep notch is made in the middle of its flat side where two steel plates are fixed. The specimen is glued to the steel plates with a thixotropic adhesive mortar containing epoxy resins. Both plates are attached to a loading platen so that they can rotate about fixing points. Different types of mixtures with different stiffness moduli, including very flexible mixtures like gap-graded and SMA mixes, as well as stiffer mixtures like high modulus mixtures, were tested. The results were compared with those obtained for the same mixtures from standard fatigue bending beam tests.

1 Introduction

Asphalt cracking is one of the most common causes of pavement distress. The cracking behaviour of asphalt concrete mixtures is difficult to analyze due to their rheological characteristics. Crack formation and propagation are caused by several factors, usually of environmental (thermal cycles and material aging) or mechanical (traffic loads) nature. These factors trigger mechanisms like top-down cracking, flexural cracking and fatigue cracking, resulting in geometrical typologies or patterns such as longitudinal cracking, block cracking, transversal cracking, fatigue cracking, among many other types described in pavement distress manuals, along with causes and remedies [1-4].

The scientific community is applying fracture mechanics concepts on quasi-brittle materials to understand the cracking behaviour of bituminous mixtures. Analytical models and experimental studies which try to simulate crack initiation and propagation are commonly used [5].

A literature search returned three tests whose main goal is to determine fracture properties of bituminous mixtures, Figure 1. The single-edge notched beam test, SE(B), provides an adequate mode I fracture thanks to its set-up and sample geometry. However, it cannot be applied to field cores due to sample shape [6]. In the case of the semicircular bending test, SCB, [5, 7, 8], sample shape makes the test suitable for both field cores and laboratory specimens and its set-up is simple, but the crack propagation with the SCB geometry creates an arching effect with high compressive stress as the crack approaches the top edge [6]. The disk-shaped compact tension test, DC(T), has a standard fracture test configuration in ASTM D 7313-07, as well as a larger sample fracture area, leading to improved tests results. On the other hand, sample preparation may weaken the area around the loading points, and moreover it is difficult to carry out the test at temperatures above 10°C.

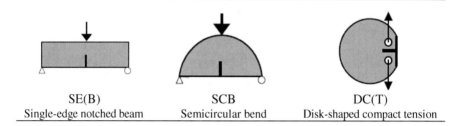

SE(B)	SCB	DC(T)
Single-edge notched beam	Semicircular bend	Disk-shaped compact tension

Fig. 1. Different specimen fracture geometries

In this research line, the Road Research Laboratory of the Technical University of Catalonia has developed a new test to evaluate cracking resistance in bituminous mixtures called Fénix test [9-11]. In this test, the dissipated energy during the process, which is a combination of dissipated creep energy and fracture energy, is calculated. The obtained values allow the determination of the resistance to cracking of bituminous mixtures. The test set-up can be seen in the following section.

2 Study

The effect of bitumen type and content on the cracking resistance of a coarse bituminous mix (G20) typically used in base courses was analyzed in a temperature range from 20 to -10°C. Three different penetration grade bitumens, i.e., 60/70, 40/50 and 13/22 with bitumen contents ranging between 3.5% and 5.5% by weight of aggregate, were selected to obtain a wide variety of stiffness indices, from flexible mixes (containing 5.5% of bitumen 60/70 at 20°C) to very stiff mixes (containing 3.5% of bitumen 13/22 at -10°C). The characteristics of bitumens are specified in Table 1 while the gradation of the limestone aggregates, which is fitted to the lower limit of the Spanish grading envelope, is shown in Table 2.

Table 1. Characteristics of bitumens

Bitumen characteristics	Unit	13/22	B40/50	B60/70
Penetration (25°C; 100 g; 5s)	0.1 mm	17	43	64
Penetration index	-	0.1	-0.2	-0.2
Ring-and-ball softening point	°C	67.3	55.9	51.7
Fraass brittle point	°C	-5	-12	-17
Ductility at 25°C	cm	15	>100	>100
Dynamic viscosity at 60°C	Pa.s	4551	651	367
Dynamic viscosity at 135°C	Pa.s	1.92	0.72	0.56
Elastic Recovery at 13°C	%	-	-	-
RTFOT Residue				
Mass loss	%	0.35	0.4	0.5
Penetration (25°C; 100 g; 5s)	% p.o.	10	23	32
Softening point increase	°C	7.5	9.5	9.6
Ductility at 25°C	cm	7	18	50

Table 2. Gradation of mixture G20

Sieve Size (mm)	25	20	12.5	8	4	2	0.5	0.25	0.125	0.063
Gradation (% passing)	100	75	55	40	25	19	10	7	6	5

Several series of Marshall specimens were prepared for each bitumen type and content and direct tensile tested by the Fénix test. The Fénix test, which was developed by the Road Research Laboratory of the Department of Transport and Regional Planning of the Technical University of Catalonia, allows evaluating the cracking resistance of asphalt mixes by calculation of the dissipated energy during the cracking process of mixtures.

The results from the test were compared with those from other tests in order to narrow the variation range of parameters defining cracking behaviour.

The Fénix test is a tensile test applied to one half of a cylindrical specimen prepared by Marshall or gyratory compaction. A 6mm-deep notch is made in the middle of its flat side where two steel plates are fixed. The plates are attached to the loading platen using two cylindrical bolts so that each plate can rotate about its fixed edge, as illustrated in Figure 2. The test is carried out under controlled displacement conditions. Displacement velocity is established at 1 mm/min and test temperature is chosen according to the environmental conditions to be simulated.

A number of parameters describing the mechanical behaviour of mixtures, such as peak load, F_{max}, displacement at peak load, ΔF_{max}, failure displacement (displacement at $F = 0.1$ kN post-peak load), Δ_R, tensile stiffness index, I_{RT}, and energy dissipated during fracture, G_D, can be determined.

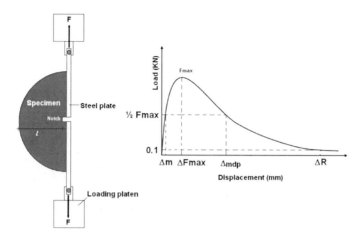

Fig. 2. Fénix test set up and load-displacement output curve

Mechanical parameters like peak load, F_{max}, displacement at peak load, ΔF_{max}, and failure displacement, Δ_R, are determined from the load-displacement curve. The tensile stiffness index, I_{RT}, is obtained using Eqn. (1):

$$I_{RT} = \frac{\frac{1}{2} \cdot F_{max}}{\Delta_m} \tag{1}$$

where I_{RT} = tensile stiffness index, kN/mm; F_{max} = peak load, kN; Δ_m = displacement before peak load at ½ F_{max}, mm.

The dissipated energy during cracking, G_D, is calculated by Eqns. (2) and (3):

$$G_D = \frac{W_D}{h \cdot l} \tag{2}$$

where G_D = dissipated energy during test application, J/m²; W_D = dissipated work during test application, kN·mm; h = specimen thickness, m; l = initial ligament length, m.

$$W_D = \int_0^{\Delta R} F \cdot du \tag{3}$$

where F = Load, kN; u = displacement, mm; ΔR = displacement at F = 0.1 kN post-peak load, mm.

3 Analysis of Results

The analysis of the effect of bitumen type and content and temperature on cracking resistance and fracture energy of mixtures tested by the Fénix test reveals that this procedure is sensitive to variation of both variables.

As an example, Figure 3 summarizes the results obtained for mixture G20 containing bitumen 60/70. Note the transition from ductile to brittle behaviour exhibited by the mixture with temperature variation. Moreover, at 5°C the mixture has higher fracture energy, i.e. the area below the curve is larger than at the other temperatures.

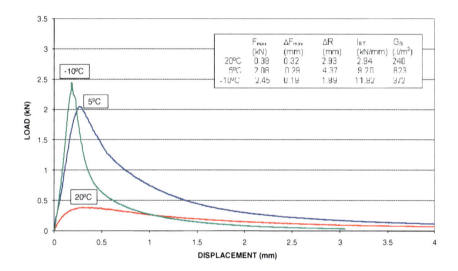

Fig. 3. Load-displacement curves at 20, 5 and -10 °C, Fénix test. Mix G20 4.5% Bitumen 60/70

It can be seen that, for harder binders, the results at low temperatures (5 and -10°C) are very similar. That is, at low temperatures, the stiffer mixtures tend to behave similarly although it is also possible that test sensitivity decreases at extreme temperatures.

Figure 4 plots the variation in the dissipated energy (G_D) with respect to the stiffness index (I_{RT}) of the mixtures for all test conditions. A solid line connects the results for mixtures at the same temperature with the same bitumen type but different content.

The results are grouped in different areas according to the temperature of the mixture and, within this area, according to the stiffness provided by the bitumen. Thus, the envelope of results at 20°C forms a sort of triangle where the more flexible mixtures, i.e. those with low I_{RT} (mixtures with bitumen 60/70), have low dissipated energy which hardly varies with bitumen content. By contrast, in stiffer mixtures, i.e. those with higher I_{RT} (mixtures with bitumen 13/22), changes in bitumen content result in significant dissipated energy variation.

The envelope of results at 5 and -10°C creates, respectively, two areas which form a triangular shape almost symmetrical to the previous one. Results at 5°C are found in the top part while results at -10°C are plotted in the bottom part. At 5°C, changes in bitumen content lead to large dissipated energy variations whereas at -10°C, the energy does not vary significantly with different bitumen contents. For the latter temperature, only the mixture containing bitumen 60/70 shows variations in the stiffness index with changing the bitumen content.

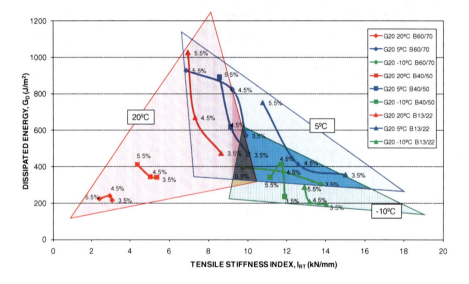

Fig. 4. Dissipated energy versus tensile stiffness index. Mix G20

The second phase of the study consisted in collecting the results for several types of semi-dense (S or AC) and gap-graded (M or BBTM) mixtures containing different penetration grade bitumens and polymer-modified bitumens tested at different temperatures, Table 3, in order to overlap the results in the previous figure and compare the fracture energy.

Two parabolic envelopes enclose the area containing many of the mixtures, Figure 5. For the same stiffness, the most crack resistant mixtures are near the upper parabola while those close to the lower parabola should not be designed. Note also that some mixtures are always near the upper parabola with temperature variation, thus exhibiting a better behaviour than those close to the lower parabola.

Table 3. Characteristics of the mixtures studied in the second phase.

Mixture	Bitumen Type	Bitumen Content (% weight of aggregate)	Testing Temperature (ºC)
S20	60/70	4.3	20, 5, -10
	40/70	4.3	20, 5, -10
	13/22	4.3	20, 5, -10
	BM-3c	4.3	20, 5, -10
AC22S MAM (1)	13/22	5.26, 6.38	20
AC22S MAM (2)	13/22	4.71, 5.82, 6.95	20
AC16S	60/70	4.71	20, 5, -5
BBTM11A	60/70	5.49	20, 5, -5
BBTM11B	BM-3c*	4.99, 5.54, 6.10	20
M10 (1)	BM-3c*	6.72	20, 5
M10 (2)	BM-3c*	6.38	20, 5
M10 (2)	RAF-AV**	6.72	20, 5
SMA	BM-3c*	6.38	20

*: polymer modified bitumen.
**: high performance polymer modified bitumen.

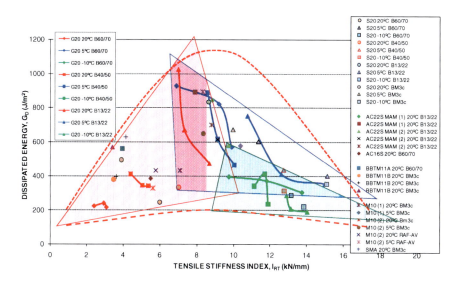

Fig. 5. Dissipated energy versus tensile stiffness index. Mixtures and test conditions

The determination of the stiffness index and dissipated energy by the Fénix test shows differences in behaviour between mixtures which, for the same stiffness, have different energies. In this case, the mixture with the highest energy has the greatest resistance to cracking. As an example, Figure 6 shows the fatigue laws obtained by a three-point bending beam test (in accordance with European standard UNE-EN 12697-24) of mixtures S20 and G20 prepared with bitumen 13/22 and tested at 20°C. Both fatigue laws are similar and it is difficult to know which of two mixtures is more resistant to cracking. However the dissipated energy of mixture S20 obtained by Fénix test is clearly higher than that of G20, Table 4, meaning that the former has better resistance to cracking than the latter.

Table 4. Parameters obtained from three-point bending beam (3PBBT) and Fénix test

Mixture Type	Flexural Modulus (3PBBT) (MPa)	I_{RT} (Fénix Test) (kN/mm)	G_D (Fénix Test) (J/m^2)
S20	11556	8.6	833
G20	9130	7.3	670

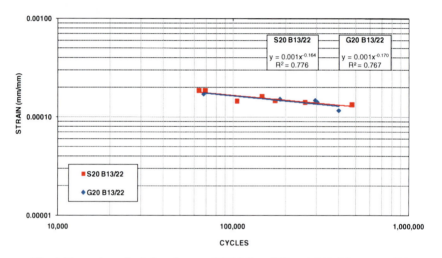

Fig. 6. Flexural tensile fatigue laws at 20°C. Mixes S20 and G20, Bitumen 13/22

Therefore, the Fénix test allows selecting the mixtures with the highest resistance to cracking during the mix design phase more easily than fatigue tests.

4 Conclusions

The Fénix test determines the cracking resistance of mixtures. Additionally, it allows differentiating the brittle and ductile response of a bituminous mixture and observing the effect of bitumen type and content, as well as test temperature, on the behaviour of the mixture.

The main conclusion drawn from the present work is that the Fénix test supplies information of the dissipated energy during the cracking process, G_D. This parameter, together with the stiffness modulus, makes it possible to establish differences in the behaviour of mixtures. Thus, it is observed that higher dissipated energy values lead to for a better response to cracking for similar moduli.

Acknowledgements. Some of the results collected during the second phase of the study were obtained from several research projects. The authors would like to make a special mention of REHABCAR project, supported by the Spanish Ministry of Science and Innovation (MICINN) within the INNPACTO 2010 program, currently under development.

References

[1] Molenaar, A.A.A.: Fatigue and Reflective Cracking due to Traffic. J. Assoc. Asphalt Paving Technol. 53, 440–474 (1984)
[2] Myers, L.A., Roque, R., Birgisson, B.: Propagation Mechanisms for Surface-Initiated Longitudinal Wheel Path Cracks. Transport Res. Rec. J. Transport Res. Board, No. 1778, 113–122 (2011)
[3] Asphalt Institute, The Asphalt Handbook, Asphalt Institute Manual Series No. 4 (MS-4), USA (2007)
[4] Koh, C., Lopp, G., Roque, R.: Development of a Dog-Bone Direct Tension Test (DBDT) for asphalt concrete. In: Proc. of the 7th Int. RILEM Symp. on Advanced Testing and Characterization of Bituminous Materials, vol. 1, pp. 585–596 (2009)
[5] Li, X., Braham, A., Marasteanu, M., Buttlar, W., Williams, R.: Effect of Factors Affecting Fracture Energy of Asphalt Concrete at Low Temperature. Road Mater. Pavement Des. 9, 397–416 (2008)
[6] Wagoner, M., Buttlar, W., Paulino, G.: Disk-shaped Compact Tension Test for Asphalt Concrete Fracture. Exp. Mec. 45(3), 270–277 (2005)
[7] Molenaar, A., Scarpas, A., Liu, X., Erkens, S.: Semi-Circular Bending Test; Simple but Useful? J. Assoc. Asphalt Paving Technol. 71, 795–815 (2002)
[8] Mull, M., Stuart, K., Yehia, A.: Fracture Resistance Characterization of Chemically Modified Crumb Rubber Asphalt Pavement. J. Mater. Sci. 37, 557–566 (2002)
[9] Pérez-Jiménez, F., Valdés, G., Miró, R., Martínez, A., Botella, R.: Fénix test: development of a new test procedure for evaluating cracking resistance in bituminous mixtures. Transport Res. Rec. J. Transport Res. Board, No. 2181, 36–43 (2010)
[10] Pérez Jiménez, F., Valdés, G., Botella, R., Miró, R., Martínez, A.: Approach to fatigue performance using Fénix test for asphalt mixtures. Constr. Build. Mater. 26, 372–380 (2012)
[11] Pérez Jiménez, F.E., Valdés, G., Botella, R.: Experimental study on resistance to cracking of bituminous mixtures using the Fénix test. In: Proc. of the 7th Int. RILEM Symp. on Advanced Testing and Characterization of Bituminous Materials, vol. 2, pp. 707–714 (2009)

Development of Dynamic Asphalt Stripping Machine for Better Prediction of Moisture Damage on Porous Asphalt in the Field

M.O. Hamzah[1], M.R.M. Hasan[1], M.F.C. van de Ven[2], and J.L.M. Voskuilen[3]

[1] School of Civil Engineering, Engineering Campus, Universiti Sains Malaysia, Malaysia
cemeor@eng.usm.my
[2] Section of Road and Railway Engineering, Delft University of Technology, The Netherlands
[3] Centre for Transport and Navigation, Delft, The Netherlands

Abstract. Stripping is a major source of pavement distress and takes place in the presence of moisture. Over the years, many laboratory tests have been proposed to evaluate moisture sensitivity of asphalt mixtures. This paper presents the development of a dynamic asphalt stripping machine (DASM) to realistically simulate stripping of porous asphalt mixtures subjected to the dynamic action of flowing water. To assess the effectiveness of the machine, two sets of specimens were prepared. One set was conditioned in the DASM by allowing water at 40°C to continuously permeate through the unextruded samples via water sprinklers at an intensity equivalent to 5400 mm/hr. The other set was stored under dry conditions at ambient temperature. Then, both sets of specimens were extruded and conditioned in an incubator before individually tested for Indirect Tensile Strength (ITS) at 20°C after 1, 3, 5 and 7 days. Resistance to stripping was evaluated from the ratio between ITS when tested wet and dry. Specimen permeability was also measure before and after conditioning. In addition, mortars that stripped from the asphalt samples were filtered on a filter material. The results showed that both ITSR and permeability reduces with conditioning time. The ITSR after 7-day conditioning was 17.2% lower than those conditioned for one day. The quantity of mortars collected on the filter material was found to increase with conditioning time. However, some stripped mortars were believed to be trapped in the mixture capillaries, and this explained the reduction in coefficient of permeability values over time.

Keywords: Flowing water action, Moisture damage, Stripping, Porous asphalt, Mortar loss.

1 Introduction

In many countries, the asphalt pavement is constantly exposed to wet conditions and high volume of water runoff due to heavy rainfall throughout the year. The prolonged exposure to water and moisture may expose the pavement to deterioration. According to Lu and Harvey [1], air voids, pavement structure, rainfall intensity and pavement age have the highest influence on moisture damage while repeated loading and cumulative truck traffic have a marginal effect.

Dawson et al. [2] mentioned that stripping was generally attributed to water infiltration into the asphaltic mixture, causing weakening of the mortar, and aggregate-mortar bond. Due to the continuous action of water and traffic loading, progressive dislodgement of aggregates could occur. Several distresses in the form of ravelling, rutting or cracking may occur in the pavement due to stripping [3]. Dawson et al. [2] stated that open-graded mixtures were deliberately designed and laid to help drain surface water. This tends to allow some water to reside more or less permanently within the mixture, contributing to the development of water-induced damage. The physical processes that had been identified as important contributors to water damage were the molecular diffusion of water through the mixture component and 'wash away' of the mortar due to the movement of water through the connected macro pores. A mechanical process that was identified as a contributor to water damage was the occurrence of an intense water pressure field inside the mixture caused by traffic loads and known as the pumping action with pressures up to 7 atm.

Caro et al. [4] mentioned that environmental conditions such as high intensity rainfall periods, high relative humidity, severe freeze-thaw cycles and other extreme environmental conditions also affected moisture damage mechanisms. These environmental conditions increased the rate and amount of moisture that could reach the material which in turn raised the damage potential. Furthermore, in-service conditions such as ageing and the dynamic loading produced by traffic had also been considered important contributors to moisture damage.

Kringos and Scarpas [5] stated that the high permeability of open graded wearing surface ensured fast drainage of water away from the road surface. However, it caused a negative effect on the material characteristics of the individual components of the asphalt and damaged the bond between the components that led to premature separation of the aggregates from the wearing surfaces due to ravelling. Kringos and Scarpas [5] also stated that damage in asphaltic mixtures could be characterised into three failure modes, namely, the washing away of mortar, damage of the mortar–aggregate interface and dispersion of the mortar. The strength of the bond between mortar and aggregates diminished in the presence of water, which, among others, could be related to the surface energies of the individual components [6].

The resistance to stripping is typically evaluated by initially immersing the specimen in the water or conditioning the specimens at a certain temperature without considering the flowing (dynamic) water action on the asphalt mixtures. However, asphalt pavements in the field are constantly exposed to the wet conditions and high volume of water run-off throughout the year, especially in the tropical monsoon climate. The asphalt pavement performance is adversely affected by stripping and unforeseen increase in maintenance budgets are often the consequence [7]. The need to unfold an understanding of the mechanism and to develop a simple but reliable test is essential. According to the Asphalt institute [8], there have been many efforts carried out in the United States in the past few years to come up with an improved laboratory test method to better predict moisture damage problem in the field.

Therefore, this paper was initiated as a contribution to the continuous effort of improving the laboratory test methods for better prediction of porous asphalt (PA) stripping in the field. Current test procedures for evaluation of moisture damage on asphalt mixtures includes Marshall immersion test, Modified Lottman test - AASHTO T283 [9] Hamburg wheel tracking test and BS EN 12697-12 [10]. In all these test methds, the action of water on the mix is primarily static in nature and involved soaking, freezing and thawing of the specimen. In the context of porous asphalt, this method does not realistically simulates the actual stripping mechanism taking place in the field where water permeates through the porous asphalt wearing course. Hence, a new machine, known as the dynamic asphalt stripping machine (DASM) was developed to simulate the dynamic action of water on PA at a laboratory scale. Nevertheless, the pumping effect during wet weather of the truck loads travelling at high speeds is not simulated.

2 Description of the DASM

The DASM is fully equipped with a water storage tank, water recycling tank, removable perforated plare, water pump and heater as well as sprinklers to simulate the rainy conditions and subject the specimens to the dynamic action of water. The schematic diagram and actual view of the DASM is shown in Figure 1.

The water recycling tank was equipped with a water heater with an accuracy of $\pm 1^\circ C$ to ensure that the temperature of the circulating water can be adjusted anywhere between ambient to $100^\circ C$. In addition, both tanks were equipped with insulator and cover to maintain a constant water temperature throughout the conditioning process. During the conditioning period, water from the storage tank was channelled through the sprinkler onto the specimen by gravitational force. The water that passed through the specimen was collected by the removable tray and drained away to the recycling tank through the drainage pipe that was fixed at the bottom of the tray. Water that was accumulated in the storage tank was channelled through the filter and pumped up to the storage tank and the process was repeated until conditioning process was completed. A filter was also located under the perforated plate to trap the mortar from the specimen which was carried away by the flowing water.

The discharge pipes are the most important component of the DASM. These pipes were fitted with ball valves, flexible hoses, sprinklers and acrylic cylinders for the conditioning purposes. The ball valve functioned as an adjuster to control the intensity of water sprinkled onto the specimen. The flexible hose was chosen to ensure the ease of handling when placing the specimens into the machine for conditioning. Meanwhile, a standard home appliance sprinkler was used to simulate the rainy condition and dynamic action of water while the acrylic cylinder was fitted to hold the specimen that was confined in the Marshall Mould in place to avoid the loss of water due to splash and control the consistency of water temperature. Subsequently, to avoid leakage of water, a rubber washer was used at every connection between UPVC adapter and PVC nut at the discharge pipes. The machine has nine discharge pipes which enabled nine samples to be conditioned and tested simultaneously.

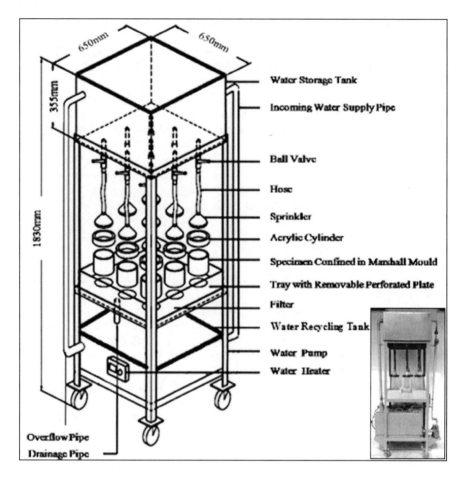

Fig. 1. Schematic diagram and the actual view (right corner) of the DASM

3 Preliminary Evaluation

Few important steps need to be considered before proceeding with the conditioning procedure of the specimens. First and foremost, it is essential to ensure that the water was free from any pollutants and the water temperature was constant throughout the conditioning process. The temperature of water used in this conditioning method was 40°C, as adopted in the immersed wheel-track test to induce stripping on wet samples [12]. Additionally, this temperature was selected to ensure the tight bonding between the specimen and mould was maintained throughout the test since the specimen (unextruded and confined in the Marshall mould) were later tested for permeability. The permeability test was carried out to evaluate the effects of stripped binder due to flowing action of water. Higher temperatures would caused the samples to slip from the mould, hence destroying the tight bond between sample and mould. A special filter was used and located in the water recycling tank to eliminate the undesired particles in the circulating water. The water from both tanks was circulated for approximately 2 hours to make sure these undesired particles were removed from water and to achieve a constant water temperature. In addition, the asphalt sample confined in the Marshall mould was used to take advantage of the strong bond between the sample and the wall of the mould.

In this assessment, the specimens were conditioned and tested at 1, 3, 5 and 7 days. Two sets of specimens for wet and dry conditions were prepared. The wet specimens were subjected to an average water intensity of 5400 mm/hr (5 mm opening of gate valve) which was approximately 12 times higher than the rainfall intensity of 432.4 mm/hr at 100 years ARI [11] to accelerate the conditioning process.

4 Materials and Tests

The aggregate material used was granite, obtained from a local quarry. A conventional bitumen 60/70 penetration grade and hydrated lime were respectively used as the binder and filler materials to prepare the PA specimen. The basic properties of all materials are summarised in Table 1. The proposed PA gradation with 14mm NMAS (Table 2) was adopted in this study. A detailed explanation on the development of the proposed gradation is available in Hasan [13]. The cylindrical specimens, prepared at 4.0% binder contents, were compacted via impact mode at 50 blows per face. The samples were then left to cool overnight at ambient temperature before conditioning process commenced.

Table 1. Materials properties

Properties	Aggregate (Crushed	Filler (Hydrated	Bitum (60/70
Abrasion Loss (%)	23.6	-	-
Aggregate Crushing Value (%)	21.5	-	-
Flakiness Index (%)	21.8	-	-
Water Absorption (%)	0.7	-	-
Polished Stone Value	51.8	-	-
Specific Gravity (g/cm^3)	-	2.350	1.030
Penetration at 25°C (dmm)	-	-	63
Softening Point (°C)	-	-	49
Ductility at 25°C (cm)	-	-	> 100

Table 2. Adopted PA aggregate gradation [13]

Sieve Size (mm)	% Passing
14	100
10	80
5	15
2.36	10
0.425	5
0.075	2

*Filler content is 2% of total aggregate mass.

Then, the set of dry specimens were stored in a closed cabinet at ambient temperature over their corresponding conditioning period. Meanwhile, the permeability test was conducted on the set of wet specimens before and after their respective conditioning period under DASM to determine the effects of water on the occurrence of mortar stripped from the mix. Part of the stripped mortar was expected to disrupt air voids continuity hence reduction in the coefficient of permeability of PA. The changes of permeability of the specimen was expressed in terms of the coefficient of permeability (k) measured using a falling head water permeameter. The permeability test was conducted on a not-extruded sample to take advantage of the strong bond between the sample and the walls of the mould.

Subsequently, both sets of specimens were extruded and placed in an incubator for 4 hours at 20°C prior for the ITS test. The dimensions of the wet (W) and dry (D) specimens were measured and air voids (Table 3) were determined. The ITS test on the dry and wet specimens were conducted on the same day and the results obtained was used to calculate the indirect tensile strength ratio (ITSR). The specimen was loaded by compressive force between two loading strips which acted parallel to the vertical diameter of the specimen. The maximum failure load was recorded and the ITS test was done in accordance with ASTM D4123 procedure [14].

Table 3. Air voids of specimens

Mix Condition	Designation	Average Air Voids (%)
Wet	WD1	24.95
	WD3	24.47
	WD5	24.37
	WD7	24.48
Dry	DD1	24.52
	DD3	24.96
	DD5	24.36
	DD7	24.39

4 Result and Discussion

4.1 Effects of Flowing Water on Stripped of Mortar and Permeability

Water that continuously permeated through the porous samples induced stripping of the mortars. Some of the mortars stripped from the aggregate were filtered through the material and retained on the DASM filter as shown in Figure 2. Figure 2(a) shows the actual initial colour of the filter while Figures 2(b), 2(c) and 2(d) show the condition of the filter after the specimens were conditioned for 3 days, 5 days and 7 days, respectively. It is hypothesed that some stripped mortars were trapped in the capillaries, disrupting air voids continuity and this explains the reduction in coefficient of permeability (k) over time in Figure 3. Another source of permeability reduction can be explained in terms of binder creep as detailed out by Hamzah et al. [15]. The measured air voids of the PA specimens used in this study is shown in Table 3.

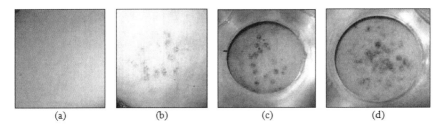

Fig. 2. Condition of filter under the perforated plate (a) Before and (b) 3 days; (c) 5 days; (d) 7 days after conditioning

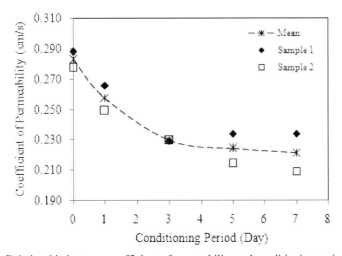

Fig. 3. Relationship between coefficient of permeability and conditioning period of PA

Table 4 shows the One-Way Analysis of Variance (ANOVA) results on the reduction of coefficient of permeability. It is confirmed that the mortar losses significantly disrupt the continuity of air voids and caused the reduction in permeability as indicated by the p-value less than 0.05.

Table 4. One-way ANOVA on the coefficient of permeability values

Source	Sum of Squares	df	Mean Square	F	p-value
Conditioned Duration	697.396	4	174.349	44.478	< 0.001
Error	19.599	5	3.920		
Total	716.995	9			

4.2 Indirect Tensile Strength

The result indicates that the ITS of the dry specimens slightly increases over conditioning time due to binder hardening over time as depicted in Figure 4. However, the ITS of the wet specimens shows otherwise due to stripping induced by the dynamic action of flowing water. The gap between the ITS of dry and wet specimens, increases as the conditioning period increases with the widest gap reaching 23% at day 7.

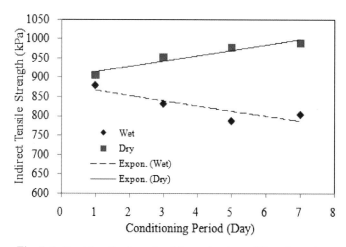

Fig. 4. Indirect tensile strength of dry and wet conditioned specimens

The indirect tensile strength ratio (ITSR) is often used to characterize the moisture susceptibility of asphalt mixtures. ITSR was calculated based on the equation given in AASHTO T283 [9]. Figure 5 shows that the ITSR of the specimens decreases as the conditioning period increases. The ITSR results indicate that this conditioning procedure significantly influences the resistance of PA specimen to stripping.

The data was further analyzed statistically using the One-Way (ANOVA) at 95% confident interval ($\alpha = 0.05$). The analysis was carried out on the ITS values for both dry and wet specimens. The ANOVA results are shown in Table 5 with the dynamic action of water has a significant effect on the indirect tensile strength of PA specimens.

Table 5. One-way ANOVA on effects of dynamic stripping on ITS of specimens

Source	Sum of Squares	Df	Mean Square	F	p-value
Condition of Specimen	67033.094	1	67033.094	45.004	< 0.001
Error	20853.080	14	1489.506		
Total	87886.173	15			

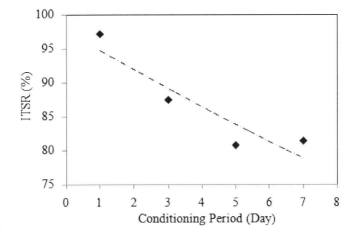

Fig. 5. ITSR at different conditioning period

5 Conclusions

A dynamic asphalt stripping machine was invented to simulate the actual stripping mechanism of porous asphalt mixtures subjected to high rainfall intensities and the dynamic action of flowing water. Based on the preliminary evaluation, the dynamic action of water has a significant effect on the indirect tensile strength ratio and permeability loss. Some stripped mortars were trapped in the capillaries, disrupting air voids continuity and this explains the reduction in mix coefficient of permeability over time. However, further study needs to be carried out to establish the relationship between the test results from the newly developed machine with other performance parameters and field (stripping) performance of porous asphalts. Nevertheless, the results of this preliminary investigation ascertained the ability of the dynamic asphalt stripping machine to evaluate the resistance to water damage of porous asphalt mixtures.

Acknowledgments. The authors would like to acknowledge the Ministry of Science, Technology and Innovation that has funded this research project through the eScience Fund that enables this paper to be written. Many thanks are also due to technicians of the Highway Engineering Laboratory at the Universiti Sains Malaysia.

References

[1] Lu, Q., Harvey, J.T.: Field investigation of factors associated with moisture damage in asphalt pavements. In: 10th International Conference of Asphalt Pavements (ISAP), Quebec, Canada, pp. 691–700 (2006)
[2] Dawson, A., Kringos, N., Scarpas, T., Pavšič, P.: Water in the pavement surfacing. In: Water in Road Structures, Geotechnical, Geological, and Earthquake Engineering, ch.5, vol. 5, pp. 81–105 (2009)
[3] Alam, M.M., Vemuri, N., Tandon, V., Nazarian, S., Picornell, M.: A test method for identifying moisture susceptible asphalt concrete mixes. Research project 0-1455: Evaluation of environmental conditioning system (ECS) for predicting moisture damage susceptibility of HMAC, The Centre for Highway Materials Research, The University of Texas at El Paso, El Paso, TX 79968-0516 (1998)
[4] Caro, S., Masad, E., Bhasin, Little, D.N.: Moisture susceptibility of asphalt mixtures, Part 1: mechanisms. International Journal of Pavement Engineering 9(2), 81–98 (2008)
[5] Kringos, N., Scarpas, A.: Raveling of asphalt mixes due to water damage, computational identification of controlling parameters. Transportation Research Record, Journal of the Transportation Research Board, 1929, 79–87 (2005)
[6] Cheng, D.X., Little, D.N., Lytton, R.L., Holste, J.C.: Moisture Damage Evaluation of Asphalt Mixtures by Considering Both Moisture Diffusion and Repeated-Load Conditions. Transportation Research Record: Journal of the Transportation Research Board, No. 1832, 42–49 (2003)
[7] Kiggundu, B.M., Roberts, F.L.: Stripping in HMA mixtures: State-of the- art and critical review of test methods, NCAT Report 88-2. National Center for Asphalt Technology (1988)
[8] Asphalt Institute, The asphalt handbook, Manual Series No. 4 (MS-4), 7th edn., USA (2007)
[9] AASHTO, AASHTO T283: Standard method of test for resistance of compacted asphalt mixtures to moisture-induced damage, 22nd edn. American Association of State Highway and Transportation Officials, Washington, DC (2002)
[10] CEN, BS EN 12697-12, Bituminous mixtures - Test methods for hot mix asphalt – Part 12: Determination of the water sensitivity of bituminous specimens. European Committee for Standardisation, Brussels (2008)
[11] MSMA, Chapter 13: Design rainfall, urban stormwater management manual, Manual Saliran Mesra Alam, Part D: Hydrology and hydraulics, vol. 4. Design fundamentals, Department of Irrigation and Drainage, Malaysia (2001)
[12] Read, J., Whiteoak, D.: The Shell bitumen handbook, 5th edn., London (2003)
[13] Hasan, M.R.M.: Studies of binder creep, abrasion loss and dynamic stripping of porous asphalt, M.Sc Thesis, Universiti Sains, Malaysia (2011)

[14] ASTM, ASTM D4123: Standard test method for indirect tension test for resilient modulus of bituminous paving mixtures,, Road and Paving Materials; Vehicle Pavement Systems, vol. 04(03). Annual Book of American Society for Testing and Materials (ASTM) Standards, West Conshohocken, PA 19428-2959, United States (1999)

[15] Hamzah, M.O., Hasan, M.R.M., Van De Ven, M.F.C.: Permeability loss in porous asphalt due to binder creep. Journal of Construction and Building Materials 30, 10–15 (2012)

Effect of Wheel Track Sample Geometry on Results

P.M. Muraya[1] and C. Thodesen[2]

[1] NorwegianUniversity of Science and Technology
[2] SINTEF

Abstract. The wheel track test can be used as a means of comparing the permanent deformation behaviour of different types of asphalt mixtures. One advantage of this test method lies in the fact that it can be applied to laboratory compacted specimens and specimens extracted from pavements in the field. However, differences in results have been noted between core samples obtained in the field and laboratory prepared samples with similar mix designs. It is thought that potential reasons for such deviations may be due to differences in the confining pressure of the sample during wheel track loading. One possible reason for such differences may lie in the fact that laboratory prepared samples are composed of rectangular asphalt slab, while field samples are circular and are encased in plaster of Paris to provide testing stability.

This paper describes a study that was performed to investigate the effects of sample geometry on wheel track test.The purpose of this study was to investigate how the specimen geometry affects the outcome of asphalt pavement rutting evaluations conducted through wheel track testing. This research is particularly applicable in Norway where more and more emphasis is placed on designing asphalts to avoid rutting and also to provide correlations between field and laboratory studies.The results of this paper indicate that the encased samples (circular) generally exhibit lower deformation than the square samples.

1 Introduction

Permanent deformation is one of the most important modes of failure in asphalt pavements. The wheel track test provides a means of assessing the permanent deformation susceptibility of asphalt mixtures in the laboratory. The test can be used to assess both laboratory compacted specimens and specimens obtained from the field. In Norway, wheel track test is used to assess the permanent deformation susceptibility of asphalt mixtures. In accordance to the Norwegian pavement specifications, the test is conducted on 200 mm diameter cored field specimens using procedure B of the NS-EN 12697-22.

A lot of wheel track testing has been conducted at the NTNU road laboratory using procedure B of the NS-EN 12697-22. These tests have been aimed at assessing the permanent deformation susceptibility of different types of asphalt mixtures. The wheel track tests have been conducted on both laboratory prepared specimens measuring 30.5 cm by 30.5 cm by 4 cm height and 200 mm diameter specimens cored from the field. However, differences in results have been noted between core samples obtained in the field and laboratory prepared samples with similar mix designs.

It is thought that one of the potential reasons for such deviations may be due to differences in the confining pressure of the sample during wheel track loading.During wheel track testing, the asphalt mixture is laterally supported around the perimeter by the edges of the mould. In case of the square specimen, this lateral support is offered by the steel edges that lie at least 15.25 cm from the centre of the sample. In case of the circular specimens, the lateral support is provided by the edge of palster of Paris surrounding the specimen. This edge lies 10 cm from the centre of the sample. The location of the edge support can affect the lateral confinement generated during loading and as a consequence affect the permanent deformation during wheel track testing. To some extent, the effect of the edge location for asphalt materials can be compared to the effect of lateral support on the bearing capacity of soils. Studies [1][3] conducted on the bearing capacity of soils under isolated footings suggest that the bearing capacity is affected by the nature of lateral support. The bearing capacity is highly dependant on the dimensions of the lateral supports.

The differences between the laboratory prepared samples and field cores for the same type of asphalt mixture may also be as a result of other factors. These include factors such as aging and degree of compaction. Since the aim of this study was to investigate the effect of specimen geometry, the other factors were kept constant and two types of specimen geometry were considered.

2 Materials and Methods

2.1 Asphalt Mixtures

Four types of asphalt mixtures were considered in this study. The mixtures included two types of dense asphalt concrete mixtures and two types of stone mastic asphalt mixtures. An overview of the specimens prepared from these mixtures is shown in Table 1. These types of asphalt mixtures were selected because they are widely used in pavement construction in Norway. Two square specimens and two circular specimens were prepared for each type of asphalt mixture. All specimens were prepared using ready made asphalt mixtures that were obtained from mixing plants and compacted to a target air voids content of 3.5%.The 3.5% air voids content was the desired level of air voids in the asphalt mixtures during field compaction. After compaction, the air voids content in the specimens was determined based on the measured dimensions of the specimens.

Table 1. Overview of specimens prepared

Label	Type of asphalt mixture	Type of bitumen	Nominal maximum size (mm)	Type of specimen	Number of specimens	Target air voids content (%)	Measured air voids content (%)	
							specimen 1	specimen 2
AC 11 70/100 Sq	Dense asphalt concrete	70/100	11	Square	2	3.5	3,7	3,9
AC 11 70/100 Cir	Dense asphalt concrete	70/100	11	Circular	2	3.5	3,4	3,6
AC 11 PMB Sq	Dense asphalt concrete	PMB	11	Square	2	3.5	3,6	3,6
AC 11 PMB Cir	Dense asphalt concrete	PMB	11	Circular	2	3.5	3,8	-
SMA 11 70/100 Sq	Stone mastic asphalt	70/100	11	Square	2	3.5	3,8	3,8
SMA 11 70/100 Cir	Stone mastic asphalt	70/100	11	Circular	2	3.5	2,9	3,4
SMA 8 70/100 Sq	Stone mastic asphalt	70/100	8	Square	2	3.5	3,7	3,6
SMA 8 70/100 Cir	Stone mastic asphalt	70/100	8	Circular	2	3.5	3,6	3,6

2.2 Specimen Preparation

The square and circular specimens were compacted in a square steel mould fitted with a collar. The internal dimensions of the mould were 30.5 cm by 30.5 cm by 2 cm while the collar had a height of 2 cm. These internal dimensions of the mould and the collar are shown inFig. 1.

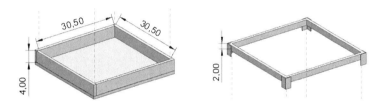

Fig. 1. Internal dimensions of the mould and collar (dimensions in cm)[4]

The dimensions of the specimens are shown inFigure 2.The specimens were composed of ready made asphalt mixtures that were obtained from mixing plants.Prior to compaction, the ready made asphalt mixtures together with the compaction mould were heated to the required temperatures for about four hours.

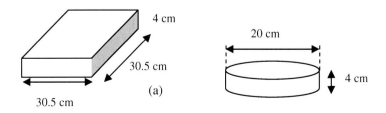

Fig. 2. Specimen dimensions (a) square and (b) circular

Both types of specimens were compacted using a procedure developed by Anton in 2007[5][5]. This procedure was developed with the aim of attaining a uniform density in the compacted specimen. In this procedure, the specimen is first compacted by hand then by a special roller compacter shown in Fig. *3*. The roller compacter consists of an arched compaction plate that rolls forwards and backwards on the surface of the specimen. If necessary, vibration can also be applied as the compaction plate rolls on the specimen.

Fig. 3. Special roller compaction equipment

Prior to hand compaction, half of the required weight of the specimen is placed evenly in a compaction mould that is fitted with a collar. This portion is then hand compacted in an orderly manner along edges of the mould after which the interior part of the specimen is compacted in a similar manner. The remaining portion of the mixture is then placed in the mould and again hand compacted in the same manner as the first portion. The mixture is then compacted to the required height using the roller compacter.

After compaction, the specimens were left to stand for a period of at least one day. The specimens to be used in the preparation of the circular specimens were then frozen, extracted from the moulds and the 200 mm diameter circular specimens cored. After coring, the surface dry density was determined and the circular specimens placed in a plastic spilt mould measuring 280 mm by 280 mm by 52.5 mm height. An illustration of this plastic mould is shown in Fig. 4. The square specimens were tested without being removed from the mould.

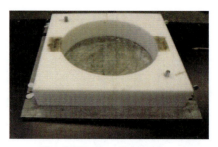

Fig. 4. Plastic split mould

2.3 Tests

The tests were performed in air at a temperature of 50^0C using a small scale test deviceshown Fig. 5 in accordance to Norwegian pavement specifications. The device consists of two wheels that can be used to test two specimens in parallel. Prior to testing, the specimens are conditioned in the test chamber at the required test temperature for a period of at least 4 hours. The wheels are then lowered on the specimen and the test started. Each wheel transmits a load of 700 N. During testing, the specimens move forwards and backwards. A forward or a backward movement constitutes of one load pass.

Fig. 5. Wheel track test device with circular sample

The rut depth is measured at a total of 27 points along the wheel track with the distance between two consecutive points being 4 mm as illustrated in Fig. 6. One measurement is taken at the centre of the wheel track and at 13 other points on either side of the track centre. The rut depth is then calculated as an average of these 27 measurements.

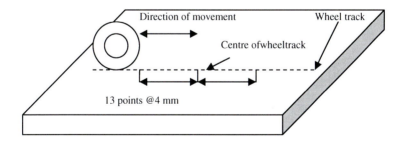

Fig. 6. Points of measurement

3 Results

Figure 7 provides an illustration of the results that were obtained for all the specimen types. From this figure it can be seen that there are apparent differences in the rut depths experienced by the SMA 11 70/100 over the course of 10,000 load cycles. This figure also provides a visual indicator of the variation level between geometry samples as well as between testing samples using the same geometry. Generally, the samples using the same mix and sample geometry yielded smaller variations than the samples which utilized differing asphalt mixes and geometries. This indicates a possible effect of geometry and mix type on deformation development.

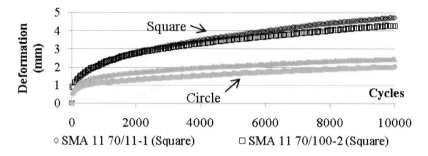

Fig. 7. Development of deformation in square and circular SMA 11 70/100 samples

3.1 Mix Type

As seen in Figure 8, the asphalt type has a clear effect on the deformations occurring on the mix. It come as no surprise to see that the AC11 PMB mix is the least susceptible to deformation, however, what is more unexpected is the extent of the variations among the mixes not using PMB. Specifically in the case of the SMA 11 sample it can be seen that the difference in specimen geometry can account for a doubling of the deformation occurring.

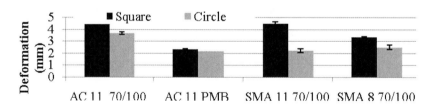

Fig. 8. Effect of asphalt type and sample geometry on deformation.

Typically the rut depth is also expressed in terms that demonstrate the depth of the slab tested. This value is known as the proportional rut depth shown and is calculated using Eqn. 1.

$$\epsilon_p = \frac{rut\ depth}{specimen\ thickness} \times 100 \qquad \text{Eqn. 1}$$

The Wheel tracking slope (WTS) is also routinely calculated and reported during wheel track testing, the WTS values is calculated using Eqn. 2.

$$WTS = \frac{d_{10000} - d_{5000}}{5} \qquad \text{Eqn. 2}$$

Where, d_{10000} and d_{5000} are the deformation after 10000 and 5000 load passes respectively. WTS is expressed in terms of mm/ (10^3 load cycle).

As seen in Figures 9 and 10 the proportional rut depth and wheel track slope were somewhat variable when the different sample geometries were used.

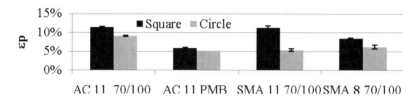

Fig. 9. Effect of asphalt type and sample geometry on proportional rut depth

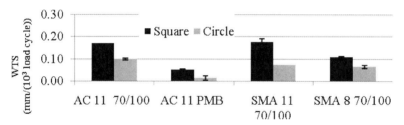

Fig. 10. Effect of asphalt type and sample geometry on wheel track slope

3.2 Effect of Sample Geometry

In Figure 11 the average values along with the standard deviation bars are shown for the asphalt samples tested in this research project. Using ANOVA (α=0.05) to analyse the data it was found that the differences between circular and square samples were not significantly different (P-value=0,053). This is somewhat surprising given the apparent differences apparent from the figure, however, it is likely due to the significant variations present.

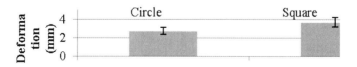

Fig. 11. Summary of effect of sample geometry on deformation

The difference in the sampe geometries was however evident when ANOVA was used to analyze the the proportional rut depth. In this case it was seen that the difference between the circular and square samples was indeed statistically significant (P-value=0,037), thus confirming the influence of sample geometry on asphalt sample rut development. As seen in Figure 12, the samples with the square geometry yielded the higher average proportional rutting.

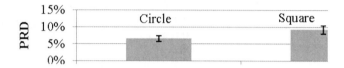

Fig. 12. Summary of effect of sample geometry on proportional deformation

Similarly to the analysis of the proportional rut depth, the analysis of the WTS indicated that the differences between the circular and the square samples were statistically significant (P-value=0,017). This indicates that the wheel track slope is indeed dependent on the sample geometry being used during the wheel track testing. As seen in Figure 13, the average WTS value was higher for square samples than for circular samples.

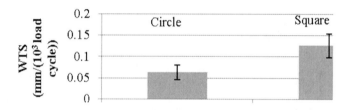

Fig. 13. Summary of effect of sample geometry on Wheel track slope

Table 2 shows the ranking of the final rut depth based on the square and the circular samples. The table shows that ranking can be affected by the specimen geometry. The AC 11 asphalt mixtures are ranked in the same order by both the square and circular samples. However, the square and the circular samples rank the SMA asphalt mixtures in a different order.

Table 2. Ranking of the observed permanent deformation

Ranking - square samples

Label	ε_p (%)	Rank
AC 11 PMB Sq	5,8	1
SMA 8 70/100 Sq	8,4	2
SMA 11 70/100 Sq	11,3	3
AC 11 70/100 Sq	11,4	4

Ranking - circular samples

Label	ε_p (%)	Rank
AC 11 PMB Cir	5,2	1
SMA 11 70/100 Cir	5,3	2
SMA 8 70/100 Cir	6,2	3
AC 11 70/100 Cir	9,0	4

4 Conclusions

The study in this paper shows that the geometry of the wheel track specimens affects the permanent deformation of asphalt mixtures. The deformation of the circular specimens was lower than that of the square specimens. In addition, the geometry also affected the ranking order for the different asphalt mixtures. This suggests that the comparison of laboratory compacted mixes and field extracted specimens should be based on specimens with the same geometry.

References

[1] Gupta, R., Trivedi, A.: Bearing Capacity and Settlement of Footing Resting on Confined Loose Silty Sands. The Electronic Journal of Geotechnical Engineering 14, Bundle A (2009)
[2] Singh, V.K., Prasad, A., Agrawal, R.K.: Effect of Soil Confinement on Ultimate Bearing Capacity of Square Footing Under Eccentric–Inclined Load. The Electronic Journal of Geotechnical Engineering 12 Bundle E (2007)
[3] El Sawwaf, M., Nazer, A.: Behavior of Circular Footings Resting on Confined Granular Soil. Journal of Geotechnical &Geo Environmental Engineering 131(3) (2005)
[4] Anastasio, S.: Laboratory verification of predicted permanent deformation in asphalt materials following climate change, Master of Science Thesis, Norwegian University of Science and Technology, Norway (2010)
[5] Morten, A.: Evaluation and Quality Assurance of a New Norwegian Mix Design System for HMA, Master of Science Thesis, Norwegian University of Science and Technology, Norway (2006)

Performance of 'SAMI'S in Simulative Testing

O.M. Ogundipe[1], N.H. Thom[2], Andrew C. Collop[3], and J. Richardson[4]

[1] University of Nottingham, UK
[2] University of Nottingham, UK
[3] De Montfort University, UK
[4] Colas, UK

Abstract. Although the use of Stress Absorbing Membrane Interlayers (SAMIs) to control reflective cracking has been proven to be effective under certain circumstances, the mechanisms involved are not agreed and design is therefore difficult for engineers to implement with confidence. This paper describes simulative laboratory tests utilizing two different SAMI systems, part of a broader project designed to throw light on the subject. The simulation was conducted to investigate the effectiveness of SAMIs under trafficking. Trafficking was carried out in the Nottingham Pavement Test Facility on a pavement consisting of a thin overlay to a cracked asphalt substrate, and this demonstrated a measurable benefit from the two SAMI systems trialed. Data analysis was then carried out and deductions made regarding those properties of SAMIs that appeared to contribute most to reflective crack resistance.

1 Introduction

Rehabilitated pavement is often plagued with the problem of reflective cracking. Traffic and thermal loadings have been identified as the major causes of reflective cracking [1-4]. Although, laboratory tests have been used successfully to evaluate the performance of materials and mixtures, it is not practicable to implement laboratory findings directly in the field without field or large scale testing being carried out, because the field conditions cannot entirely be replicated or simulated in the laboratory. To bridge the gap, it is necessary to carry out field or accelerated pavement testing. Accelerated pavement testing is generally defined as the application of wheel loads to specially constructed or in-service pavement to determine response and performance under a controlled and accelerated accumulation of damage in a short period of time [5]. In this study, the Nottingham pavement test facility was used to evaluate the performance of SAMIs under traffic loading.

SAMIs are interlayers designed to dissipate energy by deforming horizontally or vertically, therefore allowing the movement (vertical/horizontal) of the underlying pavement layers without causing large tensile stresses in the asphalt overlay. Barksdale [6] defined a stress-relieving interlayer as a soft layer that is usually thin and is placed at or near the bottom of the overlay. He stated further that the purpose of the soft layer is to reduce the tensile stress in the overlay in the vicinity of the crack in the underlying old layer and hence "absorb" stress.

The application of stress-relieving systems at the interface between the overlay and the old pavement surface reduces the shear stiffness of the interface. Debondt [1] proved using theoretical analysis that the reduction of shear stiffness allows slip of the interface, thereby isolating the overlay from the stress concentration of the crack tip.

2 The Pavement Test Facility

The pavement test facility (PTF) is made up of the following: reaction beams that provide the necessary reaction for any lateral position of the loading frame and the main beam; the load carriage used to mount the guide bearings and wheel loading assembly; the cable system which consists of 8mm cable wound around a 150 mm drum; the hydraulic system which consists of a hydraulic power pack (oil pump), hydraulic motor and a servo valve; and the feedback transducers and electronic control system used to monitor the carriage speed, carriage position and the wheel load [7]. A schematic of the pavement test facility is shown in Figure 1. The wheel movement is controlled by the hydraulic motor which pulls the cable (steel ropes) in both directions (forward and backward). It was designed to apply a load magnitude of up to 12kN at a maximum speed of 14.5 km/hr. The PTF pavement has length, width and depth of 5.0 m, 2.4 m and 1.5m, respectively.

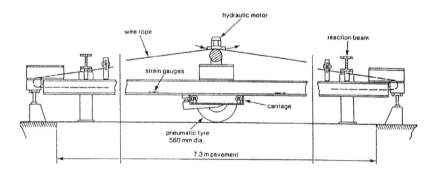

Fig. 1. Schematic of the Pavement test facility (side view) [8]

3 Material Properties

The materials required for the construction of the pavement are clay subgrade, crushed rock subbase material, 10 mm asphalt concrete with 40/60 penetration grade bitumen for the base and surface (overlay) layers and Fibredec® A and B as SAMIs. The SAMIs are prepared by sandwiching 60mm of glass fibre stands at a rate of $120g/m^2$ between two layers of bitumen emulsion and 6 mm aggregates spread at a rate of 8kg/m² are compacted onto the sandwiched glass fibres. Fibredec A is prepared with an ordinary bitumen emulsion while Fibredec B is prepared with polymer modified emulsion. The viscosities of both emulsions at 25°C, 30°C and 40°C are shown in Table 1.

Table 1. Viscosity of bitumen emulsion used in SAMIs

Bitumen emulsion	Viscosity (Pa.s) @ 25°C	Viscosity (Pa.s) @ 30°C	Viscosity (Pa.s) @ 40°C
Ordinary bitumen emulsion	0.700	0.580	0.390
Polymer modified bitumen emulsion	0.184	0.194	0.180

3.1 Subgrade and Subbase Layers

The strength of the subgrade and subbase layers was determined using the Dynamic Cone Penetrometer (DCP). The DCP has an 8kg weight dropping through a height of 575 mm and a 60° cone having a diameter of 20 mm. The result of the DCP test is shown in Figure 2. The California bearing ratio (CBR) was determined from the DCP data using the software UK DCP version 3.1 described by Done and Piouslin [9]. Also, the stiffness of the sections was calculated from equation 1 reported by Powell et al [10]. The California bearing ratio and the stiffness values are shown in Table 2. The subgrade has average CBR and stiffness of 1.33% and 21MPa, respectively, while the subbase has average CBR and stiffness of 18.33% and 113MPa, respectively.

$$E = 17.6 CBR^{0.64} \qquad (1)$$

Where,
E = Elastic modulus; and
CBR = California bearing ratio

Table 2. CBR and stiffness of subgrade and subbase, derived from DCP tests

Sections	Subgrade		Subbase	
	CBR (%)	Stiffness (MPa)	CBR (%)	Stiffness (MPa)
1	1	17.6	21	123.52
2	2	27.43	17	107.89
3	1	17.6	17	107.89
Average	1.33	21	18.33	113

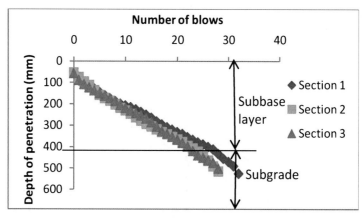

Fig. 2. DCP test results

3.2 Base and Surface (Overlay) Layers

The base and surface layers as earlier stated were made of 10 mm asphalt concrete with 40/60 penetration grade bitumen. Specimens for testing were prepared and compacted at 130°C into a mould 305 mm × 305 mm × 130 mm to a thickness of 60 mm using a roller compactor. Five cores of diameter 100 mm and thickness 40 mm were cored from each slab. The indirect tensile stiffness modulus (ITSM) test and indirect tensile fatigue test (ITFT) were carried out. The procedures for the indirect tensile stiffness method (ITSM) test and indirect tensile fatigue test (ITFT) are described in British/EN Standards [11, 12], respectively. The air voids and ITSM results at 10°C, 20°C and 30°C are shown in Table 3. The fatigue line presented in Figure 3 has a slope of 0.215. This shows that the mixture has a good fatigue characteristic.

Table 3. Indirect stiffness modulus test and air void results (nearest hundred)

Specimens	1	2	3	4	5	Average
Air voids(%)	8.00	7.79	7.97	7.50	7.54	7.76
ITSM at 10°C	8100	7500	7200	7300	7700	7560
ITSM at 20°C	4700	4300	3600	4300	4300	4240
ITSM at 30°C	2300	2200	1900	2000	2200	2120

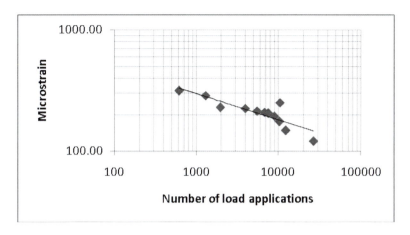

Fig. 3. Fatigue line of the asphalt

4 PTF Pavement Construction

The existing granular materials in the PTF track were removed to expose the clay subgrade (with a capping of fine sand). Crushed rock subbase material was spread and compacted with a vibrating plate in three layers to a thickness of 400 mm as shown in Figure 4(left). The first and second layers of the subbase were compacted to a thickness of 130 mm, while the third layer was compacted to a thickness of 140 mm. The 10mm asphalt concrete with 40/60 penetration grade bitumen was laid and compacted using a pedestrian roller to achieve a thickness of 60 mm for the base as shown in Figure 4(right).

To create the crack, the pavement was divided into sections. A transverse crack was created at the centre of each section (simulating an existing pavement). The cut thickness was about 5mm (thickness of the blade). Also, to study the situation where cracks are closely-spaced in the field, cracks were created at 200 mm from the end and at the end of each section. The diagram of the cuts and the PTF base layer with the cuts is shown in Figure 5.

Fig. 4. PTF subbase and base layers

The SAMI layers for sections 1 and 3 were Fibredec® A and B, respectively. Fibredec® A and B were prepared by sandwiching 60 mm glass fibre strands between two layers of bitumen emulsion and 6 mm aggregates were compacted on them using a vibrating plate. Ordinary bitumen emulsion was used to prepare Fibredec® A, while polymer modified emulsion was used for Fibredec® B. Section 2 was given no treatment (Control). A layer of sandwiched glass fibre strands is shown in Figure 6. The surface layer was made of 10 mm asphalt concrete with 40/60 penetration grade bitumen. The asphalt was prepared and compacted at an average temperature of 130°C using a pedestrian roller. The pavement structure of sections 1, 2 and 3 is shown in Figure 7.

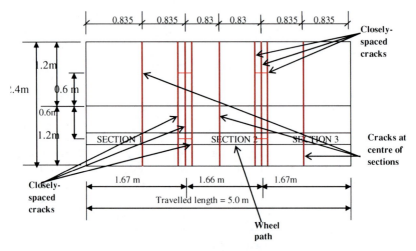

Fig. 5. Simulated cracks in PTF sections

Fig. 6. PTF SAMI

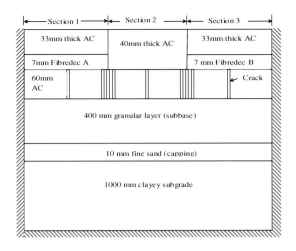

Fig. 7. Pavement structure for sections 1, 2 and 3

5 PTF Instrumentation and Trafficking

The measurements taken in this study are the numbers of wheel cycles to the first appearance of cracks and to failure. The number of wheel cycles to the first appearance of cracks was taken when a crack was first noticed on the overlay, while the full opening of cracks (cracks open and close as the wheel passes) was chosen as the failure criterion. The wheel path was painted white to monitor the appearance of cracks on the surface layer (Figure 8). The pavement was trafficked using a 9.6kN wheel load at an average speed of 3km/hr. The number of wheel repetitions as the wheel load moved forward and backward was logged with the use of an electronic counter. A digital thermocouple was used to monitor the room temperature during the test.

Fig. 8. Wheel path painted white

6 Test Results

The average room temperatures in the morning, noon and evening during trafficking were 22.7°C, 25.9°C and 26.9°C, respectively. The numbers of wheel cycles to first appearance of cracks and to failure are presented here.

6.1 Number of Wheel Cycles to Failure

The number of wheel load applications to the first appearance of cracks and failure are shown in Figure 9. The figure shows that the crack appeared first in the control section 2 with no SAMI. This indicates that both sections 1 and 3 with Fibredec® A and B, respectively performed better than section 2 with no SAMI. They had lives 2.92 and 1.93 times that of the section with no SAMI, respectively. Also, section 1 with Fibredec® A performed better than section 2 with Fibredec® B. This was in agreement with the results of a small-scale wheel tracking test carried out as part of a broader project on reflective cracking [13]. The bitumen emulsion used in Fibredec® A is thought to aid the SAMI's performance. The results show that the SAMIs in this study were able to retard reflective cracking. The crack resistance of the SAMIs is governed by their ability to dissipate energy by deforming vertically within their elastic limit, which is achieved by using material of lower stiffness and with a lower shear stiffness between the SAMIs and overlay.

To investigate a situation where cracks are closely spaced, cracks were simulated in the base layer by cutting three transverse cracks 200mm apart as shown in Figure 5. The numbers of wheel cycles to first appearance of cracks on the overlay and to failure are presented in Figure 10. It was observed that cracks appeared at the surface shortly after trafficking started. This was thought to be due

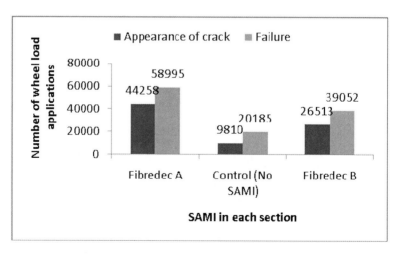

Fig. 9. Number of load applications to the first appearance of cracks and failure for the cracks at centre

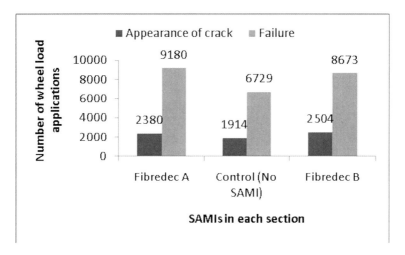

Fig. 10. Number of load applications to the first appearance of cracks and failure for the cracks closely-spaced

to the fact that the surfacing layer (overlay) was laid and compacted in three parts, thereby creating joints in the overlay close to the closely-spaced cracks in the base layer. However, this condition was the same for the three sections; therefore, the comparison of the three sections was still possible.

Figure 10 shows that section 1 with Fibredec® A performed slightly better than section 3 with Fibredec® B. For the cracks at the centre of each section in the base layer, it was observed during the test that for the SAMI sections cracks appeared first in the overlay, about 100 mm away from the simulated crack in the base. These are thought to be top-down cracks because a final crack always appeared right above the crack in the base. Attempts were made to cut cores from each section of the pavement and, unlike the cores from section 2 (control), the cores from sections 1 and 3 split into two, confirming the low shear stiffness of the overlay-SAMI interface.

7 Conclusions

The study shows that the both Fibredec® A and B are able to retard reflective cracking under traffic loading. The crack resistance of the SAMIs under traffic loading is due to the softness of the interlayer allowing the deformation of the interlayer therefore dissipating energy that would have aided the propagation of cracks, and the slip between the overlay and the SAMI, isolating the overlay from stress concentration around the crack region.

Also, the study shows that in the PTF, the section with Fibredec® A performed better than the one with Fibredec® B. This is the case because of the bitumen emulsion used in Fibredec® A.

Lastly, the findings show that the performance of the overlay is influenced by the introduction of SAMIs. Also, the effectiveness of the SAMIs depends on factors such as SAMI's composition, stiffness and the interface stiffness.

References

1. Debondt, A.H.: Anti Reflective Cracking Design of (Reinforced) Asphaltic overlays. PhD Thesis Delft University of Technology, Netherlands (1999)
2. Von Quintus, H.L., Mallela, J., Weiss, W., Shen, S.: Techniques for Mitigation of Reflective Cracking. Applied Research Associates, Champaign, IL, USA. Interim Report AAPTP 05-04 (2007)
3. Abe, N., Maehara, H., Maruyama, T., Ooba, K.: In: Abd El Hahim, A., Taylor, D., Mohamed, H. (eds.) Proceedings of the 4th RILEM Conference on Reflective Cracking in Pavements Research in Practice, Ottawa, Canada, pp. 464–474 (2000)
4. Palacios, C., Chehab, G.R., Chaignon, F., Thompson, M.: In: Al-Qadi, I., Scarpas, T., Loizos, A. (eds.) Proceedings of 6th International RILEM Conference on Pavement Cracking, Chicago, USA, pp. 721–729 (2008)
5. Saeed, A., Hall, J.W.: Accelerated Pavement Testing: Data Guidelines. National Cooperative Highway Research Program. NCHRP Report 512 (2003)
6. Barksdale, R.D.: Fabrics in Asphalt Overlay and Pavement Maintenance. In: National Cooperative Highway Research Program Synthesis of Highway Practice 171. Transportation Research Board. Washington, DC (1991)
7. Brodrick, V.B.: The Development of a Wheel Loading Facility an. In: Situ Instruments for Pavement Experiments. MPhil Thesis, University of Nottingham, Nottingham, UK (1997)
8. Brown, S.F., Brodrick, V.B.: Transportation Research Record 810, 67–72 (1981)
9. Done, P., Piouslin, S.: UK DCP 3.1 User Manual: Measuring Road Pavement Strength and Designing Low Volume Sealed Road using the Dynamic Cone Penetrometer. Department for International Development (DFID), Project Record No R7783 (2006)
10. Powell, W.D., Potter, J.F., Mayhew, H.C., Nunn, M.E.: The Structural Design of Bituminous Roads. Report LR1132. Transport and Road Research Laboratory, Crowthorne, UK (1984)
11. BSI, Method for Determination of the Indirect Tensile Stiffness Modulus of Bituminous Mixture. British Standard Institution, London, UK, DD 213 (1993)
12. BSD, Method for Determination of the Fatigue Characteristics of Bituminous Mixtures using Indirect Tensile Fatigue. British Standard Draft, London, UK, DD ABF 1993 (2003)
13. Ogundipe, O.M.: Mechanical Behaviour of Stress Absorbing Membrane Interlayers. PhD Thesis, University of Nottingham, Nottingham, UK (2011)

Towards a New Experimental and Numerical Protocol for Determining Mastic Viscosity

Ebrahim Hesami, Denis Jelagin, Björn Birgisson, and Niki Kringos

Division of Highway and Railway Engineering, Transport Science Department,
KTH Royal Institute of Technology, Sweden

Abstract. The rheological characteristics of mastics, or filler-bitumen mixtures, as a component of asphalt mixtures have a significant effect on the overall in-time performance of asphalt pavements such as low temperature cracking, fatigue and rutting behaviour. Viscosity is one of the rheological characteristics which is influenced by the physico-chemical filler-bitumen interaction. In this study, after reviewing some of more often used theories for calculating the viscosity of suspensions, a framework for calculating the viscosity of mastics is presented. This framework aims at covering the entire range of filler concentrations that is found in mastics. Also, a procedure for measuring viscosity mastic from dilute to high concentration mastic using a vane rotor viscometer is introduced. The paper is presenting the first experimental results and discusses the effect of the shape of the investigated fillers on the measured viscosity of the mastics.

1 Introduction

Research has shown that a large part of the mineral fillers that are included in asphalt mixtures are embedded inside the bituminous phase. As such, it is creating the asphalt mortar, often referred to as the mastic phase or the filler-bitumen mixture [1,2,3]. In this paper, fillers are considered as the mineral fraction that is smaller in size than 0.063mm. It can therefore be said that it is not the bitumen but the mastic phase that is binding the aggregates together, ensuring thus a stable skeleton for the needed stress-transfer. In addition to binding the aggregate skeleton, mastics also influence many of the other important asphalt mixture properties, such as the overall stability of the mixture, air void distribution, bitumen drain-down during transport, its workability during the laying process and the overall in-time performance of the pavement [1,3]. The effect of the mastic stiffness and its binding strength on the asphalt mixture fracture resistance at low temperature, fatigue resistance at intermediate temperature, and permanent deformation resistance at higher temperature can be noticeable [3]. Therefore, to understand the properties of asphalt mixtures and its resistance against failure mechanisms such as cracking, it is important to study the mastic properties as well as the mastic-aggregate bond [2].

Because the mineral filler fraction has a very high surface area in comparison to the coarser aggregates in the mixtures, the physio- chemical interaction between bitumen and filler becomes an important parameter in the mixture performance. The shape of the fillers, their size and size distribution, the nature of their surface texture, their adsorption intensity and the chemical composition of the fillers are all parameters that can potentially have a significant effect on the long-term performance of the entire mixture.

In addition to these parameters, the mixing and compaction temperature in the design and production of asphalt mixtures is also very important. For determining these temperatures, currently, the viscosity of bitumen at different temperature is used. This procedure, unfortunately, is only valid for equiviscous bitumen. For other types of bitumen, for example polymer modified, this method no longer applies. This is another reason why mastic properties should have a more dominant role in the current asphalt design procedures.

Viscosity of mastic is a parameter that can give a good indication of the mastic behaviour, incorporating some of the above described phenomena. This means, however, it is important to have 1) proper tests which are capable of measuring the viscosity of mastics over a wide range of concentrations and 2) fundamental insight into how these filler-bitumen interactions act as a function of concentration, shear rate, temperature, moisture, strain rate history and other parameters.

2 Research Aims

In this research, focus is placed on developing an understanding of the mastic viscosity as well as a reliable method to measure it. For a given binder, the filler concentration, filler properties and the physio-chemical and mechanical interaction between the filler and bitumen are important parameters that control the viscosity of mastic. To develop an understanding of the effect of fillers on the viscosity of mastic, in this research first a theoretical framework was developed based on suspension theories. At elevated temperatures, mastic can be categorized as a suspension; where bitumen acts as the viscose part of the suspension and the fillers act as the particles immersed in the bitumen. The aim of this framework is to be able to calculate the viscosity of mastic at elevated temperatures including fillers parameters, such as size, shape, surface chemistry and porosity.

Measuring the relative viscosity versus different filler concentrations gives insight into the filler bitumen interaction. Given the risk of filler sedimentation at lower and intermediate concentrations in mastics, measuring its viscosity at these concentrations can be quite challenging however.

3 Calculating the Particle Effect

Many efforts have been made over the past 10 decades to understand and describe the effect of particles on the flow characteristic of various fluids. In general, for a given liquid and a certain type of particles, suspensions with low particle concentrations have a lower viscosity compared to suspensions with higher concentrations. Higher particle concentration also means that particles come closer to each other and their interaction distances become more dominant. Eventually, with even further increased concentration, particles will come into contact and produce an inherent frictional force. To describe the viscosity of mastics, the model should be able to handle this entire range which gives quite a complex combination of phenomena to deal with. In the following a short overview is given of the efforts done to describe the particle effect in fluids and how this can be related (or not) to mastics.

3.1 Historical Perspective

Asphalt mastics are generally made by mixing asphalt binder (i.e. bitumen) with certain percentages of mineral fillers. As such, it can be treated as a suspension, in which bitumen is the fluid phase and the fillers are the particles. Over the years, suspension viscosity has received a lot of attention due to its practical importance. Einstein addressed the viscosity solution in his paper in 1906. In his theory, particles are positioned far enough from each other, such that no interaction between the particles will occur [4]. Einstein equation is shown by:

$$\eta_r = 1 + \eta' \cdot \phi \qquad (1)$$

where η_r is a relative viscosity of suspension, ϕ is the particle concentration and η' is the intrinsic viscosity which is empirically related to the particle physical characters such as size, shape and rigidity and also particle interaction with the interstitial fluid.

Einstein in 1906 derived his equation by solving the dissipation of energy for very dilute suspensions and found a value of 2.5 for η' for spherical rigid particles. His derivation was followed by a lot of research efforts to obtain a number for η' in different ways. For most filler particles η' was found to be in the range of 2.4 - 4.9 [5]. Most applications, however, treated the intrinsic viscosity as a curve fitting parameter that is related to the maximum concentration [5,6] following:

$$\eta' = \frac{2}{\phi_m} \qquad (2)$$

where ϕ_m is defined as the maximum concentration which the viscosity of the suspension tends to infinity [5,6,7]. In the present study a new definition for the maximum concentration has been defined which will be explain in this paper. Due to different values of ϕ_m depending on the binder and particle types, the amount of η' is also varying.

The Einstein equation is the basis of most theories, which are dealing with the calculation of viscosity of suspensions. Many researchers have tried to extend the Einstein equation for higher percentages of particle concentration. Rutgers in 1962

[8] and later on Thomas in 1965 [9] showed none of these models are able to calculate the viscosity of suspensions with high concentrations. Frankel and Acrivos in 1967 [10] calculated the viscosity of high concentration suspensions. They assumed hereby that the velocity of the particles is the same as the average velocity of the fluid and that suspensions behave as Newtonian fluids on a macroscopic scale. In their calculation they neglected the existence of any boundary effects.

Frankel and Acrivos solved the viscous dissipation energy for two neighboring particles with thin hydrodynamic flow between them. From this, they calculated the energy for multiple particles, and suggested the following equation for calculating viscosity.

$$\mu_r = C' \left\{ \frac{(\phi/\phi_m)^{\frac{1}{3}}}{1-(\phi/\phi_m)^{\frac{1}{3}}} \right\} \tag{3}$$

For determining the constant C' they used the Simha cage method [11]. In this method they calculate dissipation energy for a particle surrounded by other particles (six particles) in the cubic configuration (figure 1) plus an influence layer around that particle. By assuming that the velocity on the boundary of the spherical cell must be equal to an equivalent surface in a pure medium, they determined $C' = 9/8$ for spherical and $C' = 3\pi/16$ for cubical particles.

Hesami et al. [12] gave a comprehensive overview of the methods, which are most often applied to calculate the viscosity of suspension in different fields of science.

Some of these theories were designed for specific suspensions such as cement past or suspensions with uni-size particles, but none of them is able to predict the viscosity of suspensions at different particle concentrations. In the following, based on the described theories in this section, a new empirical framework is proposed that can be utilized for asphalt mastics.

3.2 Developed Framework

Some of the previously described theories are designed for low concentrations, such as the Einstein equation, and some others actually more suited for high concentrations, such as the Frankel equation. These equations could therefore be used as the boundary asymptotes for the viscosity as a function of concentration, Figure 1. Due to Frankel's definition of maximum concentration, however, at a certain concentration this asymptote tends to infinity. Obviously this can never be an accurate representation of the viscosity of the type of suspension to granular materials at a given temperature we are considering in this paper. With increased concentration there may be a sharp increase of viscosity, but it would never go to infinity and, as such, would diverge from the Frankel asymptote.

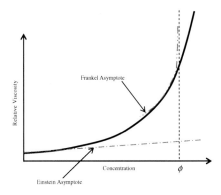

Fig. 1. Relative viscosity versus concentration bounded by two asymptotes

As discussed earlier, asphalt mastic can be considered as a suspension with different concentration, in which the filler is immersed in the bitumen. In this suspension, the filler consists of particles with different geometries and sizes that are surrounded by a layer of bitumen. This layer around the particle may in fact be divided into two sub-layers: (i) the layer of bitumen around the particle that is partially influenced by the particle and does not have the same flow characteristics as the 'free' binder that is farther away from the particle and (ii) the layer of bitumen that has adhered to the particle and as such is acting as though it has become part of the particle itself. The first layer is here referred to as the "partially influenced bitumen layer (ε)", the latter is here referred to as the "adsorbed asphalt layer (δ)". Particles with the adsorbed layer will be considered to have a new radius and will be referred to as "effective particles" [1].

The distance between the two effective particles is noted as h. This distance h is zero if the two effective particles contact each other and infinite for very dilute suspension.

An overall measure of h in a mixture can be found from [13]:

$$h = 2r\left[(\phi/\phi_m)^{\frac{1}{3}} - 1\right] \qquad (4)$$

where ϕ_m is the maximum concentration and the mixture does not have any free binder [14]. Resistance to the flow comes from particle to particle contacts and makes for a continuum network of friction. The value of the maximum concentration depends on size distribution, shape and type of aggregate and also binder characteristics such as stiffness of binder. Accurately measuring or calculating the maximum concentration according to the above definition is of paramount importance for calculating the viscosity of asphalt mastic.

Under these conditions the resistance to the flow comes from the frictional reaction between particles and the behaviour of material is predictable in the frictional regime.

When the frictional force become dominant, the distance between particles is equal or less than 2δ. Due to the direct contact of effective particles, most resistance to flow comes from the friction. Hence for calculating the viscosity in this regime this paper proposes the following equation:

$$\eta_r = \left(\frac{\delta}{r} - h_r\right) N_c * C_1 \tag{5}$$

where h_r is a relative distance between particles and is governed by:

$$h_r = \frac{h}{r} \tag{6}$$

where r is the average weighted radius of the particles.

In equation (10), the term of $(\delta/r - h_r)$ shows the strength of contact. N_c is the number of particles and it shows the number of contacts which are producing the friction force and C_1 is the friction coefficient of the whole particles structure.

The mastic with concentrations less than maximum concentration behaves in the hydrodynamic regime. In the hydrodynamic regime viscous behaviour of suspensions can be described with two equations, Einstein and Frankel, depending on the percentage of particle concentrations. For mastics with low concentrations the distance between particles is much bigger than the filler radius, so the effect of particles on each other is negligible and the Einstein equation is valid. By increasing the percentage of particles, the distance between particles becomes smaller and the particles start to affect each other. If the distance between particles is 2δ up to 2ζ, the mastic is still in the hydrodynamic regime; however significant interaction of particles increases the viscosity of mastic sharply. In this manner the particles come closer to each other but the friction interaction does not occur. The Frankel equation can explain the viscosity behaviour of mastic very well under these conditions.

Even though the Einstein and Frankel asymptotes can approximate the very low and very high conditions rather well, the transition part between these extreme limits remain without definition. For this reason an equation that can calculate the viscosity of the mastic in the transition part was derived as:

$$\eta_r = C \left(1/h_r\right)^n \tag{7}$$

Figure 2 shows a comparison between the predicted viscosity and the measured data. As can be seen the framework was quite well able to capture the viscosity over the wide range of concentration [12].

3.3 Particle-Binder Interaction

In the presented framework, the Einstein coefficient, the Frankel coefficient, the maximum concentration, C and n from the transition equation are determined by matching the framework to the experimental data. For adopting this framework to an independent theoretical framework, the model parameters need to have a physical meaning, and should serve as input parameters to predict the resulting viscosity. Important parameters for this are: shape, size, size distribution, surface texture, surface porosity, filler-bitumen chemical reaction, filler potential to bitumen adsorption, polarity of fillers and filler agglomeration. This will be investigated further in subsequent research. Additionally to developing the model framework, being able to measure the viscosity is equally important for validation as well as practical implementation into practice.

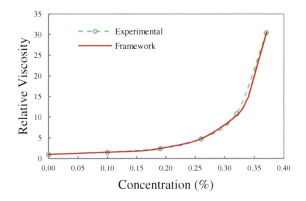

Fig. 2. Comparison between experimental data and calculated viscosity by framework

4 Development of the Experimental Procedure

Being able to accurately measure the viscosity of mastics with different fillers and filler concentration at various temperatures is an important but also challenging task, as described in the previous sections. To be able to make fundamental links between the measured mastic viscosity and physio-chemical parameters of the fillers, additional tests may also be needed. In the following, however, the developed test procedure to measure the viscosity of pure bitumen (low viscosity) up to high concentrated mastic (high viscosity) at different shear rate with high control to applying shear is explained.

4.1 Sample Preparation and Test Description

The weight of volume for a specific concentration was calculated by measuring the density of filler and bitumen at the temperature at which the mastic viscosity will be measured. All mastics where mixed at the same temperature (140 °C), to reduce the effect of the mixing operation. This temperature was chosen such that it would be appropriate for mixing the bitumen and fillers at all percentages. Mixing of the bitumen with the fillers was done with a mechanical high shear mixer. During this process attention was placed on creating a homogeneous mastic and avoiding filler agglomeration as much as possible. To prevent adding air bubbles into the mix, the filler was gradually spread in the bitumen during the mixing.The mastic was kept in an oven at 140 °C for 1 to 2 hours to give the samples time to release any remaining air bubbles. To ensure a homogeneous mixture, the mastic was mixed again at the relevant test temperature before pouring mastic into the cup of the viscometer.

The used rotational viscometer must be able to apply accurate shear rate from very low to very high for low viscose materials to high viscose materials. The geometry and the gap between inner and outer cylinder was chosen very carefully to avoid the influence of the boundaries on the measurements. In addition to a boundary bias, practically this is also important because the inner cylinder or rotor must be able to go inside the mastic even for very stiff mastic as well as prevent slippage on the wall of the inner cylinder. For this reason the vane rotor was chosen in this study, Figure 3-a. In addition to the vane rotor, an outer cylinder with small grooves on the inner wall of the cup was used, figure 3-b. The gap between rotor and cup wall was chosen to be large enough that the largest particle can move easily in the gap and narrow enough to avoid creating the so called 'plug' (or zero shear) zone.

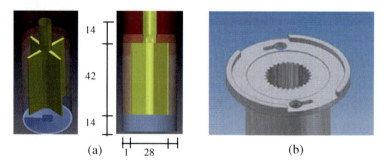

Fig. 3. a) The Vane shaped inner cylinder rotor and its dimensions in millimeter. b) The grooved outer cylinder (cup)

As the viscosity is determined by the slope of the shear stress vs. shear rate graph, a continues ramp procedure was used, in which the shear rate increases or decreases constantly in equal time steps. The device had to be able to change the shear rate very fast to have equilibrium shear rate in the measuring steps. By controlling the shear rate, the Rheometer adjusted the velocity of vane rotation and measured torque required to rotate.

4.2 Description of Fillers

In this study three types of filler were used to make the mastics: two types of silica base filler, M10 and M600 and a fly ash filler. The silica fillers have the same chemical composition but different physical characters particularly in size and specific surface area. The M10 and M600 are produced from sand after crushing by a ball mill, which creates regular aggregate shape with sharp edges, figure 4. Fly ash is generally collected from the flue gases by use of electrostatic precipitation. The fly ash producing processing will not change the shape of the particles that much, they will be round glassy particles, but it may have an influence on the size distribution [15], so care should be taken in the origin of the fly ash.

To keep the influence of the bitumen as a constant, a standard Nynas bitumen 70/100 was used as the binder for all mastics. The three types of fillers were used at concentrations from 5 to 50 percent by volume of filler to the total volume.

4.3 Test Results and Discussion

In Figure 4 the measured relative viscosity of the three mastics as a function of filler concentration is shown.

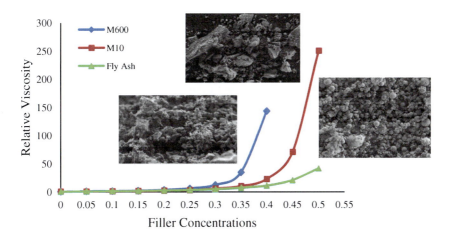

Fig. 4. Relative viscosity vs. Filler concentration for different fillers

As can be seen, the relative viscosity of the mastics increases sharply at a certain filler concentration. This concentration could be specific for each bitumen and filler combination and could perhaps be controlled by changing the shape and size of the particles. To study the shape of the particles, the fillers where investigate with a Scanning Electron Microscope (SEM), Figure 4. From these it can be seen that the flakiness of the silica based particles are similar. However M600 is finer than M10 and also has more surface area. This could partially explain that the effect of M600 on the relative viscosity is larger than M10 and the relative viscosity of mastic with M600 increases more rapidly by increasing the filler concentration compares to the mastic with M10.

From Figure 4 is can be seen that the fly ash mastics had less relative viscosity for the same particle concentrations as M600 and M10. The SEM scans of the fly ash particles show the very round nature of this filler. It could be hypothesized that due to the round shape of the Fly Ash particles they can roll more easily within the bitumen when exposed to a shear deformation. Also, the critical concentration for mastic with Fly Ash is higher than the other mastics which confirm that this mastic can have more filler with less resistance to flow.

5 Conclusions and Recommendations

In this research, focus was placed on developing an understanding of the mastic viscosity as well as a reliable method to measure it. For a given binder, the filler concentration, filler properties and the physio-chemical and mechanical interaction between the filler and bitumen are important parameters that control the viscosity of mastic. In this paper, a theoretical framework was demonstrated that is capable of capturing the range of concentration relevant for mastics. Furthermore, a developed test procedure to measure the viscosity of pure bitumen (low viscosity) up to high concentrated mastic (high viscosity) at different shear rate with high control to applying shear was described.

To demonstrate the developed methods, in this study three types of filler were used to make the mastics: two types of silica base filler, M10 and M600 and a fly ash filler. To keep the influence of the bitumen as a constant, a standard Nynas bitumen 70/100 was used as the binder for all mastics. The three types of fillers were used at concentrations from 5 to 50 percent by volume of filler to the total volume. From the measured viscosity it was clearly shown how the shape and size distribution of the filler particles influence the viscosity build-up of the mastic types.

Future aims of this research will include expanding the developed framework and test methodology to a more comprehensive protocol in which mastic viscosity can be easily determined, including fundamental fillers parameters, such as size, shape, surface chemistry and porosity.

References

[1] Buttlar, W.G., Bozkurt, D., Al-Khateeb, G.G., Waldhoff, A.S.: Journal of the Transportation Research Board 1681, 157–169 (1999)
[2] Anderson, U.D., Bahia, H.U., Dongre, R.: ASTM. stp 1147, 131–153 (1992)
[3] Chen, J., Kuo, P., Lin, P., Huang, C., Lin, K.: Materials and Structures 41, 1015–1024 (2008)
[4] Einstein, A.: Annales de physique 19, 289–306 (1906)
[5] Lesueur, D.: Advances in Colloid and Interface Science, vol. 145(1-2), pp. 42–82 (2009)
[6] Mooney, M.: Journal of colloid science 6(2), 162–170 (1952)
[7] Heukelom, W., Wijga, P.W.: AAPT, 418–437 (1971)
[8] Rutgers, I.R.: Rheologica Acta 2(4), 305–348

[9] Thomas, D.G.: Journal of Colloid Science 20(3), 267–277 (1965)
[10] Frankel, N.A., Acrivos, A.: Chemical Engineering Science 22(6), 847–853 (1967)
[11] Simha, R.: Journal of Applied Physics 23(9), 1020–1024 (1952)
[12] Hesami, E., Jelagin, D., Kringos, N., Birgisson, B.: Construction and Building Materials (accepted, 2012)
[13] Coussot, P.: Rheology of pastes, suspensions, and granular materials. Wiley (2005)
[14] Shashidhar, N., Needham, S.P., Chollar, B.H., Romero, P.: AAPT, 222–251 (1999)
[15] SS-EN 450-1, Fly ash for concrete - Part 1: Definition, specifications and conformity criteria (2005)

Interference Factors on Tests of Asphalt Biding Agents Destinated to Paving Works Using a Statistic Study

Enio F. Amorim[1], Antônio C. de Lara Fortes[2], and Luís F. M. Ribeiro[3]

[1] Professor of the Federal Institute of Education, Science and Technology of Mato Grosso (IFMT), Campus - Cuiabá, Civil Construction Department, Cuiabá, Mato Grosso, Brazil
[2] Graduating in Civil Work Technological Control at Federal Institute of Education, Science and Technology of Mato Grosso (IFMT), Campus - Cuiabá, Civil Construction Department, Cuiabá, Mato Grosso, Brazil
[3] Professor of University of Brasília (UnB), Civil and Environmental Engineering Department, Brasília, Federal District, Brazil

Summary. The asphalt binder characterization tests, with the purpose of being used in paving works, consist an excellent instrument of quality evaluation for these asphalt binders in order to guarantee the construction of higher durability works. However, it is possible observing innumerous factors during the tests which can affect, in a significant way, a more specific evaluation by each one of the proposed tests. Based on this context, the present paper presents an analysis of the principal factors that can interfere directly in the results of the characterization tests of asphalt binder, and, which can harm the performance and durability of asphaltic paving. The methodology adopted is based in a sampling of results obtained for the main characterization tests, petroleum asphaltic cement - CAP 50/70, accomplished by 08 different work teams, where a study formulated by statistic tools was applied, highlighting the main trusted results to an effective qualitative evaluation of the binders tested in paving works. In the end of this paper, the possible interference which can compromise the analysis of binders in significant ways will be shown.

1 Introduction

The asphaltic pavement, when flexible, is a structure with viscoelastic behavior, having as its main characteristic the subjection of dynamic and concentrated loads. At the transition of the load, the asphalt should deform itself and come back to its original form, showing its elastic behavior. In the other hand, when the acting load is found in elastic state, the asphalt should be able to support it, however showing a viscous behavior. Generally, because of repetitive efforts, these pavements suffer of a physic phenomenon usually called fatigue. So, materials resistant to this phenomenon have been more and more researched, with the intention of making even more durable pavement structures.

In terms of the use of asphalt binders, a way to evaluate the quality of these materials, when being used in road works, is directly related to its technologic control accomplished by tests. Thereby, doing critical analysis of the obtained results, a qualitative evaluation of the binders applied in road works can be obtained, and, establish a projection to the durability of the work as a whole thing. However, the reliability of the test results will be proved by studies using statistic tools. In this way, it might be observed possible mistakes during the tests, as the experience of operator, equipment malfunction, values of dispersion, and other problems.

Based on this context, the present paper highlights the main interference factors found in usual asphalt binder tests, focusing on paving works based on statistic parameters.

2 Work Method

Initially, for this study a data collection obtained through laboratory tests was considered, accomplished by 08 all-student team from the Federal Institute of Education, Science and Technology of Mato Grosso, in asphalt paving subject, from graduating course of civil work technologic control, where usual characterization tests with petroleum asphalt cement - CAP 50/70 were accomplished, following the Brazilians naming and test standards.

The work was organized through the ordering all the data, which will be presented later, using statistic parameters as in the following descriptions:

2.1 Percentile Determination

The present parameter is about a method of organize the data, putting the values into a 0 to 100 variation scales, in percentage form. In that scale, the values are shown in increasing order, so they can be compared to other data. This kind of parameter can be easily obtained through the Equation 01.

$$p = \frac{x-1}{n-1} .100\% \tag{1}$$

Where:
x = number from the desired order; n = quantity of organized values;
p = the percentile related to 'x'.

2.2 Quartile Determination

It consists in another method of organizing the data in increasing order, starting with 05 points, with four equal parts, using a scale going from 0% to 100%. Knowing each 25% reached in the scale represents a division. For a better understanding of it, Figure 1 shows the representation given by a quartile.

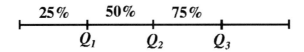

Fig. 1. Divisions of data through the quartile tool

Where:

Q1 = 1st quartile, where values lower than 25% are located;
Q2 = 2nd quartile, where values lower than 50% are located;
Q3 = 3rd quartile, where values lower than 75% are located.

To calculate the quartile of any data, it can be used the same equation used to calculate the percentile. So, to obtain the different quartiles, the equations 02, 03 and 04 are used for each one of the percentages.

* p = 25% - corresponding to ¼ of the sample.

$$p_{1/4} = \frac{n+3}{4} \qquad (2)$$

* p = 50% - corresponding to ½ of the sample.

$$p_{1/2} = \frac{n+1}{2} \qquad (3)$$

* p = 75% - corresponding to ¾ of the sample.

$$p_{3/4} = \frac{3(n+1)}{4} \qquad (4)$$

The quartile tool classifies the data in order to know how many of them have inferior values than the required percentile or quartile.

2.3 Average Arithmetic Determination

Also known as average, it is obtained through the sum of all numeric values, and then dividing the previous result by the quantity of values.

2.4 Median Determination

This parameter refers to the value which divides the organized data in two equal parts, in other words, from 0% to 50%, and from 50% to 100%.

2.5 Mode Determination

This parameter refers to the most common value within the organized data. Groups can have one, two, more than two or even no mode at all.

2.6 Standard Deviation Determination

The standard deviation is a dispersion of the organized values based on the average of this data, For a better understanding, there are two must-know definitions:

> **Average absolute deviation (DAM)**

This is the sum of all the deviations, related to its average, divided by the number of values, in the attempt of obtaining an absolute value.

> **Variance**

Defined as the sum of the squares of the deviances related to its average, divided by the quantity of values.

By knowing these previous definitions the Equation 05 can be built, for the calculation of the standard deviation of a data group.

$$\sqrt{\frac{\sum_{i=1}^{n}(x_i - \bar{x})^2}{N-1}} \qquad (5)$$

Where:

$\sum_{i=1}^{m}(x_i - \bar{x})^2$ = sum of the squares of the deviances related to its average;

N = number of values in the group.

3 Results and Discussions

3.1 Evaluation of the CAP 50/70 Tests

3.1.1 Analysis of the Statistic Parameters in the Penetration Tests (mm/10)

The statistic parameters obtained from the penetration test, based on the following data sorting, can be found on Table 1.

Table 1. Statistic parameters obtained from the CAP 50/70 penetration test

Penetration Test (mm/10)					
Order	Percentage Order	Percentile	Data (mm/10)	Division by quartile	
				1st quartile	54.75
1	0%	53	53	2nd quartile	58
2	14%	53.994	54	3rd quartile	62.5
3	29%	54.995	55	Average	58.625
4	43%	55.996	56	Median	58
5	57%	59.988	60	Mode	-
6	71%	61.996	62	Variance	22.268
7	86%	63.998	64	Standard deviation	4.718
8	100%	65	65	Average Absolute Deviation	4.714

For this test, it should be noted that the CAP 50/70 presents its values in the penetration test within the range of 50 to 70 mm/10 (ABNT NBR 6576 [1] and ASTM D5 [2]). So, analyzing the results in the data sorting, along with the obtained parameters, it can be concluded that every value is within the mentioned range. It can also be noted, that there are no value repetition, if the mode parameter does not exists. With the calculated value for the standard deviation, the values obtained in the orders 1, 7 and 8 should be discarded, once they are too far from the average found when the standard deviation is applied. However, even obtaining different values in the penetration test, all values can still be found within the desired range of 50 to 70 mm/10, which do not affect, in a significant way, the quality evaluation of the CAP 50/70 mm/10 in this test.

3.1.2 Softening Point (^{o}C)

The Brazilian standard developed by the Petroleum National Agency (ANP, 2005 [3]), proposes that the CAP 50/70 must have a minimum softening point of 46° C. The European standard (EN 12591/2009 [4]) proposes that the same CAP should be within the range of 46°C to 54°C. The parameters obtained for the softening point test (AASHTO T53 [5]) can be found on Table 2.

Analyzing the values obtained through this test, it can be noted that for the obtained values, only one student group managed to archive an acceptable result, 47.5° C, which was shown in the 8th order on Table 2. A hypothesis explaining the variations might be the operator change during the flame (fire) control and the condition of the ring material (deformable material). It is also possible noting that the average, median and mode have very close values, which sets the value of 44° C being considered the real value of the tested sample.

Table 2. Obtained parameters from the softening point test for the CAP50/70

Softening Point Test (°C)					
Order	Percentage order	Percentile	Data (°C)	Division by quartile	
				1st quartile	43.375
1	0%	41	41	2nd quartile	44
2	14%	42.988	43	3rd quartile	45.125
3	29%	43.498	43.5	Average	44.188
4	43%	43.998	44	Median	44
5	43%	43.998	44	Mode	44
6	71%	44.998	45	Variance	3.638
7	86%	45.499	45.5	Standard deviation	1.907
8	100%	47.5	47.5	Average Absolute Deviation	1.553

3.1.3 Density

Table 3 presents the parameters obtained from the density test (ABNT NBR 6296 [6] and ASTM D70 [7]).

Table 3. Obtained parameters from the density test for the CAP 50/70

Density Test					
Order	Percentage order	Percentile	Data	Division by quartile	
				1st quartile	1.004
1	0	0.991	0.991	2nd quartile	1.006
2	0.142	1.002	1.002	3rd quartile	1.012
3	0.285	1.004	1.004	Average	1.007
4	0.428	1.005	1.005	Median	1.006
5	0.571	1.006	1.006	Mode	-
6	0.714	1.009	1.009	Variance	8.74E-05
7	0.857	1.019	1.019	Standard deviation	0.009
8	1	1.02	1.020	Average Absolute Deviation	0.007

On Table 3 it can be observed that the CAP 50/70 has presented density values slightly above the water density, for almost every group. Possible variation in these results might have happened due to flaws in weighting the pycnometer or balance reading, lack of picnometer drying in any of its weightings and/or flaws related to the proportion of the materials inside the pycnometer. About the analysis done through statistic parameters, the values obtained by all groups have presented very close magnitude orders, in a way that the value divergences do not compromise in an expressive way, the quality of the density tests. Besides, references described by BERNUCCI et al, 2010 [8], report that the CAP 50/70 presents an overall density value within 1.00 to 1.02, which were the values obtained by every group, except the group related to the 1st order who has presented a density value of 0.991.

3.1.4 Saybolt-Furol Viscosity – (135°C)

For this test, the Brazilian standards specify that the CAP 50/70, at the temperature of 135°C, should present viscosity of, at least, 141 Saybolt-Furol seconds (SSF). Next, Table 4 presents the obtained values from this test for temperature at 135°C (ABNT NBR 14950 [9] and ASTM E102 [10]).

Table 4. Data obtained from the Saybolt-Furol viscosity test for CAP 50/70 - 135°C

Saybolt-Furol Viscosity Test - (135°C)					
Order	Percentage order	Percentile	Data (SSF)	Division by Quartile	
				1st quartile	143.75
1	0%	129	129	2nd quartile	168
2	14%	130.988	131	3rd quartile	176.75
3	29%	147.915	148	Average	164.375
4	43%	165.928	166	Median	168
5	57%	169.988	170	Mode	-
6	71%	174.99	175	Variance	793.41
7	86%	181.993	182	Average Deviation	28.167
8	100%	214	214	Average Absolute Deviation	24.321

Observing the data values of Table 4, it can be noted that the first 2 groups are out of the Brazilian standards, once they show values lower than 141 SSF. In the other hand, analyzing the obtained values, it becomes possible observe that the variation presented by the groups reflects that a series of factors which could have interfered in the test. Among others factors which could have caused the variations, the possibility of cracking in the tested sample, reading errors, temperature measure precision and the heating form of the studied binder should be highlighted. So, doing a global analysis of the data, the most reliable data are the ones presented by the groups 4, 5, 6 and 7.

3.1.5 Saybolt-Furol Viscosity - (150°C)

For this temperature the Brazilian standards specify that the Saybolt-Furol viscosity should be of, at least, 51 SSF. Table 5 presents a summary of the obtained values from the test, as well the results found for the statistic parameters.

Table 5. Data obtained from the Saybolt-Furol viscosity test for CAP 50/70 - 150°C

Saybolt-Furol Viscosity Test - (150°C)					
Order	Percentage order	Percentile	Data (SSF)	Division by quartile	
				1st quartile	69
1	0%	59	59	2nd quartile	81
2	14%	68.94	69	3rd quartile	94
3	14%	68.94	69	Average	81.125
4	43%	79.956	80	Median	81
5	57%	81.994	82	Mode	69
6	71%	92.978	93	Variance	219.268
7	86%	96.996	97	Standard deviation	14.808
8	100%	100	100	Average Absolut Deviation	13.571

Analyzing the data on the Table 5, it appears that every value is under the Brazilian standards. On the other hand, observing the statistic parameters, although theoretically some of them would be discarded, it would have no effect on the quality analysis of the studied CAP, when it's supposed to be used in road works.

3.1.6 Saybolt-Furol Viscosity - (177°C)

In the case of a 177°C temperature, the Brazilian recommendations impose that the CAP 50/70 should present a viscosity within 30 to 150 SSF. So, analyzing Table 6, up next, it can be observed that the first three values are out of imposed range, in a way the CAP50/70 would not be usable in paving works. Analyzing the statistical elements, it appears that almost every value should be acceptable. However, it is possible observing that the analyzed CAP is almost unusable, according to the Brazilian standards. The main interference factors are the same as the ones described on the 135° C tests.

Table 6. Data obtained from the Saybolt-Furol viscosity test for CAP 50/70 - 177°C

Saybolt-Furol Viscosity Test - (177°C)					
Order	Percentage order	Percentile	Data (SSF)	Division by quartile	
				1st quartile	29
1	0%	24	24	2nd quartile	31
2	14%	28.97	29	3rd quartile	32.25
3	14%	28.97	29	Average	30.25
4	43%	29.996	30	Median	31
5	57%	31.994	32	Mode	29
6	57%	31.994	32	Variance	9.071
7	86%	32.999	33	Average deviation	3.012
8	86%	32.999	33	Average Absolut Deviation	2.57

3.1.7 Thermal Susceptibility Index (IST)

This index relies on the tests of penetration and softening point and, according to the Brazilian standards for being an acceptable material in paving works, the values for the CAP 50/70 should be within the range of -1.5 to +0.7. Table 7 presents a summary for the obtained values of IST according to the data bank used in this study, as well the statistic treatment given to these values.

Table 7. Statistic parameters obtained for the CAP 50/70 IST

Thermal Susceptibility Index					
Order	Percentage order	Percentile	Data	Division by Quartile	
				1st quartile	-2.613
1	0%	-3.45	-3,45	2nd quartile	-2.335
2	14%	-2.953	-2,95	3rd quartile	-2.27
3	29%	-2.502	-2,5	Average	-2.439
4	43%	-2.390	-2,39	Median	-2.335
5	57%	-2.280	-2,28	Mode	-2.28
6	57%	-2.280	-2,28	Variance	0.345
7	86%	-2.240	-2,24	Average Deviation	0.588
8	100%	-1.42	-1,42	Average Absolut Deviation	0.452

In this case, most values do not meet the Brazilian standards, except the 8th order group. However, given the incidence of the calculated parameters, the 8th order value (-1.42) would not be considered reliable, once it is too far from the representative parameters, such as average, mode and median. The hypothesis about what could have affected the obtained values are the same ones described in the tests of penetration and softening.

4 Conclusions

At the end of this paper, it can be conclude that the use of statistic elements is a very essential tool to evaluate the reliability of results obtained from laboratory tests. These elements show an overall quality evaluation of the materials, an evaluation which generates more adequate answers to the use of these materials in paving works, as in the case of the asphalt binders, for example. It is also important highlighting that the gathering of no representative data can interfere directly in the quality of the work, in a way that can speed up the occurrence of pathologies like fissures, fatigue and/or excessive plastic deformations, which come from the inadequate use of a poor quality binder. Referring to the analysis through statistic parameters calculated with the data obtained from different student groups, performing the main CAP50/70 characterization tests, highlights the softening point, Saybolt-Furol viscosity, density and IST tests, it can be said that significant differences were generated and that some of the values had to be discarded due to not having a desired level of reliability. In these analyses, the only test that did not presented any problem in its qualitative evaluation was the penetration test. In this context, possible hypothesis of what could have caused the gathering of different data in each one of the tests is directly related to the following descriptions:

*operational problems: reading errors; testing procedure out of the proposed technical standards; strict criteria during the preparation and completion of the tests
*equipment malfunction: calibration problems; worn out equipment; internal dust
*material problems: cracked sample, modified sample, not representative materials.

References

[1] ABNT - Associação Brasileira de Normas Técnicas, NBR 6576. Determinação da penetração de materiais betuminosos, ABNT. Rio de Janeiro/RJ, p. 3 (1998)
[2] ASTM - American Society for Testing and Materials. D5. Standard Test Method for Penetration of Bituminous Materials, p. 4, ASTM (2003)
[3] ANP - National Petroleum, Natural Gas and Biofuels Agency. Resolution no. 19, vol. 7. ANP, Rio de Janeiro/RJ (2005)
[4] NBN EN 12591. Bitumen And Bituminous Binders - Specifications For Paving Grade Bitumens, p. 22. BSI (2009)
[5] AASHTO - American Association of State Highway and Transportation Officials. T53. Standard Method of Test for Softening Point of Bitumen, p. 6. AASHTO (2011)
[6] ABNT - Associação Brasileira de Normas Técnicas. NBR 6296. Produtos betuminosos semi-sólidos - Determinação da massa específica e densidade relativa, p. 5. ABNT, Rio de Janeiro (2004)
[7] ASTM - American Society for Testing and Materials. D70. Standard Test Method for Specific Gravity and Density of Semi-Solid Bituminous Materials, p. 3. ASTM (1997)
[8] Bernucci, L.B., Motta, L.M.G., da Soares, J.B., Ceratti, J.A.P.: Pavimentação Asfáltica - Formação Básica para Engenheiros, p. 525. Petrobrás/ABEDA, Rio de Janeiro (2010)
[9] ABNT - Associação Brasileira de Normas Técnicas. NBR 14950. Materiais betuminosos - Determinação da viscosidade Saybolt Furol, p. 9. ABNT, Rio de Janeiro (2004)
[10] ASTM - American Society for Testing and Materials. E102. Standard Test Method for Saybolt Furol Viscosity of Bituminous Materials at High Temperatures, p. 3. ASTM (2003)

Development of an Accelerated Weathering and Reflective Crack Propagation Test Methodology

Ken Grzybowski[1], Geoffrey M. Rowe[2], and Stan Prince[1]

[1] PRI Asphalt Technologies, Inc.
[2] Abatech, Inc.

Abstract. The development of reflective cracking mitigation techniques depends on the proper evaluation of different technologies from geo-synthetic inter-layers to highly modified thin lift overlays. A methodology has been developed to investigate these technologies using a novel Accelerated Pavement Weathering System (APWS) which exposes the pavement structure to the combination of temperature, moisture and UV radiation that a pavement will experience in service. This methodology is based on well documented and widely used accelerated weathering methods used by other industries to determine the durability of various materials and systems to environmental exposure. Currently, most pavement accelerated testing is based on load-associated stresses instead of temperature; no system has been developed to-date that accurately reproduces the effect of temperature, moisture, and UV radiation on a pavement structure. The APWS can be a powerful conditioning tool to evaluate different pavement materials, pavement systems, improve modelling and product performance.

This conditioning methodology has been used in conjunction with a newly developed test method that can measure the resistance to reflective cracking using a modified Asphalt Pavement Analyzer test. The Reflective Cracking Resistance Test (RCRT) was developed to run both on laboratory-prepared samples, as well as cores taken from the field. The method is ideally suited to measure the crack propagation through an interlayer or overlay as function of loading cycles. This method was used to evaluate pavement systems before and after accelerated weathering in the APWS.

This paper focuses on the results of this study and the practicality of this type of testing to understand different technologies to mitigate crack propagation. Specifically, several different pavement structures were prepared which consist of conventional overlays, overlays with geosynthetic membranes, and 4.75 mm thin lift overlay. An overview of the methodology and summary of the testing of these systems is presented in this paper.

1 Introduction

The aging of asphalt binder is significantly affected by the cyclic actions of temperature, ultraviolet radiation and water acting upon exposed surfaces and interconnected voids. Current methods of aging asphalt binder, for example the

Rolling Thin Film Oven Test (RTFOT) and Pressure Aging Vessel (PAV), were developed without incorporation of these combined parameters. These methods produce bulk samples that are then evaluated for using various asphalt binder test methods without consideration of aggregate-asphalt binder interactions.

The combined effect of UV light, water, and thermal cycling have not been studied in depth for paving materials, but have been a subject of extensive research and use in other industries, such as roofing materials, paints, plastics, automobiles, and coatings (1, 2). In pavements, a daily variation of heat and radiation occurs relative to the position of the sun and the atmospheric conditions. It is well known with asphalt pavement materials that aging that occurs in reality varies with intensity from the surface of the pavement structure, with higher aging towards the top and lower aging with depth (3). The amount of natural sunlight (providing the source of UV light) and water on the asphalt pavement system is hypothesized as having a significant effect on the aging of the asphalt binder, and consequently, the material's properties (they become harder and more brittle). To demonstrate change to material and performance by APWS conditioning properties, the ability of four pavement systems to resist crack propagation have been evaluated. The simulation of aging using UV light coupled with cyclic (simulated rain) water and subsequent testing is described below.

The development of improved test methods for aging pavements will assist in further model development and provide a tool for assessment of materials being built into road pavements, pavement architectures, and pavement designs.

2 Test Method Development

Two novel methods and apparatus were used in this work for specimen conditioning and testing. The first of these produced aged material with a similar gradient of aging that would be expected on a road pavement, whereas the second provides a realistic assessment of crack resistance using a modification to the Asphalt Pavement Analyzer (formally the Georgia Wheel Tracking Device originally developed by Dr. Lai (4, 5)) modified with the Hamburg loading configuration (6).

2.1 Accelerated Pavement Weathering System

The Asphalt Pavement Weathering System (APWS) was designed by PRI Asphalt Technologies, Inc. (PRI) to accommodate "full depth" pavement specimens for natural accelerated pavement weathering providing the missing link in pavement analysis. The APWS provides flexibility to be used for a wide-range of specimen types, shapes and sizes (see Figure 1). The APWS has controllable cycles to simulate most environmental conditions and is fully monitored allowing both "climate" and specimen data to be collected and recorded continuously throughout the conditioning. The APWS allows specimens to be weathered from the top surface down, simulating the natural aging of pavement; years of in-service exposure and weathering can be simulated in only a few months.

The APWS allows the recording and monitoring of the each variable of the test, including: water, temperature and light (UV exposure). These effectively enable controllable cycles for rain, sunlight (UV exposure) and temperature. Lamps containing quartz discharge tubes with tungsten filaments which provide a mix of radiation similar to natural sunlight (UVA, UVB, visible and infrared radiation). The watering system features the ability to control duration, time between, and volume of watering (0 to 1.0" per hour) intervals. This allows simulation of real-world conditions that may occur over several seasons in a short time span. The watering system also allows control over type of water used (deionized, salt, fresh, etc.) to even further reflect in-service conditions. A monitoring system records up to ten different sensors 24 channels) throughout the weathering process. Data can be collected for ambient temperature, individual specimen temperature (top, internal and/or bottom), humidity, water volume, etc. The APWS' unique flexibility enables it to simulate a wide variety of special weathering conditions (pavement exposure climates). These conditions can be customized to reflect individual specimen needs. Settings can be designed to mimic the in-service environment of a product, eliminate multiple variables in research, or be "user defined."

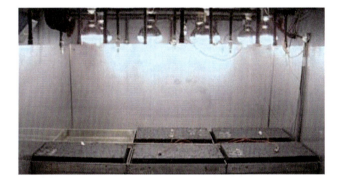

Fig. 1. Photograph of APWS with asphalt slabs in place

In a previous evaluation of a typical PG 64-22 binder, the aging developed suggests that 3,000 hours is typical of that obtained by PAV aging on a bulk sample. This is subjective to a certain extent since the aging developed in a PAV device does not consider any formation of an aging gradient within a pavement system, but rather tests a bulk sample of material. If the aging obtained is compared, the functional form of the hyperbolic relationship as suggested by Mirza and Witczak (3), a good fit to the data is obtained, as illustrated in Figure 2. This demonstrates a consistency between the aging profile produced in the APWS and the data collected by Mirza and Witczak for the calibration of the models used within the models incorporated into the newer AASHTO pavement design methods (7).

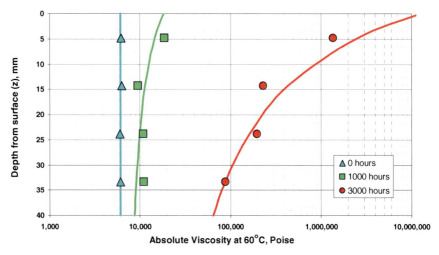

Note: 0-hours data represents binder as supplied, all others are recovered from samples cut from aged slab

Fig. 2. Aging profiles obtained showing actual test data from PRI's APWS test device and compared to fitted function in format used by Mirza and Witczak (3) and implemented in the MEPDG

2.2 Reflective Cracking Resistance Test

The Reflective Cracking Resistance Test (RCRT) is conducted in the APA. Rut resistance tests have been implemented in many of the State agency laboratories. Consequently, the development of a device for crack resistance testing is advantageous since it capitalizes on a well utilized piece of equipment. The cost of the modifications required for use of this device for the Reflective Cracking Resistance Test (RCRT) is relatively minor when compared to other test configurations.

The asphalt mixture specimens were fabricated in the PMW Slab Compactor (Figure 3) and compacted to obtain the required density and % air voids. The PMW Slab Compacter was manufactured by Precision Machine and Welding Company and enables the production of a slab of asphalt material. A specimen is typically produced in two layers and the interface between the two layers is used to test different treatments such as grids, SAMI's, fabrics and others solutions to limit and/or prevent reflective cracking. After cooling, a six inch core is taken from the slab and the sides trimmed. A notch is sawn in the beam after fabrication to ensure that the crack propagation occurs in a position which can be monitored by a digital camera. This camera is programmed to take photographs of the assembly at pre-determined intervals so that crack growth can be monitored throughout the test. Optionally, on one side of the slab 4 linear variable displacement transducers (LVDT's) can be attached (see Figure 4) to monitor the displacement changes that occur on the side of the specimen whereas the other side is painted white for observations of cracking using a simple crack width gauge (Figure 5).

Fig. 3. PMW Slab Compactor   **Fig. 4.** Positioning of LVDTs on test specimen

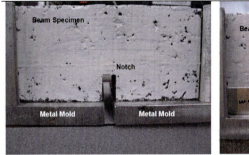
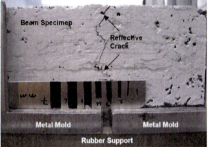

Fig. 5. Slab before (left) and after (right) crack reflection (scale in mm)

3 Experimental

The objective of the experimental work was to assess the efficiency of different pavement reflective crack resistant technologies in the RCRT device with before and after realistic accelerated aging associated with cyclic water and UV effects.

3.1 Materials

Summary information on the mixture design and slab construction is given in Table 1. The objectives of this test program was to compare the standard control pavement with an option containing a fabric (fiberglass/polyester material in fairly wide use) along with plant-produced new thin lift technology employing SBS modified binders (HiMA, 7.5% SBS content blends, a Kraton technology) with and without RAP (referenced as PmB-A and PmB-B respectively). Information on the nature of the formulations is available upon request. The discussion presented in this paper has been limited to general observations concerning the material properties to highlight the efficacy of accelerated weathering and reflective crack resistance test method.

Table 1. Summary of mix design and slab construction

Slab Ref.	Control	Fabric	PmB-A	PmB-B
Top layer Mix design size Aggregate Binder	SP 12.5 Trap Rock PG 64-22	SP 12.5 Trap Rock PG 64-22	SP 9.5 Dolomite PG 76-34	SP 9.5 Dolomite PG 76-34
Bottom layer Mix design size Aggregate Binder	SP 12.5 Trap Rock PG 64-22			
RAP %	0	0	0	25
Tack Coat ID (between layers) and rate (gal/yd^2)	Emulsion CRS-2 0.02	PG asphalt PG 64-22 0.19	Emulsion CRS-2 0.02	Emulsion CRS-2 0.02
Asphalt, % (top/bottom layers)	5.0 / 5.0	5.0 / 5.0	6.8 / 5.0	6.6 / 5.0
Lift Thickness, in. (top/bottom layers)	1.0 / 1.5	1.0 / 1.5	1.0 / 1.5	1.0 / 1.5
Sample Total Thickness, in.	2.5	2.5	2.5	2.5

3.2 Aging in the APWS

Weathering was conducted by using a 1,500-hour period in the APWS. Air temperature was allowed to follow the ambient air, a pavement surface temperature similar to summer conditions was selected, and UV light was applied with water cycles. Slabs were rotated three times at 375-hour intervals in the device to ensure that each of the four specimens received the same exposure. Figure 6 shows an excellent excerpt of records maintained for the control slab which is similar to the other three slabs for a 12-hour period. The cyclic peaks and valleys in the temperature are a result of thermal shock (water application) to the surface of the slabs. It should be noted that the materials being evaluated in this study are all typically used in a climate with a high temperature which results in a SUPERPAVE™ PG 64 grade being specified. Consequently, the upper temperature was maintained in this region for this preliminary study. While for this study some variation exists in the temperature profile, it should be noted that this can be controlled, if required, to more precise values.

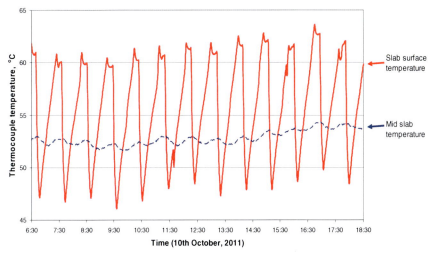

Note: Preselected parameters: Cycle: 51-minutes light only + 9-minutes light + rain; Pavement Surface Temperature: similar to typical Florida summer (PG 64 climate).

Fig. 6. Temperature records for control slab

3.3 Testing

The testing was conducted in triplicate for each of the four pavement systems evaluated both before and after aging in the APWS. The crack propagation from the notch cut was monitored along with the cyclic data collected from the instrumentation attached to the side of the specimen along with visual observations using the photographic records.

4 Results and Discussion

The individual and average results for the four series of experiments are summarized in Table 2 and plotted in Figure 7 to Figure 10. Some interesting observations can be made by inspecting the data in these figures.

Table 2. Summary of RCRT test data

Before or after APWS status	Number of wheel passes for crack to propagate to surface			
	Control	Fabric	PmB A	PmB B
Before	1856	3202	3363	1405
After	734	783	2787	1130

For the control, the performance is significantly poorer after aging, as expected, with the lowest initial performance. The life after conditoning in the APWS was just 40% of the orginal life.

For the fabric-modified pavement system, an enhanced initial performance is observed, but changes significantly after aging with the aged performance being very close to the control. This is suggestive that hardening of the binder towards the surface as illustrated in Figure 2, coupled with other aging factors, are dominating the results and effectively overcoming the benefits of the fabric reinforcement over an extended aging period. This is not to dissimilar for the field behavior for these materials which appear to have early life advantages over other systems, but does suggest they may be less effective at crack prevention as the pavement structure undergoes top-down aging. The time taken for a crack to propogate to the surface was reduced from 3,202 applications to 783 applications, a 76% reduction.

The two trials with the plant-produced new thin lift technology employing HiMA SBS PmB's, provide interesting contrasts. The material containing the 25% RAP addition performed somewhat poorly, not being too different from the aged control and fabric systems. However, for both of these blends, the performance is similar before and after conditioning in the APWS. The best performing system in this evaluation was the PmB blend (Ref A) which contained high polymer content and no RAP. This system had essentially the same performance before and after conditioning in the APWS.

The thin lift overlay without RAP exhibited initial RCRT results of 3,363 wheel passes, and after APWS accelerated weathering 2,787 passes giving a loss of 17%. The RCRT loss was 20% with the RAP blend but this mix had a low RCRT value before conditioning closer to the control mix which can be attributed to the RAP in the mix.

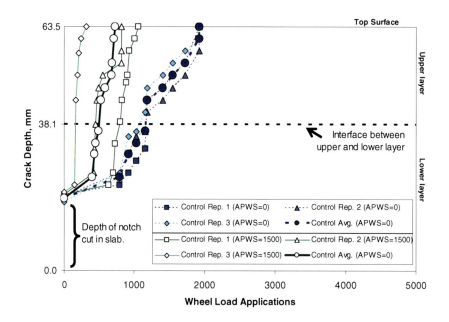

Fig. 7. Indivual and average data for "Control"

Development of an Accelerated Weathering and Reflective Crack 133

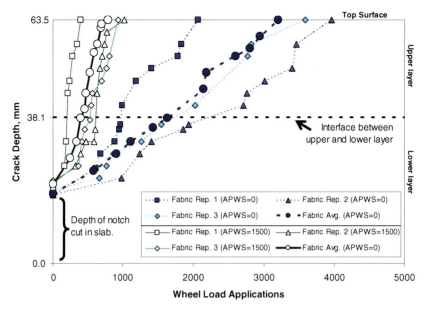

Fig. 8. Indivual and average data for "Fabric"

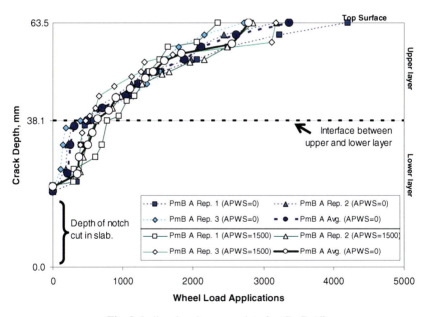

Fig. 9. Indivual and average data for "PmB A"

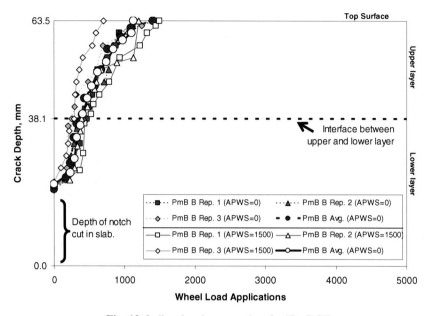

Fig. 10. Indivual and average data for "PmB B"

5 Conclusions

Test devices have been developed that enable accelerated aging of pavement specimens using similar environmental factors that would be found in a real world situation including UV light, heat and water effects in controllable cycles as an Asphalt Pavement Weathering System (APWS). In addition, an existing APA wheel track test (AASHTO T 340) device has been modified to develop a Reflective Cracking Resistance Test (RCRT) using a moving wheel load.

- Preliminary data collected using the APWS suggests that the aging profile obtained is similar to that found in real pavements when compared to the aging profile that would be expected by researchers such as by Mirza and Witczak (3). Further development of this device is currently underway to refine the inputs with a view to be capable of accurately modelling a series of climatic environments.
- Pavement performance, as evaluated with the RCRT of the unconditioned and conditioned pavement samples, exhibited significantly different results between a fabric and a competitive technology, an effective test method to evaluate RCR technologies.
- The RCRT results, after APWS accelerated aging, demonstrates the critical importance of including accelerated aging as an assessment for pavement materials and pavement performance.

Further study and comparisons with "real world" data would help with the understanding of the importance of employing the APWS in material and pavement evaluation and research.

References

[1] Xing, L., Taylor, T.J.: Correlating Accelerated Laboratory, Field and Thermal Aging TPO Membranes. J. ASTM Intl. 8(8) (2011), doi:10.1520/JAI103743
[2] Burns, R.S.: Roofing Assembly Simulated Heat and Light Test. J. ASTM Intl. 8(8) (2011), doi:10.1520/JAI103740
[3] Mirza, M.W., Witczak, M.W.: Development of a Global Aging System for Short and Long Term Aging of Asphalt Cements. Journal of the Association of Asphalt Paving Technologists 64, 393–430 (1995)
[4] Lai, J.S.: Development of a Simplified Test Method to Predict Rutting Characteristics of Asphalt Mixes Research Project 8502, Final Report, Georgia Department of Transportation, Atlanta (1986)
[5] Lai, J.S.: Evaluation of the Effect of Gradation of Aggregate on Rutting Characteristics of Asphalt Mixes Project No. 8706, Georgia Department of Transportation (August 1988)
[6] Aschenbrener, T.: Evaluation of Hamburg Wheel-Tracking Device to Predict Moisture Damage in Hot Mix Asphalt. In: Transportation Research Record 1492, TRB. National Research Council, Washington, DC, pp. 193–201 (July 1995)
[7] ARE, Inc. Guide for Mechanistic-Empirical Design of New and Rehabilitated Pavement Structures, Final Report, Part 2. Design Inputs, Chapter 2 Material Characterization. National Cooperative Highway Research Program, Transportation Research Board, National Research Council (2004)
[8] ASTM, D 4798/D 4998m, Standard Practice for Accelerated Weathering Test Conditions and Procedures for Bituminous Materials (Xenon Arc Method), vol. 04. ASTM International, Conshocken, PA, Section Four, Construction (2011)
[9] ASTM, D 4799-08, Standard Practice for Accelerated Weathering Test Conditions and Procedures for Bituminous Materials (Fluorescent UV, Water Spray and Condensation Method), vol. 04. ASTM International, Conshocken (2011)
[10] ASTM, G 141, Guide for Addressing Variability in Exposure Testing of Non-metallic Materials, vol. 14(04). ASTM International, Conshocken (2011)
[11] ASTM D, G 151, Practice for Exposing Non-metallic Materials in Accelerated Test Devices that Use Laboratory Light Sources, vol. 14(04). ASTM International, Conshocken (2011)

The Use of Ground Penetrating Radar, Thermal Camera and Laser Scanner Technology in Asphalt Crack Detection and Diagnostics

Timo Saarenketo[1], Annele Matintupa[2], and Petri Varin[2]

[1] Ph.D., Managing director, Roadscanners Oy, Finland
[2] MSc. Civil Engineering, Roadscanners Oy, Finland

Abstract. The amount and types of pavement distress have been one of the main indicators for the pavement quality in most of the pavement management systems. However, existing commercial techniques for locating these distresses and evaluating their severity have proven to be insufficiently reliable. Over the last few years new technologies have been developed and tested to provide more accurate and repeatable pavement distress mapping results.

This paper presents a summary of authors' experiences with three relatively new technologies which have provided very promising results in pavement distress surveys. All three techniques, ground penetrating radar (GPR), thermal camera and laser scanner are based on longer electromagnetic wavelengths than visual light and thus their advantage is detection of cracks inside the pavement that cannot be seen by human eyes. Over the last few years these techniques have become both fast and accurate enough to make them viable as field survey tools and at the same time improvements in data processing and storage capabilities have enabled the use of these techniques.

Ground penetrating radar has traditionally been used to measure pavement thickness but the data can also be used to evaluate pavement quality. Three dimensional GPR imaging has provided new interesting information about the formation of transverse and longitudinal cracks especially at the sites where pavement thickness varies substantially in transverse directions. Dielectric value analysis using different antenna frequencies has also proven to work well in detecting salt related cracking in asphalt. Analysis of amplitudes and frequency response analysis can also be used to detect moisture related problems in asphalt. Additionally, segregation can be detected with GPR.

The recent results from testing high precision and fast thermal cameras have provided interesting new possibilities in detecting pavement distress. Top down cracking, for instance, seems to generate slightly beneath the pavement surface and these cracks can be seen with thermal cameras before they become visible to the human eye. Thermal camera analysis also shows the effect of water pumping through the pavement due to heavy vehicle loading.

Finally this paper presents the latest results from emission analysis of laser scanner data to detect different types of pavement distress.

1 Introduction

Because of the nature, location and size of different kinds of distress in asphalt pavement, their reliable mapping has been a major challenge. Traditionally distress mapping has been carried out visually directly from a moving car or based on a digital video from the pavement, but these methods have been expensive and not always reliable and repeatable enough. That is why numerous new automated or semi automated technologies have been developed and tested over recent years with varying success.

The testing has been carried out with three relatively new technologies, ground penetrating radar (GPR), thermal camera and laser scanner along side with old and traditional techniques like falling weight deflectometer (FWD), profilometer and visual evaluations. The new non-destructive techniques have provided very promising results in pavement distress surveys. One of the advantages is that all of these techniques are based on longer electromagnetic wavelengths than visual light, which means that their ability to penetrate a pavement surface is better than the human eye. Thus they can be used to detect cracks also inside the pavement. Ground penetrating radar and thermal cameras can also be used to detect moisture in the asphalt which is often the main reason for the distress. Over the last few years new 3D GPR technology has provided detailed structural information concerning the bound and unbound layers and this information has also produced interesting findings regarding the reasons for pavement damages.

Even though there are still many improvements needed, GPR, thermal camera and laser scanner techniques are already fast and accurate enough to be used in routine pavement diagnostics surveys.

Due to the somewhat complex nature of these technologies they have not become routine pavement survey tools among most road agencies but their benefits are clear and major savings can be achieved when they are used properly.

2 GPR, Thermal Camera and Laser Scanner Technologies, Techniques and Equipment

A ground penetrating radar (GPR) survey is a non-destructive survey method that provides continuous information of pavement structure and its quality. The ground penetrating radar technique has been used in traffic infrastructure surveys as early as in the mid 1970s, initially mainly in tunnels and on bridge decks [1]. The GPR transmitter/receiver antenna, mounted on the front of the car or on a special trolley, transmits an electromagnetic pulse into the media (asphalt and road structures). The transmitted electromagnetic wave penetrates the road structures and is reflected from interfaces in the media that have different electrical properties like the boundaries between different structural layers or differences in moisture content. The reflected waves are collected by the receiver. The control unit measures the time difference between the transmitted and received pulse (travel-time) and its amplitude. The amplitude is displayed as a function of the travel time. When the measurements are made over sequential points, a continuous profile of the media can be displayed [2].

Like the GPR, the thermal camera survey is also a non-destructive survey method. Using the thermal camera the surface temperature of the road can be mapped. The presumption is that the surface temperature of the damaged areas is different from the surface temperature of the areas with no damage. This is caused by an anomaly in the water or air content in the material. The thermal mapping method is based on the solution to the heat transfer equation for a thick target with an instantaneous surface heat flux, which is [3]:

$$T(x,t) = \frac{q}{\sqrt{4\pi k \rho}} \exp\left(\frac{-x^2}{4\alpha t}\right) \qquad (1)$$

where T is temperature, x is depth beneath the surface, k is thermal conductivity, ρ is density, c is heat capacity, α is thermal diffusivity, t is time and q is the surface heat flux. For a semi-infinite solid approximation, the surface temperature is proportional to the inverse square root of time and inversely proportional to the thermal inertia, P, which varies as $(k\rho c)^{1/2}$.

Laser scanning is a NDT-technique where the distance measurement is based on the laser beam travel time from the laser scanner to the target and back. When the laser beam angle is known and beams are sent in different directions from a moving vehicle with a known position, it possible to make a three dimensional (3D) surface image, point cloud, of the road and its surroundings. In a point cloud with millions of points, every point has an x, y, z coordinate and also some reflection or emission characteristics. Earlier authors have tested the thermal camera and laser scanner methods in research conducted through the European Union's ROADEX project [4].

Figure 1 presents the survey van, which is equipped with a GPR 400MHz ground coupled antenna (in front of van) and thermal camera and laser scanner on the roof of the van.

Fig. 1. A survey van equipped with GPR, thermal cameras and laser scanner.

3 GPR and Asphalt Crack Diagnostics

In pavement diagnostics GPR provides valuable information concerning pavement and unbound layer thickness and how this can be related to damages in pavement. Figure 2 presents a case from Highway 4 in Finland where transverse cracks and longitudinal cracks are concentrated in the road sections where the pavement thickness has the greatest deviation in transverse and longitudinal directions. The top profile presents a single 1.0 GHz GPR longitudinal profile. The profile in the middle presents pavement (red) and unbound base thicknesses from 9 longitudinal profiles of the two lane road and the lowest profile presents a contour map of pavement thickness where blue colour represents thin pavement and red colour thicker pavement. Black lines represent transverse cracks and red lines the location of longitudinal cracks.

The GPR technique can be used to detect road sections or areas with cracks in the asphalt. In this technique the dielectric value of the asphalt surface is measured using an air coupled antenna surface reflection technique and sections with problems with asphalt cracking can be differentiated since they present as highly deviating dielectric values of asphalt surface. In good quality and uniform asphalt the deviation of dielectric value is very small. Figure 3 presents an example from an asphalt covered dike in the Netherlands where the deviation of dielectric value of a recently repaired asphalt section is much smaller compared with the problematic asphalt section. Cracks and salt in asphalt are reflected and refracted in a different way compared to good quality.

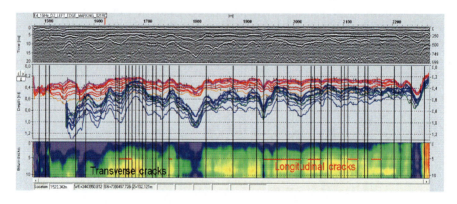

Fig. 2. An example of relationship between pavement thickness deviation and cracks.

Finally with high frequency ground coupled antennas it is also possible to detect individual cracks in the asphalt if the sampling density (scans/m) is high enough. Figure 4 presents an example from Scotland from an asphalt covered concrete road built resting on peat. Measurement was done with a 1,5 GHz ground coupled antenna.

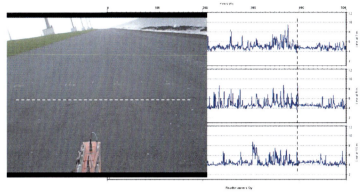

Fig. 3. The deviation of dielectric values of an asphalt covered dike in the Netherlands.

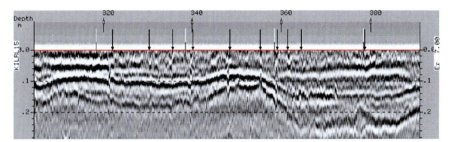

Fig. 4. Cracks in an asphalt covered concrete road built on peat in Scotland.

4 Thermal Cameras and Pavement Distress

The tests with thermal cameras have shown that a thermal camera effectively reveals cracks in pavement that cannot be seen with the naked eye. Figure 5 presents a case where a top down crack can be seen in asphalt in thermal camera data but not in digital video. The crack can be seen on the side of the inner wheel path. This also indicates that top down cracks generate slightly beneath the pavement surface and, only after reaching the pavement surface will they start to penetrate down through the pavement.

Fig. 5. An example of a crack that cannot be seen by human eye (left photo) but with a thermal camera it can be seen clearly (right photo).

Thermal camera analysis also shows the effects of water pumping through the pavement due to heavy vehicle loading. The pumping water cools down the wheel paths which can be seen with the thermal camera. Figure 6 presents an example of the relation between this pumping effect and the formation of top down cracks in highway 4 in Northern Finland. Still photo (left) and thermal image (right) showing top down cracking and water being pumped through the pavement by heavy vehicle loading. Darker colours present cooler temperatures and light colours present warmer temperatures. The data collection was done in spring when frost was thawing and water is released from melting ice. The wheel paths are much cooler than the areas outside and between them. The formation of top down cracking could be located in those sections where the temperature difference was high.

The problem with high precision digital thermal cameras is that they are not fast enough to be considered as a high speed pavement data collection method. Another problem is that the data collection cannot be carried out during daytime because direct sunlight affects the results too much and, as such, thermal camera surveys are always performed after sunset.

Fig. 6. Example of top down cracking and water being pumped through the pavement by heavy vehicle loading.

5 Laser Scanners

Laser scanner technique is quite a new method in the area of road condition surveys. The development of the application has been quite rapid and, at present, there are several techniques on the market for pavement distress mapping. Laser scanners can be used not only to "count the cracks" but they have also proven to be an excellent tool in pavement diagnostics i.e. finding the reason for the pavement failures.

The changes in pavement surface topography compared to normal shape indicate well those areas with deformation problems or frost problems in the road. Usually these deviations in the road surface are not easy to discern visually. The use of the rainbow map makes it considerably easier to visualize these deviations. These maps show road surface topography and its deviations and damage. Each colour in rainbow colour palette scale represents a 30 mm change in surface level. An optimal road surface with two sided crossfall should resemble a perfect V-shape and in sections with straight crossfall it should present as straight lines. Figure 7 presents an example of damages which developed as a result of a clogged private road exit culvert. In this road section the location of an area of uneven frost heave, caused by a clogged exit road culvert, can be detected. Cracks are formed in sections where the rainbow lines are not straight and continuous.

Fig. 7. An example rainbow view of laser scanner point cloud data (usually figures are presented in colours).

The laser scanner technique can also be used in frost heave evaluations. The measurements have to be performed twice; when the frost heave is at its maximum level and when all frost has thawed. A comparison of these two results is used to calculate the frost heave. Figure 8 presents an example of a case, where the formation of longitudinal crack can be related to differential frost heave. In this road section the longitudinal "frost" crack, shown by arrows, has been formed along the edge of maximum frost heave area of 0.11 m.

Fig. 8. An example frost heave view of laser scanner point cloud data.

Results of laser scanner survey data can be presented as emission maps. It effectively presents different features in the asphalt. Patches and cracks, for example, stand out distinctly. Figure 9 presents an example of an emission map measured from a ROADEX test site in Ohtanajärvi in Northern Sweden. The patches, cracks and road markings can be seen very clearly from the picture.

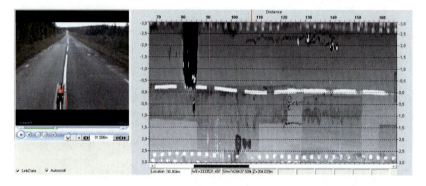

Fig. 9. An example of still photo and a laser scanner emission map.

The problems related to laser scanner surveys is that they are quite sensitive devices which may lead to errors. Dust, heavy rain and fog create disturbances such that the laser beam will reflect incorrectly back to the receiver. Therefore it is important to be cognizant of the weather conditions and similar factors when measuring.

6 Conclusions

The thermal camera method is a fast and relatively cheap method. This method can provide information about the pavement surface conditions such as cracking and structural problems like pumping. The ground penetrating radar method can be applied to study asphalt thickness and deviation of dieletric values. The laser scanning method provides information detailing the surface shape of the road and its surroundings. These three methods together can provide a good non-destructive toolkit for pavement diagnostics.

Key words in modern, cost effective and sustainable road condition management are **focus** and **preventative maintenance**. Research conducted through the European Union's ROADEX project has shown that by focusing on problem road sections, finding the reasons for their problem and selecting the optimum rehabilitation measures based on the diagnosis rather than just repairing the symptoms can deliver cost savings up to 40 %. Using this technique the lifetime of a road becomes longer which further increases the profitability of the investments.

The use of modern road survey and diagnostics techniques are vital before this new policy can be implemented. Proper use of the new NDT techniques, GPR, thermal cameras and laser scanner technique can provide guidance to the exact location of problem areas and also to the sources of the damages. However, when cracks have appeared in the pavement surface, a great part of the pavement strength has already been lost. That is why the early symptoms of these damages should be identified using the new road survey technologies and after that preventative maintenance measures can be taken to prevent these damages from growing further.

References

[1] Morey, R.: Ground Penetrating Radar for Evaluating Subsurface Conditions for Transportation Facilities. Synthesis of Highway Practose, vol. 255, National Cooperative Highway Research Program, Transportation Research Board. National Academy Press (1998)
[2] Saarenketo, T.: Electrical properties of road materials and subgrade soils and the use of ground penetrating radar in traffic infrastructure surveys. Acta Universitas Ouluensis, A471. Oulu University Press, Oulu (2006)
[3] Del Grande, N.K., Durbin, P.F.: Delamination detection in reinforced concrete using thermal inertia. In: Nondestructive Evaluation of Bridge and Highways III, Newport Beach, CA, USA, March 3-5, vol. 3587, pp. 186–199. Lawrence Livermore National Laboratory (LLNL), California (1998)
[4] Matintupa, A., Saarenketo, T.: New Survey Techniques in Drainage Evaluation – Laser Scanner and Thermal Camera. The ROADEX IV project (2011)

Asphalt Thermal Cracking Analyser (ATCA)

Hussain Bahia[1], Hassan Tabatabaee[2], and Raul Velasquez[3]

[1] Professor, University of Wisconsin-Madison, USA
[2] Graduate Research Assistant, University of Wisconsin-Madison, USA
[3] Research Associate, University of Wisconsin-Madison, USA

Abstract. The Asphalt Thermal Cracking Analyser (ATCA) is a device that can simultaneously test two asphalt mixture beams while undergoing selected thermal history. The first beam is unrestrained and thus the change of its length with temperature can be used to obtain glass transition temperature (T_g) and coefficients of thermal expansion or contraction. The second beam is restrained at the ends and can be used to measure the thermal stress build-up as a function of time and temperature. The measures of length change and stress in beams can be used to get a comprehensive evaluation of the low temperature performance including change in strain, stress as a function of time and temperature.

The ATCA is considered an important advancement when compared to other existing thermal or mechanical cracking tests in which either thermal stress or moduli are measured, while making assumptions about coefficients of contraction and ignoring glass transition change. These assumptions are believed to cause serious errors in estimating thermal stresses.

Nine asphalt mixtures obtained from field sections in Minnesota, USA were used for the development of the ATCA. It is shown that the device can be used to estimate cracking temperature and strength. Further, the relaxation modulus of mixes can be directly estimated by solving the convolution integral using the measured thermal stress and strain from the restrained and unrestrained specimen, respectively. The ATCA can also be used to investigate response of asphalt mixtures to thermal cycles and isothermal conditions.

1 Introduction

Thermal cracking is widely recognized as a critical failure mode for asphalt pavements. Due to its importance a reliable test method capable of capturing asphalt material response to environmental loading as function of both time and temperature is needed. The current low temperature specifications in the US and other countries rely heavily on measuring asphalt mixture properties obtained under mechanical loading in the Indirect Tensile (IDT) creep and strength tests. However, these tests have limited capabilities in terms of simulating thermal cracking. These limited capabilities have been recognized and to address them a test method such as the Thermal Stress Restrained Specimen Test (TSRST) has been used. Furthermore, in conventional thermal cracking tests, linear viscoelastic concepts are used to infer mechanical response to thermal loading while making assumptions about coefficients of contraction and ignoring glass transition change.

The TSRST standardized system was developed under SHRP A-400 contract by Jung and Vinson [1]. In this test, as the temperature drops the specimen is restrained from contracting thus inducing tensile stresses. The results from TSRST are the cracking temperature and cracking strength due to a single low temperature event. This test method has been extensively used in the past to investigate thermal cracking performance. Research performed by Monismith et al.[2], Arand [3], Vinson et al. [4], Romero et al.[5], Sebaaly et al. [6], Chehab et al. [7], Sauzéat et al. [8], and Velasquez et al. [9], among others showed that TSRST can be used to evaluate the susceptibility of asphalt mixtures to low temperature cracking.

This paper covers the development of the Asphalt Thermal Cracking Analyser (ATCA), which is a significant improvement of the current TSRST. In addition to measure cracking temperature and cracking strength, the ATCA measures thermal strain during cooling, which allows for the direct estimation of relaxation modulus, the glass transition temperature (T_g), and the coefficients of thermal contraction/expansion above and below T_g. Further, in addition to conventional single event thermal loading, the ATCA allows for application of thermal cycles and isothermal conditions to investigate thermal fatigue and physical hardening, respectively.

2 Materials

For the development of the ATCA, loose asphalt mixtures were collected from field pavement sections in Minnesota, USA and compacted using the Superpave Gyratory compactor targeting design parameters provided by MnROAD engineers. The asphalt binders and aggregates used for preparation of mixes represent typical materials placed in pavements in the USA. Asphalt binders placed in these sections were modified with commonly used chemical and polymer additives.

3 Development of the Asphalt Thermal Cracking Analyser (ATCA)

In an effort to address issues in existing low temperature testing setups, a device was developed that simultaneously tests two asphalt mixture beams; one unrestrained, and the other with restrained ends. The unrestrained beam is used to measure the change in length with temperature, and consequently the glass transition temperature (T_g) and the coefficients of expansion/contraction above and below Tg (i.e., $α_l$ and $α_g$). The restrained beam is used to measure the induced thermal stress build-up due to restraining conditions in the sample. The two beams tested in the ATCA are obtained from the same asphalt mixture gyratory compacted sample or core, and both are exposed to the same temperature regime. The system is schematically shown in Figure 1.

During the development of the device, numerous obstacles and challenges were faced in order to achieve acceptable results. To address the problem of adhesive de-bonding of the epoxy and the metal end pieces (Figure 3(b)) used for the restrained beam, the end pieces were initially sand blasted, which temporarily solved the issue. But with the build-up of grime in the fine sand blasted texture over time the problem reoccurred. Thus a much coarser texture was applied to the surfaces as shown in Figure 3(d), permanently solving the deboning issue.

Another significant issue observed was the apparent softening and relatively low failure stress in the material. This issue was studied significantly and at the end, two simple factors were found to be responsible for the observed behaviour: (a) the unintentional application of torsion to the beam while screwing in the end plates to the frame, and (b) loading eccentricity due to slight misalignment of the end plates as well as insufficient support of the beam leading to sagging. The first problem was resolved by filling the gap between the edge of the end plates and the bottom of the chamber with a metal spacer, effectively preventing any torsional movement in the beam (Figure 3(a)). The second issue was resolved by modifying the gluing setup by placing the plates on a rail and using a set of guide rods to insure the plates are placed completely parallel and aligned (Figure 4(b)). Furthermore, a support platform with adjustable height using a set of screws was designed and used to insure complete support of the beam midsection during the tests. Metal rollers were included on the platform to prevent any friction between the platform and beam, possibly affecting the stress and strain fields through the beam (Figure 3(c)).

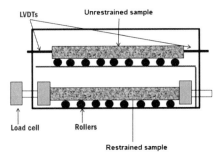

Fig. 1. Illustration of the Asphalt Thermal Cracking Analyser (ATCA)

The unrestrained and restrained samples are produced from one Superpave gyratory compacted sample. Using a masonry saw, four prismatic beams of 5 by 5-cm in cross section and 15 cm long are cut from 17 cm gyratory samples. Two of these beams are sawed in half to produce four 7.5 cm blocks. By gluing a 7.5 cm block to each end of the two 15 cm blocks, two 30 cm beams are produced (Figure 4). The effect of gluing was assessed by a set of comparative tests on one-piece and glued beams and was found to be insignificant.

As both beams are produced from the same sample and both are exposed to the same thermal history, the stress build-up, glass transition temperature, α_l and α_g can be used to get a comprehensive picture of the low temperature performance of the asphalt mixture. It is recognized that the air void content on outer edge of gyratory compacted samples may vary from that of the rest of the sample. Most of the outer area is cut off and removed during sample preparation. Nonetheless, the potential implications of this affect will be further assessed by the authors.

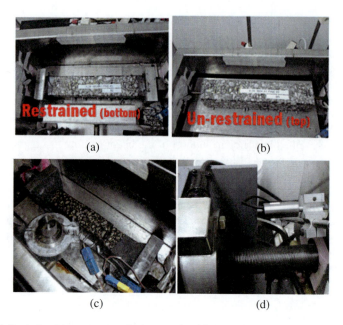

Fig. 2. (a) Restrained beam setup, (b) unrestrained beam setup, (c) restrained beam at failure, and (d) load cell and LVDT's

4 Experimental Capabilities of ATCA

The main outputs of the ATCA system are the measurements of thermal strain (Figure 5(a)) and stress (Figure 5(b)) during cooling. Many experiments, such as thermal cycling with isothermal steps (Figure 5(c)) and measurement of thermal stress relaxation are possible with the ATCA. One such test is a thermal stress relaxation experiment in which the chamber temperature is reduced to a predefined low temperature at a controlled cooling rate (0.1 to 1 °C/min) continuously monitored using temperature probes within the chamber and the core of the beams. The temperature is kept at the predefined temperature for prolonged periods, between 2 to 10 hrs, and the stress build-up in the restrained specimen as well as thermal strain in the unrestrained sample are measured continuously. The results are used to plot curves of thermal stress as a function of core temperature and test time during the extended isothermal condition (Figure 5(d)).

Results from the ATCA can also be used to calculate relevant low temperature material properties, most notably, the relaxation modulus. The relaxation modulus convolution integral can be solved numerically by directly measuring and inputting the parameters from the ATCA. Both sides of the equation are differentiated in the time domain to eliminate the integral. Thermal stress (i.e., restrained beam) and strain (i.e., unrestrained beam) data are used to solve for

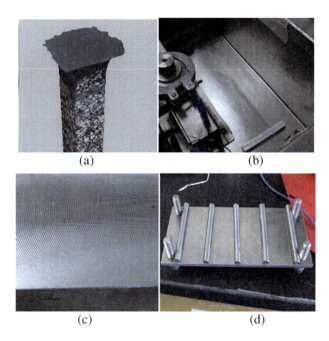

Fig. 3. Design revisions and challenges: (a) adhesive failure between epoxy glue and end piece, (b) metal spacer placed under restrained beam end pieces to prevent torsion, (c) re-textured surface of end piece to improve glue adhesion, and (d) adjustable height beam support platform with rollers

relaxation modulus. An example of ATCA results and the calculated relaxation modulus curve are shown in Figure 6.

As discussed, the ATCA can be used to fully determine the thermo-volumetric properties of asphalt mixtures. Glass transition temperature measurements in cooling and heating were obtained for the asphalt mixtures described in Table 1. The results were fitted using a mathematical relationship used by Bahia and Anderson [10] to determine the location of the glass transition temperature as well as the coefficients of contraction above and below the glass transition region. The results are plotted in Figure 7.

Another test that can be performed using the ATCA is thermal cycling. Figure 8 shows a typical example for thermal cycling testing. The asymmetric stress behaviour during cooling and heating is believed to be due to the asymmetry in the rate of build-up and reduction of time-dependent strain (i.e., physical hardening) in asphalt mixtures. This asymmetry is due to the gradual increase of the time-dependent strain rate based on the proximity of the temperature to the glass transition temperature during cooling. During heating the time-dependent strain is not differentiated from the temperature-dependent strain, thus the total accumulated potential thermal strain is decreased proportional to the coefficient of

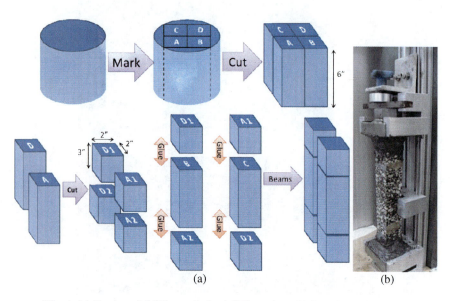

Fig. 4. (a) Cutting of SGC sample for ACTA testing. (b) Sample gluing setup

thermal expansion. This concept was used to develop a model for thermal stress calculation that takes into account glass transition behaviour and time-dependent strain (i.e., physical hardening), as discussed elsewhere [11].

Observation of ATCA results show that the thermo-volumetric response of the unrestrained samples did not significantly change from cycle to cycle. Figure 8(c) shows the thermal strain in an asphalt beam prepared with the WI binder. The temperature was cycled between +30°C and -70°C three times. No significant change in coefficients of thermal contraction/expansion and the glass transition temperature was observed from one cycle and the next. However, the heating and cooling strain curves in each cycle differ from each other, as discussed previously. It is also observed that the cooling and heating curves will deviate more significantly when cooled to temperatures well below the glass transition temperature. The trend and magnitude of contraction is very similar in all three cycles, reinforcing the idea of complete recoverability of physical hardening after each heating cycle.

Figure 9 shows thermal cycles for MnROAD Cell 33 with isothermal conditioning in last cycle for the restrained sample. It can be seen that the area of the loop (i.e., hysteresis) decreases after each cycle. Furthermore, the area of the loop significantly decreases when the specimen is subjected to isothermal conditioning at the end of the cooling step. These results indicate the importance of taking into account isothermal conditioning (i.e., physical hardening) when estimating thermal cracking susceptibility of asphalt mixtures and it is discussed in detail in other publications by the authors [11, 12].

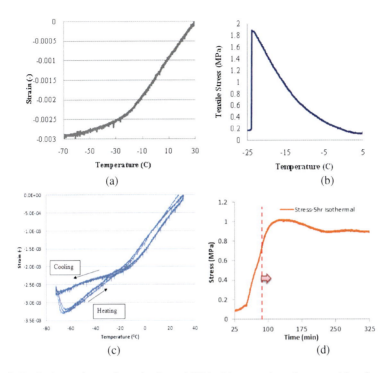

Fig. 5. Typical experimental results from ATCA, (a) measuring glass transition for asphalt mixtures, (b) measurement of stress build-up and fracture, (c) thermal cycling, and (d) measuring stress response during isothermal conditioning

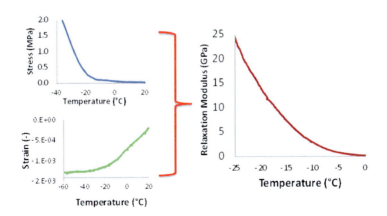

Fig. 6. ATCA results and calculated relaxation modulus curve

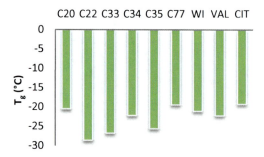

Fig. 7. Glass transition temperature of asphalt mixtures during cooling

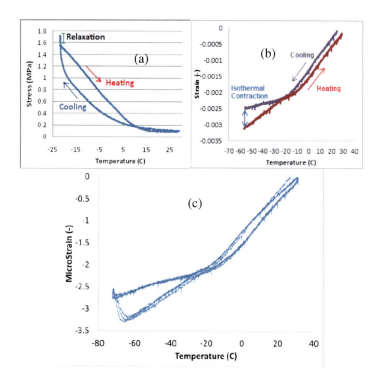

Fig. 8. (a) Thermal stress, (b) thermal strain for a full thermal cycle, (c) thermal strain in asphalt mixture beam (WI) in 3 consecutive cycles

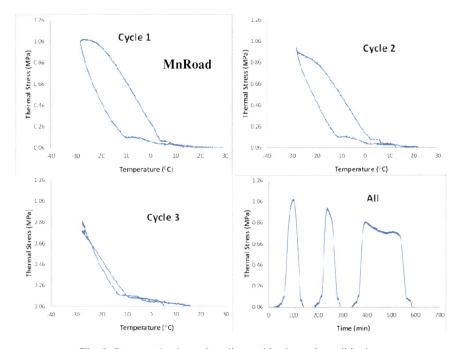

Fig. 9. Stress under thermal cycling and isothermal conditioning

5 Conclusions

A device to measure thermo-viscoelastic behaviour of asphalt materials for low temperature cracking characterization has been developed. The Asphalt Thermal Cracking Analyser (ATCA) can simultaneously test an unrestrained and a restrained asphalt mixture beam to obtain strain and stress changes during cooling and heating.

The results can be used very effectively to estimate glass transition temperature (T_g), linear coefficients of thermal expansion/contraction, and thermal stress build-up as a function of temperature and time. Based on the wide range of experimental procedures possible using the ATCA, the following important observations can be made:

- Results from the ATCA can be used to calculate directly without any assumptions the relaxation modulus. The relaxation modulus convolution integral can be solved numerically by directly measuring and inputting the strain and stress responses from the ATCA.
- ATCA measurements show that the assumption of thermal strain as a linear function of temperature is not accurate at temperatures approaching the glass transition region. Thermal strain in asphalt mixtures is greatly dependent on the glass transition region, cooling rate, and the isothermal conditioning time.

- The thermo-volumetric response of the unrestrained asphalt mixtures does not significantly change during thermal cycling. However, the strain response in each cycle differ significantly between heating and cooling due to the asymmetry of the time-dependant strain rate near and below the T_g.
- Measurements of thermal stress using the ATCA indicate the importance of taking into account isothermal conditioning time (i.e., physical hardening) when estimating thermal cracking susceptibility of asphalt mixtures. Strain and stress are highly dependent on thermal history. The claims that physical hardening does not affect asphalt mixtures cracking and stress build up cannot be supported by data collected in this study.

Acknowledgements. This research was sponsored by Federal Highway Administration National Pooled Fund Study TPF-5(132): "Investigation of Low Temperature Cracking in Asphalt Pavements Phase-II" and the Asphalt Research Consortium (ARC), which is managed by FHWA and WRI. This support is gratefully acknowledged. Authors would like to acknowledge contributions of Dr. Codrin Daranga, and Dr. Menglan Zeng, to the design of the ATCA. The results and opinions presented are those of authors and do not necessarily reflect those of the sponsoring agencies.

References

[1] Jung, D.H., Vinson, T.S.: In: Strategic Highway Research Program. SHRP A-400, Washington, DC (1994)
[2] Monismith, C., Secor, G., Secor, K.: Journal of the Association of Asphalt Paving Technologists 34, 248–285 (1965)
[3] Arand, W.: In: Proceedings of the 4th Int. Symp. on Mechanical Tests for Bituminous Mixes Characterization, Design and Quality Control. RILEM, Budapest, Hungary, pp. 68–84 (1990)
[4] Vinson, T.S., Kanerva, H.K., Zeng, H.: Strategic Highway Research Program SHRP-A-401, Washington, DC (1994)
[5] Romero, P., Youtcheff, J., Stuart, K.: Transportation Research Record: Journal of the Transportation Research Board 1661, 22–26 (1999)
[6] Sebaaly, P., Lake, A., Epps, J.: Journal of Transportation Engineering, 578–586 (2002)
[7] Chehab, G., Kim, R.: Journal of Materials in Civil Engineering, 384–392 (2005)
[8] Sauzéat, C., Di Benedetto, H., Chaverot, P., Gauthier, G.: In: Proceedings of Advanced Characterization of Pavement and Soil Engineering Materials, Athens, Greece, pp. 1263–1272 (2007)
[9] Velasquez, R., Gibson, N., Clyne, T., Turos, M., Marasteanu, M.: In: Proceedings of 6th RILEM Int. Conference on Cracking in Pavements. RILEM, Chicago, Illinois, pp. 405–414 (2008)
[10] Bahia, H.U., Anderson, D.A.: Journal of the Association of Asphalt Pavement Technologists 62, 93–129 (1993)
[11] Tabatabaee, H.A., Velasquez, R., Bahia, H.U.: Transportation Research Record: Journal of the Transportation Research Board (accepted for Publication, 2012)
[12] Bahia, H.U., Tabatabaee, H.A., Velasquez, R.: Submitted to the 5th Eurasphalt & Eurobitume Congress, Istanbul, Turkey (2012)

Using 3D Laser Profiling Sensors for the Automated Measurement of Road Surface Conditions

John Laurent[1], Jean François Hébert[1], Daniel Lefebvre[2], and Yves Savard[3]

[1] Pavemetrics Systems inc., Canada
[2] INO (National Optics Institute), Canada
[3] Ministère des Transports du Québec (MTQ), Canada

Abstract. In order to maximize road maintenance funds and optimize the condition of road networks, pavement management systems need detailed and reliable data on the status of the road network. To date, reliable crack and raveling data has proven difficult and expensive to obtain. To solve this problem, over the last 10 years Pavemetrics inc. in collaboration with INO (National Optics Institute of Canada) and the MTQ (Ministère des Transports du Québec) have been developing and testing a new 3D technology called the LCMS (Laser Crack Measurement System).

The LCMS system was tested on the network to evaluate the system's performance at the task of automatic detection and classification of cracks. The system was compared to manual results over 9000 km and found to be 95% correct in the general classification of cracks.

1 Introduction

The LCMS is composed of two high performance 3D laser profilers that are able to measure complete transverse road profiles with 1mm resolution at highway speeds. The high resolution 2D and 3D data acquired by the LCMS is then processed using algorithms that were developed to automatically extract crack data including crack type (transverse, longitudinal, alligator) and severity. Also detected automatically are ruts (depth, type), macro-texture (digital sand patch) and raveling (loss of aggregates). This paper describes results obtained recently regarding road tests and validation of this technology.

2 Hardware Configuration

The sensors used with the LCMS system are 3D laser profilers that use high power laser line projectors, custom filters and a camera as the detector [1,2]. The light strip is projected onto the pavement and its image is captured by the camera (see figures 1 and 2). The shape of the pavement is acquired as the inspection vehicle travels along the road using a signal from an odometer to synchronize the sensor acquisition. All the images coming from the cameras are sent to the frame grabber to be digitized and then processed by the CPU. Saving the raw images would imply storing nearly

30Gb per kilometer at 100 km/h but using lossless data compression algorithms on the 3D data and fast JPEG compression on the intensity data brings the data rate down to a very manageable 20Mb/s or 720Mb/km. The critical specifications for the LCMS system can be found on table 1.

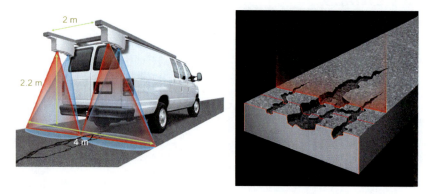

Fig. 1. LCMS on an inspection vehicle (left), laser profiling of cracks (right)

Table 1. LCMS Specifications

Nbr. of laser profilers	2
Sampling rate (max.)	11,200 profiles/s
Vehicle speed	100 km/h (max)
Profile spacing	Adjustable
3D points per profile	4096 points
Transverse field-of-view	4 m
Depth range of operation	250 mm
Z-axis (depth) accuracy	0.5 mm
X-axis (transverse) resolution	1 mm

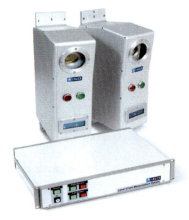

Fig. 2. Photo of the LCMS system (sensors and controller)

The LCMS sensors simultaneously acquire both range and intensity profiles. The figure 3 illustrates how the various types of data collected by the LCMS system can be exploited to characterize many types of road features. The graph shows that the 3D data and intensity data serve different purposes. The intensity data is required for the detection of lane markings and sealed cracks whereas the 3D data is used for the detection of most of the other features.

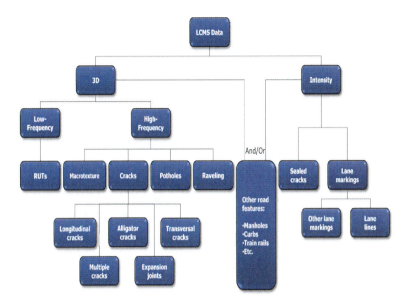

Fig. 3. Data analysis library diagram

3 Intensity Data

Intensity profiles provided by the LCMS are used to form a continuous image of the road surface. The first role of the intensity information is for the detection of road limits. This algorithm relies on the detection of the painted lines used as lane markings to determine the width and position of the road lane in order to compensate for driver wander. The lane position data is then used by the other detection algorithms to circumscribe the analysis within this region of interest in order to avoid surveying defects outside the lane. Highly reflective painted landmarks are much easier to detect in 2D since they generally appear highly contrasted in the intensity images. With the proper pattern recognition algorithms, various markings can be identified and surveyed. Figure 4 shows the results of the different types of images (intensity, range, and 3D merged image) that can be produced from the LCMS data.

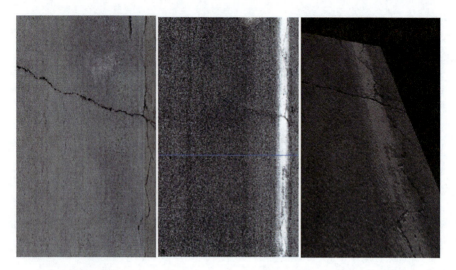

Fig. 4. LCMS data type – Range (left) – Intensity (center) – 3D merged (right)

4 3D Range Data

The 3D data acquired by the LCMS system measures the distance from the sensor to the surface for every sampled point on the road. The previous image (above left) shows a range data image acquired by the sensors. In this image, elevation has been converted to a gray level. The darker the point, the lower is the surface. In a range image the height can vary along the cross section of the road. The areas in the wheel path can be deeper than the sides and thus appear darker this would correspond to the presence of ruts. Height variations can also be observed in the longitudinal direction due to variations in longitudinal profiles of the road causing movements in the suspension of the vehicle holding the sensors. These large-scale height variations correspond to the low-spatial frequency content of the range information in the longitudinal direction. Most features that need to be detected are located in the high-spatial frequency portion of the range data. The figure 5 shows a 2m (half lane) transverse profile where the general depression of the profile corresponds to the presence of a rut, the sharp drop in the center of the profile corresponds to a crack point and the height variations (in blue) around the red line correspond to the macro-texture of the road surface.

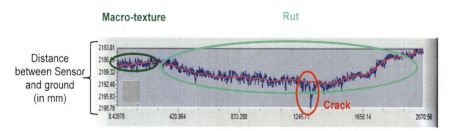

Fig. 5. LCMS (half lane) 2 m transverse profile showing ruts, cracks and texture

5 Macrotexture

Macrotexture is important for several reasons, for example it can help estimate the tire/road friction level, water runoff and aquaplaning conditions and tire/road noise levels produced just to name a few. Macrotexture can be evaluated by applying the ASTM 1845-01 norm [3]. This standard requires the calculation of the mean profile depth (MPD). To calculate the MPD, the profile is divided into small (10cm) segments and for each segment a linear regression is performed on the data. The MPD is then computed as the difference between the highest point on the profile and the average fitted line for the considered portion. MPD is the only way possible to evaluate texture using standard single point (64 kHz) laser sensors. The LCMS however acquires sufficiently dense 3D data to not only measure standard MPD but also to evaluate texture using a digital model of the sand patch method (ASTM E965) [4] as shown on figure 6.

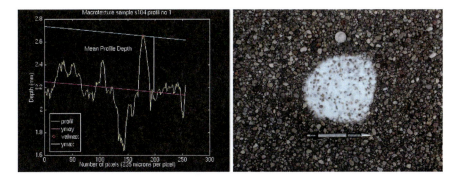

Fig. 6. MPD vs sand patch

The digital sand patch model is calculated using the following proposed Road Porosity Index (RPI). The RPI index is defined as the volume of the voids in the road surface that would be occupied by the sand (from the sand patch method) divided by a surface area. The digital sand patch method implemented allows texture to be evaluated continuously over the complete road surface instead of measuring only a single point inside a wheel path. The RPI can be calculated over any user definable surface area but LCMS reports by default the macro-texture values within the 5 standard AASHTO bands as illustrated on figure 7 (center, right and left wheel paths and outside bands).

Results show (see figure 8) that RPI measurements using the LCMS are highly repeatable as shown by road tests on several Alabama test sections and that RPI closely matches MPD measurements collected by standard texture lasers over a wide range of texture values.

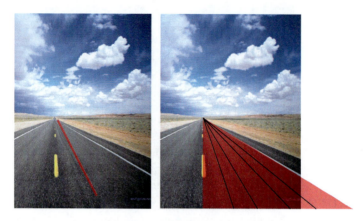

Fig. 7. MPD vs digital sand patch (RPI)

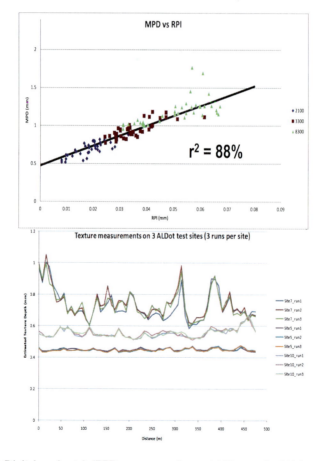

Fig. 8. Digital sand patch (RPI) accuracy and repeatability results (Alabama tests)

6 Raveling

Raveling is the wearing away of the pavement surface caused by the dislodging of aggregate particles and loss of asphalt binder that ultimately leads to a very rough and pitted surface with obvious loss of aggregates. In order to detect and quantify raveling conditions a Raveling Index (RI) indicator is proposed. The RI is calculated by measuring the volume of aggregate loss (holes due to missing aggregates) per unit of surface area (square meter). With the LCMS the high resolution of the 3D data allows for the detection of missing aggregates. Algorithms designed to specifically detect aggregate loss were developed in order

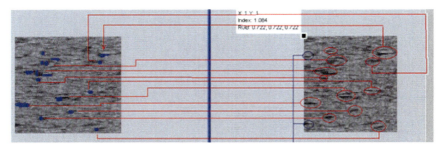

Fig. 9. Example of the automatic detection of aggregate loss in range images

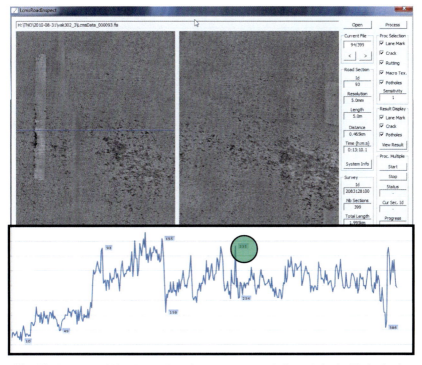

Fig. 10. Example of high RI road section on porous asphalt roads in the Netherlands

to evaluate the RI index automatically. The figures 9 demonstrate the results of aggregate detection (in blue) on range images. Figure 10 show an example of a high RI rated road section measured on porous asphalt roads in the Netherlands. Finally, the results of a repeatability test (3 passes) also on road sections in the Netherlands are shown on figure 11.

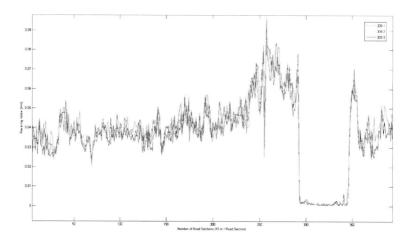

Fig. 11. Repeatability of RI measurements (3 passes) on road sections in the Netherlands

7 Cracking

Detecting cracks reliably is far more complex than applying a threshold on a range image. As mentioned previously the 3D profile data needs to be detrended from the effects of rutting and vehicle movements. Macrotexture is also a problem; road surfaces have very variable macrotexture from one section to the next and even from one side of the lane to the other. For example, on roads with weak macrotexture we can hope to detect very small cracks which will be harder to detect on more highly textured surfaces. It is thus necessary to evaluate and to adapt the processing operations based on the texture and type of road surface. Once the detection operation is performed, a binary image is obtained where the remaining active pixels are potential cracks. This binary image is then filtered to remove many of the false detections which are caused by asperities and other features in the road surface which are not cracks on the pavement. At this point in the processing, most of the remaining pixels can correctly be identified to existing cracks, however many of these crack segments need to be joined together to avoid multiple detections of the same crack. After the detection process, the next step consists in the characterization of the cracks. The severity level of a crack is determined by evaluating its width (opening) typically cracks will be separated in low, medium and high severity levels. The cracks also need to be grouped into two main categories: Longitudinal and transverse cracks. Furthermore, transverse cracks are further divided into complete and incomplete types and joints need to be classified separately. Longitudinal cracks are further refined into three sub-categories: simple, multiple and alligator.

The LCMS system was used by the MTQ to survey nearly 10,000km of its road network. In order to validate the system an independent 3rd party under the supervision of the MTQ was mandated to manually qualify the crack detection results of the LCMS system over the entire survey. To do this each 10m section was visually analyzed and the results were categorized in 3 classes (Good, Average and Bad). A forth class (NA) was used when for when it was not possible to correctly evaluate a section. Figure 12 shows an example of crack detection results on a 10m pavement section. Transverse cracks are identified with a bounding box. Regions in red indicate high severity cracks (15mm+) and light blue and green represent low severities (less than 5mm). Table 2 shows the results of the compilation of the manual evaluation. The final results are deemed excellent by the MTQ as the overall 'Good' rating reaches 96.5%. Repeatability tests were also conducted on several MTQ test sections and the results shown on the figure 13 also demonstrate very repeatable crack detection results on these sections.

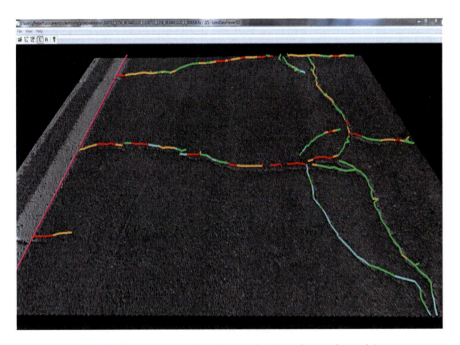

Fig. 12. Example crack detection results (severity = color code)

Table 2. 10,000 km automatic vs manual survey results

District #	Total (10 m sections)	Results (manual classification)							
		Number of images (10 m sections)				Proportion (%)			
		Good	Average	Bad	NA	Good	Average	Bad	NA
84	35288	34144	310	144	690	96,8	0,9	0,4	2,0
85	4243	4101	53	51	38	96,7	1,2	1,2	0,9
86	147903	144040	516	1520	1827	97,4	0,3	1,0	1,2
87	149926	138453	1170	5728	4575	92,3	0,8	3,8	3,1
88	189097	183010	1064	2002	3021	96,8	0,6	1,1	1,6
89	125003	121835	442	2015	711	97,5	0,4	1,6	0,6
90	123653	116930	2980	2434	1309	94,6	2,4	2,0	1,1
91 & 92	215513	213142	197	956	1218	98,9	0,1	0,4	0,6
Total	990626	955655	6732	14850	13389	96,5	0,7	1,5	1,4

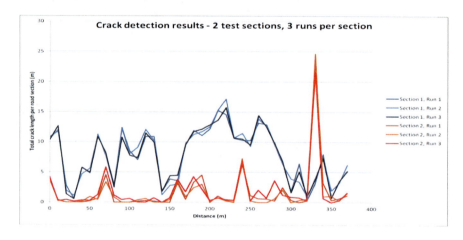

Fig. 13. Repeatability results (3 passes) on two MTQ road sections

8 Conclusions

We have presented a road surveying system that is based on two high performance transverse 3D laser profilers that are placed at the rear of an inspection vehicle looking down in such a way as to scan the entire 4m width of the road surface with 1mm resolution. This configuration allows the direct measurement of many different types of surface defects by simultaneously acquiring high resolution 3D and intensity data. Examples of different algorithms and results were shown using the 3D data to detect cracks, ruts, evaluate macro-texture and to detect raveling while the intensity data was used for the detection of lane markings.

The LCMS system was tested at the network level (10000 km) to evaluate the system's performance at the task of automatic detection and classification of cracks. The system was evaluated to be over 95% correct in the general classification of cracks.

A Road Porosity Index (RPI) was proposed as a model to measure the equivalent of a digital sand patch. The digital sand patch (RPI) method implemented allows texture to be evaluated continuously over the complete road surface and within each of the five AASHTO bands.

A Raveling Index (RI) indicator calculated by measuring the volume of aggregate loss (holes due to missing aggregates) per unit of surface area (square meter) was proposed. This indicator was shown to allow the quantification of the amount of raveling present and was shown to be highly repeatable.

References

[1] Laurent, J., Lefebvre, D., Samson, E.: Development of a New 3D Transverse Profiling System for the Automatic Measurement of Road Cracks. In: Proceedings of the 6th Symposium on Pavement Surface Caracteristics, Portoroz, Slovenia (2008)
[2] Laurent, J., Hébert, J.F.: High Performance 3D Sensors for the Characterization of Road Surface Defects. In: Proceedings of the IAPR Workshop on Machine Vision Applications, Nara, Japan (2002)
[3] ASTM E1845 - 09 Standard Practice for Calculating Pavement Macrotexture Mean Profile Depth, Active Standard ASTM E1845 Developed by Subcommittee: E17.23
[4] ASTM E965 - 96, Standard Test Method for Measuring Pavement Macrotexture Depth Using a Volumetric Technique, Active Standard ASTM E965 Developed by Subcommittee: E17.23 (2006)

Pavement Crack Detection Using High-Resolution 3D Line Laser Imaging Technology

Yichang (James) Tsai[1], Chenglong Jiang[1], and Zhaohua Wang[2]

[1] Gerogia Institute of Technology, School of Civil and Environmental Engineering
[2] Georgia Institute of Technology, Center for Geographic Information Systems

Abstract. With the advancement of 3D sensor and information technology, a high-resolution, high-speed 3D line laser imaging system has become available for pavement surface condition data collection. This paper presents preliminary results of a research project sponsored by the U. S. Department of Transportation (DOT) Research and Innovation Technology Administration (RITA) and the Commercial Remote Sensing and Spatial Information (CRS&SI) technology program. The objective of this paper is to validate the capability of 3D laser pavement data gathered during an automated pavement survey. An experimental test, using continuous profile-based laser data collected from Georgia State Route 80 and 275, was conducted to evaluate the performance of 3D line laser imaging technology. Based on the experimental results, the 3D laser pavement data are robust under different lighting conditions and low-intensity contrast conditions and have the capability to deal with different contaminants on a pavement's surface. It can support an accurate crack width measurement, which will contribute to further crack classification task. The 3D laser pavement data have a good capability to collect cracks that are greater than 2mm wide; however, the data resolution limits the detection of hairline cracks to approximately 1mm. The findings are crucial for transportation agencies to use when determining their automated pavement survey policies. Recommendations for future research are discussed in the paper.

1 Introduction

Pavement surface distress measurement is an essential part of a pavement management system (PMS) for determining cost-effective maintenance and rehabilitation strategies. Visual surveys conducted by engineers in the field are still the most widely used means to inspect and evaluate pavements, although such evaluations involve high degrees of subjectivity, hazardous exposure, and low production rates. Consequently, automated distress identification is gaining wide popularity among transportation agencies.

As early as 1990, Haas and Hendrickson [1] presented a general model of pavement surface characteristics that integrates multiple types of sensor information to simplify automated pavement distress survey. For the past two decades, many researchers have developed automated pavement crack detection and evaluation methods, which is an important component of automated pavement distress surveys. Most studies [2–5] employed 2D intensity-based imaging systems to provide the input for crack detection algorithms. Due to the mechanism of data acquisition, the performance of crack detection algorithms is severely hampered in the presence of shadows, lighting effects, non-uniform crack widths, and poor intensity contrast between cracks and surrounding pavement surfaces [6]. Xu et al. [7] used artificial constant lighting, such as LED lighting, that prevents the impact of shadows during data collection; however, the beam width of the LED lighting was not thin enough to provide sufficient depth solution, and it also did not provide a good solution to the pavements with low-intensity contrast. Some researchers [8] attempted to use 3D stereovision or photogrammetric systems for pavement surface reconstruction, but they were still in the experimental stage, and the resolution was operationally limited. Therefore, the challenge for automated pavement crack surveying persists in spite of all the research work that has been carried out to improve image acquisition techniques by minimizing defects [9].

With the advances in sensor technology, a 3D line-laser-imaging-based pavement surface data acquisition system has become available. The Laser Crack Measurement System (LCMS) [10] can collect high-resolution 3D continuous pavement profiles for constructing pavement surfaces. The objective of this paper is to validate the capability of 3D laser pavement data to detect cracks in support of subsequent crack classification. The paper is organized as follows. This section reviews related research on automated pavement crack surveying and identifies the objective of this study. Section 2 briefly introduces the 3D line-laser-imaging system for pavement data collection. Section 3 demonstrates the advantages of 3D laser pavement data under different conditions. Section 4 presents the validation results of the crack width measurement results using 3D laser data. The last section concludes the findings and makes recommendations for future research.

2 3D Laser Technology for Pavement Surface Data Collection

This section briefly introduces the 3D line-laser-imaging-based data acquisition system for pavement surface data collection. A sensing vehicle integrated with two high-performance laser profiling units is shown in Figure 1. Each profiling unit consists of a 3D laser profiler (laser line projector), a custom filter, and an area scan camera as the detector. The profiling unit employs the existing concept of structured light and triangulation. The profiler projects the laser light stripe onto an object's surface, and its image is captured by the area scan camera. Based on the pre-calibrated positions and angles of projector and detector and also the deformation of the laser line projected on the object, the elevation of the object surface can be calculated. In this case, a complete 3D-dimensional set of points of the pavement surface can be acquired.

Fig. 1. A sensing vehicle integrated at Georgia Tech

Each laser profiling unit produces 2,080 data points per profile. Integrating two profiling units, the sensing system uses 4,160 3D data points to cover a 4-meter pavement width, which is usually sufficient for single normal road lane (as shown in Figure 2). Therefore, the resolution in transverse direction is about 1mm. In addition, the acquired 3D laser profile has been designed to have a 12-degree clockwise tilt angle to the pavement's transverse direction in order to ensure that the laser profiles can intersect with transverse cracks. The resolution is 0.5mm in elevation direction. The highest resolution in longitudinal direction (driving direction) depends on the vehicle's driving speed. In the integrated sensing vehicle, Distance Measurement Instrument (DMI) and the accompanying encoder are used to coordinate the vehicle driving and 3D pavement data collection. The system can collect transverse profiles at 4.6mm intervals at a speed of 100km/h. The high allowable operational speed also provides the potential for the system to conduct pavement data collection under highway condition.

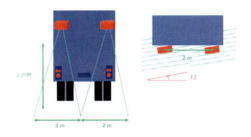

Fig. 2. LCMS system and projection of laser [10]

3 Advantages of 3D Laser Pavement Data

Since a 3D laser profile uses the range (elevation) information to describe pavement surface, it has several advantages compared to traditional techniques. Unlike a 2D digital image, the range data based on a 3D laser profile is hardly influenced by different lighting conditions. Poor intensity contrasts and contaminants like oil stains will also not interfere with the segmentation algorithms using the acquired range data (which is equivalent to 3D laser data in this paper). This section demonstrates the advantages of 3D laser pavement data.

3.1 Pavements under Different Lighting Conditions and Low Contrast Conditions

Our previous study demonstrated the robustness of 3D laser pavement data under different conditions [6]. The experimental data in this test were collected on Georgia State Route 80. The crack detection employed the dynamic optimization algorithm [11], which was originally developed for medical image processing. Figure 3 shows three different lighting conditions on the roadway. The different lighting conditions definitely led to different intensity appearances of the roadway surface, and this introduced challenges for crack detection using a traditional 2D intensity image. However, in the collected 3D laser pavement data, there were no distinctive differences among the three conditions, and the crack detection gave robust results (Figure 4).

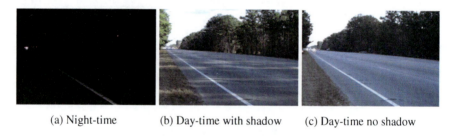

(a) Night-time (b) Day-time with shadow (c) Day-time no shadow

Fig. 3. Examples of three lighting conditions [6]

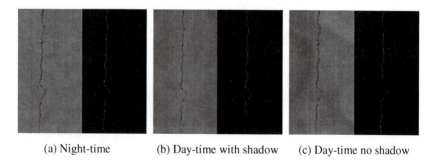

(a) Night-time (b) Day-time with shadow (c) Day-time no shadow

Fig. 4. 3D laser data and corresponding crack detection results for three lighting conditions [6]

Figure 5 shows the pavement sample with low-intensity contrast. The pavement cracking in a traditional road image (Figure 5 (a)) was hard to differentiate due to the pattern of the surrounding surface. This challenge was tremendously reduced by employing the 3D laser pavement data, and the crack detection also gave reliable result.

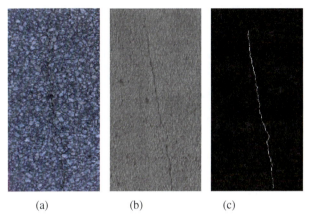

(a) (b) (c)

Fig. 5. Pavement sample with low-intensity contrast: (a) the roadway image; (b) 3D laser data; (c) crack detection results [6]

3.2 Pavements with Oil Stains and Other Contaminants

Oil stains usually appear to be darker than the surrounding area on a pavement surface. In a traditional 2D digital image, crack pixels have similar characteristics. Therefore, it is difficult to differentiate pavement cracking from contaminants, such as oil stains. However, benefiting from the features of a 3D laser profile, oil stains are no more distinctive in range data.

This experimental test used a representative pavement sample to present the advantages of the 3D laser technique under the influence of oil stains and other contaminants. The data used in this test were collected on Georgia State Route 275 between Milepost 0 and Milepost 1. There's an intersection and a gas station around this section of road. The pavement surface is usually not clean and is affected by contaminants; however, considering that there is not too much traffic along SR 275, the pavement surface condition is relatively good. Figure 6 demonstrates the selected sample in which contaminants, such as oil stains, influence the pavement's appearance. Figure 6 (a) shows the intensity image of the pavement sample. Different contaminants influence the appearance of the pavement, including some oil stains around the image center, dark strips caused by tire marks, camera discoloration, and lane marking. Figure 6 (b) shows the 3D laser data of this sample. Although contaminants, such as oil stains, have a distinctive intensity, they do not have an obvious elevation change on the surface. Therefore, they have been eliminated in the range data; however, cracking can still be clearly observed because it usually has a sharp elevation drop compared to the surrounding area.

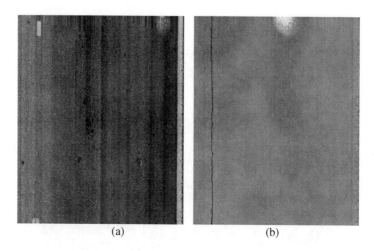

Fig. 6. Pavement images with oil stains and other contaminants

4   Validation of Crack Width Measurement Using 3D Laser Pavement Data

Almost all the existing automatic crack evaluation studies focus on using 2D intensity images as the input. Compared to the traditional 2D digital image technique, the emerging 3D laser technique can provide a more accurate width measurement as well as the crack depth information. Crack width is a common and important crack classification factor in most DOTs' pavement evaluation protocols, especially when differentiating severity levels. It is also crucial information to determine pavement maintenance operations, such as crack sealing/filling. However, crack width has rarely been used in the past crack classification studies. Considering the characteristics of the 2D intensity image, the accuracy of crack width measurement (measured pixel by pixel) is limited; even for a high-resolution image, crack width measurement is still influenced by other factors that also influence crack detection, such as lighting conditions and pavement contaminants (e.g. oil stain). The 3D laser technique provides an opportunity to measure crack width more accurately.

In this experimental test, a total of 12 locations were selected from State Route 275 between Milepost 1 and Milepost 2 for crack width measurement. Figure 7 shows an example of location selection for crack width measurement. The left image is the range data with detected crack map overlaid, and the automatically measured crack width information using 3D laser data is labelled beside the corresponding crack elements. The right one is the intensity image with crack map overlaid, and the selected 12 locations are marked for reference.

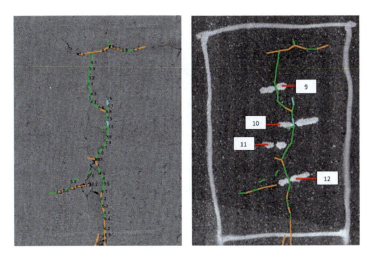

Fig. 7. Example of location selection for crack width measurement

Table 1. Crack width measurement validation results for 12 locations

Location No.	Detected Crack Width (mm)	Manually Measured Crack Width (mm)	Absolute Difference (mm)
1	3.5	3.5	0
2	2.8	3	0.2
3	4	3.5	0.5
4	Not detected	1.5	N/A
5	Not detected	1	N/A
6	3.8	3	0.8
7	Not detected	1	N/A
8	3.1	3	0.1
9	4.8	4	0.8
10	2.9	3	0.1
11	Not detected	1	N/A
12	4	5	1
Avg.	3.6	3.5	0.4

In order to validate the automatic crack width measurement accuracy for the 12 selected locations, the ground true crack widths were manually measured in the field. The validation results are shown in Table 1. Based on the results, the 3D laser pavement data have a good capability to collect cracks which are greater than 2mm wide; however, it is difficult to detect hairline cracks with widths of approximately 1 mm. This is reasonable considering the characteristics of the laser technique and, also, the resolution of LCMS system (4 mm at driving direction and 1 mm at transverse direction). Hairline cracks are also difficult to detect using traditional digital intensity images. Actually, based on Georgia Department of

Transportation (GDOT) pavement survey practice, hairline cracks, which usually occur in low severity level load cracking, do not have significant influence on the deducted value computation; in other words, although a hairline crack is also a pavement distress issue, it does not have a large influence on the current pavement condition survey results. For the detected cracks, the automatically measured width information is relatively consistent with a field measured ground truth. The maximum absolute difference is about 1mm, and the average absolute difference is about 0.4 mm. The results show the capability of using the accurate crack width information for further crack classification tasks.

5 Conclusions and Recommendations

Developing an automated pavement cracking survey has gained strong interest among transportation agencies. However, a fully automated pavement crack detection system based on a 2D intensity-based imaging data acquisition system still remains a challenge. The emerging 3D line-laser-imaging-based data acquisition system has the potential to collect more robust pavement surface data under different conditions. The objective of this study is to demonstrate the feasibility of using a 3D laser-based pavement data acquisition system to conduct automated pavement cracking surveys. A sensing vehicle was integrated by our research group at Georgia Tech, and the LCMS system was employed as the 3D pavement data acquisition system. The experimental tests were conducted on actual pavements in Georgia, and different pavement data were collected from State Routes 80 and 275. The major findings of the paper are summarized as follows:

1. The 3D laser pavement data are robust under different lighting conditions and low-intensity contrast conditions;
2. The 3D laser pavement data have the capability to remove the interference of contaminants on pavement surface, such as oil stains, tire marks, discoloration caused by the camera lens, incomplete lane marking, etc.
3. The 3D laser pavement data can support an accurate crack width measurement, which is critical for further crack classification and evaluation. Based on the experimental results, the absolute average difference between detected crack widths and manually measured ground truths was 0.4 mm. The cracks with widths equal to and greater than 2mm can be detected. However, hairline cracks (with crack width around 1mm) cannot be detected by the system, mainly due to the currently available resolution of laser data.

The preliminary experimental tests show that the emerging 3D laser pavement data acquisition technique is promising for conducting automated pavement crack surveys; it can be operated at highway speed with high data resolution, and it has several outstanding advantages over the traditional intensity-based pavement data acquisition. The following are recommendations for future research:

1. The paper is based on the results of preliminary experimental tests, and a comprehensive field test is still needed. The comprehensive test will include the pavement data with different kinds of cracks (e.g. transverse cracks, block cracks, short cracks, sealed cracks, and cracks filled with dirt) and severity levels, different asphalt pavement surfaces with various textures (e.g. dense graded, open graded pavement, and chip seal), and different roadway conditions (e.g. patches).
2. The experimental evaluation was based on the visual inspection of data and crack detection results, but there is still a need to introduce a quantitative and objective method to further evaluate the 3D laser pavement data. Kaul et al. [9] introduced the scoring method based on buffered Hausdorff distance to quantitatively evaluate the performance of different crack detection/segmentation algorithms [12], and it can be used as a good tool in further research.
3. Different crack detection/segmentation algorithms still need to be tested with the 3D laser pavement data. Most of the existing crack detection algorithms (e.g. dynamic optimization algorithm used in this study) are based on a 2D intensity image, and further experimental tests need to be done to demonstrate their performance with the emerging 3D laser data. Considering the unique features of 3D laser pavement data, it may also be necessary to improve the existing methods or propose new crack detection methods that will fit better with the different data.
4. Current experiments are in the automated crack detection stage. In order to conduct an automated pavement crack survey, crack classification and measurements still need to be further studied. The existing studies on crack classification are limited. With the high accuracy of crack width measurement and additional crack depth measurement, the 3D laser pavement data have the potential to conduct the crack classification tasks, even for different crack severity levels (which are also considered in the state DOTs' pavement survey manuals).

Acknowledgements. The work described in this paper was sponsored by the US Department of Transportation RITA program (RITARS-11-H-GAT). The authors would like to thank the assistance provided by Mr. Caesar Singh, the program manager of US DOT. The views, opinions, findings and conclusions reflected in this presentation are the responsibility of the authors only and do not represent the official policy or position of the USDOT, RITA, or any State or other entity.

References

[1] Haas, C., Hendrickson, C.: Computer-based Model of Pavement Surfaces. Transportation Research Record (1260), 91–98 (1990)

[2] Cheng, H.D., Chen, J., Glazier, C., Hu, Y.G.: Novel approach to pavement cracking detection based on fuzzy set theory. Journal of Computing in Civil Engineering 13(4), 270–280 (1999)

[3] Wang, K.C.P.: Designs and implementations of automated systems for pavement surface distress survey. Journal of Infrastructure Systems 6(1), 24–32 (2000)

[4] Lee, B.J., Lee, H.D.: Position-invariant neural network for digital pavement crack analysis. Computer-Aided Civil and Infrastructure Engineering 19(2), 105–118 (2004)
[5] Huang, Y., Tsai, Y.: Enhanced Pavement Distress Segmentation Algorithm Using Dynamic Programming and Connected Component Analysis. Transportation Research Record (2011) (accepted for publication)
[6] Tsai, Y., Li, F.: Critical Assessment of Detecting Asphalt Pavement Cracks under Different Lighting and Low Intensity Contrast Conditions Using Emerging 3D Laser Technology. Journal of Transportation Engineering (2011) (accepted for publication)
[7] Xu, B.: Summary of Implementation of an Artificial Lighting System for Automated Visual Distress Rating System. Presented at Transportation Research Board Annual Meeting (2007)
[8] Hou, Z., Wang, K.C.P., Gong W.: Experimentation of 3d Pavement Imaging through Stereovision. In: Proc. of International Conference on Transportation Engineering, pp. 376–381 (2007)
[9] Kaul, V., Tsai, Y.J., Mersereau, R.M.: Quantitative Performance Evaluation Algorithms for Pavement Distress Segmentation. Transportation Research Record (2153), 106–113 (2010)
[10] Laurent, J., Lefebvre, D.: Development of a New 3d Transverse Laser Profiling System for the Automatic Measurement of Road Cracks. Presented at the 6th Symposium on Pavement Surface Characteristics (2008)
[11] Alekseychuk, O.: Detection of Crack-Like Indications in Digital Radiography by Global Optimisation of a Probabilistic Estimation Function, PhD Thesis, BAM-Dissertationsreihe, Band 18 (2006)
[12] Tsai, Y., Kaul, V., Mersereau, R.M.: Critical Assessment of Pavement Distress Segmentation Methods. Journal of Transportation Engineering 136(1), 11–19 (2010)

Detecting Unbounded Interface with Non Destructive Techniques

Jean-Michel Simonin[1], Cyrille Fauchard[2], Pierre Hornych[1], Vincent Guilbert[2], Jean-Pierre Kerzrého[1], and Stéphane Trichet[1]

[1] LUNAM Université, IFSTTAR, Route de Bouaye, CS4,
 F-44344 Bouguenais Cedex, France
[2] CETE Normandie Centre 10 chemin de la Poudrière BP 241
 76121 Le Grand Quevilly Cedex

Abstract. The French road network has been built more than 30 years ago, and consists mainly of bituminous pavements. Some of them have also been maintained several times by thin overlays. On these pavements, a lot of damage such as potholes and alligator cracking has been observed, in particular after periods of heavy rain or freeze/thaw. Frequently, this type of damage is assumed to be linked with interface debonding between these overlays and the old pavement, associated with moisture effects. To detect such damages, some non destructive techniques (NDT), as electromagnetic techniques (GPR, step-frequency radar or infra-red) or as mechanical techniques (from static deflection measurements to seismic wave propagation methods), appear as promising approaches. This paper compares two differents NDT to detect debonding during an experiment carried out on the large pavement fatigue carrousel of IFSTTAR in Nantes.

The tests presented in this paper are performed on a 15m long pavement section, consisting of 3 bituminous layers over a granular subbase. Several types of defects have been included at the interface between the two base layers or at the interface between the base layer and the wearing course. Debonded areas of different size and form have been created artificially, using different techniques (sand, Teflon or kraft paper). The construction has been done by a road construction company, using standard road works equipment.

At the start of the experiment, different NDT techniques (Colibri, step frequency radar) are used to detect the different geometrical characteristics of artificial defects. This allows comparing the capability of each technique to detect such damages.

Keywords: Accelerated Pavement Testing, debonded interface, step frequency radar, dynamic investigation, non destructive tests.

1 Introduction

The French road network has been built more than 30 years ago, and consists mainly of old bituminous pavements. Some of them have also been maintained several times by thin overlays (less than 8cm). On these pavements, a lot of damage such as potholes and alligator cracking has been observed these last years, in particular after periods of heavy rain or freeze/thaw. Frequently, this type of damage is assumed to be linked with interface debonding between these overlays and the old pavement, associated with moisture effects. These debondings have a large influence on the residual life of the pavement [1], and thus their early detection is a very important issue for pavement maintenance.

To detect such interface damages, some non destructive techniques (NDT), as electromagnetic techniques (ground-penetrating radar, step-frequency radar or infra-red) or as mechanical techniques (from static deflection and radius of curvature measurements to seismic wave propagation methods), appear as promising approaches. They could also be efficient to detect and survey internal cracks. This paper compares 2 different NDT to detect debonding on a test site built for an experiment carried out on the large pavement fatigue carrousel of IFSTTAR in Nantes.

2 Description of the Test Site

The test site has been built on the pavement fatigue carrousel of IFSTTAR (figure 1). It is a large scale circular outdoor test equipment, unique in Europe by its size (120m long) and loading capabilities (maximum loading speed 100 km/h, loading rate 1 million cycles per month). Contrary to most Accelerated Testing Pavement equipments, it is able to test pavements up to failure in a few weeks. The machine comprises a central motor unit and 4 arms that can be equipped with different wheel configurations. The circular test track can be divided in several different test sections, loaded simultaneously. The width of the test track (6m) allows to apply traffic loads on the same track at two different radii.

Fig. 1. The pavement fatigue carrousel of IFSTTAR

This study is part of a full scale experiment started on the test track in 2008 with low traffic pavements. The major part of the ring has been built by a private company as part of a research contract. The ring was divided into several sectors to test some alternative road construction techniques. A 15m long part of the ring has been used for by IFSTTAR for its own research work. On this section, it was decided to include debonded interface, in a pavement structure representative of the national road network. The main objective was to test and compare different non destructive techniques. A second objective was to survey the evolution of the defects.

This 15m long pavement section, consists of 3 bituminous layers (2x0.10m thick base layers, and 0.06m thick wearing course), over a 0.20m thick unbound granular sub base. It is close to the 19m radius of the ring and is 4m wide. The wheel path is 1.60m wide.

The subgrade is a sand 0/4 from the Missillac quarry and it could be decomposed in two layers. The first layer (2.2 m) is composed by a sand present on the site since 2001 with a good bearing capacity (about 120 MPa). The second layer, about 0.35m, is constituted of new clayey sand, added to raise the initial level of the soil, for the pavement structure. This layer is of poor quality (35 MPa).

The unbound granular material (UGM) is a 0/20 mm crushed gneiss aggregate, in conformity with the French classification GNT B2C2 [2]. It can be noted that the UGM has relatively high fines content (9.9 %), close to the maximum of French specifications.

The characteristics of the bituminous materials are given in table 1. The wearing course is a 0/10 mm bituminous concrete (HMA1), and the base and subbase consist of a 0/14 mm base course asphalt material (HMA2 & HMA3 base).The two mixes are made with aggregates of the same origin (Brefauchet quarry) and the same binder (35/50 grade pure bitumen).

Table 1. Main characteristics of the bituminous materials

	Content (%)	
Size (mm)	HMA1 course 35/50	HMA2 & HMA3 base 35/50
10/14 aggregate	7.3	31.7
6.3/10 aggregate	33.9	12
2/6.3 aggregate	26.1	25
0/2 sand	31.7	31.3
Filler	1	
Bitumen 35/50	5.7	4.1

During the construction, optical fibre sensors have been placed at the interfaces between bituminous layers, at the beginning of the section, to evaluate strain measurements obtained with these sensors, and to compare them with classical strain gages for bituminous layers (PE1, PE2 & PE3). Several objects have also been included in the pavement (figure 2): Wood (D4) and Teflon objects (T2, T3, T5 & T10) placed at the interface between wearing course and base layer, outside the wheelpath. These objects have been put in place to test different infra-red detection techniques. Other rectangular objects have been included to simulate debonded interfaces, at different depth. Table 2 gives the level, size and type of each object.

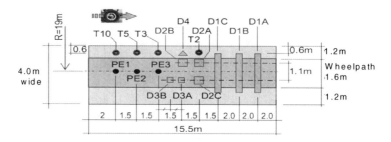

Fig. 2. Interface defects before wearing course construction

Kraft paper is assumed to represent a sliding interface without damage of the pavement layers. Sand simulates a sliding interface with an alteration of the material at the interface. Teflon has been used to represent a high level of damage of the interface, with a large change of mechanical properties, but also of thermal properties of materials. During the construction, sand and Teflon are easy to set-up. Kraft papers are more difficult to maintain at the right place, and may be moved or destroyed by the finisher.

Table 2. Characteristics of the artificial debonded areas

Name	Level (m)	Size (mxm)	Type
D1A	-0.06	1x2.5	Kraft paper
D1B	-0.16	1x2.5	Kraft paper
D1C	-0.16	1x2.5	Sand
D2A	-0.06	0.5x0.5	Kraft paper
D2B	-0.16	0.5x0.5	Kraft paper
D2C	-0.16	0.5x0.5	Sand
D3A	-0.06	0.2x0.2	Kraft paper
D3B	-0.06	0.2x0.2	Teflon

3 Pavement Investigation

The experiments presented compare 2 new non destructive techniques: dynamic investigation with the Colibri apparatus and high frequency radar measurements, with a Step Frequency Radar (SFR). Several longitudinal and transversal profiles have been investigated with the 2 systems. The paper presents only a longitudinal profile to compare the results obtained with the 2 NDT techniques.

3.1 Dynamic Investigation Method

3.1.1 Principle of the Method

Roadways constitute continuous structures on which the complex Frequency Response Function (FRF) [3] can be measured. The Colibri apparatus is a mobile automated measurement device, which applied this method. It uses a hammer to apply a dynamic solicitation (shock s(t)) to the road surface and measures the vertical surface response (vertical acceleration x(t)) close to the solicitation with an accelerometer placed at 0.10m from the impact (figure 3) . It deduces the inertance frequency response function, $(A(f))$, which is the ratio between a harmonic acceleration response and the harmonic force [3]. It is calculated at each test point in a broad frequency range.

For a healthy structure, the shock generates vibrations of the whole pavement. When a structure includes a defect (interface or crack), low frequency vibration modes appear which correspond to the vibration of a part of the structure (above the delamination or close to the crack). The inertance modulus estimated for the delaminated structure is higher than the one of the healthy structure. It increases at each eigen frequencies. Thus, a difference of inertance could be observed in a sensitive frequency band. This band and particularly the lowest frequency, depends on the characteristics of the defect (extension, depth, nature).

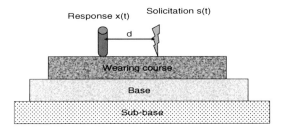

Fig. 3. Principle of dynamic investigation method

Application to Pavement Investigation

Application to pavement investigation consists in measuring the inertance function all along a roadway section. Then, the process aims at comparing the FRF modulus by defining a reference FRF representative of the healthy structure and at identifying FRF which are significantly different from this reference function. It has to be noted that the reference function is related to the investigated roadway. So, we suppose that measurements are recorded on a homogeneous structure (materials and layer thicknesses). The variations observed are then representative of the presence of damages which lead to a softer structure.

For each measurement point, i, a spectral analysis of a series of tests leads to calculate the inertance, $A(f,i)$, and the coherence function, $\gamma(f,i)$, between the pavement response and the applied solicitation. These functions depend on the frequency, f, and on the measurement point. The coherence function estimates the dependence of the output signal compared to the input signal. It is a real value ranging between 0 (no dependence) and 1 (full dependence). A minimum threshold of coherence (usually 0.8) is chosen to validate or not the calculation of the inertance. This threshold can be adapted according to the studies. For each frequency and each measurement point, FRF is validated if the coherence value is higher than this threshold. Thereafter, the analysis is restricted to the population of validated measurements.

On a homogeneous zone, data are then processed in 2 steps:

> Estimation of a reference function representative of the healthy structure;
> Calculation of a normalized damage.

To estimate the reference function modulus, it is assumed that a part of the tests was carried out on a healthy zone. This could be done voluntarily by investigating an un-trafficked zone such as an emergency lane. In practice we usually consider the set of modulus, $|A(f_k,i)|$, measured at a fixed frequency, f_k. The reference value at this frequency, $|A_{réf}(f_k)|$, is defined as a percentile of selected population. We usually adopt the percentile 20 which allows obtaining a low value representative of the healthy structure and eliminating abnormal measurements. The set of reference values are used to build the reference transfer function representative of the healthy structure, $|A_{réf}(f)|$.

Inertance modulus increases with frequency. We normalize the FRF modulus $|A(f_k,i)|$ using the modulus of the reference function. For each frequency and each measurement point, we calculate the damage, $D(f_k,i)$ according to Eqn. (1). This value varies between 0 and 1. The matrix D represents the damage on the road section for the different frequencies. It can be presented as a "damage mapping" where:

> The X-coordinate is the abscissa along the road section;
> The Y coordinate is the frequency band;
> The level of gray (or colors) represents the level of damage.

$$D(f_k,i) = 0 \text{ if } |A(f_k,i)| < |A_{réf}(f_k)| \text{ and } 1 - \frac{|A_{réf}(f_k)|}{|A(f_k,i)|} \text{ if } |A(f_k,i)| \geq |A_{réf}(f_k)|$$
(1)

3.1.2 Colibri Apparatus

The Colibri apparatus (figure 4) is a complete automated system for dynamic investigation of pavements. It includes:

> A hammer with a force-cell to measure the shock application;
> An accelerometer placed at 0.10m from the impact to measure the surface response and held by a spring mass system;
> An optical sensor to control the level of the shock.
> An engine to move up the hammer and produce the shock
> An electric jack enabling the lowering and raising of the system on the road surface as well as the placement of the sensors when Colibri is mounted on a vehicle;
> Electronics and data acquisition systems and a computer to manage the measurement sequences and to store the data.

During a measurement sequence, the system is positioned on the road surface. The engine moves up the hammer at the appropriate level. Then the hammer falls down and applies a wide-band dynamic impulse to the pavement. Signals (force and acceleration) are recorded by the computer. Usually, the test is repeated 3 times to have a good signal processing. If the system is mounted on a vehicle, the computer controls the vertical displacements of the Colibri system using the electric jack to make automatically measurements at different points.

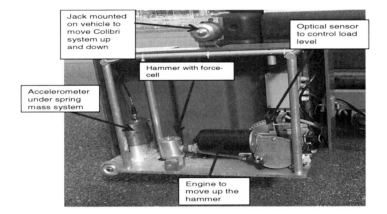

Fig. 4. The Colibri prototype mounted on a vehicle

3.1.3 Application on the Test Site

The Colibri prototype has been used to investigate several road profiles above the different debonded areas. Measurements have been made every 0.05 m along each road profile. Each measurement includes 3 impacts. Then, the system is moved manually to the next measurement point.

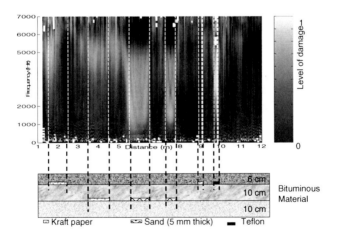

Fig. 5. Results of dynamic investigation method applied on longitudinal profile

Figure 5 shows the analysis of a longitudinal profile from D1A to D3B. The picture shows the damage level as a function of the distance (X) and the frequency (Y). The level of gray represents the damage level from 0 (black) to 1 (white). The debonded areas are indicated in the graph below. They can also be located on the picture. The results show that the dynamic investigation method is able to detect damage areas, but also to locate them and make a difference, in terms of damage level, between a debonded interface (kraft paper) and a higher damage level with a bad cohesion of the materials at the interface. The frequency range affected is also related with the type of debonding [4,5]. The lower is the first frequency sensitive to the defect, the larger is the debonded area or the lower is the material cohesion at the interface. Other work shows that the method is also sensitive to reflective cracking [5].

3.2 Step Frequency Radar (SFR)

3.2.1 Principle of the Method

The measurement of thickness of asphalt layers is usually performed with a Ground Penetrating Radar (GPR). This system consists of a pulse generator in time domain, that generates the signal emitted towards the studied road layers, and that receives the reflected part, via one or several antennas, placed above (horn antennas) or just on (dipole antennas) the road surface. The actual limit of GPR is the resolution of the thickness measured that depends on the permittivity of the medium, and on the central frequency of the used antenna, that is currently limited to 2.5GHz. In order to improve the resolution, Inverse Fast Fourier Transform (IFFT) properties show that the signal generation of frequencies over a wide band (frequency domain) is equivalent to a pulse generation in time domain. The Step Frequency Radar technique allows the use of high frequencies in a large band so that the equivalent pulse in time domain, after an IFFT, offers the capability of measuring very thin asphalt layers less than 2 cm. The system used here is made up of a Vector Network Analyzer (VNA) that emits monochromatic waves in the band [1.4GHz-15GHz], corresponding to the band pass of the Exponential Tapered Slot Antennas (ETSA) used for this study. The equivalent central frequency, after IFFT, is about 7.5GHz.

3.2.2 Short Background on the Detection of layer interfaces

We consider the two HMA layers media with a thin debonded zone between these two layers. This defect could be detected as far as the emitted electromagnetic field from the antenna has a wavelength of the order of half the thickness and the surface area of the defect. The vertical and horizontal resolutions are deduced from Eqn (2).

$$r_v = \frac{\lambda}{2} \text{ and } r_h = \sqrt{\frac{\lambda^2}{16} + \frac{\lambda_z}{2}} \qquad (2)$$

Where λ is the wavelength of the emitted electric field that is parallel to the surface (Transverse Electric mode) and z is the height of the antenna above the surface. For instance, for a HMA layer with a permittivity of 5 and a thickness of 5cm, and an ETSA radiated in the [1.4GHz – 20GHz] band at a height of 15cm from the HMA surface, a debonded or damaged interface of 3 mm to 4.5cm of thickness, and of 2.6cm to 10cm of horizontal length, is detectable.

This is a theoretical approach. Actually it is better to consider the limit of 7.5GHz given by the central frequency of the impulsion emitted from the ETSA, that leads to a potential detection of less than 1cm of thickness, and a horizontal dimension of less than 4cm.

3.2.3 Application to the Test Site

The measurements were carried out on the pavement fatigue carrousel (Ifsttar testing facility), on the pavement section described in the first part of this paper. The SFR system can be implemented for in-situ testing, as a GPR system. The results presented concern the detection of two defects: the D2C (sand interface) and D3B (Teflon interface) presented on Figure 2. We also propose a characterization of the D3B defect in terms of nature, by the measurement of its permittivity. This is possible because there is a significant permittivity difference between the Teflon and bituminous materials. It is less feasible for other natures of interfaces, with sand for instance.

For an accurate acquisition, an automatic motorized bench (Figure 6) controls the horizontal displacement of the antennas above the HMA surface. The results presented here are related to the monostatic case: only one antenna acts as the emitter and receiver. Software specifically developed synchronizes the motor displacement and the data acquisition.

Fig. 6. Picture of motorised bench with two ETSA over the pavement

Figure 7 shows the radargram (or B-scan) of the SFR measurement on the D2C defect. The sand interface is clearly detected between the HMA2 & HMA3 layers. Its length is about 60 cm and its thickness (less than 0.7cm) can be estimated by considering the travel times between its top and its bottom (less than 0.2ns) and an approximated permittivity of 5 in the middle. Nothing allows its characterization in terms of nature because its permittivity is not estimated: the first calculation conducted is currently insufficient and some improvements must be carried out.

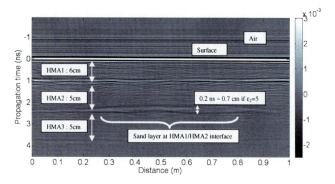

Fig. 7. B-scan obtained with the SFR system from monostatic measurement on the D2C (sand) defect

The SFR measurements realized just above the defect D3B is presented on Figure 8. The Teflon interface is clearly identified by a displacement of the HMA1/HMA2 interface position in time and by a variation of the reflected amplitude.

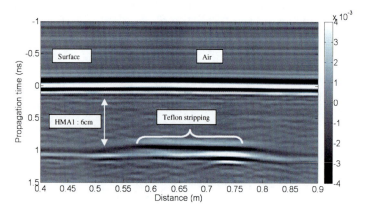

Fig. 8. B-scan obtained with the SFR system from monostatic measurements on the D3B (Teflon) interface

Permittivities of the HMA layers are estimated according to the classical approach [6, 7 & 8]. It consists in measuring the amplitude A_{HMA} in time domain of the reflected pulse, and comparing it with the amplitude A_{fmp} of the total reflection on a flat metal plate. The plane wave approximation is assumed and the HMA electrical conductivity is neglected. The reflection coefficient R_{01} (sign positive) between the air and HMA1 and R_{12} (sign dependent on dielectric contrast) between HMA1 and HMA2 are directly proportional to the ratio of these respective measurements:

$$R_{01} = \frac{A_{HMA1}}{A_{fmp}} \quad \text{and} \quad R_{12} = \frac{A_{HMA2}}{A_{fmp}} \quad (3)$$

The permittivities of the first and second layer are:

$$\varepsilon_1 = \left(\frac{1+R_{01}}{1-R_{01}}\right)^2 \quad \text{and} \quad \varepsilon_2 = \varepsilon_1 \left(\frac{1-R_{01}^2+R_{12}}{1-R_{01}^2-R_{12}}\right)^2 \quad (4)$$

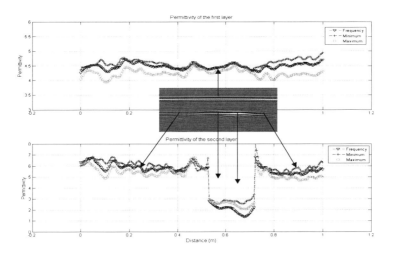

Fig. 9. Results of estimated permittivity for the two HMA layers. Presence of Teflon decreases the permittivity value of the second interface.

The permittivity of the HMA1 layer and the HMA2 layer are presented on Figure 9 for each position of the antenna above the surface, with a step of displacement of 1cm. The value for the HMA2 layer shows a strong variation where the Teflon stripping is located. The best estimated permittivity value of 2.3 +/- 0.5 is close to the theoretical one (2.1). It is performed with the permittivity calculation conducted in the frequency domain of the local Inverse Fourier Transform of the reflected pulse on the Teflon. The estimation based on the calculation conducted in the time domain of the local reflected pulse on the Teflon, by both considering the minimum or the maximum of the impulsion, are less accurate. On figure 9 (bottom) the permittivity of the teflon presents variations of a factor of about 1.4. These could be attributed to the presence of voids at the interface between the Teflon and HMA2 layer.

4 Conclusion

The paper presents the application of 2 NDT methods to detect debonding inside pavements. These methods are based on different physical phenomena. The first one is based on a dynamic mechanical approach including frequency and statistical analysis. The second one is based on the propagation of high frequency electromagnetic waves. These methods allow to detect and compare different defects on a test site. The dynamic investigation method is easy to apply on test site. Data are automatically processed. The method allows to detect and locate debonded areas and also internal cracking (not detected usually by other pavement investigation methods). The method shows a clear difference of response depending on the type of debonded interface. The Colibri device can be used to conduct investigations on real pavements in safe conditions, due to the use of a vehicle, and automated measurement device. The measurement rate can be several hundred meters per day (depending on the measurement interval). It could be used to investigate short sections at the project level to detect internal damage early, before it is visible on the surface, or to follow it after application of an overlay. The SFR technique is easily carried out on test site. The principle of measurement is the generation of very short electromagnetic impulsions in the time domain such that very thin layers (less than 2 cm) can be detected (debonded or damaged interface). Depending on the permittivity of the investigated media, the thickness and the nature of the defect (type of material, presence of voids or water) can also be estimated. The nature of the defect is easier to detect if it is located close to the surface, and between two bituminous layers (meaning two layers with similar permittivities). Compared with classical GPR, the measurement speed of SFR is lower presently, due to limitations from capacity of the electronics and data acquisition system. However, this can be improved with the rapid progress of the capacity of these systems. With this evolution, a similar measurement speed to classical GPR could probably be attained (about 40 km/h).

References

[1] Savuth, C.: Auscultation structurelle des chausses mixtes: détection des défauts d'interface à l'aide de la déflexion, Ph D. INSA de Rennes, France (2006)
[2] NF EN 13285, unbound mixtures - Spécifications (French standard NFP 98 129) (December 2010)
[3] Ewins, D.-J.: Modal testing: theory, practice and application, 2nd edn. Research studies press LTD, Letchworth (2000)
[4] Simonin, J.-M.: Contribution à l'étude de l'auscultation des chaussées par méthode d'impact mécanique pour la détection et la caractérisation des défauts d'interface. Thèse de doctorat. INSA, Rennes (2005)
[5] Simonin, J.-M., Lièvre, D., Dargenton, J.-C.: Bearing Capacity of Roads, Railways and Airfields. In: Proc. Intern. Conf., Balkema, vol. 1, pp. 459–466 (2009)
[6] Chew, C.W.: Waves and fields in inhomogeneous media. Van Nostrand Reinhold, New York (1990)
[7] Spagnolini, U.: IEEE Transaction on Geoscience and Remote Sensing 35(2) (1997)
[8] Fauchard, C., Dérobert, X., Côte, P.: NDT&E International 36, 67–75 (2003)

New Field Testing Procedure to Measure Surface Stresses in Plain Concrete Pavements and Structures

Daniel I. Castaneda[1] and David A. Lange[2]

[1] Graduate Student at the University of Illinois at Urbana-Champaign
[2] Professor at the University of Illinois at Urbana-Champaign

Abstract. Plain concrete pavements are subject to premature cracking due to the formation of residual stresses, structural confinement, and loss of subgrade support. These factors diminish the loading capacity of pavements and can result in cracking when combined with stresses attributable to curling and wheel loads. It is, thus, advantageous to quantify the unloaded stress state of pavements in order to modify its rated capacity and prevent premature cracking that, in some cases, necessitates costly repair or replacement. A research program was developed to craft a new field testing procedure capable of measuring the stress state in cantilevered concrete beams and in-situ slabs. The experimental results of this testing procedure showed that the stress state at the concrete surface could be fully quantified. Finite element modelling of the concrete beams and in-situ slabs further corroborates that the surface stresses in pavements can be viably measured.

1 Introduction

Uncontrolled cracking of concrete pavement slabs at airports is a persistent problem that is affected by a complex combination of live loads, dead loads, built-in residual stress, structural confinement, and loss of subgrade support. Evaluation of field problems is hampered by the lack of practical methods to assess stress of the in-place concrete material. The purpose of this study was to develop an experimental procedure to reliably quantify the stress state in concrete pavements with minimal disruption to airside operations.

Previous efforts to quantify the stress in concrete pavements date as far back as 1979 when A. M. Richards adapted an 'overcoring' technique from the geotechnical field [1]. The testing method, he reported, produced unreliable strain measurements and could not adequately describe the stress state in the concrete material. In more recent years, the US FAA National Airport Pavement Testing Facility (NAPTF) adapted a standardized testing method of measuring the residual stress in metal beams to concrete beams [2]. In this standardized method, a small through-hole is cored in the vicinity of a strain gage affixed onto the web surface of a structural beam [3]. The change in strain is correlated to a change in stress state by use of the Kirchhoff relations for the stress-field about a hole in an infinite plate [4]. The researchers at NAPTF modified the procedure by instead observing the change in strain atop the surface of a concrete beam subjected to a cantilever load when varying diameter partial-depth cores were made in the vicinity. Their preliminary results indicated that the stress state could be partially quantified [2]. The current study is a partnership between the University of Illinois at Urbana-Champaign (UIUC) and the NAPTF to further develop this test method.

2 Experimental Program

Research was undertaken by the authors at UIUC to validate the testing procedure developed by the NAPTF wherein a cantilever concrete beam was partially cored along its top surface. A modified testing procedure was developed at UIUC that improved the expediency and reliability of the measurements by instead using a hand-operated circular saw to cut around the affixed strain gage. This method proved to be more reliable, thus the testing regiment was expanded to include in-situ concrete slab testing where similar, successful results were observed.

2.1 Materials and Testing Equipment

Concrete beams measuring 15.24 cm in width, 15.24 cm in height and 86.36 cm in length were cast alongside cylinders measuring 10.16 cm in diameter and 20.32 cm in height. The beam dimensions closely matched those used by the researchers at the NAPTF providing for ample workspace when coring 7.62 cm and 10.16 cm diameter cores along the top surface of the beam. A typical concrete mix design is shown in Table 1.

Table 1. Typical concrete mix design

	SG	SSD (kg/m^3)
Cement	3.15	70.7
CA	2.72	230
FA	2.70	154
Water	1.00	29.7

The specimens were designed with a water-to-cement ratio of 0.42 and were allowed to cure for 7 days under plastic sheathing. After 7 days, the specimens were de-molded and transferred to a moist-cure room where the relative humidity was maintained at 100% and the temperature at 23 °C through 28 days. At 28 days and beyond, the cylinders were tested for their strength and stiffness while the beams were tested for their surface stress induced by cantilever loading.

The beams were positioned onto a 48.9 kN capacity MTS loading frame table as depicted in Fig. 1. In this position, 22.86 cm of the left-most underside of the beam was rigidly anchored by four threaded bars positioned on either side of the beam. Two cross-bars were tightly secured using bolts along the threaded bar to apply a downward, securing force onto the concrete beam. A point load was positioned atop the beam at a location 58.42 cm rightward from the supported underside edge. This load was distributed evenly across the width of the beam by a cylindrical steel bar and a 10.16 cm square steel plate cushioned by a rubber gasket. A clearance of 5.08 cm separated the point load and the right-most edge of the beam specimen.

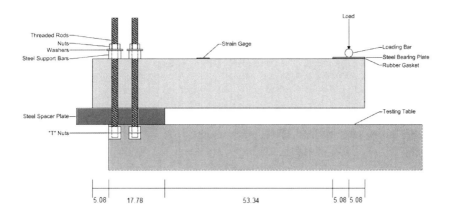

Fig. 1. Drawing of the cantilevered beam test setup and linear dimensions (cm)

Foil strain gages of 30 mm length were firmly epoxied onto the top surface of the beam at varying distances from the point load and capably measured the tensile strains produced by the application of 1.95 kN of force at a loading rate of 11.12 N/sec.

A jointed concrete pavement test strip was available for testing at the Advanced Transportation and Research Engineering Laboratory (ATREL) in Rantoul, IL. This road strip had been previously subjected to millions of equivalent single axle loads, ESALs, in the span of weeks [5]. A large traffic simulator remained in place and was used as a reaction frame for a hydraulic bottle-jack to be placed in-line with the underside of the traffic simulator's immobilized wheel carriage and the center-edge surface of the jointed pavement. This load was distributed by a 30.48 cm square steel plate. A 133 kN capacity load cell was situated between the hydraulic bottle-jack and wheel carriage and measured applied loads upwards of 90 kN at an approximate load rate of 740 N/sec.

A hand-operated electric, circular saw was outfitted with a 17.78 cm diameter masonry saw blade and was used to cut linear notches of varying depths into the beam and in-situ slabs surrounding the strain gages orthogonal to the directions of predicted principal stresses. High-precision digital calipers measured the depth of the cuts made into the concrete. During and after cutting, strain readings were monitored and recorded using a portable strain data indicator device.

2.2 Notching of Cantilever Concrete Beams

The affixed strain gage measured tensile strains atop the concrete beam as a cantilever load of 1.95 kN was applied inducing calculable moments between 99.06 kN-cm and 89.15 kN-cm depending on the center-location of the strain gage. The strains observed in the cantilever beams were converted into stresses after using the appropriate stiffness value from testing of the concrete cylinders. An electric circular saw was used to cut linear notches in the concrete beam perpendicular to the direction of bending. These notches were made at varying distances away from the center of the affixed strain gage and their depths were incremented by 1.27 cm to a total depth ranging from 5.08 cm to 6.35 cm. These nominal depths were later measured during testing using high-precision calipers.

Initially, the concrete beams were singly notched at one side of the affixed strain gage in the direction towards the end load. In later testing, symmetric double notches were employed. The short time of passing the saw cut along the width of the concrete beam did not necessitate cooling water as had been necessary in coring operations. As such, the strain gage was simply protected by a coating of polyurethane and flush guides (see Fig. 2).

Fig. 2. Saw cutting of cantilever concrete beam

2.3 Saw-Cutting of In-Situ Jointed Plain Concrete Pavements

One strip of 15.24 cm thick plain concrete pavement road test strip was available for testing at the ATREL facility located in Rantoul, IL. The 3.66 m width road strip had been sawn-cut into 1.83 m sections in both lateral and longitudinal directions producing a midline along the full length of the road strip. Many of the pavement sections had suffered extensive cracking from previous investigations;

however, six slabs were identified as suitable for testing based on having little to no surface cracking. In order to simplify the scope of the problem, an industrial saw cutter was used to wholly separate each of these six slabs from its adjoining slabs in order to reduce aggregate bridging and associated load transfer. Thus, each slab was separated from three neighbouring slabs while the fourth edge was always a free-edge and required no saw-cutting.

The placement of the traffic simulator and road strip allowed for symmetric center-edge loading across all six slabs. A 30.48 cm-square steel plate was positioned along the interior edge adjacent to the midline in order to distribute the bottlejack load and ensure similar stress states across the tests. A rectangular rosette of strain gages were affixed onto the surface of the pavement at distances of either 0.61 m or 0.91 m from the midline edge of the slab in the direction away from the applied load (see Fig. 3). A 133 kN capacity load cell was situated between the bottle jack and the traffic simulator.

Fig. 3. Bottle jack positioned at center-edge of slab

3 Results and Discussion

The observed strain measurements for singly notched beams were markedly similar to those observed for those concrete beams which had been cored. However, those results only partially quantified the state of stress requiring an imprecise extrapolation of the stress in the material. Doubly notching the concrete beams more fully quantified the state of stress when the notch depth-to-spacing ratio was approximately 0.4.

3.1 Cantilever Concrete Beams

Figure 4 depicts the strain behaviour of a cantilever beam as saw cuts at a distance of 4.32 cm from the center of the strain gage are made. This distance offers the closest saw cut without risking damage to the affixed strain gage. The first set of cuts is made between the strain gage and the end load at depths of 0.64 cm, 1.27 cm, 2.54 cm and 3.81 cm. The beam is unloaded and re-loaded to observe its diminished response. A second, symmetric set of cuts cut is made at the opposite end of the strain gage. Unloading and loading the cantilever beam shows that the induced strain by the end load is completely diminished by this second set of cuts. This is also readily apparently when the strain recovery closely matches the previous stable value. This unchanging strain value suggests that the strain gage is fully isolated from the applied load meaning that the discrepant strain magnitude in the concrete material is due to the strain induced by residual stresses and relieved by saw-cutting. In this case, for a calculated Young's modulus of 46,778 MPa, the estimated residual stress is 1.6 MPa in tension – a significant fraction of concrete's tensile strength. Table 2 summarizes the results of additional concrete beam tests.

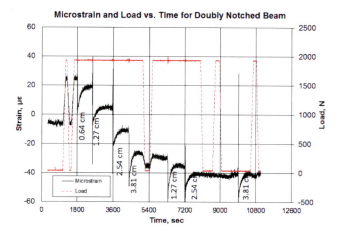

Fig. 4. Strain results for doubly notched cantilever beam

Table 2. Summary of results for doubly notched cantilever beams

Beam	Difference in Unloaded Strain, µε	Young's Modulus, MPa	Estimated Minimum Residual Stress, MPa
N4	16	46891	0.75
N5	19	44051	0.84
N6	34	46691	1.59
N7	40	42348	1.69
N8	29	44099	1.28

3.2 In-Situ Jointed Plain Concrete Pavements

The promising results from doubly notched concrete beams justified adapting the procedure to full-scale testing on in-situ concrete slabs where two sets of parallel cuts perpendicular to each other could measure the surface strain relaxation in two nominal directions. A third reference gage oriented at 45 degrees was also employed in order to resolve the orientation of the principal stresses. The application of an edge-center load produced discernable strains – compressive in the direction lateral to the load and tensile in the direction longitudinal to the load. A summary of the test results are tabulated in Table 3. Similarly, these results depict the full strain relaxation of material stresses suggesting that the discrepant strain again quantifies the material residual stress.

Table 3. Summary of results for in-situ slabs

Slab	Notch Spacing, cm	Lateral Strain, με	Reference Strain, με	Longitudinal Strain, με	Estimated Minimum Residual Stress, MPa
1	7.62	48	56	45	1.74
2	7.62	0	10	23	0.65
3	7.62	21	8	20	0.97
4	7.62	34	4	0	1.06
5	10.16	17	22	13	0.66
6	10.16	-24	-26	-13	-0.82

3.3 Finite Element Analysis

The simple geometry of the beams and slabs lent themselves to be analysed using finite element software. Four major variations of the beams were modelled in PATRAN as T6 (linear strain triangle) elements: a plain beam, a singly notched beam, a doubly notched beam with notch spacing 8.89 cm, and a doubly notched beam with notch spacing of 12.7 cm. The singly notched beams were modelled with notch depths of 1.27 cm, 2.54 cm, and 3.81 cm while the doubly notched beams also included a notch depth of 0.64 cm. similarly, the slabs were modelled as Hex8 elements with 1.27 cm element lengths. The notch spacing for the modelled slabs were maintained as square dimensions with side lengths of 5.08 cm, 7.62 cm, 10.16 cm and 12.7 cm. The depth of the four slab saw cuts were equal in their respective models increasing in depth from 1.27 cm to 11.43 cm in 2.54 cm increments.

Figure 5 depicts the von Mises stress distribution of the cantilever concrete beam for an unnotched configuration as well as singly and doubly notched beams at depths of 2.84 cm and notches separated by 8.89 cm. The upper surface of the beam is in tension while the lower surface is in compression. The left-most ends of the beams are fixed along their underside to mimic the boltage to the MTS frame as previously described in Figure 1. Figure 6 depicts the von Mises stress distribution for a slab with four 6.4 cm deep cuts spaced 7.62 cm apart.

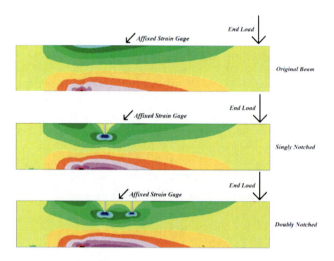

Fig. 5. Profile view of stress fields of plain, singly and doubly notched beams

Fig. 6. Plan view of stress field of slab before and after 6.4 cm quadrilateral cut

Figure 7 depicts the modelled stresses relieved in a slab for a square area of side length 7.62 cm at a distance of 0.61 m and 0.91 m from the applied load as a third order best-fit polynomial function. Vertical range bars are also included and depict the range of stress relieved based on the distance of the element from the saw cut edge to the interior of the square area. Experimental data points from the slab testing are superimposed on the graph to better demonstrate the correlation.

Conservatively, the point at which the modelled function first relieve 100% of the induced stress denotes the minimum notch depth-to-spacing ratio (D/S) required to fully identify the residual stresses. Ratios above this minimum ratio are indicative of a "hinging" effect where the geometry of the slab has changed sufficiently as to no longer be adequately characterized as a slab in bending, but a material solely acted upon at its connection with the interior of the slab. Based on the location of the strain rosette, a minimum notch D/S ratio of 0.4 to 0.45 is estimated to be sufficient in fully relieving the surface stresses in a concrete material.

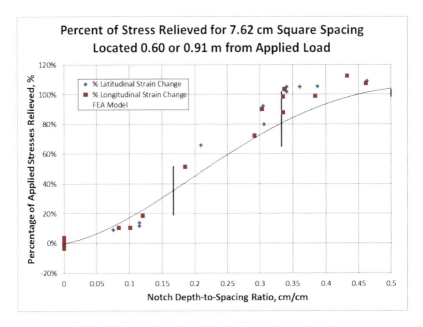

Fig. 7. Comparison of modelled and experimental results

It is also found that the notch spacing should not exceed 12.7 cm since it appears that the effect of strain relief is rendered moot as very deep cuts are required and boundary conditions of the isolated square area are effectively approaching infinite conditions. In the other extreme, the notch spacing cannot be made smaller than the length of the affixed strain gage. However, strain gages smaller than the maximum aggregate size should be avoided in order to better capture the homogeneity of the concrete.

4 Conclusions

Measuring stresses in concrete through coring has been problematic due to the incomplete identification of stresses leading to imprecise extrapolations and the use of cooling water (resulting in the creation of local moisture and heat expansion) [2]. As presented in this study, sawing cuts into the concrete around a strain gage or strain rosette can completely quantify the state of total stress in a concrete material. Additionally, if the stresses due to the applied loads, moisture and thermal gradients, and other known stresses are subtracted, the discrepant stress can be categorized as the residual stress.

While the new testing procedure is simple in its execution and subsequent analysis, it is important to note the limitations. The procedure is a direct measurement of the surface stresses at a discrete point along a large slab. As such, it is not indicative of the stress state in all locations. It is prudent, then, for users to carefully consider the location of testing in their assessment of the slab conditions.

Moreover, the experimental result is valid only for the time of testing during which it was performed since various parameters affecting the concrete (like daily thermal cycling and seasonal wetting and drying) are in effect. This new sawcutting procedure to quantify residual stresses in concrete can be thoughtfully and carefully employed; and used as a tool among others to better characterize the state of stress in concrete. To expand upon the findings of this study, future studies should consider the effect of larger concrete pavement thicknesses among other variables.

Acknowledgements. Financial support for this project was made possible with funds from the US Federal Aviation Administration through the FAA Center of Excellence in Airport Technology (CEAT), and fellowship funds from the UIUC College of Engineering.

References

[1] Richards, A.M.: Trans. Res. Rec. (713), 9–15 (1979)
[2] Guo, E.H., Pecht, F., Ricalde, L., Barbagallo, D., Li, X.: In: Proceedings of the 9th International Conference on Concrete Pavements, San Francisco, CA (2008)
[3] ASTM E387, Standard Test Method for Determining Residual Stresses by the Hole-Drilling Strain-Gage Method. ASTM International (2008)
[4] Timeshenko, S.P., Goodier, J.N.: Theory of Elasticity, 3rd edn. McGraw-Hill (1970)
[5] Cervantes, V., Roesler, J.: In: IL Center for Trans Series 09-053, Rantoul, IL (2009)

Strain Measurement in Pavements with a Fibre Optics Sensor Enabled Geotextile

Olivier Artières[1], Matteo Bacchi[2], Paolo Bianchini[1], Pierre Hornych[3] and Gerrit Dortland[1]

[1] TenCate Geosynthetics France, Italy and The Netherlands
[2] Impresa Bacchi, Italy
[3] IFSTTAR, France

Abstract. A new sensing solution based on the combination of a technical geotextile and fibre optics measurement technologies has been developed for strain and temperature measurement in pavement. This monitoring system has been evaluated in the laboratory with a 4-points fatigue device. Our results show a very high sensitiveness of the sensor enabled geotextile to be able to detect strain smaller than 10 micro-strain and fast dynamic movements with frequencies up to 1000 Hz. Installation trials have been carried out in different locations with conventional road paving equipment, in both asphalt and concrete pavement applications. The response of this sensing solution to traffic loads is very good and makes of this technology a powerful tool for road ageing assessment, analysis and maintenance.

1 Introduction

A lot of work has been done in the past years to understand the ageing process of road pavements and to develop solutions to decrease or eliminate cracks and structural deformations due to traffic and climatic fatigue. A very common solution is the use of technical paving textiles both to reinforce the base of the overlays and create a watertight continuous bituminous liner to avoid water penetration deep into the structure. However, even if these solutions are effective in increasing the lifetime of the structure, there remain big issues for roads designers, contractors and owners such as evaluating the in-situ performance of the structure, selecting the best techniques depending on the road's subsoil and structure and assessing the actual ageing of the pavement to predicting the period and the scale of maintenance operations.

Monitoring of road pavements is not easy as the environment is very aggressive for the sensors: hot temperature and high compaction stresses. Classical techniques use electro-mechanical sensors. The rate of damage of these sensors is very high: it is common to lose 20 to 50% of the sensors installed just after the compaction process. Also these sensors are big compared to the size of the cracks they aim to monitor which creates scale and border effects detrimental to their detection capabilities.

This paper presents a new sensing solution based on the combination of a technical geotextile and fibre optics measurement technologies that has been developed for strain and temperature measurement in pavement. After introducing the principles of this sensing solution, its performance both in laboratory and in scale 1:1 real conditions are described.

2 The Monitoring Solution Based on the Fibre Optics Sensor Enabled Geotextile

Fibre optics have been widely used for many years in civil engineering applications, specialty pipelines, structural health monitoring systems and hydraulic works applications such as concrete and earth dams, levees and dikes.

By embedding optical fibres onto a geotextile fabric (Figure 1), TenCate GeoDetect® is an innovative sensor enabled geotextile that enhances the performance of the fibre optics sensors when applied in contact with soil, concrete or asphalt: the geotextile fabric creates an excellent anchoring interface with the surrounding media. With the geotextile being securely anchored in the asphalt or concrete, and the strong connection between the optical fibre and the geotextile, even very small soil strains can be detected. This friction interface also facilitates the transfer of movements from the geotextile to the fibre optic line. Moreover, and when necessary, high tensile stiffness and reinforcement properties can be included to the sensor enabled geotextile. Different fibre optic sensing technologies can be embedded, such as Fibre Bragg Gratings which measures very narrow optical index changes written at given locations inside the optical fibre line for point specific measurements, or Brillouin and Raman technologies which provide distributed measurements at any point along the optical fibre up to 50 kilometres in length. Fibre optic sensing technologies are able to measure very precisely parameters such as temperature or strain under static or dynamic conditions. The monitoring solution includes the fibre optics sensor enabled geotextile, the instrumentation equipment and data acquisition software (Figure 1). Different monitoring strategies may be incorporated into the design. Either periodic or continuous monitoring can be used as an early warning system.

In comparison to existing monitoring systems made of numerous individually wired sensors, this solution measures continuously up to hundreds of points along the full length of the structure with a single instrumentation configuration. It can provide deformation location with a spatial resolution of less than 0.5 m in some cases. Once installed, the sensor enabled geotextile communicates the strain and temperature data to the system's instrumentation equipment. Strain lower than 0.01% can be measured, and with the proper software, changes in temperature can be monitored at 0.1°C. The optical sensing technology requires no sensor calibration prior to the measurement; temperature compensation may be necessary for amplitudes higher than 10°C.

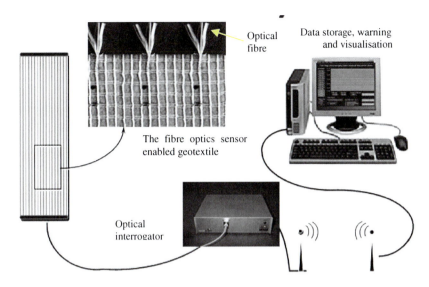

Fig. 1. The TenCate GeoDetect® system components

3 Laboratory Tests

Preliminary trials have been carried out at the Impresa Bacchi laboratory in Carpiano, Italy, to check the resistance of the sensor enabled geotextile to hot temperature, to monitor the stresses due to the compaction of the bituminous concrete and assess its sensitiveness in measuring strain in a concrete asphalt layer.

3.1 Preparation of the Specimen

Specimen asphalt beams were built into a box 40 cm x 30 cm, with the following layers from the top to the bottom: 3 cm layer of bituminous concrete, the sensor enabled geotextile impregnated with a bituminous emulsion, 3 cm layer of bituminous concrete (Figure 2), and compacted. To create the beams we used a Dyna-Comp, pneumatic roller compactor with a maximum vertical force of 30 kN (Figure 3). The roller compactor provides a pneumatically powered means of compacting slabs of asphaltic material in the laboratory under conditions, which simulate in-situ compaction. Slabs produced measured 300 mm by 400 mm and 50 mm thick.

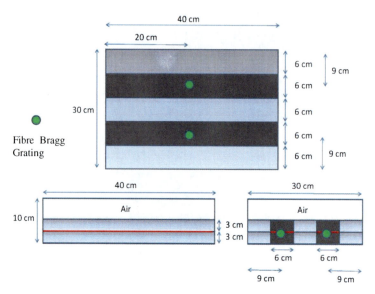

Fig. 2. Top view and cross-sections of the concrete asphalt blocks, each producing two specimen beams with one strip of sensor enabled geotextile (red) embedding one optical line and one Fibre Bragg Grating in the middle of the beam

Fig. 3. Compaction of the second layer of bituminous concrete

The precise depth of a slab can be preset enabling the user to compact a certain mass of material to a selected volume thus providing a target mix density. Several compaction cycles were tested with about 3 passes for each to achieve normal compaction strength. Different levels of vertical force can be selected up to approximately 30 kN. As the width of the roller is 300 mm, the compaction effort of the largest static site roller can be reproduced. The strongest cycle started with 3 kN, then 5 kN, then 10 kN, then 19 kN.

The sensors reacted very well to the different passes which are clearly visible from the strain measurements. The maximum strain measured was about 1,3 % (Figure 4). From these blocks are cut 2 beams. A total of 8 beams were produced. Each of them included a 40 cm x 15 cm sensor strip, embedded with one optical fibre line which contained one Fibre Brag Grating in the middle (Figure 2). The first part of the experiment shows the resistance of the sensor to the installation, compaction stresses and to hot asphalt temperature. The temperature of the bituminous concrete was 140°C during placement on the sensor enabled geotextile strip. No damage was observed on the optical fibre or on the FBG during the placement.

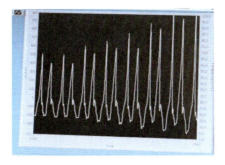

Fig. 4. The strain curve measured during compaction with an increasing strength. The peaks corresponding to the passes are visible.

3.2 Tests with the Bending Machine

The beams were tested into a 4 points "Nottingham" bending test machine (Figure 5). The model is an IPC Global UTM - 25 based on a 25 kN capacity hydraulically-driven load-frame equipped with the 4 Point Bend Apparatus, a stand-alone system for four-point fatigue life testing of asphalt beams subjected to repeated flexural bending. The cradle mechanism allows for backlash free rotation and horizontal translation of all load and reaction points. Pneumatic actuators at either end of the cradle centre the beam laterally and clamp it. Servo-motor driven clamps secure the beam at four points with a pre-determined clamping force. Haversine loading is applied to the beam via the built-in digital servo-controlled pneumatic actuator.

In our case we tested the beam with several loading curve (Sinusoidal, haversine, triangular, rectangular, etc).

A total of 4 beams were tested. The sensor strips were connected to a FBG interrogator with a dynamic acquisition frequency from 1 to 100 Hz, depending on the fatigue cycle chosen. As a result, all cycles were accurately measured with the sensor enabled geotextile, even when very low vertical amplitude of 10 µstrain was applied (Figure 5). For each beam, at the end of the loading cycles, a pseudo-static normal loading was applied to reach a deflection of 1 cm in the middle of the beam: the corresponding 0.2% horizontal strain was measured by the sensor.

 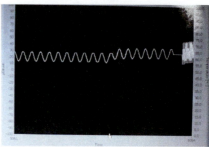

Fig. 5. The specimen beam inside the four points bending machine (left) and the strain curve resulting of sinusoidal bending cycles producing strain variation of 10 micro strains amplitude (right)

4 In-Situ Trials

4.1 The IFSTTAR Accelerated Pavement Test Facility

After these laboratory evaluations, a scale 1:1 test sections was built with an asphalt pavement and monitored with the fibre optics sensor enabled geotextile. It was installed in the fatigue carousel at IFSTTAR (LCPC) in Nantes (Figure 6).

Fig. 6. The IFSTTAR Accelerated Pavement Test facility

This Accelerated Pavement Test facility was built for the study of full scale experimental pavements submitted to heavy traffic levels. This major facility became operational in 1984. It makes it possible to reproduce in less than a week up to a full year's truck traffic load supported by a heavily trafficked pavement, with load speeds capable of reaching 100 km/hr. The site comprises three 110-m long rings with an average radius of 17.50 m and a width of 6 m. It is possible to position the loads at different radii of rotation depending on the arm length. The loads may be adjusted between 45 kN on a single wheel and 135 kN on either a three-axle configuration with single wheels or a double axle with two wheels each. The facility experimental site consists in three test rings, with its central motorization and four arms being movable from one ring to another. Further description is given in [1, 2].

4.2 Tested Structure and Installation of the Sensor Enabled Geotextile

Two sensor enabled geotextile strips have been installed bellow a 8 cm thick asphalt concrete layer with classical road construction equipment (Figure 7). One strip is installed in the direction of the traffic, the second strip is installed perpendicular to the road. This second strip embedded one optical line with three FBGs spaced 1 m apart along the line. The sensors were monitored before, during and after installation. No damage was observed during this operation. Installation creates the highest stress, between 600 and 2000 µstrain, part of it due to temperature increase (Figure 8). A slow relaxation of the strain values have been observed since the completion of the test.

4.3 Results

The first measurements took place 4 months after the installation on June 16, 2011. The configuration of the 4 arms was the following: three arms were equipped with single axles with dual wheels (12.00 R 20.0 tires), loaded at 50 kN, and the fourth arm was equipped with a single axle with a super single wheel, (455/55 R22.5 tire) also loaded at 50 kN.

Fig. 7. Pouring asphalt concrete on top of the sensor enabled geotextile strips (left) and compaction (right)

The response of the fibre optics sensors was measured for 11 different lateral positions of the wheels, spaced 105 mm apart. The tests were made at a loading speed of 40 km/h, and an average pavement temperature of 28°C. Figure 9 shows an example of the transverse strain measured by one sensor enabled geotextile strip for the wheels position n°8, where the dual wheels pass near the centre of the wheel path. We can observe that mainly FBG2 is strained, the FBG3 measuring some small negative strain (contraction). At this position, FBG2 is just below the centre of one of the dual wheels, and the measured strain under dual wheel is maximum. FBG3 located 1m apart measures only a small strain. The dual wheels produce a maximum positive transverse strain (in extension) close to 400 μstrain. At this same position, the single wheel does not pass on top of the sensor, and produces only a small negative strain.

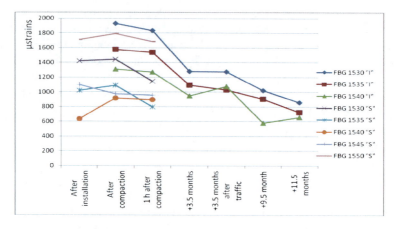

Fig. 8. Relative strain measured by the Fibre Bragg Gratings during and after installation in the asphalt pavement. Zero value: just before installation.

In comparison, figure 9 shows the response of a strain gage, used as classical instrumentation on the test track to measure strains in asphalt layers. Again, the figure shows the transverse strains under the passage of the 4 rolling wheels, for position 8. Under the dual wheels, the maximum transverse strain level is 300 μstrain. Even if strain gage and FBG are both local measurements, it is difficult to compare their measured strain values directly: strain gages measure a local strain under a given position of the wheel (over a length of about 10 cm), while the geotextile may transfer a part of strain to the optical fibre, which can react even if the load is not applied on top. However, it can be seen that the quality of the signal is satisfactory (Figure 10), and that the shape and amplitude is similar to the strain gage response (Figure 11). Figure 10 also shows that measurements made at different ageing times of the pavement are consistent. But further work is necessary to define how the signal could be interpreted, to obtain meaningful strain values.

Strain Measurement in Pavements with a Fibre Optics Sensor Enabled Geotextile 209

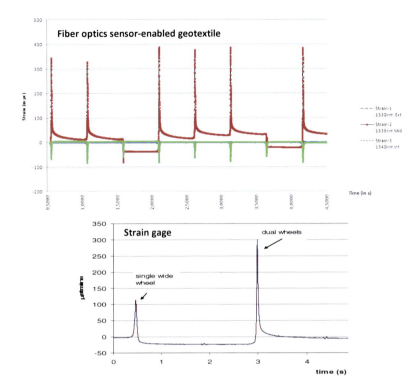

Fig. 9. Comparison of the transverse strain measured at the base of the bituminous layer under the 4 rolling wheels with the fibre optics sensor enabled geotextile (+3.5 months) and strain gages

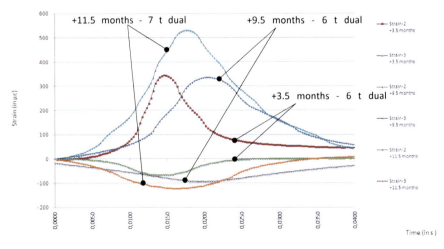

Fig. 10. Strain curves measured with the fibre optics sensor enabled geotextile resp. +3.5, + 9.5 and 11.5 months after installation.

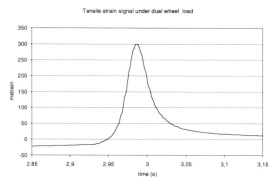

Fig. 11. Strain curve measured with strain gages

5 Conclusion

The fibre optics sensor enabled geotextile has proven to be a good monitoring system to measure strain into concrete asphalt road pavements. By adopting specific installation procedures, its survivability to installation stresses, high temperature and compaction, is very good compared to other sensor devices. It can be applied directly during the road construction thus measuring the real state of the road. The sensitiveness measured both in laboratory and on site are a few micro-strains. The sensor requires no calibration after installation that makes it very easy to handle. This sensor technology has been measuring for more than one year different pavement structures in the field which makes it very reliable for mid and long term maintenance data acquisition, even on damaged areas up to 5% strain.

References

[1] Gramsammer, J.C., Kerzreho, J.P., Odeon, H.: The LCPC's A.P.T. Facility: Evaluation of Fifteen Years of Experimentations. In: Proceedings of the 1st International Conference on Accelerated Pavement Testing, Reno, Nevada, October 18-20, pp. 18–20 (1999)

[2] IFSTTAR The Accelerated Load Testing Facility (2011),
http://www.lcpc.fr/en/presentation/moyens/manege/index.dml

Evaluating the Low Temperature Resistance of the Asphalt Pavement under the Climatic Conditions of Kazakhstan

Bagdat Teltayev and Evgeniya Kaganovich

JSC "Kazakhstan Highway Research Institute", Kazakhstan

Abstract. The report presents the analysis of the results of investigations of the low temperature resistance of viscous air-blown bitumens used in the construction and repair of asphalt pavements in Kazakhstan. The non-conformity has been established between the test results by the Superpave method characterizing the low-temperature crack resistance and the actual appearance of such cracks in the pavements during the first service years. This non-conformity can indicate that the influencing factors, particularly the character of pavement cooling, are not taken into account properly. Typical cases of air temperature lowering based on the meteorological data as well as the change of tensile stresses in the asphalt concrete for these cases have been analyzed with applying the linear theory of viscoelasticity. It has been found that the time and temperature of the first crack appearance substantially depend not only on the minimum design temperature but also on the character of its attaining.

1 Introduction

According to the results of asphalt pavement diagnostics, low temperature cracking is the basic type of deformation for the most part of the Kazakhstan territory with minimum values of pavement temperatures from -28°C to - 46°C. Crack formation takes place after the first winter with 15 – 50 m spacing depending on the region of road operation, asphalt concrete composition and road pavement structural features, with the appearance of new cracks and corresponding decrease in their spacing during the 2nd and 3rd service years.

It is well-known that the crack resistance of asphalt concrete is mainly governed by low temperature properties of bitumen. Initially, the conformity of the low temperature properties of bitumen (Fraas brittle temperature after ageing) with the minimum design pavement temperature was taken as a criterion of the bitumen conformity to the climatic service conditions. However, in the course of investigation it was found that the reliability of this criterion was insufficient because of a significant divergence in the test results obtained with devices of various design or different manufacturers. Therefore, at the present time, the Superpave method with the use of a bending beam rheometer (BBR) has been recognized a more preferable one for determining the low temperature characteristics of binders.

2 Experimental Studies of Low Temperature Properties of Bitumens

The following five most-used bitumens were chosen for studies: grade BND 60/90 of the Alma-Ata (TOO "ABZ-1"), Pavlodar (PNKhZ) and Omsk (ONPZ) refineries and grade BND 90/130 of the Alma-Ata (TOO "ABZ-1") and Pavlodar (PNKhZ) refineries. The bitumens were studied in a bending beam rheometer (BBR) in compliance with ASTM D 6648 – 08 [1] after ageing in a rolling thin film oven test (RTFOT) [2] and in a pressurized ageing vessel (PAV) [3]. The research was carried out at temperatures in the range from - 12°C to - 34°C. On the basis of the test results, curves of bitumen creep $\varepsilon(t)$, stiffness $S(t)$, and index of stress relaxation rate $m(t)$ at different temperatures have been built and analyzed.

As a result of the analysis, it has been established that all the bitumens are characterized by increased deformability and decreased stiffness with an increase in load application time and a rise in temperature. At the same time, as to the relaxation capacity, bitumens of various grades and of different manufacturers behave differently at different temperatures.

Figures 1 – 4 present the stiffness modulus and the index of stress relaxation rate versus temperature at loading time 60 s.

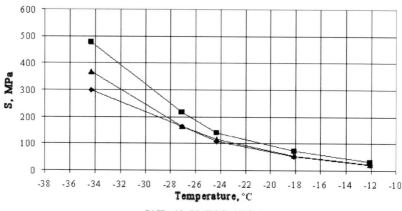

Fig. 1. Stiffness/temperature relationship for TOO "ABZ-1" and ONPZ bitumens after ageing (RTFOT + PAV)

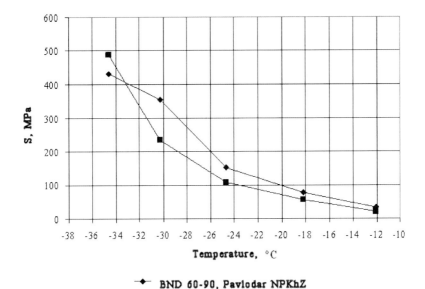

Fig. 2. Stiffness/temperature relationship for PNKhZ bitumens after ageing (RTFOT + PAV)

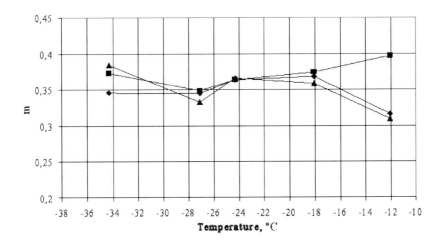

Fig. 3. Relationship between index of stress relaxation rate and temperature for TOO "ABZ-1" and ONPZ bitumens after ageing (RTFOT + PAV)

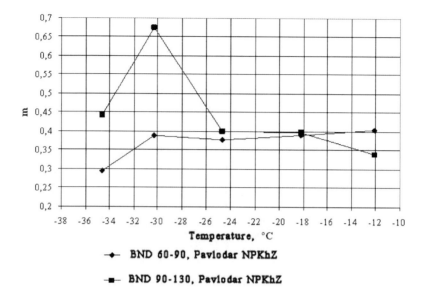

Fig. 4. Relationship between index of stress relaxation rate and temperature for PNKhZ bitumens after ageing (RTFOT + PAV)

The analysis of these relationships shows that at a temperature of - 12°C the stiffness of all bitumens is practically the same. With a further decrease in temperature the stiffness grows according to the non-linear relationship. The temperature/ stiffness curves for bitumens of BND 60-90 grade from the Alma-Ata and Omsk refineries practically coincide in the temperature range from - 12°C to - 27°C but further lowering in temperature brings about a more rapid growth in stiffness of the Omsk bitumen.

The prevailing opinion presently is that the bitumens of less viscous consistency (with consistency evaluation by a penetration value at + 25°C) are more stable against temperature cracking. However the data from Figures 1 and 2 indicate that bitumens of different manufacturers can manifest quite different (and sometimes opposite) properties. So, in the temperature range of -12°C to - 32°C bitumens of BND 60/90 and BND 90/130 grades from the Pavlodar refinery show properties that can be explained on the basis of usual notions: the stiffness of BND 60/90 bitumen is higher than that of BND 90/130 bitumen but with a further decrease in temperature, bitumen of BND 90/130 grade becomes stiffer. The stiffness/temperature relationship for the bitumens from the Alma-Ata refinery is of expectable character.

The relationship between the temperature and relaxation rate index m for bitumens has also a complicated character. So, in the temperature range from -12°C to -18°C index m is different for the bitumens of various refineries and grades. However, in the temperature range from -18°C to -24°C index m is relatively constant and equal to 0.36-0.40. With a further decrease in temperature the relaxation capacity of bitumens from different refineries changes variously. So, up to a temperature of - 34°C, index m for the bitumens from the Alma-Ata and Omsk refineries changes in a relatively small interval while that for the bitumens from the

Pavlodar refinery changes substantially. For bitumen of BND 90/130 grade from the Pavlodar refinery a value of index m grows from 0.4 to 0.67 in the temperature range of -24°C to - 30°C and then begins to fall sharply reaching a value of 0.44 at -34.5°C. Bitumen of BND 60/90 grade of this refinery keeps practically a constant value of index m at temperatures of -12°C to - 30°C. A subsequent decrease in temperature results in a substantial drop in a value of the relaxation rate.

The above analysis shows that the mechanical behavior of the investigated bitumens at low temperatures is complicated, and the peculiarities of each of them should be taken into consideration when designing the asphalt concrete compositions.

As it is known, the Superpave specifications require that for a loading duration of 60 s, the modulus of bitumen stiffness S does not exceed 300 MPa at the minimum design temperature and the value of the index of stress relaxation rate is not less than 0.3 [4]. According to zoning of the Kazakhstan territory as to the asphalt pavement operational temperatures [5], the minimum design temperatures have been established for the Republic's regions, which are as follows: -28, -34, -40, and -46°C. The results of bitumen BBR tests have shown that in compliance with the Superpave method, the bitumens tested meet the requirements of low-temperature resistance at all design temperatures except -46°C. At the same time, the results of road diagnostics indicate that the temperature cracks occur universally in the asphalt pavements. Such discrepancy between the experiment results and in-situ data can be indicative of the fact that the influencing factors are not considered adequately. In these studies, a character of pavement cooling, particularly the rate of temperature decrease, has been taken as such factor.

3 Prediction of the Crack Formation Indices for Various Regimes of Cooling

The determination of arising tensile stresses and evaluation of a possibility of the appearance of low temperature cracks in the asphalt road pavement at several one-time prolonged drops of ambient air temperatures have been performed with the use of a design model.

It is well known that the mechanical properties of asphalt concretes as viscoelastic materials depend on the duration of load application and temperature [6,7]. Therefore, according to the linear theory of viscoelasticity [8] the temperature stresses in the asphalt pavement are defined by Boltzmann-Volterra integral:

$$\sigma_T(t) = \int_0^t E(t-\tau) d\varepsilon_T(\tau) \qquad (1)$$

where $E(t)$ – function of asphalt concrete;

t – time under consideration when stress $\sigma_T(t)$ is determined;

τ – time preceding t;

$\varepsilon_T(t)$ – relative deformation at time τ.

The asphalt concrete pavement is considered as a layer of thickness h, infinite in length in the horizontal directions, that overlies the continuous homogeneous foundation. The temperature of the asphalt pavement will be characterized by its value at the pavement surface, i.e. the air temperature. This is based on the considerations that irrespective of the actual temperature distribution through the pavement depth, cracks can appear at points of the asphalt pavement surface when the limiting thermal stress values are achieved. On further lowering in air temperature with a rate of 1 -2°C/hour the side surfaces and the top of a newly developed crack have a temperature practically equal to the air temperature, and the stresses that cause a further growth of the crack will be governed by a temperature value in the vicinity of the top of the crack already developed.

The temperature cracks appear when there is no possibility of free deformation in the horizontal directions at temperature lowering. Free deformation of any section of the continuous asphalt pavement is hindered by its adjacent sections. Under such conditions a value of unrealized relative thermal deformation of any section of the continuous asphalt pavement is estimated by formula:

$$\varepsilon_T(\tau) = \alpha[T(\tau) - T(\tau = 0)] \qquad (2)$$

where α – coefficient of linear thermal deformation, 1/°C;
T – temperature, °C.

Functions of the asphalt concrete relaxation at temperatures of +10°C to -40°C, determined using the M. W. Witczak model [8], have been approximated with Prony series [9]:

$$E(t) = E_0 + \sum_{i=1}^{m} (E_i \cdot e^{-t/\tau_i}), \qquad (3)$$

where
  E_o, E_i – constants defined by the least-squares method
  τ_i – preset values of relaxation time (10^{-12}, 10^{-11}, ... 10^7)
  m – number of exponential functions taken to be equal to 20.

At the present time, in Kazakhstan, the asphalt concrete mixture is conventionally prepared using 5% of limestone powder and about 5.5% of bitumen. Thereby, the volumetric indices of the asphalt concrete V_a and V_{beff}, defined by method [10], are approximately as follows: V_a=4.75% and V_{beff}=10.65%. In computations the average value of the coefficient of linear thermal deformation of the asphalt concrete has been accepted, which is equal to $\alpha = 3.3\sqrt{10^{-5}}$ 1/°C. The values of relaxation time τ_i and Prony series coefficients E_i in equation (3) are given in Table 1.

Since the rheological properties of asphalt pavement characterized by relaxation function E(t) in equation (1) change with time due to the change in temperature T(t), true time t is replaced by reduced time $\xi(t)$ determined from formula:

$$\xi(t) = \int_0^t \frac{dt}{a_T[T(t)]} \qquad (4)$$

where $a_T[T(t)]$ – function of temperature-time superposition.

Function of temperature-time superposition is defined by the following expression:

$$a_T[T(t)] = \frac{t(T)}{t(0)} \quad (5)$$

where t(T), t(0) – values of the duration of load application at temperatures of T (°C) and 0 °C, respectively.

Logarithm of function $a_T[T(t)]$ in the range of temperature variation from +10°C to -40 °C has been presented by a second-degree polynomial:

$$\log a_T[T(t)] = a_1 + a_2 \cdot T + a_3 \cdot T^2 \quad (6)$$

where a_1, a_2, a_3, – regression parameters, equal to 4.507; -0.219; 2,783.10^3 respectively.

Table 1. Values of relaxation time τ_i and Prony series coefficients E_i

Serial number of series	Relaxation time τ, s	Prony series coefficient E, MPa	Serial number of series	Relaxation time τ, s	Prony series coefficient E, MPa
0	-	20.337	11	10^{-2}	2.584·10^3
1	10^{-12}	1.081·10^3	12	1.1	1.819·10^3
2	10^{-11}	1.319·10^3	13	1	1.135·10^3
3	10^{-10}	1.731·10^3	14	10	636.952
4	10^{-9}	2.174·10^3	15	10^2	329.097
5	10^{-8}	2.673·10^3	16	10^3	161.878
6	10^{-7}	3.159·10^3	17	10^4	78.362
7	10^{-6}	3.557·10^3	18	10^5	38.732
8	10^{-5}	3.759·10^3	19	10^6	19.230
9	10^{-4}	3.670·10^3	20	10^7	12.273
10	10^{-3}	3.255·10^3	-	-	-

The results of the analysis of environment air temperature variation under the climatic conditions of North Kazakhstan on the basis of meteorological station data have shown that during the cold periods of the year there occur long-term temperature drops with a rate of 0.93 to 1.67 °C/hour. Thereby, three typical cases of air temperature lowering can be distinguished: 1) initial temperature $T_0 = T(t=0)$ is about -5 to -7 °C; air temperature decreases with a rate of about K ≈ 1 °C/hour; 2) T_0 ≈ -17 to-18°C, K ≈ 1 °C/hour; 3) T_0 ≈ -2 to-3°C, K ≈ 1,6 to 1,7 °C/hour. The duration of temperature drop is 33, 18 and 12 hours for the first, second and third cases, respectively.

Figures 5 and 6 present plots of the tensile stress change with time in the asphalt pavements built with these bitumens for the above three cases of ambient air temperature lowering. As can be seen, as the temperature decreases according to the linear law there occurs a non-linear growth of stresses with time in the asphalt pavement. Among the three cases of pavement cooling the most dangerous one is the last when the rate of temperature drop is the highest. The least dangerous case is the first one when the rate of temperature drop is the lowest and cooling starts with a negative temperature that is relatively lower in an absolute value. It should be noted that in all the cases of pavement cooling under consideration, the temperature stresses were higher for the asphalt concretes with BND 60/90 bitumen as should be expected. It is necessary to point out that when using BND 90/130 bitumen, cooling in the second and third cases brought about practically the same stresses during the first six hours of cooling. Thus, it can be said that the most dangerous cases from the viewpoint of low temperature cracking in the asphalt pavements are those when the bitumens of more viscous consistency (BND 60/90) are used and when the cooling begins at lower temperature and at higher rates.

Having the data on the change of stresses with time and on the strength of asphalt concretes over the low-temperature range, it is possible to predict the time when the first cracks appear in the continuous asphalt pavement. The strength of asphalt concretes has been determined at a temperature of -10 °C by a correlation formula from "Guide" [9], and it has been assumed that the asphalt concrete strength does not change in the range of low temperatures from -5°C to - 25°C. Determined under such conditions, the design values of strength for the studied asphalt concretes with BND 60/90 and BND 90/130 bitumens were 2.54 MPa and 3.17 MPa, respectively. The values of probable time and temperature of the crack appearance in the initially continuous asphalt pavement are given in Table 2. In conformity with the results of the above analysis of the stress growth in the pavement with time for the three considered cases of cooling, it has been found that in the cases when cooling starts at a relatively low temperature (case 2) and at the highest rate (case 3), the temperature cracks appear earlier than in the case of prolonged cooling with a relatively low rate (case 1). Thereby, in cooling cases 2 and 3 the probable time of crack appearance is the same and equal to 3 hours while in case 1 it is 4.5 hours. In cases 2 and 3 of temperature decrease, the probable values of the crack appearance time are the same and equal to 3 and 4 hours for the asphalt pavements built with BND 60/90 and BND 90/130 bitumens, respectively. In case 1, the cracks appear in 4.5 and 7.0 hours in the asphalt pavements with BND 60/90 and BND 90/130 bitumens, respectively.

The temperature of crack appearance is in the range of -9.4°C to -21.7°C. It has been established therewith that the highest temperatures of crack appearance are characteristic for the case of pavement cooling with the high rate (case 3) while the lowest ones are characteristic for the case of pavement cooling started at low temperatures (case 2). The values of the crack appearance temperature obtained for the asphalt pavement with BND 90/130 bitumens are lower by 1.0 – 2.3 °C as compared with those obtained for the asphalt pavement with BND 60/90 bitumen.

Thus, it can be noted that for the asphalt pavements built with bitumens the viscosity of which differ by the one-two order in the range of low temperatures from 0 °C to -20 °C, the time when the first cracks appear can differ by 1.0 – 2.5 hours depending on the conditions of lowering the ambient air temperature.

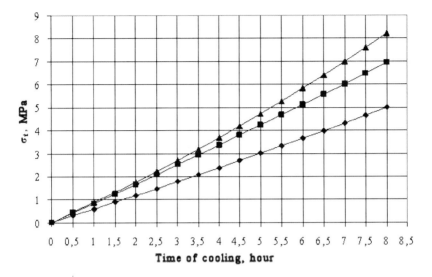

Fig. 5. Tensile stress growth in the asphalt pavement (BND 60/90) during cooling:
♦ -T_0 = -6.1 °C, k= 0.93 °C/hour;
■ -T_0 = -17.6 °C, k = 1.03 °C;
▲ -T_0 = -2.8 °C, k = 1.66°C/hour

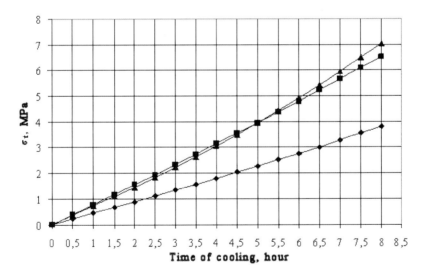

Fig. 6. Tensile stress growth in the pavement (BND 90/130) during cooling:
♦ -T_0 = -6.1 °C, k= 0.93 °C/hour;
■ -T_0 = -17.6 °C, k = 1.03 °C;
▲ -T_0 = -2.8 °C, k = 1.66°C/hour

Table 2. Indices of asphalt pavement temperature cracking

Predicted indices of crack appearance	Bitumen grade	Case No.		
		1	2	3
Time t_{cr}, hour	BND 60/90	4.5	3.0	3.0
	BND 90/130	7.0	4.0	4.0
Temperature, T_{cr}, °C	BND 60/90	-10.3	-20.7	-7.8
	BND 90/130	-12.6	-21.7	-9.4

4 Conclusion

1. Under the conditions of the sharp-continental climate in Kazakhstan with minimum pavement temperatures of -28°C to -46°C the study of low temperature properties of bitumens is of great importance for predicting the state of asphalt pavements during operation.

2. Determining the rheological characteristics of bitumens with a bending beam rheometer points to the complicated character of their behavior in the temperature range from -12°C to -34°C, which influences the bitumen choice with regard to the climatic conditions.

3. The results of investigating bitumens by the Superpave method show that they correspond to the climatic conditions of the most part of the Kazakhstan territory while according to the road diagnostics data the temperature cracks occur practically universally in the asphalt pavements. This is indicative of a probability that the influencing factors are not adequately taken into account.

4. In this work an attempt has been made to simulate the tensile stress development in the asphalt pavement for three typical cases of temperature lowering. The most unfavorable combinations of the temperature and the rate of temperature drop have been established.

References

[1] ASTM D 6648-08: Standard Test Method for Determining the Flexural Creep Stiffness of Asphalt Binder Using the Bending Beam Rheometer, BBR (2008)
[2] ASTM D 2872-08: Standard Test Method for Effect of Heat and Air on a Moving Film of Asphalt (Rolling Thin-Film Oven Test) (2008)
[3] ASTM D 6521-08: Standard Practice for Accelerated Aging of Asphalt Binder Using a Pressurized Aging Vessel, PAV (2008)
[4] Performance Graded Asphalt Binder Specification and Testing. Superpave Series, vol.1, Asphalt Institute (1999)
[5] Teltayev, B., Kaganovich, E.: Bitumen and asphalt concrete requirements improvement for the climatic conditions of the Republic of Kazakhstan. In: Pre-Proceedings of the XXIVth World Road Congress, Mexico (2011)

[6] Huang, Y.H.: Pavement Analysis and Design, 2nd edn. Pearson Education, Inc., Upper Saddle River (2004)
[7] Papagiannakis, A.T., Masad, E.A.: Pavement Design and Materials. John Wiley & Sons, Inc., New Jersey (2008)
[8] Tschoegl, N.W.: The Phenomenological Theory of Linear Viscoelastic Behavior. An Introduction. Springer, Berlin (1989)
[9] ARA, Inc, ERES Consultants Division: Guide for Mechanistic-Empirical Design of New and Rehabilitated Pavement Structures. Final Report. NCHRP Project 1-37 A. Transportation Research Board of the National Academies, Washington, DC (2004)
[10] The Asphalt Handbook. MS-4, 7th edn. Asphalt Institute (2007)

Millau Viaduct Response under Static and Moving Loads Considering Viscous Bituminous Wearing Course Materials

S. Pouget[1], C. Sauzéat[2], H. Di Benedetto[2], and François Olard[1]

[1] EIFFAGE Travaux Publics
  Research & Development Department
  8 rue du Dauphiné BP 357, F-69960 Corbas Cedex, France
[2] Université de Lyon
  Ecole Nationale des Travaux Publics de l'Etat, Vaulx-en-Velin, F-69120, France
  CNRS, FRE 3237, Département Génie Civil et Bâtiment
[3] Rue Maurice Audin, Vaulx-en-Velin, F-69120, France

Abstract. This paper deals with the influence of viscous bituminous wearing courses materials on orthotropic steel deck bridges. These researches are part of a French national project "Orthoplus", which is briefly introduced. The approach to take into account the surfacing behavior and to develop calculation tools is explained. First the behavior of the different bituminous constituent materials is investigated. A linear viscoelastic modeling is proposed with a rheological model, previously developed at the Civil Engineering and Buildings Department ("DGCB") of University of Lyon / ENTPE. This model is implemented in a Finite Elements software, which enables simulation of any transportation structures considering viscous behavior of bituminous materials.

In order to validate these developments, the highest bridge in the world, the Millau Viaduct (in the south of France), is studied. In-situ measurements are especially carried out on the bridge. Static and moving loads at two different constant speeds (10 km/h and 50 km/h) are applied using a normalized truck. Steel structure of the bridge is instrumented in order to access strain (and then stress) level. Focus is made on comparisons between experimental strain data and simulations results using Finite Elements Method (FEM). In particular, necessity to take into account viscous properties of the bituminous materials to determine response of the whole bridge structure is emphasized.

The accurate estimation of high strain level in steel structure (using FEM calculation) could allow life time calculation of orthotropic bridges which are particularly sensitive to fatigue phenomena.

1 Introduction

This study is a part of a 2.5 million euros French national project called "Orthoplus: Advanced engineering of orthotropic decks and their wearing courses for a global optimization of their life-cycle" [1]. The so-called Orthoplus project is lead by a consortium of 7 public and private partners: SETRA, EIFFAGE Travaux Publics, EIFFEL, LCPC, Arcadis, CTICM and the ENTPE (c.f. acknowledgements).

The most famous example of such orthotropic deck bridge is the Millau Viaduct (*Figure 1*), known as the highest bridge (240m high) and the longest multiple cable stayed bridge (2460m long) in the world. The Millau Viaduct was financed and built by the Eiffage Group within the framework of a concession arrangement. Its subsidiary, "Compagnie Eiffage du Viaduc de Millau" is the concession operator of the structure for 75 years since the end of 2004. The Millau Viaduct is the subject of special investigation in the "Orthoplus" project. Some in-situ experiments are carried out and are used to validate the theoretical developments. In the project, three other smaller structures, with different scale, are also studied (in-situ or in-laboratory) to develop and validate the calculation tools. Orthotropic is the contraction for "orthogonally anisotropic". Such bridges have a complex behavior [2-8]. Both geometry of the structure and very high flexibility of steel plates induce severe stress and strain fields in the surfacing, hence durability issues for wearing courses. The structural role of the wearing courses on orthotropic steel deck bridges is usually neglected during the design process. One objective of "Orthoplus" project is to evaluate this role, considering usual surfacing composed of bituminous materials.

In this paper, the linear behavior of each bituminous constituents of the deck are first determined experimentally and modeled using previous developed models which have already proved their accuracy. These models have been implemented in commercial Finite elements software in order to simulate multiple structures [9]. Secondly, one in-situ orthotropic structure, the Millau Viaduct is investigated, experimentally and with Finite Elements simulations.

Fig. 1. View of the Millau Viaduct (France) –the highest bridge and the longest multiple cable-stayed bridge in the world

2 Materials Behavior and Modeling

In this paper, the structure and the constituent materials of the Millau Viaduct are chosen as the reference (*Figure 1*). The deck is composed of a steel plate (thickness is between 12mm for fast lane and 14mm for slow lane) reinforced with steel stiffeners. The surfacing is composed of a 3mm thick sealing sheet (Parafor

Pont®) and a 70mm thick bituminous surfacing layer (Orthochape®). The sealing sheet made with bituminous mastic ensures the water protection of steel and the perfect bonding of bituminous surfacing on the steel.

First, the behavior of each constituent material is determined. In this paper, only the linear behavior is considered (small strain domain). No non-linearity (fatigue, permanent deformations, cracks) was taken into account. Thus, the steel is considered as isotropic linear elastic with classical parameters values (E=210GPa and ν=0.3). In the following, we focus on the bituminous materials behavior (Orthochape® and ParaforPont®).

2.1 Advanced Viscoelastic Characterization of Bituminous Materials

The first considered material, Orthochape®, is a bituminous mix made with a continuous 0/10mm aggregates grading and with 5.5% by weight of the aggregates of Orthoprène®, a polymer modified bitumen. Complex modulus tests in tension/compression have been performed to characterize its linear visco-elastic behavior in the small strain domain ($\varepsilon < 10^{-4}$). The test principle is described in *Figure 2*. It should be noticed that radial strain measurement (ε_r) are added to the classical axial stress (σ_z) and strain measurements (ε_z) (*Figure 2*). Measured sinusoidal signals (*Figure 2*) are expressed in complex form (equation (1)).

$$\begin{cases} \sigma_z^* = \sigma_{0z}.e^{j(\omega t)} \\ \varepsilon_z^* = \varepsilon_{0z}.e^{j(\omega t+\phi_{\varepsilon_z})} \\ \varepsilon_r^* = \varepsilon_{0r}.e^{j(\omega t+\phi_{\varepsilon_r})} \end{cases} \quad (1)$$

where σ_{0z}, ε_{0z} and ε_{0r} are the amplitude and 0, ϕ_{ε_z} and ϕ_{ε_r} the phase lags of respectively the axial stress, axial strain and radial strain.

The complex Young's modulus E* and the complex Poisson's ratio ν* are then obtained using equations (2) and (3). They are defined with their norm and phase angle, respectively |E*| and ϕ_E for the complex Young's modulus and |ν*| and ϕ_ν for the complex Poisson's ratio.

$$E^* = \frac{\sigma_z^*}{\varepsilon_z^*} = |E^*|e^{j\phi_E} = \frac{\sigma_{0z}}{\varepsilon_{0z}}.e^{j(\phi_{\varepsilon_z})} \quad (2)$$

$$\nu^* = -\frac{\varepsilon_r^*}{\varepsilon_z^*} = |\nu^*|e^{j\phi_\nu} = -\frac{\varepsilon_{0r}}{\varepsilon_{0z}}.e^{j(\phi_{\varepsilon_r}-\phi_{\varepsilon_z})} \quad (3)$$

With the postulated hypothesis of isotropy, the 3D linear viscoelastic behavior is completely determined by E* and ν* [10].

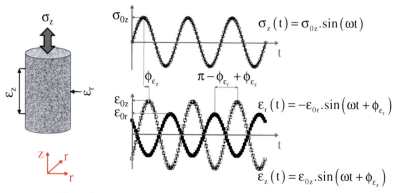

Fig. 2. Schematic explanation of the complex modulus test

Measurements were made at 9 different temperatures (from -30°C to 50°C), sweeping 7 frequencies from 0.01Hz to 10Hz. From experimental data, the Time Temperature Superposition Principle (TTSP) is considered as a first approximation (some discrepancy appears at very low frequencies and/or for high temperatures, due to the effect of polymers in the Orthoprene® bitumen). The complex Young's modulus E* and the complex Poisson's ratio ν* master curves are plotted at a reference temperature (T_{ref}) of 10°C in *Figure 4*. The classical WLF law (William, Landel and Ferry) [11] is used to fit the shift factor a_T (equation (4)).

$$\log(a_T) = -\frac{C_1(T - T_{ref})}{C_2 + T + T_{ref}} \quad (4)$$

with $C_1=35$ and $C_2=218$ for the Orthochape® mix.

The a_T curve is also plotted in *Figure 4* as a function of temperature. This newly proposed representation enables values of complex modulus or Poisson's ratio (norm and phase angle) to be easily obtained for any temperature and frequency, with only one figure. The use of this figure consists of:

- Step 1: read the desired temperature T on the right axis and join the a_T curve
- Step 2: read the corresponding equivalent frequency on the horizontal axis (this value corresponds to the a_T value for the desired temperature T).
- Step 3: read the value on the master curve. This value corresponds to the parameter at the desired temperature and for a frequency of 1Hz.
- Step 4 (not shown): to obtain the value for any frequency, keep the previously found frequency range constant (a_T value) and drag it to the desired frequency (upper limit). Perform again Step 3 with the lower limit of this range.

For the Parafor Pont® sealing sheet, the same procedure was used. As mastic, a modified great-sized rheometer was used [12]. The complex shear modulus was determined. The complex Young's modulus was obtained assuming a constant real value for the Poisson's ratio, 0.45. The classical WLF law parameters are $C_1=28$ and $C_2=203$. *Figure 5* presents the complex modulus master curve and the a_T curve for the Parafor Pont® material.

2.2 Isotropic Linear Visco-Elastic Modeling: The DBN Model

A 3D Isotropic Linear Visco-Elastic (ILVE) modeling is proposed to be used for the finite element analysis. The DBN (Di Benedetto – Neifar) model is used. This model has been developed at the University of Lyon / ENTPE. It is an attempt to describe with a unique formulation the complex elasto-visco-plastic behavior observed for different types of bituminous materials. More details on DBN model are given in [10], [13-16]. The general analogical form of the DBN model consists in an assembly in series of a linear isotropic elastic body and "n" elements having a body of EP type in parallel with a linear isotropic dashpot. It is noteworthy that the number of chosen elements "n" is a free choice for the user. If small amplitude loadings are applied, the observed behavior is linear and the obtained DBN asymptotic form is presented *Figure 3*. It consists of a generalized Kelvin Voigt model of "n" elements each having a Young's modulus E_i, a Poisson's Ratio ν_i and a viscosity η_i. The calibration of the constants (E_i, ν_i, η_i) is made from an optimization procedure in the frequency domain. Due to the lack of space, chosen constants are not given. Simulation with DBN model is also presented *Figure 4* for Orthochape® mix and *Figure 5* for ParaforPont® sealing sheet. This linear visco-elastic model has been implemented in a Finite Elements software and results from calculations on the Millau Viaduct are presented in the following.

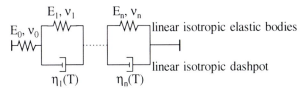

Fig. 3. Asymptotic expression in the linear domain of the DBN model (equivalent to generalized Kelvin-Voigt model)

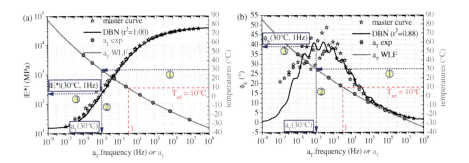

Fig. 4. Experimental master curves and DBN models simulation (20 elements) for the bituminous mix Orthochape® plotted at a reference temperature T_{ref} equal to 10°C, with the norm |E*| and the phase angle ϕ_E of the complex Young's modulus E* and the norm |ν*| and the phase angle ϕ_ν of the complex Poisson's ratio

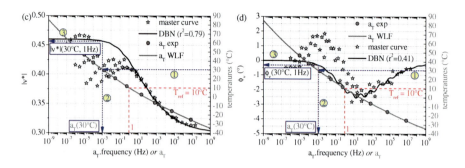

Fig. 4 .*(Continued)*

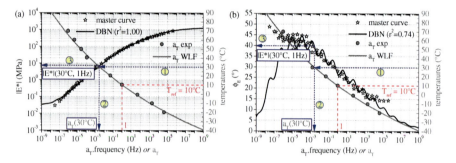

Fig. 5. Experimental master curves and DBN models simulation (20 elements) for the bituminous sealing sheet Parafor Pont® plotted at a reference temperature T_{ref} equal to 10°C. Left: norm of the complex Young's modulus E*; Right: phase angle of the complex Young's modulus E*

3 In Situ Measurements: The Millau Viaduct

The main orthotropic structure studied in "Orthoplus" project is the Millau Viaduct. In this paper, in-situ experiments are described. Some Finite Elements analyses were performed and compared with experimental data, to show the ability of the developed calculation tools.

3.1 Experimental Study

Experiments consist in loading the Millau viaduct and measuring strain in the steel structure with a net of gauges.

The objective was to load the slow and emergency lanes. The loading was applied by a truck and its trailer of 38.1 tons distributed on 5 axles (*Figure 6*). Each axle was precisely weighed and the footprint of the tires was recorded. Ambient temperature was between 12.1 and 12.3°C. Two longitudinal locations (same as for loading) were chosen for investigating the strain of the steel deck (underside). Some gauges were stuck under the deck around stiffeners n°6 and 7 to obtain information on the stress field around welding (*Figure 8*). Some bi-directional gauges are also stuck below the deck between stiffeners n°5, 6, 7 and 8 (*Figure 8*). Others gauges are used but not detailed here.

Measurements were performed under two loading cases:
- Static case :
  The truck was positioned precisely. Location of the most loaded axle (second one of the tractor truck, with two twin wheels) is indicated by dx and dy (Figure 7). This axle was located on 2 different longitudinal position, dy, firstly, over a crossbeam (dy=16.7m) and secondly in the middle of 2 crossbeams (dy=14.6m). 9 different transversal positions dx were tested (Figure 7). One twin wheel of the axle was thus located over the stiffener n° 6, 7 and 8. Transversal strain (ε_{xx}) are presented in Figure 8 for one position of the most loaded axle, located in the middle of 2 crossbeams (dy=14.6m), at transversal position dx=1.77 m. Each data point represents the average value of strain after stabilization.
- Moving case :
  As for static case, location of the most loaded axle (second one of the tractor truck, with two twin wheels) is indicated by dx in transversal way (Figure 7). Longitudinal position dy is not needed. Same 9 transversal positions dx as for static case were tested. Tests were performed at two constant speeds (10 and 50 km/h) to underline viscous effects. Longitudinal and transversal strain (ε_{xx} and ε_{xx}) are presented in Figure 8 for one speed (10km/h) and at transversal position dx=1.75 m.

Measurements are compared with calculations results.

3.2 Analysis with FEM Calculations

Finite Element calculations (FEM) are performed using "COMSOL" software in 3D to simulate experimentations on the Millau viaduct. Due to the complexity and the size of the structure, some simplifications are necessary (*Figure 7*):

- 6 similar elements are considered longitudinally (25m). Extremities are clamped.
- slow and emergency lanes are modeled (4.8m), which represents 8 stiffeners.
- crossbeams are supported on one extremity (near stiffener n°1) to represent actions of the rest of the bridge structure.
- steel plate, sealing sheet and bituminous mix surfacing are modeled with 3D brick elements while crossbeams and stiffeners are modeled with 2D shell elements. Mesh is refined around the stiffeners n° 6, 7 and 8, inducing 3.12 millions degrees of freedom.
- wheel loads are modeled by rectangular loaded surface (20cm x 30 cm).

Steel is assumed to be isotropic linear elastic materials having a Young modulus E=210000 MPa and a Poisson's ratio ν=0.3. Wearing course behavior, introducing temperature and viscous properties, is considered. This viscous behavior is introduced by the isotropic linear viscoelastic DBN model with calibrated 20 elements from complex modulus tests.

In this section, the tension is considered as positive. Moreover, perfect bond is assumed between layers. Some simulation results are presented in *Figure 8*. Discontinuities appear in the curve, which are due to the stiffeners presence and the way they are modeled with 2D shell elements. Comparisons with experimental data show rather good agreement, taking into account the errors in locating truck and gauges.

Fig. 6. Pictures of the Millau Viaduct during the experimental measurements. (a) loading truck during weighing; (b) loading truck on the Millau Viaduct; (c) stiffeners and crossbeam with gauges

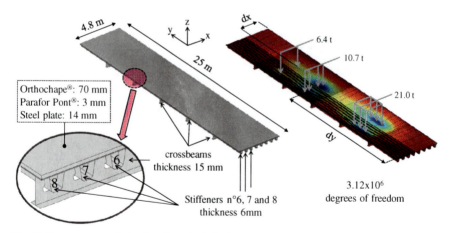

Fig. 7. Geometry, mesh, load and vertical displacement field of the Millau Viaduct struc in the Finite Element Code

3.3 Analysis of the Effect of Vehicle Speed on Bridge Response

Maximum transversal strains (ε_{xx}) for static and moving cases (*Figure 8*) plotted in Figure 9 as a function of truck speed "v". Results are given for a gau located under steel plate between stiffeners n°6 and 7 for an average dx = 1.77 m

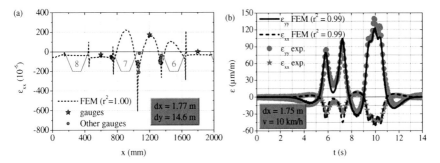

Fig. 8. Comparison between measured strains and FEM strain calculations. (a) static case for dx = 1.77 m and dy = 14.6 m; (b) moving case at 10 km/h for dx = 1.75 m and for one gauge located under stiffener n°7 between two crossbeams

This example (Figure 9) shows result also observed for other gauges. It allows to synthesize the results (experimental and calculation) obtained on the Millau Viaduct. Viscous calculations (ILVE case) allow a good approximation of strain evolution in the steel deck at any vehicle speed.

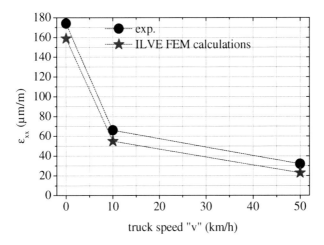

Fig. 9. Transversal strains ε_{xx} measured and calculated under the steel plate for different speeds (0, 10 and 50 km/h) and for dx approximately equal to 1.77m.

4 Conclusion and Outlook

In this paper, the approach to take into account the bituminous surfacing in the behavior of orthotropic structure is explained. First, a 3D experimental characterization of bituminous materials is carried out. The linear behavior, supposed to be isotropic, was completely defined and a previously developed model was calibrated and used. Its ability to simulate the behavior of each material is shown. This model was implemented in a Finite elements code allowing simulation of any structure. Secondly, to validate this approach, some comparison between structure tests and simulation are carried out. Experimental study carried out on Millau viaduct –the higher bridge in the world– shows the importance of taking into account viscous and thermal effects, induced by the bituminous constituent on the structure behavior. This requires to correctly measure and model the behavior of bituminous material. Comparisons between experimental data and simulations are encouraging. Yet, in-situ measurements give a great amount of results, and some complementary analyses are required to validate the simulation tool.

Acknowledgments. The authors wish to thank the French "National Research Agency" for supporting this research. Messrs. Dune and Gallice from the Millau Viaduct Company and Mr Servant are also greatly acknowledged for allowing the measurements on the Millau Viaduct and of course the 7 public and private partners of the Orthoplus project.

- Arcadis
- CTICM: Centre Technique Industriel de la Construction Métallique
- EIFFAGE Travaux Publics: public works subsidiary company of the EIFFAGE group
- EIFFEL: steel subsidiary company of the EIFFAGE group
- ENTPE: Ecole Nationale des Travaux Publics de l'Etat
- LCPC: Laboratoire Central des Ponts et Chaussées
- SETRA: Service d'Etudes sur les Transports, les Routes et leurs Aménagements.

References

[1] Le Quéré, C.: Revue Travaux, vol.(843), p. 4 (2007) (in French)
[2] Huurman, M., Medani, T.O., Scarpas, A., Kasbergen, C.: In: International Conference on Computationnal & Experimental Engineering, Corfu (2003)
[3] Medani, T.O.: PhD, Delft University of Technology, p. 280 (2006)
[4] Seim, C., Ingham, T.: Transportation Research Record 1892, 98–106 (2004)
[5] Fanjiang, G.-N., Ye, Q., Fernandez, O.N., Taylor, L.R.: Transportation Research Record 1892, 69–77 (2004)
[6] Wolchuk, R.: Transportation Research Record 1688, 30–37 (1999)
[7] Hulsey, J.L., Yang, L., Raad, L.: Transportation Research Record 1654, 141–150 (1999)
[8] Connor, R.J., Fisher, J.W.: Transportation Research Record 1696, 100–108 (2000)
[9] Pouget, S.: PhD ENTPE-INSA, p. 254 (2011) (in French)
[10] Nguyen, H.M., Pouget, S., Di Benedetto, H., Sauzéat, C.: European Journal of Environmental and Civil Engineering 13(9), 1095–1107 (2009)
[11] Ferry, J.D.: scoelastic Properties of Polymers, p. 672. John & Sons (1980)
[12] Pouget, S., Sauzéat, C., Di Benedetto, H., Olard, F.: Road Materials and Pavement Design 11(Special Issue), 111–144 (2010)
[13] Neifar, M., Di Benedetto, H.: Road Materials and Pavement Design 2(1), 71–96 (2001)
[14] Olard, F., Di Benedetto, F.: Journal of AAPT 74, 791–828 (2005)
[15] Di Benedetto, H., Neifar, M., Sauzeat, C., Olard, F.: Road Materials and Pavement Design 8(2), 285–316 (2007)
[16] Tiouajni, S., Di Benedetto, H., Sauzéat, C., Pouget, S.: Road Materials and Pavement Design, 34 (2011); accepted
[17] Pouget, S., Sauzéat, C., Di Benedetto, H., Olard, F.: Materials and Structures 43(3), 319–330 (2010)

Material Property Testing of Asphalt Binders Related to Thermal Cracking in a Comparative Site Pavement Performance Study

A.T. Pauli, M.J. Farrar, and P.M. Harnsberger

Western Research Institute, 365 North 9th Street, Laramie Wyoming 82072, USA

Abstract. Chromatographic fractions of asphalt binders, where some of the binder materials were obtained during the construction of a comparative performance field site constructed in Rochester, Minnesota, USA, where transverse (thermal) crack survey data had been obtained annually since construction were investigated in terms of compositional properties, specifically average molecular size of the wax-oil fraction. This compositional property measured or previously reported for eight SHRP asphalts was subsequently compared to binder dilatometric properties previously reported for the same materials. This compositional property was also compared to transverse (thermal) crack survey data for four binders. Results from these investigations suggested that binders of high wax content where the wax is of a comparatively higher average molecular size may lead to more extensive thermal cracking.

1 Introduction

Structuring in asphalt binders involving the crystallization of paraffin and microcrystalline waxes is suspected to potentially lead to thermal fracture of asphalt pavements [3-5]. A well studied mechanical response influenced by this type of structuring phenomenon is referred to as isothermal physical (reversible) age hardening, which is thought to contribute to thermal cracking of pavements constructed in cold climates [1-5]. This type of response, originally observed by rheological analysis (i.e., creep compliance testing) of sub-ambient isothermally conditioned asphalt samples, has led to speculation as to the nature of the changes in the molecular and/or phase composition of asphalt under this type of low temperature conditioning.

Bahia and Anderson [1] reported that as pavement temperatures approach the asphalt's glass transition temperature a collapse of free volume in an asphalt's composition, particularly changes in intermolecular configurations analogous to polymer microphase rearrangements, may be responsible for observed physical hardening effects. These researchers further reported that circumstantial evidence showed a link between asphalt wax content and physical hardening effects but also observed a relationship between asphalt wax content and molecular weight suggesting to these investigators that more research of this phenomenon was warranted.

Claudy *et al.* [3], in the same time period, reported that differential scanning calorimetry (DSC) and thermo-microscopic, including polarized light and phase contrast techniques, studies supported the hypothesis that crystallizing moieties in asphalt were partially responsible for the physical hardening phenomena observed by Bahia and Anderson [2, 1]. DSC studies reported by these researchers showed that at temperatures below 0°C, where heat capacity increases more rapidly, the onset of a glassy state develops, as measured by a glass transition temperature. It is further discussed by these researchers that at temperatures above 0°C heat flow versus temperature-change plots exhibit peaks or enthalpy events in DSC curves corresponding to changes in the hydrocarbon matrix generally known to correspond to melting and crystallization events. Thus, conditioning involving isothermal storage of asphalt materials at sub-ambient temperatures (i.e., at -15°C for 1 to 8 days), followed by DSC heating scans exhibited shifts in melt/crystallization peaks and glass transition temperatures in DSC, particularly after 24 hours of conditioning, suggesting to these investigators that molecular reorientation occurs with time. These researchers further speculated that the volume shrinkage mechanism proposed by Bahia and Anderson [2, 1] was due to *"coalescence to form crystalline domains or amorphous domains in the asphalt solvent phase"*, and could be described by phase separation or a spinodal decomposition process. To investigate their hypothesis samples of SHRP asphalt AAG-1, characterized as low in wax content, were doped with different concentrations of n-paraffins and tested with DSC and thermal microscopy techniques. It was observed that glass transition temperatures decreased relative to an initial value of -11.4°C measured for neat asphalt AAG-1 when 3% by mass of $C_{20}H_{42}$ (T_g = -29.2°C), $C_{24}H_{50}$ (T_g = -20.4°C), $C_{28}H_{56}$ (T_g = -15°C) and $C_{33}H_{66}$ (T_g = -13°C) were present in the sample. Glass transition temperatures for samples containing 3% by mass of $C_{36}H_{74}$ (T_g = -10.8°C) and $C_{40}H_{82}$ (T_g = -12.5°C) were observed to be close to that of the original neat material. Photomicrographs further showed more prominent structuring in wax-doped materials compared to the original un-doped asphalt. Claudy *et al.* [3] further suggested that crystalline fraction precipitation was not the complete story, and that a need for additional research was warranted to explain the role of molecular mobility in the proposed physical hardening mechanism. They concluded by saying that asphalt at low temperatures may be *"akin to a gel"*.

Recently, Hesp *et al.* [4, 5] have considered how morphological and chemical properties of asphalts influence physical or reversible age hardening. In these investigations X-ray diffraction, optical microscopy, and mass spectroscopy techniques were employed to study asphalt binders derived from pavement trial sites (test tracks) where the materials under consideration were observed to exhibit thermal cracking. The physicochemical data obtained on asphalt materials by these methods were compared with mechanical properties determined by extended freeze-time bending beam rheometry (EBBR) [5]. A significant finding from these studies showed that EBBR thermal cracking temperatures, T_{crack}(measured at an m-value of 0.35), changed (increased) with extended periods of thermal conditioning of asphalt binder samples (i.e, stored at -10°C and -20°C up to 14 days), when compared with data obtained for samples thermally conditioned from much shorter time periods (i.e., 1 hour). Cracking severity field data thus correlated more favorably with T_{crack}(m-value = 0.35) values of 3-day condition data compared to cracking temperatures measured for much shorter time conditioned materials. In this work, asphalts with higher wax content corresponded to materials which showed greater changing (shifting) thermal cracking temperatures after extended thermal conditioning. Optical microscopy images of asphalt morphology further revealed that waxy asphalts which exhibited a coarse-grain structure were particularly susceptible to reversible age hardening. These researchers further hypothesized that asphaltene structuring (studied by X-ray diffraction which provides information on aromatic ring structure and size) could also contribute to physical hardening, particularly in the air-blown waxy asphalts.

The studies referenced here seem to suggest that the occurrence of higher concentrations of wax in asphalts may partially contribute to thermal cracking in pavements constructed in cold climates. In particular, wax gelation has been proposed as one mechanism by which molecular structuring may occur with these materials [3]. Wax gelation, if it occurs in asphalt, is likely controlled by several compositional factors including wax concentration, type (molecular weight, n-, iso-, or cyclic nature), interaction with other constituents in the asphalt (asphaltene, resins, naphthene aromatics, saturates) and conditioning factors including cooling rate and duration of isothermal storage.

The present paper reports on studies aimed at characterizing binder composition and physico-chemical properties determined for asphalt binder materials of approximately the same PG-grade collected from comparative performance pavement sections where thermal cracking performance data, specifically transverse thermal cracking, was available. Correlations are also considered, derived from historical data reported for the eight SHRP core asphalts to make the case that variations in binder composition and/or physicochemical properties relate to variations in mechanical properties, and hence to performance behavior. Finally, a hypothesis is presented that contends that higher molecular weight molecules associated with the wax-oil fraction of waxy asphalt binders potentially gel to form a microstructure as pavement temperatures decrease resulting in binders which are susceptible to shrinkage and eventual thermal cracking of the pavement.

2 Experimental

2.1 Chromatographic Fractionation

Material fractions were derived from asphalts employing SARA (Saturates, Aromatics, Resins, Asphaltenes) chromatography [6, 7]. To prepare samples for SARA chromatography asphaltenes were separated from maltenes by precipitation in isooctane (2,2,4-trimethyl pentane, HPLC-grade, Fisher Scientific). In this procedure the precipitating solvent was combined with asphalt in a 50:1 (mL:g) ratio, stirred on a hotplate with refluxing for 30 min, and filtered using a 10-15μm fritted glass filter after settling overnight. The asphaltene filter cakes were washed with additional solvent and dried in a nitrogen purged vacuum oven at 80°C. Maltenes were recovered from the eluate by rotoevaporation and dried to constant weight on an oil bath at 120°C. Corbett (SARA) separations [7] were performed on maltene materials based on a published ASTM method, ASTM D4124-09 [6].

2.2 Thermo Gravimetric Analysis of SARA Fractions: Apparent Molecular Mass Determination

Thermogravimetric analysis (TGA) was utilized to measure vaporization temperature distributions for SARA fractions separated from asphalt [8]. This method employs a TA Instruments Q5000IR thermogravimetric system with auto sampling capabilities. In this procedure 10.0 mg of sample is weighed into a 100-μl Pt pan (TA #957207.904) and analyzed using a heating rate of 20°C per minute. Values in percent weight loss are recorded as a function of time at the specified heating rate. Weight percent per time data is converted to derivative mass change per temperature as reported. Two events (a period of material vaporization and a burn off of non-volatiles) are typically observed.

Mass or weight loss, dM versus vaporization temperature distributions, was determined as the peak or maximum value, T_{max} (K), from a five-parameter Weibull distribution function. Saturates were observed to fit reasonably well with this function while naphthene aromatic materials fit less well due to their more complex distribution. In both cases the Peak vaporization temperatures, T_{max} (K), were utilized as a measure of the average molecular size of particles comprising selected SARA fractions.

3 Results and Discussion

3.1 Analysis of SHRP Asphalt Data: Asphalt Dilatometry, Wax Content and Molecular Mass of IEC Neutrals

Before discussing the results reported in the present study, correlations were sought relating historical data (e.g., mechanical-dilatometric properties and physico-chemical properties) previously reported for the eight SHRP core asphalts. During the SHRP program several different chromatographic methods [9], SARA (saturates, aromatics, resins, asphaltenes) chromatography being one example, were adopted to characterize "chemical" classes of material types present in SHRP asphalts. Ion exchange chromatography (IEC) [9] was another technique adopted to separate "chemically" defined molecular species from an asphalt based on polarity and/or acid/base functionality. This separation method, in addition to strong acid and base species, also produced amphoteric species (weak acid/base functionality), and a neutral oil referred as an IEC neutral fraction. Waxes were subsequently separated from this neutral oil material utilizing a cold solvent precipitation technique. Many of these SHRP asphalts and their material fractions were further characterized in terms of physicochemical, mechanical and rheological properties. Based on the literature reviewed in the introduction, physicochemical, mechanical and rheological properties of the IEC neutrals fraction may give insight into an asphalt's propensity to resist and/or be susceptible to thermal fatigue given that this material solidifies at the lowest temperature compared to other material fractions. This material phase of an asphalt would then be responsible for retaining the flow property of a binder prior to solidification during cooling processes. Table 1 lists a some compositional/physicochemical [10] and mechanical properties [2] presumed to be pertinent to physical hardening tendencies as discussed in the introduction section, specifically wax content and molecular weight, here of the wax-oil phase or IEC neutrals material.

From the data reported in this table it is readily observed that wax content and molecular mass (i.e., measured via vapor phase osmometry [10]), trend with each other and with dilatometric properties of whole asphalt (i.e., absolute change in specific volume @ -15°C [2]), asphalt AAM-1 being the predominant outlier. On the basis of these findings we thought it of value to conduct additional studies to verify the historic molecular mass data reported for the SHRP asphalt.

Table 1. Absolute change in specific volume @ -15°C, number average molecular mass of IEC neutrals, percent wax per whole asphalt extracted from IEC neutrals, and the percent crystalline material determined by DSC.

Asphalt	[a] Absolute Change in specific volume @ -15°C (mL/g)	[b] VPO Number Average Molecular Mass of IEC Neutrals	[b] % Wax/Asphalt Extracted from IEC Neutrals	[b] % Crystalline Material (DSC)
AAA-1	0.0010	620	1.4	0.4
AAB-1	0.0016	680	4.7	2.3
AAC-1	0.0020	770	7	2.7
AAD-1	0.0008	580	1.9	0.6
AAF-1	0.0017	710	5.7	1.7
AAG-1	0.0007	650	1.2	0
AAK-1	0.0012	610	3.4	0.4
AAM-1	0.0021	1200	26	2.8

[a] Bahia [2]; [b] Robertson et al. [10].

3.2 Apparent Molecular Mass of Asphalt Wax-Oil Fraction Materials and Dilatometric Properties of SHRP Asphalt

Investigations were conducted to characterize the apparent molecular mass of wax-oil phase materials, in this case SARA saturate and naphthene aromatic material fractions derived from asphalts, including the eight SHRP core asphalts and four additional asphalts derived from a comparative performance site where thermal crack survey data were reported. Mechanical data testing was also conducted with the four comparative performance site materials. All twelve asphalts were initially separated into SARA fractions following a published ASTM procedure [6]. SARA saturate and naphthene aromatic fractions were then characterized individually in terms of their apparent molecular size distribution by vaporization temperature measurements utilizing thermogravimetric analysis (TGA). This method was selected for convenience given that in simple homologous series systems, aliphatic hydrocarbon systems for example, molecular mass and boiling point temperature are found to be functionally related. Hence, TGA analysis should be a simple approach to generally characterize a molecular size distribution as compared to alternative approaches including gel permeation chromatography or vapor-phase osmometry.

Table 2 lists material mass-balance data and mass fraction averaged peak vaporization temperatures, T_{max} (K), measured for SARA generated wax-oil phase material fractions, designated here as saturates plus naphthene aromatics, quantified by

$$T_{max} = \left\{ \frac{(SAT\%)}{(SAT\% + NA\%)} \right\} \cdot T_{max}(SAT) + \left\{ \frac{(NA\%)}{(SAT\% + NA\%)} \right\} \cdot T_{max}(NA) \quad (3)$$

Vaporization temperatures, T_{max}(SAT+NA), or simply T_{max} reported in Kelvin-units are assumed to be a measure of the size or apparent molecular mass of the hydrocarbon molecules (e.g., n-paraffins, branched alkanes and alkyl-naphthenic species) which comprise these two fractions. Thus, for TGA analysis only dispersive forces should account for the intermolecular interactions among these types of molecules as they dissociate into a vaporous state indicative of a mass loss of the bulk sample as measured by TGA. This combined wax-oil fraction material is further presumed to be similar in molecular composition to IEC neutral materials, given the absence of heteroatom content or polar functionality.

Table 2. Wax-oil peak vaporization temperatures, T_{max} (K) (i.e., saturates and naphthene aromatics) determined for 8 asphalts based on relative SARA percentages

Asphalt	%SAT per wax-oil	%NA per wax-oil	T_{max}(SAT), K	T_{max}(NA), K	T_{max}(SAT+NA), K
AAA-1	34%	66%	577	679	644
AAB-1	28%	72%	651	703	688
AAC-1	37%	63%	672	720	702
AAD-1	40%	60%	535	588	567
AAF-1	26%	74%	641	722	701
AAG-1	25%	75%	546	675	643
AAK-1	32%	68%	648	695	680
AAM-1	23%	77%	704	748	738

The plot depicted in Figure 1 shows that T_{max} values, with the exception of asphalt AAD-1, correlate with dilatometric properties (i.e., change in specific volume, ($\Delta v = \Delta(1/\rho)$), reported in Table 1 [2, 1]. As reported by Bahia and Anderson [1], other properties than wax content, molecular mass of asphalt for example, may also potentially contribute to the dilatometric properties of asphalt binders. We thus hypothesized that some form of structuring associated with the wax-oil phase of an asphalt may involve the previously proposed gelation mechanism [3]. This gel structuring may further be associated with higher molecular weight n-paraffins and may also involve immobilization of other continuous phase materials (e.g., branched alkanes, naphthene and or aromatic species with alkyl side chains) as a function of decreasing temperature. To test our hypothesis TGA testing of SARA fractions was conducted on materials obtained during the construction of a comparative performance field site where transverse (thermal) crack survey data had been obtained annually since construction for validation purposes.

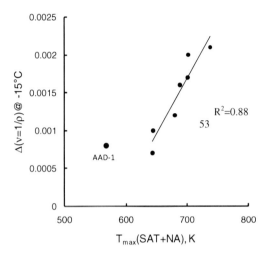

Fig. 1. Correlation plots relating wax-oil vaporization temperature and average molecular weight to dilatometric properties (absolute value in change in specific volume measured at -15°C [2]) of 8 SHRP asphalts

3.3 Distress Survey Data of Comparative Performance Site Asphalts Compared to Apparent Molecular Mass of Asphalt Wax-Oil Fraction Materials

Asphalts originating from different crude sources derived from materials collected at a comparative performance site constructed in Rochester, Minnesota were fractionated to produce a wax-oil material which was subsequently characterized by TGA to measure T_{max} values. At these sites the construction specifications of pavement sections were held constant with the exception that the source of the asphalt was varied from one section to the next within the site. Performance data gathered from these sites, (e.g., distress survey) were correlated with predictive models and test protocols developed as part of ongoing studies to validate research results.

The Rochester comparative performance site was constructed in Olmsted County on county road 112 in August 2006. The project consisted of three lifts of HMA over a reclaimed base. Five performance sections were constructed. Within each section two 500-ft monitoring sections were established. The project used four different asphalt sources designated MN1-2, MN1-3, MN1-4, and MN1-5. Samples of these asphalts were collected during construction. Two of the five sections were constructed with a polymer (Elvaloy) modified asphalt (MN1-1 and MN1-2). The difference between the two sections is that the MN1-1 section included 20% RAP. The HMA used to construct the MN1-1 section was also the project HMA. Mix designs were performed for each section.

Seasonal climate data based on the weather station nearest the project (Rochester International Airport) were utilized to report extreme temperatures. The low temperature for the first winter (2006/2007) was reported to be -29.4°C, the low temperature for the second winter (2007/2008) was reported to be -28.3°C and the low temperature for the third winter (2008/2009) was reported to be -32.2°C [11]. Temperatures lower than -28°C occurred on five occasions over a three year period. The low temperature grades required for the Rochester location at confidence levels of 50%, 94% and 98% are -28, -28, and -34, respectively [12]. Pavement condition monitoring was performed by WRI personnel just after construction on October 3, 2007, at which time no distress was observed in any of the 500-ft. monitoring sections.

Figure 2 depicts a bar graph of normalized transverse crack length (linear feet, L-ft) reported in the WRI 2008 and MNDOT 2009 surveys of the Rochester comparative performance site for sections constructed with each of the four binders. Higher normalized transverse crack length values were observed in the third performance year compared to year two.

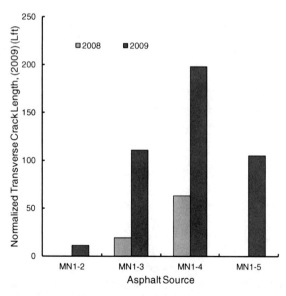

Fig. 2. Bar graph of normalized transverse crack length (linear-ft) reported for four comparative performance site asphalts based on the WRI 2008 and MNDOT 2009 surveys conducted in Rochester, MN

Sections constructed with asphalt MN1-4 showed the highest normalized transverse crack length values, while sections constructed with MN1-2, reported to be a polymer modified asphalt, showed almost no reported cracking for the same time period. Sections constructed with asphalts MN1-3 and MN1-5 showed intermediate transverse crack length values.

Wax-oil phase fractions chromatographically separated from materials collected from the Rochester, MN comparative performance site asphalts were then characterized in terms of their mass fraction averaged peak vaporization temperature (T_{max}). Table 3 lists mass fraction averaged peak vaporization temperatures, T_{max} (K), measured for wax-oil material and normalized transverse crack lengths (linear-ft) reported for four comparative performance site pavement sections. With the exception of the polymer modified asphalt, MN1-2, T_{max} (K) values trend by increasing with normalized transverse crack lengths. These findings helped to support our hypothesis that differences in compositional properties of the wax-oil phase of an asphalt binder potentially result in differences in performance.

Table 3. Peak vaporization temperatures determined for wax-oil fractions (i.e., saturates and naphthene aromatics) and the mass fraction averaged T_{max} (K) of the wax-oil determined for four comparative performance site asphalts obtained by TGA

Asphalt	%SAT/ wax-oil	%NA/ wax-oil	T_{max}(SAT) K	T_{max}(NA) K	T_{max}, K	Normalized TC Length (L-ft)
MN1-2	32%	68%	638	694	676	11.3
MN1-3	33%	67%	613	678	657	110.6
MN1-4	34%	66%	635	705	681	197.7
MN1-5	32%	68%	583	665	639	105.2

So, in regard to wax content being solely responsible for physical hardening, and hence thermal fatigue, Claudy et al. [3] actually observed that glass transition temperatures decreased relative to an initial value of -11.4°C measured for neat asphalt AAG-1 when 3% by mass of $C_{20}H_{42}$ (T_g = -29.2°C), $C_{24}H_{50}$ (T_g = -20.4°C), $C_{28}H_{56}$ (T_g = -15°C) and $C_{33}H_{66}$ (T_g = -13°C) were doped into AAG-1 but also observed that glass transition temperatures for samples containing 3% by mass of $C_{36}H_{74}$ (T_g = -10.8°C) and $C_{40}H_{82}$ (T_g = -12.5°C) were observed to be closer to that of the neat material. These results seemed to show that lower molecular weight wax spiked into AAG-1 could have actually softened the material as suggested by the lowering of the glass transition temperature, while intermediate or higher molecular weight wax spiked into this same asphalt affected the glass transition to a much lesser extent. By speculating that Claudy et al. [3] may have potentially observed glass transition temperatures to increase if higher molecular weight wax (i.e, higher than $C_{40}H_{82}$) would have been tested in the manner they employed, we observed that molecular mass of the wax-oil materials also had a significant effect on dilatometric properties based on our investigation of the historic SHRP data.

4 Conclusions

In this paper we have formulated a hypothesis suggesting that binders of comparatively higher wax content, where the wax is of a comparatively higher average molecular weight, may form a more rigid gel-structure (also considered by Claudy *et al.* [3] and Hesp *et al.* [4]), which impacts the dilatometric properties of the whole binder. To test this hypothesis historical data, specifically reported for the eight SHRP core asphalts, was analyzed and revealed that a strong correlation exists between higher wax content, higher average molecular mass of the solvent or wax-oil phase of an asphalt and asphalt binder dilatometric properties. To further support our hypothesis asphalts which differed based on crude source gathered from a comparative performance field site (Rochester, Minnesota) where thermal crack survey data was reported, were also investigated in terms of average molecular size of their wax–oil phase. Results from these studies also showed that asphalts with wax-oils of higher apparent molecular mass correlated with materials which cracked more with the exception of one polymer modified asphalt. These findings have led to a working hypothesis that gel structuring, if operative in asphalt, is likely controlled not only by the presence of wax but by the molecular mass of the wax.

Acknowledgements. The authors gratefully acknowledge the Federal Highway Administration, U. S. Department of Transportation, for their financial support: Contract No. DTFH61-07-D-00005 and DTFH61-07-H-00009. The authors would also like to thank G. Forney and J. Beiswenger and for their contribution to this work in performing the laboratory experiments.

Disclaimer. This document is disseminated under the sponsorship of the U.S. Department of Transportation in the interest of information exchange. The United States Government assumes no liability for its contents or use thereof. The contents of this report reflect the views of Western Research Institute, which is responsible for the facts and the accuracy of the data presented herein. The contents do not necessarily reflect the official views or the policy of the United States Department of Transportation. Mention of specific brand names of equipment does not imply endorsement by the United States Department of Transportation or by Western Research Institute.

References

[1] Bahia, H.U., Anderson, D.A.: ACS Division of Fuel Chemistry Preprints 37(3), 1397–1404 (1992)
[2] Bahia, H.: Dissertation, University of Michigan (1991)
[3] Claudy, P., Letoffe, J.M., Rondelez, F., Germanaud, L., King, G.N., Planche, J.P.: ACS Division of Fuel Chemistry Preprints 37(3), 1408–1426 (1992)
[4] Hesp, S.A.M., Iliuta, S., Shirokoff, J.W.: Energy & Fuels 21, 1112–1121 (2007)
[5] Zhao, M.O., Hesp, S.A.M.: Int. J. Pavement Eng. 7(3), 199–211 (2006)
[6] ASTM D4124-09: In: Annual Book of ASTM Standards, Road and Paving Materials; Vehicle-Pavement Systems, Section 4, vol. 4(3), pp. 381–388, ASTM International, West Conshohocken (2011)
[7] Corbett, L.W.: Analytical Chemistry 41, 576–579 (1969)

[8] Goodrum, J.W., Siesel, E.M.: Journal of Thermal Analysis and Calorimetry 46(5), 1251–1258 (1996)
[9] Branthaver, J. F., Petersen, J. C., Robertson, R. E., Duvall, J. J., Kim, S. S., Harnsberger, P. M., Mill, T., Ensley, E. K., Barbour, F. A. Schabron, J. F.: SHRP-A-368, Binder Characterization and Evaluation, vol. 2, Chemistry. Strategic Highway Research Program, National Research Council, Washington, DC (1993)
[10] Robertson, R.E., Branthaver, J.F., Harnsberger, P.M., Petersen, J.C., Dorrence, S.M., McKay, J.F., Turner, T.F., Pauli, A.T., Huang, S.-C., Huh, J.-D., Tauer, J.E., Thomas, K.P., Netzel, D.A., Miknis, F.P., Williams, T., Duvall, J.J., Barbour, F.A., Wright, C.: Fundamental Properties of Asphalts and Modified Asphalts: Interpretive Report, FHWA-RD-99-212. U. S. Department of Transportation, Federal Highway Administration, McLean, VA (2001)
[11] National Oceanic and Atmospheric Administration National Weather Service, http://www.nws.noaa.gov/sitemap.php
[12] LTPPBind, Version 2.1, developed for the Federal Highway Administration, http://www.fhwa.dot.gov/research/tfhrc/programs/infrastructure/pavements/ltpp/install.cfm

Influence of Differential Displacements of Airport Pavements on Aircraft Fuelling Systems

A.L. Rolim[1], L.A.C.M. Veloso[2], H.N.C. Souza[3], P.L. de O. Filho[1], and L.V. de A. Monteiro[1]

[1] INFRAERO - Brazilian Airports Infrastructure Company
[2] UFPA- Federal University of Para, Brazil
[3] IME - Military Institute of Engineering, Brazil

Abstract. This paper reviews an analytical and experimental investigation of an airport pavement after the occurrence of a ductile rupture on a pipe from an aircraft hydrant fuelling system. In order to identify the causes of the rupture, numerical analysis were carried out to evaluate the vertical and horizontal displacements of the pavement imposed on hydrant pit boxes. Parameters that could validate the numerical model were obtained by means of experimental analysis of the pavement. Therefore, information such as thermal expansion of the airport pavement due to weather conditions and to aircraft jet blast was evaluated. Also, the effects of a high loading caused by aircraft wheels and other airport service vehicles on the pavement, and settlement or movement of adjacent apron were considered.

1 Introduction

1.1 Background

In October 2008, a leakage of aircraft fuel (JET-A1) was found at a thermal expansion joint located between an apron concrete pavement and a Pit box ofanaircraft fuelling system atthe International Airport of Belem, Brazil.The leakage was caused due to a rupture of a hydrant pipeelbow. In January 2009, a report was issued by the IPT (Technology Research Institute) assuring that the crack of the metal pipe occurred by a mechanism of ductile fracture due to bending stresses [1].

After this report, an "in-situ" investigation started with the support of the Federal University of Para. The purpose of this research was to determine the possible causes of the rupture, based on experimental tests at the same location of the failure, analysing the pavement mechanical behaviour and its correlation with the problem.

An excavation was performed after the leakage in order to drain the fuel and extract the fractured pipe section,as shown in Figure 1a.

Fig. 1. (a)Pit box of the aircraft fuelling system and (b) the fractured elbow [1]

2 Experimental Tests Design

2.1 Apron Layout

The test was performed on an airport apron that consists of a concrete pavement20 meters wide by 40 meters long,composed by 64 adjacent slabs. The pavement layout is shown in Figure 2.

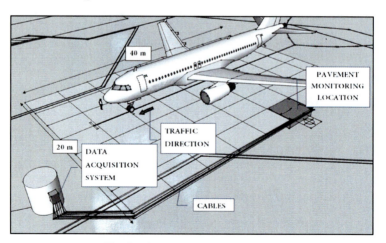

Fig. 2. Airport rigid pavement layout

2.2 Instrumentation

The main objective of the instrumentation was to obtain accurate data in order to evaluate the structural airport pavement behaviour under aircraft and service vehicle loadings and weather conditions.Experimental tests werecarried out during regular airport operation.Ten strain gages wereplacedon the surface of the pavement to measure strain values. These gages were connected to a data acquisition system that recorded the strains as the aircraft wheel passed over the pavement, and it was enabled to record a 10 Hz frequency of data sampling.

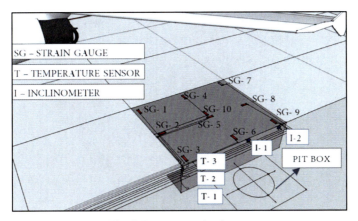

Fig. 3. Instrumentation layout

The pavement temperature was measuredusingequipment enabled to receive the analogical signal of three thermo-resistive sensors and send the collected data already as a digital signal by means of binary bases, containing IP address, and data function with error analysis to the Airport network infrastructure.Therefore, it was possibleto evaluatea real time data collection remotely, at the engineering office located eight hundred meters far from the airport operational area.

Finally, the rotation of the plate was monitored using two inclinometers installed in the lateral surface of the pavement, one for each direction, as shown in Figure 3.

3 Pavement Modelling

3.1 Pavement Model Overview

Finite element software SAP2000 was used to model the influence of aircraft loadings and weather conditions over the concrete pavement and also to verify if the displacement caused by its mechanical behaviour would be able to cause the aircraft fuelling system pipe rupture.

The pavement material properties considered in this study were the same adopted in its original design. The concrete pavement had compression strength of 42 MPa and 38 cm deep. A three-dimensional concrete pavement section was analyzed, Figure 4. The pavement section was placed on a 10 cm-deep concrete base, with 15 MPa compression strength and subgrade layer represented by Winkler foundation [2]. Solid elements were used to mesh the model with the nodes coincident with strain gages locations.

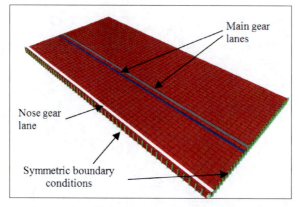

Fig. 4. Pavement Model

Also, one type of aircraft, Airbus 320 (A320), was investigated, which represents the highest aircraft load over the apron. It was considered a quasi-static analysis to simulate the aircraft passage on the pavement. Table 1 shows the main characteristics considered in this model for this aircraft, according to Airbus Manual [3].

According to the data exhibited on Table 1, it was considered concentrated moving loads of 113.1 kN and 125.8 kN corresponding to nose and main gear tires on the lanes showed in Figure 4.

Table 1. Aircraft characteristics

	A320
Number of passengers	180
Maximum takeoff weight	75900 Kg
Maximum ramp weight	75500 Kg
Nose gear tire pressure	1.103×10^6 N/m²
Main gear tire pressure	1.227×10^6 N/m²

In this study the temperature gradient was assumed to be bilinear throughout the pavement thickness, one linear gradient distribution from top to mid and other from mid to bottom considering the measured temperatures of a typical day. The temperature gradient was applied in the FE model in order to take into account weather conditions effects, as shown in Figure 5.

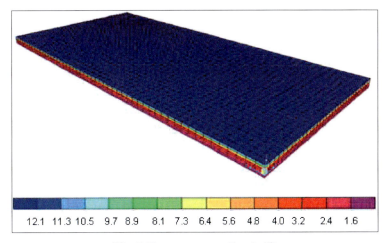

Fig. 5. Temperature gradient in °C

3.2 Pipe Model Overview

Finite element software ANSYS was used to estimate the influence of plates movement over the pipe. After the validation of the pavement model, values of displacement were obtained and applied to the pipe at the same direction of the thermal dilatation of the pavement influence.

The uniaxial element PIPE16 was used for the straight section of the pipeline considering tension-compression, torsion, and bending capabilities. The PIPE18 was used for the elbow, because this type of element considers a curved element, the geometry, the pipe diameter, radius of curvature, wall thickness, flexibility factors, internal fluid density and corrosion thickness allowance [4, 5], Figure 6.

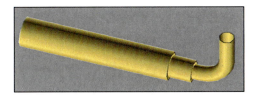

Fig. 6. Pipe modelling

The pipe section and properties adopted to build the model were the same presented at the fuelling system project and IPT report. The straight part of the pipe section measured approximately 0,17m of outside diameter and after it a reductionwas considered and the elbow section measuring 0,09 m of outside diameter.

4 Results and Analysis

4.1 Experimental Results

All tests were performed during one week beginning on 22 July 2011. After it, the collected data was organized. Figure 7-a shows the temperature measured at the pavement surface, mid-section and base, while Figure 7-b shows the surface and surface-base gradient temperatures measured at the same period.

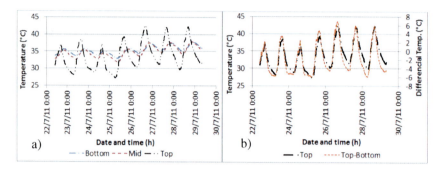

Fig. 7. Pavement measured temperatures from 22 to 28 July 2011. (a) Surface, mid and base pavement temperatures. (b) Surface and surface-base gradient

It is possible to notice the cyclic behaviour of the temperatures during one day period (24 hours). Both base and mid temperatures are in phase but slight out of phase compared with surface temperature, Figure7-a. Comparing the surface and surface-base gradient temperatures, Figure7-b, they are in phase and the temperatures variations have the same magnitude, approximately 12°C.

Although the pavement monitoring was not performed at the same period of the fuel leakage, the climate in the north of Brazil is equatorial, and the ambient temperature suffers small variations along the year. Figure 8 presents the daily ambient temperature variation measured in Belem International Airport at the period of the pipe failure, October 2008.

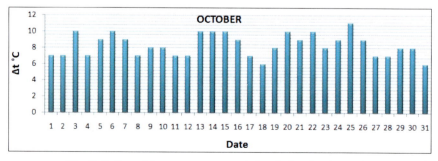

Fig. 8. Daily ambient temperature variation during October 2008

As expected, the maximum daily ambient temperature variation at the period of the pit box accident has approximately the same magnitude of the surface temperature pavement and surface-base gradient measured during the pavement monitoring.

Figure 9-a shows the strain measured near to the pit box and the Figure9-b shows the correlation of a typical measured strain and the surface-base gradient temperatures. In the Figure9-a, the strain gage 6 (SG-6) signal presented a different behaviour compared to the two other signals measured along the same direction.

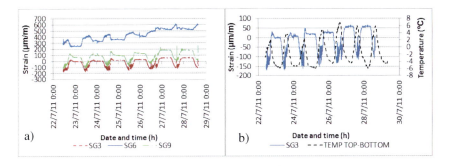

Fig. 9. Measured strain close to the pit box and surface-base gradient temperatures

The SG 6 was the strain gage fixed on the closest concrete surface to the pit box. An inspection evidenced that the concrete surface had been repaired with grout after damages caused by the pit box excavations. In the Figure 9-b, the measured strains also presented a cyclic behaviour. However, it is not in phase with the surface-base gradient temperature. Certainly, it is related with the complex thermal behaviour of the pavement.

A thermographic camera was used to measure the temperature on the pavement surface due to the aircraft jet blast. The jet blast influence on the pavement surface temperature is related with aircraft operation and type. Usually, the aircraft stops on the apron with engines turned off. Figure 10 shows the great observed pavement heating due to the jet blast where the surface temperature around the aircraft turbine reached more than 150° C.

Fig. 10. Thermal influence of Aircraft turbine

Despite the high localized temperature just below the turbine of the aircraft, analysing the measured strains, at the periods of aircraft arrival, it was not possible to identify large increase in measured strains due to the jet blast.

4.2 Numerical Analysis

After evaluating the experimental results to perform a calibration of the FE model, it wasobtained a lateral displacement of approximately 0.7 mm, as shown in Figure 11.

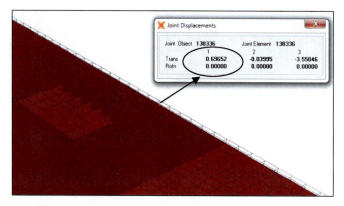

Fig. 11. Lateral displacement of the pavement (0.697mm)

The pipe bending stresses were evaluated considering the maximum lateral displacement of the concrete pavement obtained with the pavement FEM model. Figure 12-a shows mechanical behaviour of the pipe considering the direction in which the pavement moves towards the pipe, and Figure 12-b shows the opposite direction of the displacement.

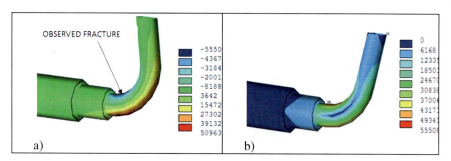

Fig. 12. Pipe stresses (N/m^2)

Figure 12-a represents the situation of the heating of pavement where it moves towards the pipe, while Fig 12-b corresponds the cooling of the pavement. Tensile stresses in the region where it was observed the fracture occur with the pavement cooling. The pipe maximum tensile stresses reached only 0.055 MPa.

5 Discussion and Concluding Remarks

According to API 1540 recommendation, the aircraft fuelling system design should consider that high loadings can be imposed on hydrant pit boxes from aircraft wheels, tugs, and other service vehicles or from settlement or movement of adjacent aprons to prevent the transmission of these loadings to hydrant risers (to which the hydrant pit valve is fitted).Each hydrant pit valve should be effectively isolated from its hydrant by means of a sealing arrangement that can accommodate both lateral and vertical differential movement[6].Also the fuel line shall enter the side of the pit; bottom entry shall not be allowed [7].

Therefore, based on the results of this study, the following conclusions have been drawn:

- The original design of the pit box does not attend to international recommendations;
- The temperature gradients of the studied pavement under traffic load caused a maximum lateral displacement of approximately 0.7 mm;
- Considering the maximum lateral displacements, the pipe stresses are many times inferior to the pipe´s strength.

Acknowledgements. This work was supported by INFRAERO and the Civil Engineering Department of the Federal University of Para. Special thanks are given to the engineer Silvio Souza for his technical support in test planning and organization. The authors would also like to express their appreciation to the engineerJacksonMarques Reis for his assistancepresented within. The contents do not necessarily reflect the official views and policies of the INFRAERO. The paper does not constitute a standard, specification, or regulation.

References

[1] IPT – Institute of Technical Research, Failure Analysis of an Elbow ofAircraft Fueling System Pipeline (Institutional Report), São Paulo – Brazil (2009)
[2] Caliendo, C., Parisi, A.: Stress Prediction Model for Airport Pavements with Jointed Concrete Slabs. Journal of Transp. Engineering 136(7), 664–677 (2010)
[3] Airbus, S.A.S.: Aircraft Characteristics for Airport Planning(2005) (Issue: September 30, 1985) (Rev. May 01, 2011)
[4] ANSYS 11, Ansys Elastic Curved Pipe Tutorial (2008)
[5] Sam Kannappan, P.E.: Introduction to Pipe Stress Analysis. John Wiley and Sons, New York (1986)
[6] American Petroleum Institute - API: Design, Construction, Operation and Maintenance of Aviation Fuelling Facilities, Energy Institute, London (2004)
[7] SAE – Aerospace: Aviation Fuel Facilities Aerospace Recommended Pratice, SAE International, USA (2006)

Rehabilitation of Cracking in Epoxy Asphalt Pavement on Steel Bridge Decks

Leilei Chen and Zhendong Qian

Intelligent Transportation System Research Center, Southeast University
Nanjing, 210018, P.R. China

Abstract. Cracking is the main distress mechanism for epoxy asphalt concrete pavements on bridge steel decks. However, there are no sealing materials and techniques for crack in bridge deck pavements. To solve this problem, a chemically cured sealant for epoxy asphalt pavement on bridge steel deck has been developed and evaluated. Viscosity test, tensile test, pull-out test, tensile bond behavior test and shear bond behavior test were conducted in the laboratory, and the newly developed sealant proved to have rather good workability, bulk performance, interfacial performance and cooperative behavior with epoxy asphalt concrete. Monotonic and fatigue beam bending tests were also employed to assess the effects of different sealing techniques. The tests results showed that the bending strengths of sealed beams are approximately the same as undamaged beams while their fatigue lives are significantly shorter than the undamaged ones. Fatigue equations of undamaged beams and sealed beams were developed based on the fatigue test results. They provide ways to predict fatigue lives of the epoxy asphalt steel deck pavements both before and after sealing.

1 Introduction

Epoxy asphalt concrete has been proved to be an excellent material for steel deck pavement. It has been widely used in the steel bridge pavements all over the world, especially in China [1]. However, investigations show that cracking is the major distress of epoxy asphalt pavements [2]. Once cracks appear on a pavement surface, the pavement structure will become discontinuous. If no appropriate treatments be taken, this may lead to further deteriorations like potholes, debonding and even the rusting of the steel deck. These may significantly reduce the serviceability and the service life of the bridges. So it is quite necessary to take measures as soon as cracks appearing in the steel deck pavement.

The cracks are handled in many ways [3-6], ranging from pavement maintenance activities, such as surface treatments and crack sealing, to full-scale pavement rehabilitation projects, like overlay and resurfacing. The cost-effectiveness analyses have also been conducted in the Strategic Highway Research Program (SHRP), in which the maintenance treatments such as crack sealing were found to be most cost-effective [7]. However, there is no cure-all sealing material and technique that can deal with all kinds of cracks, the sealing material and technique are still being researched all over the world.

Over the past two decades, a new generation of highly modified crack sealants has been introduced to the market [8]. These products essentially fall into three families based on their chemical composition and manufacturing process: cold-applied thermoplastic materials, hot-applied thermoplastic materials and chemically cured thermosetting materials [9]. However, for the lack of effective evaluation, in many cases, the exact cause for sealant failure remains unknown, successful sealant installation can't be repeated [10]. In addition, the sealants mainly aim at treating the cracks in highway pavement. Sealant products for steel deck pavement can rarely be found.

As to sealing technique, according to the researches and the practical experiences, sealants can be placed into cracks in numerous configurations, and these configurations can be grouped into four categories: flush-fill, reservoir, over band and combination (reservoir and overband) [9]. However, few studies about the sealing techniques for crack in steel deck asphalt pavement could be found. The following questions remain to be solved: (a). whether the configurations can fit the crack in steel bridge pavement or not, and (b). which sealant configuration is most effective when using in steel deck pavement.

Cracks appear frequently in the steel deck pavement in China due to heavy overloads and severe environment conditions. However, only since recent years has the importance of maintenance of steel deck pavements been recognized, most crack treatments are copies of the highway pavement as mentioned above. The differences between structural condition and the stress mode of the highway and the steel deck pavement lead this copy to certain failure. In this case, the sealant product for cracks in steel deck pavement is urgently needed, and suitable techniques should also be investigated to find out an effective way to install the crack sealants.

3 Objective

This paper presents a development and assessment of a crack sealant for epoxy asphalt pavement on steel bridge deck. The effect of different crack sealing techniques used on steel deck epoxy asphalt pavement is also evaluated to find a suitable sealing material and technique for crack in steel bridge deck pavement.

4 Materials and Method

4.1 Sealant Development

The chemically cured thermosetting materials have been widely used in recent years for their good performances. Epoxy asphalt concrete is also a thermosetting material which is mixed with epoxy asphalt and high quality aggregates. Considering the similar performances and the good compatibility with epoxy asphalt mixture, the epoxy resin was selected to develop a chemically cured thermosetting sealant.

Selection of the raw material. The primary adhesive and curing agent are two main compositions of a chemically cured sealant. Based on the requirement of the raw material and the results of market investigations, liquid bisphenol-a type epoxy resin was selected as the primary adhesive due to its low initial viscosity. Meanwhile, mixed amines were selected as the cure agent, since no single amine could lead to a good effect. A chemical analysis was conducted to the selected primary adhesive and curing agent, the results showed the epoxy value of the primary adhesive was 0.43, which means a small initial viscosity. The active hydrogen equivalent of the cure agent was ranging from 110 to 120, which will be used to calculate the mixing percentage of the epoxy resin and the amines as presented below.

Mixing percentage. The mixing percentage of the primary adhesive and the cure agent can be estimated by Eqn. (1):

$$M_{ca} = AHE \times EV \tag{1}$$

Where M_{ca} is the mass portion of the cure agent in 100g primary adhesive, AHE is the hydrogen equivalent of the cure agent and EV is the epoxy value of the primary adhesive. The M_{ca} was calculated to be 47.3 to 51.6 according to Eqn (1).

To determine a more accurate mixing percentage of the cure agent and the primary adhesive, the tensile test was employed following ASTM D638-08. Tensile specimens with M_{ca} ranging from 40 to 60 were prepared and tested at 23°C after curing, as shown in Figure 1. The test results were listed in Table 1.

(a) The tensile test specimens   (b) The tensile test equipment

Fig. 1. The tensile test of the developed sealant

Table 1. Tensile Test Results of the Sealant with Different Mixing Percentages

Technical Indexes	Mca / mass portion				
	40	45	50	55	60
Tensile strength/MPa	7.5	9.0	10.2	10.5	10.0
Fracture elongation /%	96.2	90.4	86.9	83.2	81.7

It can be found from Table 1 that the tensile strength reaches the peak value when the mass portion is 55. However, the fracture elongation of the specimens decreases with the rising of the mass portion. On the other hand, the tensile strength and the fracture elongation don't vary a lot when the mass portion of the cure agent changing from 50 to 60. So taking a comprehensive consideration of the tensile strength, the fracture elongation and the operation convenience, the mass portion of the cure agent is determined as 50. It means that the mass ratio of the primary adhesive and the cure agent is 2:1.

4.2 Epoxy Asphalt Mixture Specimens

Materials Preparation. Two main materials are involved in the test. The details of preparing for both materials and specimen are introduced below.

The binder used in the test is 2910-type local epoxy asphalt, which is composed of two components marked as A and B. Component A is the epoxy resin while component B consists of petroleum asphalt and curing agent. The basic information of the material is given in Table 2.

The basalt and the limestone powder for steel bridge pavement are selected as aggregate based on the practical engineering. The max aggregate diameter is 13.2mm. The gradation curve for the aggregate is shown in Figure 2.

Table 2. Technical Index of 2910-Type Local Epoxy Asphalt

Technical Indexes	Measured Value	Criteria	Test Method
Mass ratio (A:B)	100:290	100:290	
Tensile strength (MPa, 23°C)	3.26	≥ 2.0	ASTM D 638
Fracture elongation (%,23°C)	242	≥ 200	ASTM D 638
Viscosity from 0 to 1 Pa·s(min)	110	≥ 50	JTJ052-2000

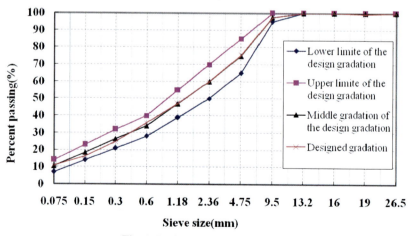

Fig. 2. Designed gradations in the test

Specimen Preparation. Based on the Marshall mixture design procedure, the optimum asphalt content was determined as 6.5%. The asphalt mixtures were shaped to slabs and cut to the beams and cubes for the tests. Firstly, the binder and the aggregate were mixed at 125°C. After reserved for 40min at 120°C, a slab specimen was shaped with roller. Then, after curing for 5h at 130°C, the slab was ready to be cut. Three replicates were prepared for each test.

4.3 Evaluation Criteria and Method of Material

Working Mechanism. The structural condition and the stress mode of the highway pavement vary a lot from those of steel deck pavement [11], as shown in Figure 3.

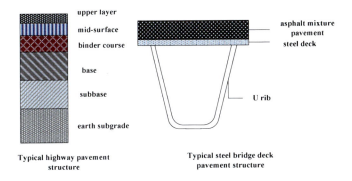

Fig. 3. Comparison between highway pavement and steel deck pavement

Because of the different structural conditions, the maximum tensile stress and strain often appear at the bottom of the surface layer in highway asphalt pavement, while the tensile stress and strain peaks often appear at the top of the surface layer in the steel deck pavement. Therefore, most highway asphalt pavement cracks are bottom up cracks due to the tensile stress at bottom of the asphalt layer and top-down cracks caused by the shearing stress on the top of the surface [12]. However, in the steel deck pavement, most of the initial cracks are top-down cracks due to the tensile stress on the top on the steel deck pavement [13]. In this case, the evaluation criteria of sealant for steel bridge pavements should be different from those for highway pavements.

Criteria. The existing crack sealant often failed either cohesively or adhesively. Cohesive failure, characterized by the fracture of the sealant in the bulk, but still adhered to the crack walls; while on the other hand, the adhesive failure, a debonding near the sealant/asphalt concrete (AC) interface, is much more common. So both the bulk properties and the interfacial properties of the sealant should be evaluated to make sure it works well. In addition, the workability is also an important index to the sealant, especially to the chemically cured sealant, it may affect the effect of the crack sealing significantly. So the assessment of the sealant referred to three main items: bulk property, interfacial property and workability.

Criteria of crack sealants for steel deck pavement were proposed, as presented in Table 3, through a comprehensive consideration of the characteristics of steel deck pavement and the performance requirements in some standards listed below:

- One-component silicone resin sealing material for concrete pavement joints (ASTM D5893-04);
- Classification and requirements for building sealants (ISO/DIS 11600-2000);
- Sealants for building joint of concrete construction (JC/T881-2001);
- Jointing sealants for concrete bridge and other concrete trafficable by vehicles (JC/T 976-2005);
- Epoxy grouting for concrete crack (JC/T 1041-2007).

Table 3. Criteria of Crack Sealants for Steel Deck Pavement

Item	Content	Criteria	Test Method
workability	initial viscosity /mPa·s	<800	ASTM D2393
	operable time /min	>30	ASTM D2393
bulk property	tensile strength /MPa	≥8	ASTM D638
	fracture elongation /%	≥60	ASTM D638
interfacial property	adhesive strength /MPa	≥3	GB/T 5210

Other Method. Besides the criteria above, there were two more tests employed here to evaluate the cooperative behavior of the sealant and the AC: tensile bond behavior test and shear bond behavior test, as shown in Figure 4. The sealant was filled into two epoxy asphalt mixture cubes at 23°C and then cured. The tensile strength and the shear strength of the specimens would be tested respectively following the test method GB/T 13477.

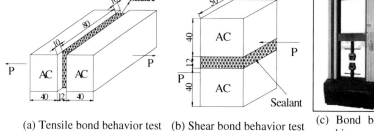

(a) Tensile bond behavior test (b) Shear bond behavior test (c) Bond behavior test machine

Fig. 4. The cooperative behavior test of the sealant and AC

4.4 Evaluation Methods of Sealing Technique

The beam bending strength test and beam bending fatigue test are adopted to compare the sealing effect of different configurations.

Firstly, some 63.5mm×50mm×381mm epoxy asphalt mixture beams were prepared and divided into two subsets: one subset was kept as is it and the other subset was cracked artificially at the midpoint and then sealed with the newly developed sealant. According to the crack conditions, the sealed beams could be also divided into two groups: the completely-fractural beams and the partly-fractural beams. The completely-fractural beams are those beams which fracture to two halves while the partly-fractural beams are the beams with a crack depth ratio of 0.5. Six types of the crack channel configurations were prepared for the effect comparison, which were 5mm×5mm, 10mm×5mm, 20mm×5mm, 5mm×10mm, 10mm×10mm, 20mm×10mm along longitudinal and vertical directions. Then the cracks were sealed with the developed sealant above. At last, after the sealant was cured, the beam bending strength test was conducted at 15°C to determine the maximum bending strength of the beams. The Four point bending fatigue test was then conducted using the Universal Testing Machine (UTM) at 15°C. Controlled-stress loading mode was selected in the fatigue test. The semi-sinusoidal wave was used at 10Hz, and 0.40、0.45、0.50 were determined as the stress ratio according to the relative research[14]. The control stress could be calculated through multiplying the maximum bending strength of the beams by the stress ratio. Three replicates were prepared for each testing.

5 Results and Discussions

5.1 Sealants

Workabilities. The viscosity test was employed to evaluate the work abilities of the sealant using a Brookfield rotational dial viscometer with a rotating speed of 100 rpm ($29^{\#}$). The results were recorded in the Figure 5. It can be found easily that viscosity of the sealant increasing with the curing time growing. The initial viscosity of the sealant is 133mPa·s, and the viscosity after curing for 30 minutes is 260mPa·s. Both of them can meet the criteria of the sealant for steel deck pavement well as listed in Table 3.

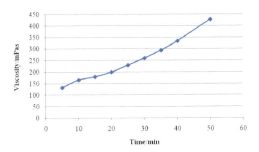

Fig. 5. The viscosity-time curve of the sealant

Bulk Performance. Studies on crack sealants have mostly focused on their bulk properties [*15*]. Manufacturers have done much to improve the bulk properties of the sealant over the years. This has resulted to sealants with good cohesion. It also can be found in this study that the sealant has a rather good bulk performance, as listed in Table 1, the tensile strength and the fracture elongation are 10.2MPa and 86.9% respectively, which could satisfy the criteria in Table 3 well.

Adhesive Performance. The interfacial property of the sealant and the epoxy asphalt mixture was examined through the pull-out test. Two groups of total six replicates were examined and the average adhesive strength is 3.28MPa. The test result is much greater than the adhesive requirement of the bonding layer material, which is 2.75Mpa [*16*], and it also meet the requirement in Table 3 well. However, on the other hand, the fractural sections in the test are all at the interface between the sealant and the epoxy asphalt mixture, indicating that the interfacial performance is still a weak point of the sealant.

Cooperative performance. The average results of the tensile and shear bond behavior are 4.152MPa and 2.852MPa respectively. The fractural sections are both at the interface between the sealant and the epoxy asphalt mixture. It is proved again that the bulk performance of the sealant is much better than the interfacial performance, so more efforts should be taken to improve the interfacial performance of the sealant in the future.

5.2 Sealing Techniques

Beam Bending Strength Test. The beam bending strength test was conducted to the undamaged beams and the sealed beams. The average bending strength of the undamaged beams is 19.14MPa. The bending strength of the sealed beams with different channel configurations don't vary a lot, the results are ranging from 18.34MPa to 19.40MPa, and the average bending strength is 18.78MPa, not much smaller than that of the undamaged beams.

A special phenomenon was observed in the tests that the fracture sections were neither at the bulk of the sealant nor at the interface of the sealant and the AC, but at the epoxy asphalt mixture near the formal crack. The reason might be that the bulk performance and the interfacial performance are larger than the bending strength of the epoxy asphalt mixture at 15°C. So it also explained why the bending strength of the sealed beams are not different from that of the undamaged beams.

Beam Bending Fatigue Test. The fatigue tests have been conducted at 15°C and the results are presented in Figure 6.

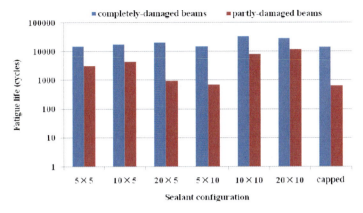

(a) Fatigue life of sealed beams with different sealant configurations

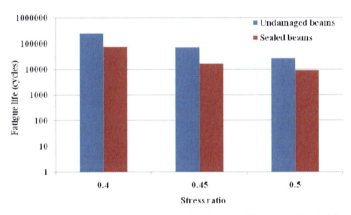

(b) Comparison between fatigue lives of undamaged beams and sealed beams

Fig. 6. Fatigue test results of different beams under 15°C

Figure 6(a) shows the test results of sealed beams with different sealing configurations. The tests have been conducted to the completely-fractural beams and partly-fractural beams, with a control stress of 8.0MPa. It can be indicated from Figure 6(a) that the completely-fractural beams have a better sealing effect than the partly-fractural beams. The reason is that the completely-fractural beams can be sealed more completely and easily. Although the crack channel may also make the sealant fill into the crack deeper, it's difficult to make sure the sealant filling into the full depth of the crack.

In addition, it also can be found from Figure 6(a) that the 10mm×10mm and 20mm×10mm sealant configurations have better sealing effects. The reason might be that the 10mm cut in depth can make the sealant filling deeper into the crack and therefore bring a better sealing effect than the 5mm depth. Meanwhile, the 10mm and 20mm cut in width may add the flexibilities of the epoxy asphalt mixture beams, and then leading to a longer fatigue life.

Figure 6(b) presents the test results of the sealed beams and the undamaged beams under different stress ratios. The sealed beams with 10mm×10mm sealant configuration were selected in the test for its good sealing effect. Figure 6(b) shows that the fatigue lives of sealed beams are significantly smaller than that of the undamaged beams. It also can be observed from Figure 6(b) that the fatigue lives of both two kinds of beams appeared linear law in the logarithmic coordinate system. So Eqn. (2) is adopted to determine the relationship between the fatigue life N_f and the control stress σ [17]:

$$N_f = K_1 \left(\frac{1}{\sigma_1} \right)^{y_1} \tag{2}$$

Assuming $K'_1 = \lg K_1$, and taking logarithm to the both sides of Eqn (2), it may transfer to Eqn. (3)

$$\lg N_f = K'_1 - c_1 \lg \sigma_1 \tag{3}$$

Where N_f is the fatigue life of the beams, σ_1 is the control stress in the fatigue test, K'_1 and c_1 are regression coefficients.

After a regressive analysis to the results in Figure 6(b), the fatigue equations of undamaged beams and the sealed beams can be determined respectively, as Eqn. (4) and Eqn. (5)

$$\lg N_f = -9.3305 \lg \sigma_1 + 69.545 \tag{4}$$

$$\lg N_f = -9.6744 \lg \sigma_1 + 71.323 \tag{5}$$

Eqn. (4) and Eqn. (5) have provided ways for the life prediction of the epoxy asphalt steel deck pavements before cracking and after sealing.

6 Summaries and Conclusions

This paper presents an assessment to a newly developed crack sealant for steel deck epoxy asphalt concrete pavement, the effect of different sealing techniques have also been evaluated. The results are listed below:

A chemically cured crack sealant for epoxy asphalt steel deck pavement was developed, and the evaluation criteria for the sealant were also determined. The workability, bulk performance, interfacial performance and cooperative performance of the developed sealant were assessed through viscosity test, tensile test, pull-out test, tensile bond behavior test and shear bond behavior test. The test shows that the developed sealant can satisfy the requirements and criteria of the steel deck pavement crack sealant well. The pull-off test, cooperative performance test and the fatigue test results all presented that the fracture section were at the interfacial between the sealant and the epoxy asphalt mixture. It can be indicated that the interface between the sealant and the epoxy asphalt mixture are the most possible position of sealing failure. So efforts should still be made to improve the interfacial behavior of the sealant and the epoxy asphalt mixtures.

The beam bending strength test was employed to evaluate the ultimate bearing capacity of the undamaged epoxy asphalt mixture beams and the sealed beams. The results show that both of the bending fractures occurred at the epoxy asphalt mixture sections, and the bending strength of the sealed beams do not vary a lot from that of undamaged ones at 15℃. The fatigue test results of sealed completely-fractural beams and sealed partly-fractural beams show that the sealing effect of completely-fractural beams is better than that of partly-fractural beams. The 10mm×10mm and 20mm×10mm sealant configurations were proved to have better sealing effects than other sealant configurations. The fatigue results of undamaged beams and sealed beams under different control modes show that the fatigue lives of sealed beams are significantly smaller than that of undamaged beams.

The fatigue equations of undamaged beams and sealed beams at 15℃ were regressed, and they have provided ways for the life prediction of the epoxy asphalt steel deck pavements before cracking and after sealing.

Acknowledgment and Disclaimer. This work was undertaken with funding from the Western Transportation Construction Technical Program for Chinese Ministry of Transport (No. 2009318000086). This funding is greatly appreciated. The opinions and conclusions expressed in this paper are those of the authors and do not necessarily represent those of Chinese Ministry of Transport.

References

[1] Gaul, R.: In : Proceedings of Selected Papers from the, GeoHunan International Conference, pp. 1-8 (2009)
[2] Qian, Z., Han, G., Huang, W., et al.: Chin. Civ. Eng. J. 85(10), 132–136 (2009)
[3] Smith, K., et al.: Publication SHRP-M/UFR-91-504. SHRP, National Research Council, Washington, DC (1991)
[4] Bullard, D., Smith, R., Freeman, T.: Publication. SHRP-H-322, SHRP, National Research Council, Washington, DC (1992)
[5] Jordan, W., Howard, I.: In: Proceedings of 89th Annual Meeting of the Transportation Research Board, Washington, DC (2010)
[6] Lee, J., Kim, Y.: In: Proceedings of 89th Annual Meeting of the Transportation Research Board, Washington, DC (2010)
[7] Smith, R., Freeman, T., Pendleton, O.: Publication SHRP-H-358, SHRP, National Research Council, Washington, DC (1993)
[8] Al-Qadi, I., Yang, S., et al.: In: Proceedings of 86th Annual Meeting of the Transportation Research Board, Washington, DC (2007)
[9] Kelly, L., Smith, A., et al.: Publication. SHRP-H-348, SHRP, National Research Council, Washington, DC (1993)
[10] Masson, J., Lacasse, M.: In: Wolf, A. (ed.) Durability of Building and Construction Sealants. RILEM, Paris. pp. 259–274 (2000)
[11] Chen, T.: Research on the cracking behavior of epoxy asphalt pavement on long span steel bridge. Southeast university, Nanjing (2006)
[12] Zhang, Q., Zheng, J., Liu, Y.: Chin. Civ. Eng. J. 25(2), 13–22 (1992)

[13] Liu, Z.: Research on Key Technology of Long-span Steel Bridges Deck Surfacing Design. Southeast University, Nanjing (2004)
[14] Ghuzlan, K., Carpenter, S.: Trans. Res. Rec. 1723, 141–149 (2000)
[15] Zanzotto, L.: Laboratory Testing of Crack Sealing Materials for Flexible Pavements. Transportation Association of Canada, Canada (1996)
[16] Huang, W.: Theory and method of deck paving design for long-span bridges. China Construction Industrial Press, Beijing (2006)
[17] Lin, G.: Research on the Fatigue Life of the Steel Bridge Deck Pavement based on the Fracture Mechanics. Southeast university, Nanjing (2006)

Long-Term Pavement Performance Evaluation

Laszlo Petho[1] and Csaba Toth[2]

[1] ARRB Group Ltd, Sustainable Infrastructure Management
[2] Budapest University of Technology, Department of Highway and Railway Engineering

Abstract. Hot mix asphalt (HMA) performance in terms of fatigue resistance is a well-developed area of pavement design worldwide. Different equipment is available to predict the fatigue performance due to the well supported technical background.Sophisticated pavement design methods utilises the asphalt performance to predict pavement performance and the relative performance comparison between different mix types is also feasible.The research work presented in this paperprovidesan analytical approach to validate laboratory fatigue tests and in-service pavement performance. Large diameter cores (320 mm) were taken from heavily trafficked heavy duty pavement structures and subsequently 2 point bending tests were performed on the cut specimens. The pavement response to loading derived from the FWD measurement was compared to the performance obtained from the laboratory fatigue tests andthe remaining life of the pavement structure was assessed.

1 Background and Scope of the Investigation

The determination of the remaining structural useful life of the pavement structures is one of the most interesting, but most difficult tasks of pavement engineering. Reliable calculation of the allowable loading of an existing roadway can be beneficial for the asset owner, since predicting the future behaviour results in reliable allocation of the resources. Economic advantages for the asset owner/manager have been proven previously [1].The general mechanistic procedure (GMP) is limited to the assessment of load associated distresses. The method uses computer software to determine critical strain responses in pavement layersresulting from the static application of a standard reference load.The critical responses assessed for asphalt materials is the horizontal tensile strain at the bottom of the layer and for subgrade and selected subgrade material it is the vertical compressive strain at the top of the layer [2,3].The GMP requires the design moduli for existing pavement layers and the subgrade to be estimated as accurately as possible, for example by back-calculation [4].Once the design moduli of the existing pavement layers and subgrade have been determined, calculations can be performed. The critical location in the pavement for the calculation of strains for an asphalt overlay is the bottom of the overlay. This approach considers that the existing asphalt layer(s) are in cracked condition and limits the stiffness of the asphalt material for pavement design purposes. This approach has been proven reliable and takes the uncertainties of the pavement design properties into consideration.

The considerations highlighted in this paper do not provide critical analysis of the above procedure, but highlights the benefits and difficulties of the other approach, where the material properties of the existing asphalt layers are not limited. In this approach, the critical strains in the asphalt layers are calculated directly from the measured deflection bowl. The bearing capacity and the pavement response to a certain wheel loading can be evaluated by different types of test equipment. In this research the Falling Weight Deflectometer (FWD) was utilised, which has a high capacity and also provides a high reliability level. The deflection bowl can be captured exactly using a high number of geophones attached to the measurement frame, which ensures that the shape of the deflection bowl can be described by reliable mathematical functions. The parameters of the mathematical functions for describing the deflection bowl provide the basis of the analysis and evaluation of the curvature of the deflection bowl. The curvature in this analysis is defined as being the reciprocal of the radius of the deflection bowl at a certain distance from the loading centre. By utilising the curvature of the deflection bowl and alsoconsideringbasic linear elastic theory the actual strain in the asphalt layers can be calculated.

In the first phase of the research work FWD tests were performed in the outer wheel path (OWP) and between wheel paths (BWP) using the approach described by Molenaar et al. [5]. Following the measurement the curvature values were calculated applyingthe new method as described in this paper. In the second phase of the research the fatigue properties of the existing asphalt layers were evaluated using a new method developed by the authors. The fatigue parameters of a hot mix asphalt can be determined by means of a cyclical fatigue test; the most common test methods are the two point bending (2PB) test on a trapezoidal asphalt specimen or the four point flexural bending test (4PB) using an asphalt beam specimen [6].The procedure of sample preparation is labour intensive for laboratory mixes; however, deriving samples from existing asphalt layers can be even more difficult and requires extra care and effort. Therefore fatigue relationships are mainly collected for new production mixes and experiences with fatigue properties are largely related to those new mixes. In order to be able to estimate the in-service fatigue properties of existing asphalt layers a new core drilling method had been developed and applied, where the cylinder of the drilling equipment is larger than usual. After extraction of these large cores trapezoid specimens can be cut out by means of a precision saw-cut machine.

Based on the GMP it is usually accepted that the existing asphalt layers do not have significant impact on the pavement structural capacity. This is explained by the hypothesis that the existing (old) asphalt layer in the reconstructed pavement structure will soon be in a cracked condition and the newly constructed asphalt layer will be the critical layer. This approach takes the uncertainty of the existing pavement into account and provides a high reliability in the pavement design process; however, the associated construction costs are also high.In existing heavy duty pavements, where the existing (old) asphalt layers are relatively thick and the asphalt layers are in relatively good condition, it can be envisaged that the existing asphalt layers are structurally sound and can greatly contribute to the overall bearing capacity of the pavement structure. When sufficient friction can be achieved between the existing and new asphalt layers, the upgraded structure will be bent together, and maximum strain will occur at the bottom of the existing asphalt layers. In this case the fatigue properties of the existing asphalt layers are highly important.

The on-site and laboratory testing described in this paper was carried out by the Budapest University of Technology, Department of Highway and Railway Engineering under the supervision of the authors in 2009 and 2011.

2 Mathematical Description of the Deflection Bowl

FWD setup usually utilises 7-10 geophones (Figure 1). The outputs of FWD testing are plots of the deflection bowl constructed from deflections at various offsets from the load centre. Mathematical functions can be applied using the measured surface deflections; these mathematical functions can be then analysed and evaluated to derive information from the behaviour of the existing pavement structure. In this investigation the curvature of the deflection bowl was determined according to the adjusted 'Witch of Agnesi' [7]. The deflection bowl can be described according to Eqn. (1):

$$D(x) = \frac{D_0 \cdot 3r^2}{(\alpha \cdot x)^\beta + 3r^2} \quad (1)$$

Where:
- D_0 = maximum deflection under the loading plate (mm)
- r = radius of the loaded plate (mm)
- α, β = coefficients which describe the shape of the bowl
- x = offset to the maximum deflection, in horizontal direction (mm)

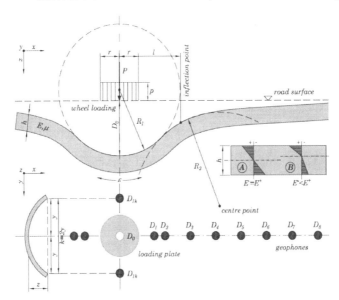

Fig. 1. Explanation of the curvature and strain distribution in the pavement structure

It is proven that the minimum radius of the curvature at the position $x_1=0$ can be calculated according to Eqn. (2). Coefficient c is introduced as the substitution of the shape coefficients α, β in order to make the equation simpler. Derivation of Eqn. (2) can be found in [8]:

$$R_1(x_1) = -1{,}5 \frac{r^2}{c \cdot D_0} \qquad (2)$$

Where:
$R_1 =$ maximum radius of the deflection bowl (m)
$r =$ radius of the loaded plate (mm)
$c =$ coefficient which describe the shape of the bowl
$D_0 =$ maximum deflection (mm)

The application of mathematical functions fitted to the measurement points of the deflection bowl ensures the practical use of other parameters as area-index, or stiffness index. Based on a detailed analysis of test results collected previously on rigid and flexible pavements revealed that the distance of the inflection point to the load centre is significantly different for different types of pavement structures. Based on this phenomenon it provides a suitable means for non-destructive analysis.The strain response to loading can be calculated directly from the geometry of the deflection bowl, without applying any difficult regression equation subject to the curvature and the asphalt thickness being known. The strain value is a relative number (ε) and can be calculated at the bottom of the pavement which has a thickness of h according to Eqn. (3):

$$\varepsilon(x) = \frac{h}{2R(x)} = \kappa(x)\frac{h}{2} \qquad (3)$$

Where:
$\varepsilon(x) =$ strain value at the bottom of the asphalt layer,ata horizontal offset x (m)
$h =$ overall thickness of the asphalt layers (m)
$R(x) =$ radius of the deflection bowl, in a horizontal offset x (m)
$\kappa(x) =$ curvature of the deflection bowl, in a horizontal offset x (m)

In this paper Eqn. (3) and the related calculations were utilised.

Estimating the Strain Response of the Asphalt Layer Based on the Deflection Bowl Parameters

FWD test series were carried out on two different heavy duty pavement structures, on the Hungarian motorway network. The applied target load was 50 kN at a contact stress of 707 kPa.

2.1 Investigation on the Highway M2

Highway M2 is a heavy duty pavement carrying heavy traffic between Hungary and Slovakia; the sectionin scopewas constructed in 1996. The road formation consists of a single carriageway which provides access for a single lane in each direction. The pavement structure consists of multiple asphalt layers on stabilised base layers as described in Table 1.

Table 1. Pavement structure, Highway M2

Layer description	Thickness (mm)
Asphalt layers	220
SAMI	5
Cement stabilised layer	250
Subgrade	infinite

One set of the test series was conducted in the outer wheel path of the northbound lane, and an other set of the test was repeated in the same lane between the wheel paths. According to the mathematical approach described earlier,the radii of the curvature were calculated and the statistical analysis of the results is tabulated in Table 2.

Table 2. Statistical analysis of the curvature radii, Highway M2

Statistical analysis, Highway M2, curvature radii	Outer wheel path (OWP)	Between wheel paths (BWP)	Difference (%)
Number of test points	35	36	n/a
Mean value (m)	1017	1112	109%
Standard deviation (m)	685	642	n/a
Coefficient of variation	0.67	0.58	n/a

Based on Table 2 it can be noted that the curvature value calculated, between the wheel paths provided higher values, which can be related to better load distribution properties. This is usually expected, because of the unloaded nature of the pavement structure between the wheel paths.After utilisation of Eqn. (3) the strain values can be estimated. The results are summarised in Table 3.

Table 3. Statistical analysis of the estimated tensile strain at the bottom of the asphalt layers, Highway M2

Statistical analysis, Highway M2, calculated strains	Outer wheel path (OWP)	Between wheel paths (BWP)	Difference (%)
Number of test points	35	36	n/a
Mean value (microstrain)	154	142	92%
Standard deviation (microstrain)	81	78	n/a
Coefficient of variation	0,52	0,55	n/a

2.2 Investigation on the Motorway M3

The pavement structure of Motorway M3 was constructed in 1979 and the road section provides connection to the Eastern part of the country and Eastern Europe. The traffic intensity is very high, with a high percentage of heavy vehicles using the roadway. The pavement structure consists of multiple asphalt layers on stabilised base layers as described in Table 4.

Table 4. Pavement structure, Motorway M3

Layer description	Thickness (mm)
Asphalt layers	303
Cement stabilised layer	200
In situ stabilisation	150
Subgrade	infinite

A FWD test was conducted on the westbound carriageway, in the outer wheel path of the slow lane and the fast lane, respectively. The slow lane was considered as the heavily loaded traffic laneand the fast lane was considered as the unloaded part of the pavement structure, since the fast lane is mainly used by light vehicles. Statistical analysis of the radii calculated from the FWD testing is provided in Table 5. The estimated strains are tabulated in Table 4.

Table 5. Statistical analysis of the curvature radii, Motorway M3

Statistical analysis, Motorway M3, curvature radii	Slow lane, OWP	Fast lane, OWP	Difference (%)
Number of test points	67	57	n/a
Mean value (m)	343	651	190%
Standard deviation (m)	154	351	n/a
Coefficient of variation	0.45	0.54	n/a

Table 6. Statistical analysis of the estimated tensile strain at the bottom of the asphalt layers, Motorway M3

Statistical analysis, Motorway M3, calculated strains	Slow lane, OWP	Fast lane, OWP	Difference (%)
Number of test points	67	57	n/a
Mean value (microstrain)	539	296	55%
Standard deviation (microstrain)	271	134	n/a
Coefficient of variation	0.50	0.45	n/a

3 Estimating the Fatigue Properties of the Existing Asphalt Layers

In the second phase of the research work fatigue properties of the existing asphalt layers were tested and evaluated.Comparing the actual strains with allowable strains of the existing asphalt layers is essential for remaining life calculation; however, it is difficult to obtain reliable allowable strains of the existing asphalt layers.

An extra-large diameter cylinder was manufactured for the normal drilling equipment andspecial pincers with curved blades were also developed for the process.It should be noted that it was sometimes difficult to extract such a large and heavy core, but the process utilised ensured that the asphalt specimens were not bent or damaged on any way.Trapezoid specimens were cut out utilising a precision saw-cut machine.

It should be noted that the loading environment is different between the pavement structure and in the laboratory testing. However, since specimens were extracted between the wheel paths, the authors consider this approach theoretically correct, since the pavement structure can be considered completely unloaded in this position. Figure 2 summarised the overall procedure of the specimen extraction and preparation. Figure 3 summarises the results of the fatigue tests performed on the extracted specimens from the pavement structure.

In-service fatigue properties of existing asphalt layers were tested using cyclical fatigue test performed using the two point bending (2PB) test on trapezoidal asphalt specimens.

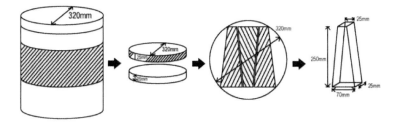

Fig. 2. Schematic process of the core drilling and trapezoidal specimen extraction

I should be noted that the fatigue functions derived from the laboratory tests are not transfer functions. Reliability factors should be utilised to relate a mean laboratory fatigue life to the in-service fatigue life at desired project reliability [2].It should also be noted that high statistical significance and data fit (R-squared) can be usually achieved by testing 18 specimens [6], which is in line with the results collected at the BME asphalt laboratory.

A reliable remaining life calculation could be performed only for the unloaded area (between wheel path and fast lane), since the asphalt specimens for 2PB testing were derived from this section of the pavement. A reliability factor of 1.5 at 90% reliabilitylevel [2] was applied. Table 7 summarises the calculated traffic loading on the Highway M2, and Table 8 on the Motorway M3.

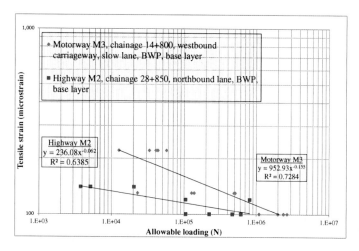

Fig. 3. Fatigue results on the Highway M2 and Motorway M3

Table 7. Calculated allowable traffic loading, Highway M2

Location	Calculated tensile strain (microstrain)	Calculated traffic loading at 90% reliability (ESA)
Outer wheel path	154	N/A
Between wheel paths	142	2.5E+06

Table 8. Calculated allowable traffic loading, Motorway M3

Location	Calculated tensile strain (microstrain)	Calculated traffic loading at 90% reliability (ESA)
Slow lane	539	N/A
Fast lane	296	2.6E+04

The calculated allowable traffic loading for Highway M2 is in the right ballpark and provides reasonable results for the pavement performance evaluation. The results for Motorway M3 show results out of the envisaged traffic loading range. However, the following should be noted:

- The pavement structure of Highway M2 and Motorway M3 are different, therefore there might be limitations on the accuracy of the strain calculation introduced in this paper.
- Mean value of the calculated strains was taken into account in the calculations, where the standard deviation was relatively high (0.45-.055). This relatively high scatter of the measured values influences the accuracy. More frequent testing would be desired to ensure that the results are statistically accurate.

- Relatively low reliability could be obtained in the fatigue test, which is probably due to the insufficient number of test specimens (14 points for M3 and 9 points for M2). According to the results collected at the BME asphalt laboratory R-squared value for 2 point bending test on trapezoidal specimens are above 0.8, if 18 specimens are used in the fatigue test.

4 Conclusions, Suggestion for Further Research

The research results presented in this paper provide an analytical tool for pavement performance evaluation. Large diameter cores (320 mm) were extracted from heavy duty pavement structures and subsequently 2 point bending tests were performed on the cut specimens. FWD tests were conducted on the loaded and relatively unloaded area of the pavement structure in order to evaluate the pavement structure in terms of the remaining life.Due to budget constrains relatively few tests were performed in this validation. However, the approach described in this paper provides a possible direction for future research work. It provides an approach for reliable pavement design forre-construction works, where the existing asphalt pavement layers can be evaluated and their relative performance could be taken into account, resulting in cost effective design solutions.

References

[1] Robinson, B., Clayton, A., Alderson, K.: Austroads, Remaining Life of Road Infrastructure Assets: An Overview, AP-R235/03, Austroads, Sydney, NSW (2003)
[2] Jameson, G.W.: Austroads, Guide to pavement technology: part 2: pavement structural design, AGPT02/10, Austroads, Sydney, NSW (2010)
[3] Claessen, A.I.M., Edwards, J.M., Sommer, P., Uge, P.: Asphalt Pavement Design – The Shell Method. In: International Conference on the Structural Design of Asphalt Pavements. Ann Arbor (1977)
[4] Jameson, G.W., Shackleton, M.: Austroads, Guide to pavement technology: part 5: pavement evaluation and treatment design, AGPT05/09, Austroads, Sydney, NSW (2009a)
[5] Molenaar, A.A., Houben, L.J.M., Alemgena, A.A.: Estimation of maximum strains in road bases for pavement performance predictions. In: Maintenance and Rehabilitation of Pavements and Technological Control, Guimaraes, Portugal, pp. 199–206 (2003)
[6] EN 12697-24, Bituminous mixtures. Test methods for hot mix asphalt. Part 24: Resistance to fatigue
[7] Scharnitzky, V.: Mathematical formula collection. Technical Books, Budapest (1989)
[8] Primusz, P., Toth, C.S.: Geometry of the deflection bowl. Revue of Roads and Civil Engineering, 18–25 (December 2009)

Structural Assessment of Cracked Flexible Pavement

L.W. Cheung, P.K. Kong, Gordon L.M. Leung, and W.G. Wong

Department of Civil and Structural Engineering,
The Hong Kong Polytechnic University, Hong Kong, China

Abstract. Highway authorities in various countries have been using different tools to determine the structural capacity of pavement for their rehabilitation programs, pavement design, research and management for more than 50 years. The earlier common tools were Benkleman beams and Deflectographs, which have been gradually phased out after the introduction of Falling Weight Deflectometer (FWD) since the 1980s. FWD is able to record pavement surface deflections in relation to a dynamic load, simulative to a moving wheel. Back-analysing surface deflection measurements enables estimation of in-situ moduli of materials in different pavement layers. Cracked or poor materials give relatively low in-situ layer moduli.

In Hong Kong, although FWD and the associated back-analysis and forward-analysis, have been in use for a number of years, the results of a recent review on FWD residual life show doubts on the correlation between the estimated residual life from FWD survey and the actual in-situ pavement performance in practice. Hence, a new method, simply making use of FWD's surface deflection measurements, to estimate the residue structural capacity is developed.

This paper presents the research on studying the use of FWD center deflection for crack identification of flexible pavement. Detailed observation of cracks on cores from 31 pavement sections were used to define the condition code. Back-analysed stiffness levels related to the center deflection were used to develop Structural Condition Index. The findings lead to the development of a simplified non-destructive structural assessment technique to determine the probability of crack existence within flexible pavement.

1 Introduction

The Hong Kong's road network consists of 2,071 kilometres of roads [1]. About HK$900 million was spent annually on roads to maintain the serviceability [3]. Expressways are inspected daily; trunk roads and other primary roads are inspected weekly. Other roads and footpaths are inspected at half yearly intervals [2]. The results of a recent review on FWD residual life show doubts on the correlation between the estimated residual life from FWD survey and the actual in-situ pavement condition. It is concluded that many roads in Hong Kong have been repeatedly repaired to various extents and by different methods for the last thirty years. Great errors are often noted from the conventional idealised assumptions of homogeneous pavement layer thicknesses and consistent material types, which are simply based on information of limited numbers of core. The inaccuracies are furthered by the use of temperature factors applied to either back-analysed stiffness or FWD deflection. Such temperature factors may work well with intact materials, but are difficult to account for pavements with a long and complicated maintenance history, or pavement with debonding and defect embedded in any of the sub-layers. In order to maximise the resource allocation, direct and more reliable pavement assignment methodology has been developed. The method engages the FWD's center deflection measurements, without the needs of pavement layer thicknesses and temperature corrections, to categorize the residue structural condition.

2 Falling Weight Deflectometer Survey

The research consisted of 31 numbers of flexible pavement section, each with an average section length of 20m. On each pavement section, FWD tests with a targeted contact pressure of 700kPa were conducted at 1 m intervals along the wheelpath and the lane center. The magitudes of the deflections of the pavement surfaces, up to 1.8m from the load centre were measured.

3 Data Analysis

To categorize the pavement condition, a simple Structural Condition Index (SCIn) based on the magnitude of the center deflection of FWD is used. SCIn values range between 0% and 100%, which attempt to provide a simple method to describe the likely structural condition of a pavement – the higher the SCIn value, the better is the pavement condition. Another advantage of using SCIn is that such index number could be easily understood not only by pavement engineers but also by any other members in the management team. The procedures of developing SCIn are shown as below:

1. Determine the probability distribution $f[D_1]$ of the FWD survey, of which D_1 is the center deflection of a FWD test;
2. Determine the cumulative probability distribution $F[D_1]$ of the center deflections of the FWD survey;
3. Determine the SCIn in percentage using the following equation:

$$SCIn = 1 - F[D_1] \qquad (1)$$

Structural Assessment of Cracked Flexible Pavement 279

3.1 Determination of Probability Distribution and Cumulative Probability Distribution of Center Deflection

Both probability distribution $f[D_1]$ and cumulative probability distribution $F[D_1]$ were determined separately for lane center and wheel path. Figure 1 and Figure 2 show the probability distribution of the center deflection at lane center and wheel path respectively. It is noted that both $f[D_1]$ are not in normal distribution and they are left-skewed. Maximum center deflection at wheel path is larger than that at lane center reflecting the effect of repeated wheel load along the wheel path. Figure 3 shows the cumulative probability distribution of center deflection at lane center and wheel path. The figures confirm that pavements on both lane center and wheel path are similar, with only slight difference in deflections between 250 μm and 500 μm.

3.2 Point FWD Test and Condition Code

Point FWD testing was performed at 56 core locations. Full core of the bituminous layer was retrieved at each of these locations. FWD testing was performed at these points before coring. These core locations covered a wide range of pavement thicknesses (from 55 mm to 436 mm), bituminous layer conditions (full-depth cracking to intact) and pavement temperatures (from 14 °C to 38 °C).

Condition Codes (from 1 to 5) aiming at categorizing the conditions of bituminous materials at the core locations were assigned to each core. An intact core with no visual defect and no cracking had a Condition Code of 5 (the best visual condition). An intact core with relatively minor defects, such as minor voiding but without cracking, had a Condition Code of 4. For those cores with one cracked sub-layer in the bituminous material, a Code of 3 was assigned to them. Cores with two cracked sub-layers had a Condition Code of 2. Those cores with the worst condition (i.e. with cracks in all bituminous sub-layers) had a Condition Code of 1.

Table 1. Definition of condition code of core

Condition Code	Definition
1	Cores with the worst condition (e.g. with cracks in all bituminous sub-layers)
2	Cores with two cracked sub-layers in the bituminous material
3	Cores with one cracked sub-layer in the bituminous material
4	Intact core with relatively minor defects, such as minor voiding but without cracking
5	Intact core with no visual defect and no cracking

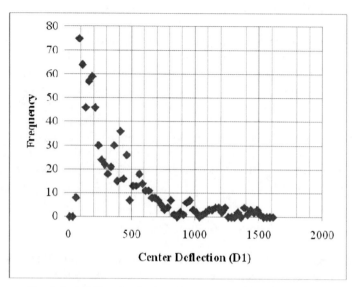

Fig. 1. Probability distribution of center deflection at lane center

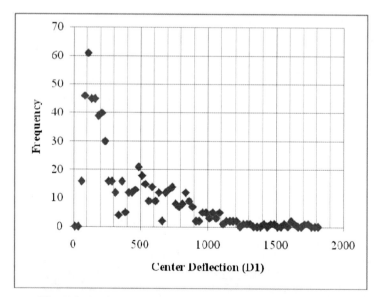

Fig. 2. Probability distribution of center deflection at wheel path

Structural Assessment of Cracked Flexible Pavement

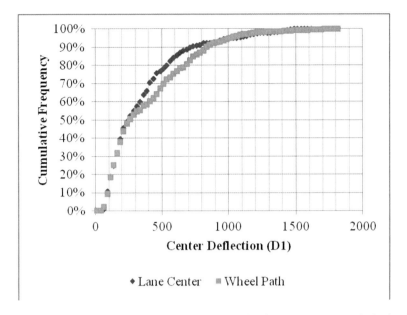

Fig. 3. Cumulative probability distribution of center deflection at lane center and wheel path

Processing of FWD data was conducted using the computer program ELMOD. ELMOD is an acronym for Evaluation of Layer Moduli and Overlay Design, which uses Boussinesq-Odemark pavement analysis approach for calculations. It back-calculated the stiffness moduli of the multilayered pavement. Method of back-calculation was based on the radius of curvature approach (Odemark-Boussunesq) used by the highway authoity in Hong Kong [4].

Figure 4 presents the relationship between the core condition and the back-analysed bituminous stiffness. As expected, the bituminous stiffness increased with reducing defect severity. When the stiffness value was above 2000 MPa, none of the cores were in the Condition Codes of 1 and 2. When the stiffness value increased to 5000 MPa, only cores of bituminous materials in the Condition Codes of 4 and 5 were noted.

Table 2. Condition code included within different stiffness region

Stiffness (MPa)	Condition	Condition Code Included
>5000	Good	4 & 5
2000-5000	Fair	3, 4 & 5
<2000	Poor	1 & 2

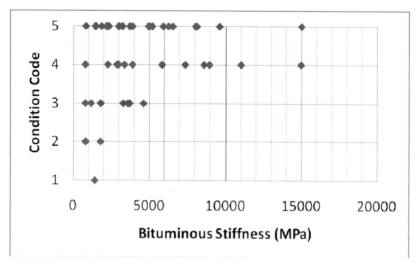

Fig. 4. Relationship between condition code and bituminous layer stiffness

3.3 Relating Center Deflection and Back-Analysed Stiffness Modulus

Figure 5 and Figure 6 show the relationship between center deflection and back-analysed stiffness modulus. In the figures, the thin solid line represents the regression line for all data points and the thick dash-dotted lines indicate the boundaries of the 95% confident interval. The D_1 deflection values corresponding to 2000 MPa and 5000 MPa at the lower 95% confident level limit would be 385 μm and 127 μm respectively for the lane center, while 363 μm and 138 μm for the wheel path (Table 3). Taking the conservative approach, the minimum deflection value of 127 μm might be used to represent good core condition (with stiffness value at least 5000 MPa) and the minimum deflection value of 363 μm might be used to represent bituminous materials at least in the fair condition (without multiple cracked sub-layer).

Table 3. Deflection at specific stiffness level

Stiffness (MPa)	Deflection		
	Lane Center (μm)	Wheel Path (μm)	Minimum (μm)
2000	385	363	363
5000	127	138	127

Structural Assessment of Cracked Flexible Pavement

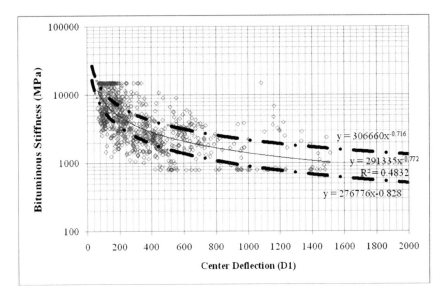

Fig. 5. Back-analysed stiffness modulus against center deflection at lane center

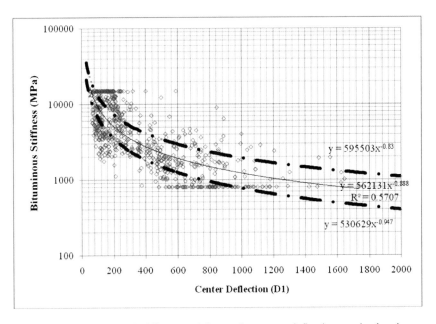

Fig. 6. Back-analysed stiffness modulus against center deflection at wheel path

3.4 Determination of Structural Condition Index

Figure 7 shows the Structural Condition Index against center deflection at lane center and wheel path. When center deflection is 127 µm, the corresponding SCIn's of both lane center and wheel path are about 75%. When the center deflection is 363 µm, the corresponding SCIn's of both lane center and wheel path are about 40%. Hence, flexible pavements having a SCIn between 75% and 100% are likely in good condition (i.e. with intact core without any cracked sub-layer); pavements of SCIn below 40% are likely in poor condition (with two cracked sub-layers or more). The pavement condition in relation with the SCIn range is summarized in Table 4.

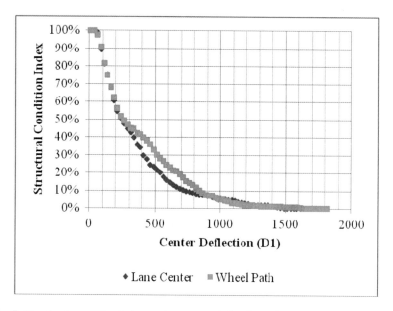

Fig. 7. Structural condition index against center deflection at lane center and wheel path

Table 4. Structural condition index

Condition of Bituminous Layer	Crack Condition	Deflection (µm)	Structural Condition Index
Good	no crack in all sub-layers	0-127	75-100
Fair	might have one cracked sub-layer	>127-363	40-75
Poor	cracks in at least two sub-layers or full depth crack	>363	0-40

4 Conclusion

A simple method using the center deflection of FWD to assess the severity of cracking in the bituminous materials of flexible pavement for the local condition in Hong Kong has been suggested. It classifies the materials in three categories (good, fair and poor). Good represents pavements likely in intact condition. Fair represents pavements having one cracked sub-layer at most. Poor represents pavements likely to have two cracked sub-layers or have full depth cracking problem.

References

[1] Highways Department, HKSAR: Hong Kong Road Network, Highways Department, Hong Kong Special Administrative Region Government, Hong Kong (2011), http://www.hyd.gov.hk/eng/major/road/road/road.html
[2] Highways Department, HKSAR, Highways Fact Sheets "Road Maintenance", Highways Department, Hong Kong Special Administrative Region Government, HK, http://www.hyd.gov.hk/eng/public/publications/factsheet/index.htm
[3] Highways Department, HKSAR: Highways Fact Sheets "Highways - Hong Kong: The Facts". Information Services Department, Hong Kong Special Administrative Region Government, Hong Kong (2010), http://www.gov.hk/en/about/abouthk/factsheets/docs/highways.pdf
[4] Research & Development Division, Highways Department, HKSAR. Guidance Notes on Backcalculation of Layer Moduli and Estimation of Residual Life Using Falling Weight Deflectometer Test Data RD/GN/027A, Highways Department, Hong Kong Special Administrative Region Government, Hong Kong (2009)
[5] Noureldin, S., Zhu, K., Harris, D., Li, S.: Non-Destructive Estimation of Pavement Thickness, Structural Number, and Subgrade Resilience Along INDOT Highways, Indiana Department of Transportation and the U.S. Department of Transportation Federal Highway Administration, United States (2005)

Comparison between Optimum Tack Coat Application Rates as Obtained from Tension-And Torsional Shear-Type Tests

Salman Hakimzadeh, Nathan Abay Kebede, and William G. Buttlar

University of Illinois at Urbana-Champaign

With the increased usage of HMA overlays in pavement rehabilitation, research on interface bonding between adjacent layers of HMA pavement has gained considerable attention. Pavement layers constructed with insufficient bonding can result in a number of pavement failure modes including slippage cracking, top-down cracking, premature fatigue cracking, and delamination. Despite the significance of interface bonding on pavement performance, selection of tack coat type and application rate is still based on experience and engineering judgment. Until now, most of the studies conducted to evaluate the bonding between different layers of HMA pavement have been based on shear-type interface tests. Considering that pavement interface failure can be attributed to both shear and tension modes, tension-type tests are also needed to truly optimize the process of selecting and designing the tack coat system. Recent studies have shown that the optimum tack coat rate determined by shear-type interface tests may be lower than the optimum tack coat rate determined through testing in tension. The purpose of this study is to make a comparison between optimum tack coat application rates as obtained from Torque Bond Test and a new tension-type interface test called the Interface Bond Test (IBT). The variables considered in this study include: test mode, tack coat type, and tack coat application rate. For the mixtures and bonding materials evaluated herein, the optimum tack coat application rate was found to be approximately twice as high for maximizing tensile bond fracture energy as compared to torsional shear. Implications and recommendations for further study are discussed.

1 Introduction

Pavement structures are composed of different HMA layers; therefore, the life and performance of the pavement depends not only on the properties of each layer such as stiffness, modulus, and fracture energy but also on the quality of bonding between adjacent layers. Whenever the adjacent layers are not completely bonded together the stress distribution, magnitude, and location of critical responses, such as tensile strain, will be significantly different as compared to when the layers are fully bonded [1]. Poor bonding between different layers of HMA can result in reduction of pavement capacity to withstand traffic and environmental loading resulting in different types of distresses that can reduce the pavement life by 40 to 80 percent [2]. Some of the distresses that have been identified by researchers include slippage cracking, premature fatigue cracking, top-down cracking, band-type reflective cracking (Figure 1), potholes, surface layer delamination, and increased difficulty to achieve compaction [1]. In practice, bonding between adjacent layers of a pavement structure is typically attempted by spray-application of a thin film of bituminous material between layers, termed 'tack coat.' Some of the important factors affecting the quality of bonding between adjacent layers of pavement structure include tack coat type, tack coat application rate, mixture type, surface cleanliness, surface texture, temperature, and moisture [3].

Fig. 1. Band-type reflective cracking on US 136 near Peoria, IL

A number of test methods have been proposed to evaluate pavement bonding, which can be categorized into three groups: direct shear tests, tension tests, and torsional shear tests [3]. Figure 2 provides a schematic representation of these interface testing modes.

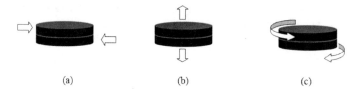

Fig. 2. (a) Direct shear, (b) Tension, (c) Torsional shear

Some of the common test methods that are in use today are the UTEP Pull-off Device (UPOD) [4], the ATacker Test [5], the Superpave Shear Tester (SST) [6], the Torque Bond Test [7], the NCAT shear test [8], the Leutner Test [9], the FDOT shear tester [10], and the direct shear test [11]. Until now, shear-type tests have been most commonly employed to evaluate bond. However, interface debonding can be attributed to both shear and tensile type failure mechanisms. Considering that interface debonding can result from both shear and tension modes, investigation on tension-type failures have gained attention recently.

It has been reported that the optimum tack coat application rate as obtained from shear type tests (such as the Direct Shear Test) is different than that which is obtained from tension type tests [12]. The purpose of this paper is: (a) to introduce a fracture energy based interface bond test (IBT) which appears to be a practical method to evaluate the bonding between adjacent layers of pavement in tensile mode, and; (b) to compare optimum tack coat application rates as obtained from tensile type test and torsional shear type tests. The scope of this study was on laboratory compacted, fine-graded asphalt mixture specimens with two different tack coat types (polymer modified and non-polymer modified) and residual application rates in the range of 0 to 0.68 L/m^2 (0 to 0.15 gal/yd^2). From these results, conclusions regarding the differences in tack coat application rates associated with optimization using either test were drawn.

2 Interface Bond Test (IBT)

A new Interface Bond Test (IBT), developed by Hakimzadeh et al., was utilized in this study [12] to evaluate tensile bond. IBT is a fracture energy based test that provides a fundamental characterization of tensile bond using a classic fracture testing mode (compact tension). One of the advantages of the test is its ability to evaluate the bonding between adjacent layers of a pavement which involves thin layers (as thin as 19 mm (3/4 in.)) of HMA. This is an important feature, as agencies are increasingly moving to thinner, high-performance overlay systems to maximize rehabilitation funds and to maintain surface properties and to preserve underlying pavement structure. In addition, the IBT generates fundamental tensile fracture data that can be readily used in computational models in order to facilitate system optimization and linkage between material properties and field performance. The test is relatively easy to perform and the specimens can be fabricated from field cores or laboratory prepared cylindrical samples. IBT specimens can be cut from a 150 mm (6 in.) diameter cylinder to the desired thickness using a water-cooled masonry saw. After cutting the edges, the notch and grooves can be fabricated using a water-cooled masonry saw with a 1 mm wide blade. The final specimen dimensions are shown in Figure 3. Additional details regarding the sample preparation and development of the IBT test have been previously reported [12].

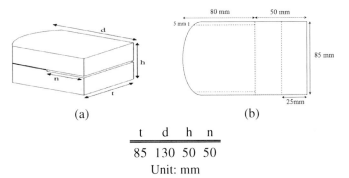

t	d	h	n
85	130	50	50

Unit: mm

Fig. 3. IBT Specimen dimensions (a) isometric view, (b) plan view

The IBT involves the application of tensile loading through pins inserted into the loading holes located in aluminum platens and the measurement of Crack Mouth Opening Displacement (CMOD) with a clip-type extensometer (Figure 4). In order to provide stable post-peak fracture, the test is controlled through a constant CMOD rate. The interface fracture energy is calculated by determining the area under the Load-CMOD curve and normalizing that quantity by the fracture surface area.

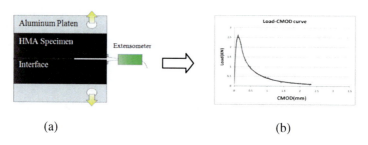

(a) (b)

Fig. 4. (a) Elevation view of experimental test setup and (b) Load-CMOD curve

3 Torque Bond Test

In order to evaluate the interface bonding in torsional shear mode, the Torque Bond Test, initially developed in Sweden for the in-situ evaluation of bond strength, was used in this study [13]. In this test, after coring the pavement to about 20 mm below the interface, torque is applied manually to the top of the overlay surface until interface failure occurs. The maximum torque that the interface can tolerate before failure occurs is measured and used to calculate the interface bond strength. In order to conduct the test in laboratory, fabricated specimens are clamped below the interface using a gripping unit. After gluing a steel plate to the top of the overlay using epoxy, a torque wrench is mounted to the steel plate and the torque is applied until the interface failure occurs [14]. The force required for failure is recorded, along with the location of the failure and temperature of the interface. The bond strength for the specimen is finally calculated using the following equation:

$$\tau = \frac{12M \times 10^6}{\pi D^3}$$

Where,
τ = interlayer bond strength (kPa)
M = peak Value of applied shearing torque (N·m)
D = diameter of core (mm)

4 Experimental Study

Five sets of specimens were used in this study, representing residual tack coat application rates of 0, 0.09, 0.23, 0.46, and 0.68 L/m^2 (0, 0.02, 0.05, 0.1, and 0.15 gal/yd^2). The tack coat materials used in this study were Trackless (non-polymer modified) and SS-1hp (slow setting polymer modified) (with asphalt residues of 58% and 61% by volume, respectively) obtained from a local emulsion supplier. A 19 mm (3/4 in.) nominal maximum aggregate size HMA mixture with 5.6% asphalt binder (PG64-22) was obtained from a counter flow drum-type plant at Open Road Paving in Champaign, Illinois. The mixture gradation is presented in Table 1.

Table 1. Mix gradation

Sieve size (mm)	25	19	12.5	9.5	6.25	4.75	2.36	1.18	0.6	0.3	0.15	0.075
% passing	100	96.1	84.2	76.8	62.5	52.9	33.9	21.2	13.3	7.7	5.3	4.5

In order to prepare the specimens in the laboratory, the loose mix was compacted into a cylinder of 50 mm (2 in.) height and 100mm (4 in.) and 150mm (6 in.) diameter for the Torque Bond Test and IBT, respectively, using a superpave gyratory compactor. This represents the bottom half of the test specimen. The tack coat material was then carefully applied using a paint brush on the surface of the specimen. Tack coat application rate was accurately controlled by measuring the weight of the specimen before and during application of the tack coat. Loose mix was then placed on the tack-coat treated specimen and was compacted again in the Superpave gyratory compactor to produce a 100mm (4 in.) tall composite specimen.

5 Results

After cutting the gyratory compacted specimens into the IBT geometry, drying to ambient moisture conditions, affixing the aluminum platens using an epoxy with 2,600 psi tensile strength, and mounting gage points onto the specimens, the specimens were placed in an environmental chamber at a conditioning temperature of -12°C for 2 hours. The IBT test then was performed with CMOD rate of 0.5 mm/min. Additional details regarding the testing temperature and loading rate have been previously reported [18]. All IBT testing was performed with an Instron 8500 servo-hydraulic load frame with an environmental chamber capable of controlling temperature to within ±0.2°C. Load was measured with a 10kN load cell and the crack mouth opening displacement (CMOD) was measured with an epsilon model 3541-0020-250-ST clip-on gage. This equipment is the same as that used to develop the ASTM D7313 DC(T) test protocol [15], or disc-shaped compact tension test, located at the Advanced Transportation Research and Engineering Laboratory (ATREL) in Rantoul, IL. Figure 5 shows the experimental setup with the loading fixtures and clip-on CMOD gage.

Fig. 5. IBT test setup and failed specimens

The Load-CMOD curves obtained from the tests were recorded and fracture energies were calculated as shown in Figure 6.

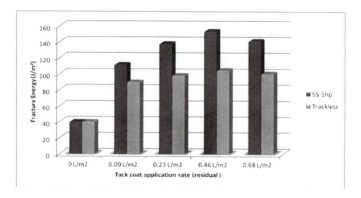

Fig. 6. Averaged interface fracture energies for different residual tack coat application rates

According to Figure 6, as the residual tack coat application rate increases, the interface fracture energy increases until it reaches to a peak value (153 and 98 J/m2 for SS-1hp and Trackless, respectively) and then decreases. Thus, the optimum tack coat application rate from the standpoint of fracture energy from tensile type interface testing occurs at a residual application rate of about 0.46 L/m^2 (0.1 gal/yd^2) for both SS-1hp and Trackless tack coat material. In addition, Figure 6 obviously shows that the SS-1hp tack coat material has superior interface fracture energy at low temperature as compared to the Trackless tack coat. Considering that the bulk material fracture energy of the same HMA mixture tested using IBT on a homogenous, notched specimen (no interface) was measured to have 301 J/m^2, it was observed that the SS-1hp and Trackless tack coat material provided about one-half and one-third of the bulk fracture energy, respectively. The Coefficients of Variation (COV) of the results range between 5 and 15% showing that the test provides acceptable repeatability. It should be noted that the obtained results are based on one HMA mixture, one testing temperature, and one loading rate. Further investigation on a wider range of HMA mixtures, testing temperatures and loading rates is recommended.

As mentioned earlier, one of the advantages of the IBT is that it represents a classic fracture type test for a quasi-brittle material system (notched specimen with controlled crack mouth opening), which can be coupled with numerical modeling to obtain a more fundamental understanding of interface debonding and to simulate debonding in actual pavement structures. Therefore, although testing bituminous material at low temperatures should not be viewed as a replacement for all other interface bond tests, it opens the door for the use of scientific tools such as fracture mechanics in the evaluation of the bonding between adjacent layers of a pavement structure.

The Torque Bond Test was performed at a temperature of 22°C. The torque wrench is manually rotated across an angle of 90° (parallel to the specimen's surface) within about 30 seconds, according to British Board of Agreement (BBA) guideline [14]. The laboratory test set-up and a typical failed specimen after testing are shown in Figure 7. The obtained peak values of applied shearing torque were then recorded and the torque bond strengths were calculated. Figure 8 shows the torque bond strength results.

Fig. 7. Torque Bond Test laboratory setup and failed specimen

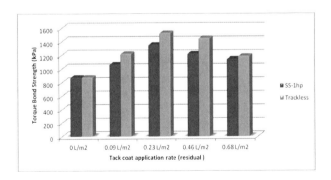

Fig. 8. Averaged torque bond strength for different residual tack coat application rates

According to Figure 8, the optimum tack coat application rates as obtained from torsional shear-type testing occurred at a residual tack coat application rate of 0.23 L/m^2 (0.05 gal/yd^2) for both SS-1hp and Trackless tack coat material. Moreover, Figure 8 suggests that the Trackless tack coat provides a better shear strength as compared to SS-1hp, which is in a good agreement with previous findings presented in the literature. The COV of the results range between 10 and 30%, indicating that the test is not as repeatable as other interface bond tests. The relatively high COV is probably due to the manual nature of load application, which produces a non-uniform loading rate. In addition, bending stresses result from the inability of the operator to apply a perfectly in-plane torsional load. Finally, the specimen rotates slightly within the clamp while torque is being applied, which can lead to variability in loading profile and loading rate.

6 Discussion

For the mixture and tack coat materials investigated herein, it was found that the tack coat application rate needed to create peak (or optimum) torsional shear strength at intermediate temperature was about half as much as that required to create optimum tensile fracture energy at low temperature. This is in agreement with previously reported findings, although this study represents the first comparison of a shear-type test to tensile fracture energy as obtained with the IBT. In addition, it is noteworthy to observe that the ranking of tack coat materials was reversed when comparing tension versus shear results. Thus, tack coat materials that provide a high shear bond strength at intermediate temperatures may not provide a high degree of tensile fracture energy at low temperatures. As another means of comparison, the COV measured in the IBT was significantly better (about half as large) as that measured in the torque bond test. On the other hand, the torque bond test has the advantage of simplicity and applicability to in-situ testing.

The results suggest that availability of shear and tensile type bond tests allow the pavement designer to better tailor tack coat type and application rate to each specific application. For the rehabilitation of uncracked or unjointed underlying pavement, and in warmer climates, maximization of interface shear properties would probably be given a higher weighting factor as compared to tensile properties. For the rehabilitation of cracked pavement structures, and in colder climates, the maximization of tensile bond properties would probably be given a higher weighting factor as compared to shear properties. An added benefit of the higher tack coat rate associated with the maximization of tensile bond fracture energy is the possibility of enhanced reflective and thermal cracking resistance of the overlay system. This is particularly true in the case of gap-graded, spray-paver applied bonded overlay systems, as reported by Ahmed et al. [16, 17]. The spray paver system combined with a gap-graded overlay mixture allows the use of even higher tack coat applications used in this study (more than twice the maximum amount used in this study), as the gap-graded mixture allows the tack coat material to wick upward into the mixture. When polymer-modified emulsion tack material is used, the fracture resistance of the overlay can more than double. Additional research is needed to determine a systematic approach (testing suite) to optimize tack coat material and application rate on the basis of providing sufficient shear and tensile properties, and in some applications, to maximize the fracture properties of the overlay itself in combination with the optimization of interface bond.

7 Conclusion and Recommendation

In this study, a new fracture energy based Interface Bond Test (IBT) was introduced. The IBT is a controlled fracture test which provides stable crack propagation and characterizes the bond in terms of interface fracture energy. The test was conducted on lab-prepared specimens produced with different tack coat application rates and tack coat materials. The IBT was then compared with a torsional shear type interface bond test (Torque Bond Test) and the obtained optimum tack coat application rates were compared. The following conclusions can be drawn based on this study:

- The IBT test is a practical method to evaluate the bonding between adjacent layers of pavement in terms of interface tensile fracture energy. The results of the laboratory study conducted herein demonstrated the ability of the test to clearly distinguish between the interface bonding of samples produced with different tack coat application rates and modified vs. unmodified tack coat material.
- The optimum tack coat application rate as obtained from torsional shear type test occurred at the residual tack coat application rate of 0.23 L/m^2 (0.05 gal/yd^2) for both SS-1hp and Trackless tack coat material, while the IBT test produced an optimum rate of 0.46 L/m^2 (0.1 gal/yd^2). This finding was in agreement with previous findings presented in the literature.
- The ranking of tack coat materials was reversed when comparing tension versus shear results. Thus, tack coat materials that provide a high degree of shear bond strength at intermediate temperatures may not provide a high degree of tensile fracture energy at low temperatures.

Considering that pavement interface failure can be attributed to both shear and tension modes, tension-type tests are also necessary to truly optimize the process of selecting and designing the tack coat system. Since interface fracture energy (as measured in the IBT) is a good indicator of bonding between adjacent layers of pavement structure, it is recommended that tension-type tests such as IBT be used along with shear-type tests in the design and control of tack coats and thin-bonded overlay systems. It is also suggested that more studies including pavement numerical modeling be performed to gain additional insight towards the interface debonding mechanisms in pavement structures and its mitigation. Additional research is underway to determine a systematic approach (testing suite) to optimize tack coat material and application rate on the basis of providing sufficient shear and tensile properties, and in some applications, to maximize the fracture properties of the overlay itself in combination with the optimization of interface bond.

Acknowledgement. The authors gratefully acknowledge the support provided by colleagues at the Illinois Center for Transportation and Road Science LLC in the course of this study. The views and opinions expressed in this paper are those of the authors who are responsible for the facts and accuracy of the data presented here and do not necessarily reflect the views and opinions of the sponsor.

References

[1] Romanoschi, S.: Characterization of Pavement Layer Interfaces, Ph.D. Dissertation, Louisiana State University, Baton Rouge (1999)
[2] Khweir, K., Fordyce, D.: Influence of Layer Bonding on the Prediction of Pavement Life. In: Proceedings of the Institution of Civil Engineering Transport, vol. 156, pp. 73–83 (2003)
[3] Leng, Z., Ozer, H., Al-Qadi, I.L., Carpenter, S.H.: Interface Bonding Between Hot-Mix Asphalt and Various Portland Cement Concrete Surfaces: Laboratory Assessment. Journal of the Transportation Research Board (2009)
[4] Eedula, S.R., And Tandon, V.: Tack Coat Field Acceptance Criterion, FHWA/TX-06/0-5216-1, Center for Transportation Infrastructure Systems, The University of Texas, El Paso (2006)
[5] Atacker, T.M.: A Tack Coat Testing Device, Operator's Guide. InstroTek, Inc. (2005)
[6] Mohammad, L., Huang, B., Raqib, M.: Influence of Asphalt Tack Coat Materials on Interface Shear Strength. Journal of the Transportation Research Board, No.1789, 56–65 (2002)
[7] Choi, Y.K., Sutanto, M.H., Collop, A.C., Airey, G.D.: Bond Between asphalt layers. Project Report to the UK Highways Agency, Scott Wilson Pavement Engineering Ltd., Nottingham, UK (2005)
[8] West, R., Zhang, J., Moore, J.: Evaluation of Bond Strength between Pavement Layers. NCAT report 05-08 (2005)
[9] Leutner, R.: Untersuchung des schichtenverbundes beim bituminosen oberbau. Bitumen 41(3), 84–91 (1979)
[10] Tashman, L., Nam, K., Papagiannakis, T.: Evaluation of the Influence of Tack Coat Construction Factors on the Bond Strength between Pavement Layers. Washington center for asphalt technology, Pullman (2006)
[11] Donovan, E.P., Al-Qadi, I.L., Loulizi, A.: Optimization of Tack Coat Application Rate for Geocomposite Membrane on Bridge Decks. Journal of the Transportation Research Board, No.1740, 143–150 (2000)
[12] Hakimzadeh, S., Kebede, N.A., Buttlar, W.G., Ahmed, S., Exline, M.: Development of Fracture Energy Based Interface Bond Test for Asphalt Concrete. Journal of Road Materials and Pavement Design (RMPD) 81 (2012)
[13] Walsh, I.D., Williams, J.T.: HAPAS certificates for procurement of thin surfacing. Highways and Transportation 48(7-8), 12–14 (2001)
[14] British Board of Agreement: Guidelines Document for the Assessment and Certification of Thin Surfacing Systems for Highways. SG3/05/234, British Board of Agreement, Watford, UK (2004)
[15] ASTM D7313– 07 a Standard Test Method for Determining Fracture Energy of Asphalt-Aggregate Mixtures Using the Disk-Shaped Compact Tension Geometry, vol. 4(3), ASTM, Road and Paving Materials

[16] Ahmed, S., Dave, E.V., Buttlar, W.G., Exline, M.: Fracture Properties of Gap & Dense Graded Thin Bonded Overlays. Journal of the Association of Asphalt Paving Technologists 79, 443–472 (2010)
[17] Ahmed, S., Dave, E.V., Behnia, B., Buttlar, W.G., Exline, M.: Fracture Characterization of Gap-Graded Thin Bonded Wearing Course. In: Proceedings of the Second International Conference Environmentally Friendly Roads (ENVIROAD), Warsaw, Poland (2009)
[18] Hakimzadeh, S., Buttlar, W.G., Santarromana, R.: Evaluation of Bonding between HMA Layers Produced with Different Tack Coat Application Rates using Shear-type and Tension-type Tests. Journal of the Transportation Research Board (2012)

Using Life Cycle Assessment to Optimize Pavement Crack-Mitigation

Ali Azhar Butt, Denis Jelagin, Björn Birgisson, and Niki Kringos

Division of Highway and Railway Engineering, Transport Science Department,
KTH Royal Institute of Technology, Sweden

Abstract. Cracking is very common in areas having large variations in the daily temperatures and can cause large discomfort to the users. To improve the binder properties against cracking and rutting, researchers have studied for many years the behaviour of different binder additives such as polymers. It is quite complex, however, to decide on the benefits of a more expensive solution without looking at the long term performance. Life cycle assessment (LCA) studies can help to develop this long term perspective, linking performance to minimizing the overall energy consumption, use of resources and emissions. To demonstrate this, LCA of an unmodified and polymer modified asphalt pavement using a newly developed open LCA framework has been performed. It is shown how polymer modification for improved performance affects the energy consumption and emissions during the life cycle of a road. Furthermore, it is concluded that better understanding of the binder would lead to better optimized pavement design, hence reducing the energy consumption and emissions. A limit in terms of energy and emissions for the production of the polymer was also found which could help the polymer producers to improve their manufacturing processes, making them efficient enough to be beneficial from a pavement life cycle point of view.

1 Introduction

Problems like low temperature cracking and fatigue cracking have always been an issue in cold regions like the European Nordic countries [1]. Cracking is in fact very common in areas having large variations in the daily temperatures and can cause large discomfort to the users due to uncomfortable rides and disturbances caused by frequent maintenance periods. It also increases the cost to the society, as often higher taxes will have to be paid to overcome increased number of maintenance actions. Improvement of crack-mitigation in asphalt pavements could therefore have a significant contribution to the society at large.

The rheological properties of bitumen have an important effect on the cracking of the asphalt mixtures, since they provide the glue of the aggregate skeleton [2]. To improve the binder properties against cracking and rutting, researchers have studied for many years the behaviour of different binder additives such as polymers [3-5]. The benefit of using polymers to modify the binder properties is well established but to quantify the long term benefit, an investigation of the effect of this modification over the entire life time of the pavement should be made. Life Cycle Assessment (LCA) tools can therefore be utilized.

Due to the depletion of resources and concerns of climatic change, LCA for different products, systems and activities have increased in popularity among researchers for the past years. LCA studies can help to determine and minimize the energy consumption, use of resources and emissions to the environment by giving a better understanding of the systems. LCAs can also purpose different alternatives for different phases of a life cycle of the system. Unfortunately, LCA has not yet been adopted by the industry or the road authorities as part of the procurement and material selection procedure. This could partly be explained due to the lack of a technical tool that accurately represents all the aspects of the pavement sector and is able to make close predictions of the in-time pavement response. For this reason, a new technical LCA framework is being developed. This paper is giving an example of the application width of such a tool for the case of prevention of low-temperature cracking in asphalt pavements.

1.1 Low Temperature Properties of Asphalt Concrete

To improve the quality of our roads and prevent pavement distresses such as cracking and rutting, certain measures can be taken. For example, improved road design, optimal use of materials or improving mixtures properties as a whole. Polymers like Styrene-Butadiene-Styrene (SBS) and natural rubbers are often used in the pavement industry to enhance the properties of the asphalt mixtures against premature damage. Polymers have the ability to create a secondary network or a balance system in the bitumen by either molecular interactions or react chemically with the bitumen [6]. Several studies have concluded that adding small amount of polymer (3-6% depending on what type of polymer is used) usually results in dispersed polymer particles in the continuous bitumen matrix and improves the properties of the binder against rutting and cracking [3-5, 7-9].

Due to heavy loads on the pavements and inefficient maintenance operations, roads sometimes deteriorate much quicker than expected. This directly leads to increased energy usage, higher cost and more emissions to the environment. It is therefore in favour of all stakeholders to optimize the efficiency of the maintenance operation over the lifetime of the pavement as much as possible. To achieve this, different case studies and possibilities are to be studied based on different design alternatives. Hence, an approach is required which could help in decision support during the lifetime of the pavement.

1.2 Development of an Open LCA Framework

LCA is a versatile tool to investigate the environmental aspect of a product, a service, a process or an activity by identifying and quantifying related input and output flows utilized by the system and its delivered functional output in a life cycle perspective [10]. Ideally, it includes processes from the cradle to the grave of a product. In the case of asphalt pavements, the cradle can be the extraction of materials and the grave can be the burial of the asphalt pavement in the sub-grade. Use of resources and environmental loads can be reduced by studying the effects and the impacts on the environment during the different phases of a road's lifetime. A new open LCA framework for asphalt pavements was recently developed by Butt et al. [11] that considers energy consumption and emissions produced during the lifetime of the pavement. The LCA framework is fed the output from pavement design tools which are then processed to quantify energy, raw materials and emissions during different phases of a road's life time. The functional unit was defined as the construction, maintenance and end of life of 1 km asphalt pavement per lane for a nominal design life.

Certain system boundaries have to be assumed while developing the LCA framework. The study was focused on the project level, therefore it was assumed that the road location was known and the use of the land for some other purpose was not considered. Furthermore, the thickness of the asphalt layer was assumed to be constant along the length of the road per functional unit. Fuel and electric energies were accumulated separately for different processes in the lifetime of a road. This assumption was necessary because electricity being a secondary energy source could only be added to the fuel energy if the electricity production energy and efficiency are known. The raw materials considered for the framework are bitumen, aggregate and additives like waxes and polymers.

1.3 Pavement Design (Mechanistic Calibrated MC Model)

A calibrated mechanistic design tool used in this study has recently been evaluated for Swedish conditions [12]. The analysis and design framework presented by Gullberg et al. [12] is an extension of the earlier work by Birgisson et al. [13], in which a framework for a pavement design against fracture based on the principles of viscoelastic fracture mechanics has been reported. One key observation regarding this approach is that each mix is evaluated based on its dissipated creep strain energy limit ($DCSE_{lim}$), which is a measure of how much damage mixture can tolerate before a non-healable macro-crack forms. In a design procedure the $DCSE_{lim}$ acts thus as a threshold between healable micro-cracks and non-healable macro-cracks. This is a threshold that has proven to be fundamental and independent of mode of loading [14].

In Romeo et al. [9], SuperPave indirect tension (IDT) tests were performed on unmodified and polymer modified asphalt mixtures. In this study it was found that the polymer modification results in a higher damage tolerance of the asphalt mixture, i.e. higher $DCSE_{lim}$. The impact of the $DCSE_{lim}$ increase on the design thickness is presently investigated with the design framework reported in Gullberg et al. [12]. All other material properties are assumed not to be affected by polymer modification.

1.4 Research Aims

In this paper, LCA of an unmodified and polymer modified asphalt pavement is performed using the newly developed open LCA framework. The effect of a polymer modification for crack resistance on the energy consumption and emissions during the life cycle of a road is investigated in this paper. The polymer production and transportation energy is also estimated in order to determine the benefit of polymer modification of asphalt pavements in terms of environmental costs.

2 LCA Case-Study

The design of the pavement section is based on the work by Almqvist [15]. The asphalt pavement thickness design is done for a lifetime of 20 years using a mechanistic calibrated pavement design model [12]. The pavement consists of a 50 mm thick wearing course above a structural course. The thickness of the structural course changes for different cases depending on the design. The base layer is 178 mm thick whereas the sub-base is 1.0 m lying on top of the bedrock. The design is done for a mean temperature of 5 °C (corresponds to Swedish climate zone 3) assuming the design equivalent single axle load (ESALs) to be 10e6.

The following three cases are analysed using the LCA framework: Simulations are performed with unmodified asphalt, SBS polymer modification and unknown polymer modification of asphalt which results in 0%, 50% and 100% increase of the $DCSE_{lim}$, respectively. SBS polymer enhances the properties of the asphalt against rutting and cracking [8-9]. For the case 2, 3.5% SBS polymer modified asphalt has been considered [9]. With the addition of 3.5% SBS to the unmodified asphalt, IDT tests have shown that the $DCSE_{lim}$ changes from 3.57 to 5.34 kJ/m^3. Hence an increase of almost 50% is achieved. For case 3, it is assumed that the unknown polymer is 3.5% by weight of the blend and provides 100% increase in the $DCSE_{lim}$. The thicknesses of asphalt layers used for the LCA are as shown in Table 1. For the analyses, the total asphalt pavement thickness has been considered containing 5.2% binder content. The construction, and bitumen and aggregates storage sites are considered to be 25, 75 and 35 km from the asphalt plant, respectively. The emissions from electricity and diesel production are as inventoried by Stripple [16]. Energy consumption data for the asphalt production was acquired from Skanska, a large Swedish contractor. It is also assumed that an increase of 17% in fuel consumption is required for polymer modification of the asphalt mixture. The functional unit (FU) defined for the study is construction of 1 km of asphalt pavement for a nominal design life. Lane width is selected to be 4 m wide.

Table 1. Asphalt pavement layer thicknesses for different cases

Cases	Description	Increase in $DCSE_{lim}$ (%)	Wearing Course Thickness (mm)	Structural Course Thickness (mm)	Total asphalt pavement Thickness (mm)
1	Unmodified asphalt	0	50	100	150
2	Unmodified asphalt with 3.5% SBS	50	50	69	119
3	Unmodified asphalt with 3.5% unknown polymer	100	50	36	86

The comparison between Case 1 and Case 2, 3 will give insight into the added benefits in terms of reduced energy and greenhouse gas (GHG) emissions when polymer is added to the asphalt against crack resistance. Based on the results of the previous studies mentioned in the above, it was found that a small percentage of polymers not only provide resistance against cracking but also allows for the reduction of the asphalt layer thickness. This decrease in thickness itself will save energy and reduce emissions in a road's life cycle, but polymers production and transportation should also be considered in this number.

2.1 Results

The results of the LCA analysis are summarized in Table 2 and Table 3. Parameters a, b, c are the unknown energy values (in GJ) for the SBS whereas d, e, f are energy values (in GJ) for the unknown polymer which are associated with the electric, fuel and transportation energies, respectively. Parameters g, h, i and j are CO_{2-eq} values (in tonnes) for the polymer production and transportation. For Case 2, SBS polymer modification of asphalt led to an increase of 50% in the $DCSE_{lim}$ which resulted in a decrease of the structural course by 31%, assuming the same service life of the pavement. For the calculation of Case 3 it was assumed that 3.5% of an unknown polymer is added in asphalt which would increase the $DCSE_{lim}$ to 100% which lead to a decrease of 64% w.r.t. case 1 and a further decrease of almost 50% w.r.t. case 2. From Table 2 can be seen that the total used energy therefore reduces from 830 GJ (Case 1) to 700 GJ (Case 2) to 508 GJ (Case 3). From Table 3 can be seen that the total CO_{2-eq} reduces from 55 (tonnes) to 47 to 34, respectively. These values, however, still do not include the energy spent and emissions created when including polymers into the process. For this reason, in the following the thresholds will be determined for these.

Table 2. LCA results from the case study

Energy Consumed	Item	Energy Consumed per ton of material (MJ/ton)	Total Energy consumed (GJ)	Case Study 1 Σ Energy (GJ)	ETE (GJ)	% Energy consumed	Total Energy consumed (GJ)	Case Study 2 Σ Energy (GJ)	ETE (GJ)	% Energy consumed	Total Energy consumed (GJ)	Case Study 3 Σ Energy (GJ)	ETE (GJ)	% Energy consumed
Electricity	Bitumen Production	252.00	18.87	99	220	5.07%	14.45	78	173	4.60%	10.44	56	125	4.58%
	Polymer Production	-	-			-	a			-	d			-
	Aggregate Production	21.19	28.93			7.78%	22.95			7.31%	16.58			7.28%
	Asphalt Production	35.28	50.80			13.66%	40.30			12.83%	29.13			12.79%
Fuel	Bitumen Production	1060.00	79.37	610	610	9.57%	60.77	527	527	8.68%	43.91	383	383	8.65%
	Polymer Production	-	-			-	b			-	e			-
	Aggregate Production	16.99	23.19			2.80%	18.40			2.63%	13.30			2.62%
	Asphalt Production	242/(281 for case 2-3)	348.48			42.01%	321.18			45.86%	232.11			45.70%
	Bitumen transported* to the asphalt plant		9.57			1.15%	7.33			1.05%	5.30			1.04%
	Polymer transported* to the asphalt plant		-			-	c			-	f			-
	Aggregate transported* to the asphalt plant		81.46			9.82%	64.62			9.23%	46.70			9.20%
	Asphalt transported* to the construction site		61.37			7.40%	48.69			6.95%	35.19			6.93%
	Laying Asphalt		3.86			0.47%	3.86			0.55%	3.86			0.76%
	Compacting Asphalt		2.27			0.27%	2.27			0.32%	2.27			0.45%
	Total Process Energy =				830			700 + (2.23 x a) + b + c				508 + (2.23 x d) + e + f		

ETE (Equivalent Thermal Energy) factor for electricity is 2.23 MJ
* Transportation distances were doubled in the calculation as loaded trucks are empty on return.
a Electric energy required to produce SBS in GJ.
b Fuel energy required to produce SBS in GJ.
c Transportation fuel energy required to produce SBS in GJ.
d Electric energy required to produce unknown polymer in GJ.
e Fuel energy required to produce unknown polymer in GJ.
f Transportation fuel energy required to produce unknown polymer in GJ.

Table 3. Resulting emissions for different case studies

	CASE STUDY 1			CASE STUDY 2			CASE STUDY 3		
Emissions to air (tonnes)	CO_2	N_2O	CH_4	CO_2	N_2O	CH_4	CO_2	N_2O	CH_4
Bitumen production	12.95	7.94E-06	2.64E-06	9.92	6.08E-06	2.02E-06	7.17	4.39E-06	1.46E-06
Polymer production	-	-	-	g'	g''	g'''	i'	i''	i'''
Aggregate production	1.94	4.93E-05	5.21E-06	1.54	3.91E-05	4.13E-06	1.11	2.82E-05	2.99E-06
Asphalt production	27.72	5.79E-04	2.45E-05	25.53	5.31E-04	2.17E-05	18.45	3.84E-04	1.57E-05
Paving	0.31	6.18E-06	1.93E-07	0.31	6.18E-06	1.93E-07	0.31	6.18E-06	1.93E-07
Compacting	0.18	3.64E-06	1.14E-07	0.18	3.64E-06	1.14E-07	0.18	3.64E-06	1.14E-07
Transportation	12.04	2.44E-04	7.62E-06	9.53	1.93E-04	6.03E-06	6.89	1.39E-04	4.36E-06
Polymer transportation	-	-	-	h'	h''	h'''	j'	j''	j'''
Σ	55.14	8.90E-04	4.03E-05	47.00	7.79E-04	3.42E-05	34.10	5.66E-04	2.48E-05
CO_2-eq	55.41			47.23 + g + h			34.27 + i + j		

2.2 Polymer Production and Transportation

The polymers production and transportation energies are not included in Case 2 and 3, which should be considered to make an objective judgement of the long term effect of the modification. For this reason, in the following the thresholds of the energy and emission limits are determined for the polymer production and transportation based on the study's cases results (Table 4).

Table 4. Beneficial bitumen modification boundaries w.r.t. energy and emissions allocation

Energy spent on polymer	(GJ/FU)	Case 1 Vs Case 2	Case 1 Vs Case 3
ETE Electricity used/FU	a, d	<40.5	<103
Fuel consumption/FU	b, e	<78	<195
Transportation Energy/FU	c, f	<9.5	<24
Total Polymer Energy/FU		**<129**	**<322**
GHGs Emissions (tonnes)			
Polymer production/FU	g, i	<8	<20.5
Polymer Transportation/FU	h, j	<0.3	<0.7
Total Process Emissions		**<8.3**	**<21.2**

It was determined that for a polymer modification that increases the $DCSE_{lim}$ to 100%, the total sum of the energy and GHG emissions spent on polymer production and transportation should be less than 322 GJ/FU and 21 tonnes $CO_{2\text{-}eq}$/FU when comparing to the case of unmodified asphalt for the modification to be beneficial from an energy point of view. When compared to the SBS polymer modified asphalt, i.e. Case 2, the total energy and GHG emissions spent on the SBS should be less than 129 GJ and 8 tonnes $CO_{2\text{-}eq}$ to be beneficial per FU.

3 Conclusions and Recommendations

Use and effect of polymers (e.g. SBS) in asphalt mixtures to enhance performance is no more new to the asphalt industry. Polymer is known to enhance the properties of the binder, making it less vulnerable against rutting and cracking. In this paper, a newly developed LCA framework was used to determine the effect on polymer usage on the energy and emissions during the pavements lifetime. It was also observed that by enhancing the properties of the binder by adding polymer, led to a thinner pavement design for the same design life.

From the case study, it could be concluded that better understanding of the binder would lead to better optimized pavement design, hence reducing the energy consumption and emissions. A limit in terms of energy and emissions for the production of the polymer was also found which could help the polymer producers to improve their manufacturing processes making them efficient enough to be beneficial from a pavement life cycle point of view. In other words: positive effects obtained due to use of additives are only beneficial when energy and emissions are lower in comparison to the unmodified asphalt in a life cycle perspective.

It can be seen from the results of all the three studies that asphalt production was the most energy consuming process. Hence, the binder properties and the use of additives like polymers should be further studied. This could help in improving the binder properties against cracking and rutting and could also help in reducing the resource consumption, energy and emissions in the asphalt mix plant. Material transportation distances should also be kept as short as possible.

Acknowledgements. The authors would like to thank Dr. Susanna Toller for her expert advice in the development of the LCA framework.

References

[1] Zeng, H.: On the low temperature cracking of the asphalt pavements, PhD thesis, TRITA-IP FR 95-7, Royal Institute of Technology, KTH, Stockholm Sweden (1995)
[2] Isacsson, U., Zeng, H.: Journal of Materials Science 33(8), 2165–2170 (1998)
[3] Lu, X.: On polymer modified road bitumens, PhD thesis, TRITA-IP, Royal Institute of Technology, KTH, Stockholm Sweden (1997)
[4] Sengoz, B., Isikyakar, G.: Journal of Hazardous Materials 150(2), 424–432 (2008)
[5] Kumar, P., Chandra, S., Bose, S.: International Journal of Pavement Engineering 7(1), 63–71 (2006)
[6] Isacsson, U., Lu, X.: Materials and Structures 28, 139–159 (1995)
[7] Kim, S., Sholar, G.A., Byron, T., Kim, J.: Journal of the Transportation Research Board (2126), 109–114 (2009)
[8] Ping, G.V., Xiao, Y.: Challenges and Recent Advances in Transportation Engineering. In: ICTPA 24th Annual Conference & NACGEA International Symposium on Geo-Trans., Paper No. S2-001, Los Angeles, CA, USA (2011)
[9] Romeo, E., Birgisson, B., Montepara, A., Tebaldi, G.: International Journal of Pavement Engineering 11(5), 403–413 (2010)
[10] Baumann, H., Tillman, A.-M.: An Orientation in LCA methodology and application. In: The Hitch Hiker's guide to LCA, Studentlitteratur, Göteborg (2003)
[11] Butt, A.A., Mirzadeh, I., Toller, S., Birgisson, B.: International Journal of Pavement Engineering (2012) under review
[12] Gullberg, D., Birgisson, B., Jelagin, D.: International Journal of Road Materials and Pavement Design (2012) under review
[13] Birgisson, B., Wang, J., Roque, R.: Addendum to Implementation of the Florida Cracking Model into the Mechanistic-Empirical Pavement Design. University of Florida, Gainesville (2006)
[14] Zhang, Z., Roque, R., Birgisson, B., Sangpetngam, B.: Journal of the Association of Asphalt Paving Technologists 70, 206–241 (2001)
[15] Almqvist, Y.: Master thesis, TRITA-VBT 11:06, Royal Institute of Technology, KTH, Stockholm, Sweden (2011)
[16] Stripple, H.: Life Cycle Assessment of Road, A Pilot Study for Inventory Analysis. IVL Swedish Environmental Research Institute, Göteborg (2001)

Preliminary Analysis of Quality-Related Specification Approach for Cracking on Low Volume Hot Mix Asphalt Roads

David J. Mensching[1], Leslie Myers McCarthy[2], and Jennifer Reigle Albert[3]

1 Graduate Assistant, Villanova University, 800 East Lancaster Avenue, Villanova, PA, USA 19085
david.mensching@villanova.edu
[2] Assistant Professor of Civil Engineering, Villanova University, 800 East Lancaster Avenue, Villanova, PA, USA 19085
leslie.mccarthy@villanova.edu
3 Assistant Professor of Civil Engineering, Pennsylvania State University – Harrisburg, 777 West Harrisburg Pike, Middletown, PA, USA 17057
jaa23@psu.edu

Abstract. During the last twenty years, efforts have been made to implement performance-related specifications (PRS) for hot mix asphalt (HMA) construction in the United States. The National Cooperative Highway Research Program (NCHRP) Project 9-22: *Beta Testing and Validation of Hot Mix Asphalt Performance-Related Specifications* created software using models similar to those in the interim American Association of State Highway and Transportation Officials (AASHTO) Mechanistic-Empirical Pavement Design Guide (MEPDG). The program predicts an effective dynamic modulus (E*) parameter to determine information pertaining to major distress types. In this study fatigue cracking (bottom-up/alligator cracking) is analyzed. A predicted life difference (PLD) is then calculated between job mix formula (JMF) and as-built conditions, resulting in an assigned pay factor. In this study, a low volume HMA roadway in rural Rhode Island was analyzed using the volumetric-based models. A sensitivity analysis was conducted by varying asphalt contents, in-situ air void targets, and dust-to-asphalt ratios to evaluate their effects on fatigue cracking levels. The aim was to assess the suitability of the software as a tool for pay factor development in Rhode Island. Based on preliminary results, results are significantly sensitive to changes in JMF target in-situ air voids. Future considerations regarding pay factor development for low volume roadway projects include: added costs or savings as a result of implementation, the development of a more simplistic method of computing pay factors, and comparing results with pavement management system (PMS) information on other comparable flexible highway pavements in Rhode Island.

1 Introduction

This paper presents the results of a preliminary analysis involving a potential pay factor specification for traditional fatigue cracking on low volume hot mix asphalt

(HMA) roadways. Through National Cooperative Highway Research Program (NCHRP) Project 9-22: *Beta Testing and Validation of HMA PRS*, a software program geared towards the comparison of performance predictions for job mix formula (JMF) and as-built HMA characteristics was created for pay factor development [1]. Prediction models based on the Witczak Predictive Equation (WPE) and internally-conceived closed-form solutions (CFS) similar to those featured in the American Association of State Highway and Transportation Officials Mechanistic-Empirical Pavement Design Guide (MEPDG) software predict service life values, which directly relate a predicted life difference between the as-built and JMF lives to a distress-specific pay factor.

This study explored software predictions for alligator cracking pertaining to a full-depth reconstruction HMA project paved in the state of Rhode Island (RI) during the 2010 construction season. Rhode Island is the smallest state in the United States and located in the northeast region between New York City and Boston. The objectives were to utilize a statistical test measure (linear regression t-test) to determine the significance of changes in three critical parameters identified during construction: asphalt content by weight (AC%), dust-to-asphalt ratio (D/A), and target in-situ air voids (AV); and to provide conclusions and recommendations to determine other testing methodologies to further analyze the software's tendencies. If the volumetric parameters proved significant, further investigation could be conducted by agencies to adjust specification limits or values in an effort to maximize predicted life for the pavement in the fatigue cracking distress mode. A research approach was devised with the intent to perform the following tasks: 1) collect pertinent project data as software inputs, 2) develop statistical significance methodology, 3) conduct a preliminary sensitivity study using three critical volumetric factors identified by the Rhode Island Department of Transportation (RIDOT), and 4) provide recommendations for additional study with the optimal goal being to better understand prediction behavior for specification enhancement.

2 Background

Development of performance-related specifications (PRS) was initiated in 1988 with the publication of NCHRP Project 10-26 and the formation of a Transportation Research Board (TRB) steering committee, which identified PRS as a matter of high-priority in the area of asphalt research. The Federal Highway Administration (FHWA) then cited PRS to be a High Priority National Area with the objective to "develop and implement specification based on effective predictors of pavement performance with appropriate incentive/disincentive clauses based on those predictors" [2]. Many efforts have been made towards developing a conceptual framework for PRS [3, 4]. In a report by Epps et al. [4], the need for prediction models using mechanistic-empirical theory integrated as part of PRS resulted in a software application entitled *HMASpec*. *HMASpec* used a life-cycle cost factor to compute a pay adjustment [5] to construction project final contract amounts.

Aside from the efforts of the NCHRP Project 9-22 and Westrack projects, there were a few studies in particular that outlined the advantages of widespread PRS implementation and efforts on development of PRS on a more localized level. The advantages of a PRS were reported to vary from minimized life-cycle costs, to the consideration of lot variability, and the incentive/disincentive system that stems from these methods [3]. These factors provide benefits to both transportation agencies and contractors, as transportation agencies can set their own performance limits, obtaining a level of confidence that the pavement will not fail over a given period of time based on predictive models. Contractors could then produce a mix that would provide substantial performance characteristics resulting in an incentive, rather than risk producing an underperforming section resulting in a disincentive payment.

Explorations of localized (state-level) PRS for HMA construction were done in a study by Buttlar and Harrell [6]. The researchers stressed the importance of a framework for the progression to more advanced types of specifications, such as end-result specifications (ERS) and PRS, citing the need for statistical quality assurance (QA) procedures and a development of performance-based pay factors.

A study done for Arizona DOT (ADOT) sought to implement a PRS plan and characterization of HMA performance to be used in models for evaluation of HMA construction jobs in the state [7]. In the report, databases for prediction of the three critical distresses (rutting, alligator cracking, and thermal fracture) were constructed, as well as a framework for the ultimate development of a field validation procedure based on the Asphalt Mixture Performance Tester (AMPT). The study demonstrated that prediction models derived through database creation yield localized calibrations that will produce relevant results. This finding has the potential to be paramount to gaining the confidence of transportation agencies and the paving industry.

Over the course of the last decade, efforts have been made to drastically change the methods in which QA and incentive/disincentive specifications are utilized in transportation agencies. Through NCHRP Project 9-22, the preliminary research-grade software utilized in this study was developed to provide agencies with a resource for the implementation of a quality-related specification. The software has the capability to predict pavement service life as a tool for assigning incentives or disincentives to contractors, based on adherence to specifications and performance standards. Industry experts have expressed the need for performance-based evaluations of asphalt mixtures as a whole, believing that acceptance quality characteristics (AQCs) should be tied more to the overall performance of a pavement as opposed to strictly volumetric-based QA protocols. With the MEPDG selected as the basis for prediction models the software was created to compare service life factors for the as-designed and as-built conditions [1]. An enhanced version of the software program is still under development in NCHRP Project 9-22A: *Field Validation of QRSS Version 1.0*. It has been designed to operate in a systematic fashion, where JMF analysis is performed before as-built comparisons on pavement performance can be made. The input to the program includes mixture volumetrics, design features, traffic, and sampling data among other items. The output includes stiffness properties and measures of performance such as predicted life difference (PLD), distress, and pay factors for the project [1]. This study will utilize E* outputs only in attempts to test the statistical significance of changes as a result of changes in AC%, D/A, or AV.

3 Project Site Description

In this study, one HMA full-depth reconstruction section was selected for a trial application of quality-related specification analysis. The project, included as a study site for NCHRP Project 9-22A, was a 1.9 kilometer (km) (1.2 mile) stretch of Rhode Island-102 (RI-102) in Foster, Rhode Island. Figure 1 shows a cross-section of the pavement profile of RI-102 based on job specifications:

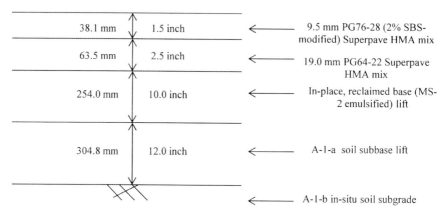

Fig. 1. Cross-section of RI-102 reconstruction project (not to scale)

This portion of the highway can be classified as a rural-principal arterial, with an annual average daily traffic (AADT) of 3,500 vehicles per day (vpd), based on information from RIDOT officials and AADT maps. RI-102 acts as a north-south throughway for western Rhode Island, traveling through several rural areas. The roadway features one lane in each direction and displays a posted speed limit and assumed design speed of 56 kilometers-per-hour (35 miles-per-hour). The RIDOT Pavement Management Systems (PMS) showed that this stretch of roadway was crack sealed in 1999, and a 0.6-km (0.4-mile) segment was fully-reconstructed in 1987, with 38.1 mm (1.5 inch) of Class I-1 HMA, 38.1 mm (1.5 inch) of a modified binder lift, 127 mm (5 inch) of a cold-recycled base, and 457 mm (18 inch) of A-1-a subbase material. The rest of the section was given an overlay treatment in 1985.

4 Test Setup

For proper manipulation of the software, the initial step towards analysis related to gathering the required inputs. Upon completion of this study stage, a matrix was created for testing, with the desired outputs being effective dynamic modulus (E*) at the JMF and as-built conditions. It is important to note that the featured construction project included two paved HMA layers. For this study, the properties of one pavement lift were changed by one variable at a time. The following JMF ranges or values were used for analysis in each lift, based on specifications obtained from RIDOT and potentially observed QA deviations [8]. This analysis matrix amounted to 16 tests per HMA lift that was varied, specifically:

- AC%: Default, Default ±0.3%, Default ±0.6%, Default ±1.0%;
- D/A: Default, 0.60, 0.80, 1.00, 1.20;
- AV: Default, 4%, 6%, 8%, 10%, 12%.

After software execution was completed, a statistical analysis procedure was initiated to determine the significance of each parameter for fatigue cracking predictions. Based on the basic assumption that the relationship between each variable, and the delta E* (ΔE*), or as-built E* less JMF E*, for each project, is linear and normally distributed, simple statistical tests were executed. The normality assumption was based off of the WPE, which played a major role in the predictions presented in this study [1]. The assumption of linearity is primarily based off of the principle that incentive/disincentive was calculated by the software in a purely linear equation. Since the output was tied directly to the PLD/pay factor and a weighted average corresponding to tonnage inputs, with no exponential factors being considered, it was assumed that the relationship between E* or ΔE* and JMF parameter is linear. In this study, a linear regression t-test was analyzed to determine the significance of the results at a given confidence level. From this information, a final determination was made regarding the level of sensitivity associated with each input variable.

4.1 Linear Regression t-Test

A linear regression t-test was used to determine whether a parameter was statistically significant with regard to ΔE*. As stated previously, since the underlying assumption is that a linear relationship exists between the dependent and independent variables (alligator cracking ΔE* and JMF parameter, respectively), a t-statistic can be computed to test a hypothesis. A simple flowchart can serve as an insightful guide as to how the following procedure was executed. Figure 2 outlines the statistical methodology applied.

Using a technique similar to the one completed by researchers at West Virginia University [9], the following linear relationship, shown in Eqn. (1), was first assumed to exist, as:

$$Y_i = \beta_0 + \beta_1 x_i + \varepsilon_i \quad for\ i = 1, 2, \ldots, n \tag{1}$$

In this equation, β_0 and β_1 are coefficients, where β_0 signifies the intercept of the line with the y-axis and β_1 dictates the slope. The standard error, ε_i, represents the scatter around the linear relationship [9].

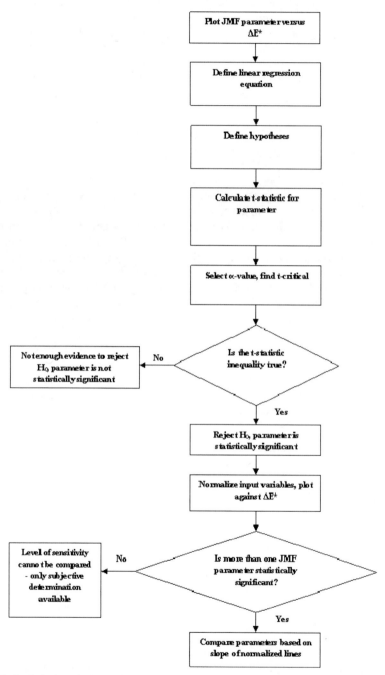

Fig. 2. Statistical analysis methodology for the evaluation of parametric impacts on fatigue cracking

Based on the values obtained during the sensitivity analysis, a regression equation will be computed along with a plot of the analysis outputs to test a hypothesis using a t-statistic. In order to achieve a level of significance for the data, a null hypothesis was derived. For this study, the null hypothesis, H_0, was that the slope, β_1, of the regression equation is zero, meaning that the variable, x_i, has no statistically significant impact on the results achieved. If the alternative hypothesis, H_1, was supported through rejection of the null, this implies that the parameter, x_i, has a statistically significant linear impact on the results, Y_i, for a given confidence level, α, which will be set at an initial value of 0.05 (95% confidence). The null and alternative hypotheses for the t-test are shown below:

$$H_0: \beta_1 = 0$$
$$H_1: \beta_1 \neq 0$$

The t-statistic was calculated by dividing the least squares estimate β_1 by the standard error for β_1, as shown in Eqn. (2):

$$t_{\hat{\beta}_1} = \frac{\hat{\beta}_1}{s(\hat{\beta}_1)} \quad (2)$$

The calculated t-statistic was compared to the t-critical value from a basic t-table for the given α-value, using a two-tailed approach, and degrees of freedom, $n - 2$, which will be the number of samples less two. If the t-statistic exceeded the t-critical value, the null hypothesis could be rejected. Eqn. (3) shows the inequality previously described:

$$|t_{\hat{\beta}_1}| \geq t(\frac{\alpha}{2}, n - 2) \quad (3)$$

4.2 Normalization Technique

Each statistically significant JMF parameter was normalized so that a comparison could be made to determine the degree of significance of each parameter. In this case, normalization occurs when the input (i.e. AC%) was divided by the mid-range value of the AC% variation, in this example, the original JMF AC% [9]. For D/A, the input values would be normalized when divided by 0.80. In the scenario regarding AV, the mid-range value for normalization would be 8.0%. The normalized input value would then be plotted on the x-axis, with ΔE* included on the y-axis. Since all three variables could potentially be plotted in one location for a particular project, the slope of this line will compare the degree of sensitivity for each JMF parameter [9]. In the case that only one parameter was found to be statistically significant for a particular project, a comparison for level of sensitivity could not be completed and a subjective determination would be made based on the slope of the normalized line. For full-depth reconstruction conditions, two separate analyses would be run, one for each constructed HMA (surface and binder) lift.

5 Discussion of Test Results

In order to obtain replicates for a specific parameter, data for each construction lot were gathered for each of the 31 software runs, resulting in a total of 165 data points. As defined in NCHRP Project 9-22A, five constant tonnage lots were created for software analysis, per lift. This allowed for the test to capture lot variation, as asphalt construction has often been shown to vary from batch-to-batch. The output, alligator cracking E*, was predicted at an effective temperature and effective frequency using a Monte Carlo simulation on the WPE based on historical standard deviations and project-specific values to represent the statistical means required for proper simulation [1]. The as-built E* was then subtracted from the JMF E* to obtain a ΔE* value. Based on preliminary software executions, ΔE* may be a contributing element in the pay factor calculation, as the degree of quality is related largely to the stiffness of the mixture. Statistical procedures were then executed on ΔE* to determine significance as attributed to a particular volumetric factor (AC%, D/A, or AV). At the time of publication, the pay factors cannot be disclosed because the software is still part of an active NCHRP research project.

For the alligator cracking module, the attributes of the HMA binder lift represent most of the prediction results. Therefore, when AC%, D/A, and AV values were varied for the HMA surface lift, there was virtually no change in ΔE*. Any changes in ΔE* were likely due to variations as a result of the Monte Carlo simulations in the software.

However, in the case of varied HMA binder lift characteristics, changes were noted. After the statistical analysis was completed, it was found that AC% is not significant to ΔE*, while D/A was very close to the significance threshold at a 95% confidence level, and AV was found to be statistically significant. Table 1 displays the t-test results for the binder layer in RI-102.

Table 1. Linear regression t-test results for RI-102 binder (19.0 mm) layer

	AC%	D/A	AV
$n - 2$	33	23	28
$t_{\hat{\beta}_1}$	1.444	-2.049	9.763
$t(\frac{\alpha}{2}, n - 2)$	2.035	2.069	2.048
Result	NOT SIGNIFICANT	NOT SIGNIFICANT	SIGNIFICANT

Upon determination of statistical significance, the normalization technique was not required since only one parameter was statistically significant. In order to evaluate the correlation present between the statistically significant variable (AV) and ΔE*, a plot of the average ΔE* at each tested AV level was constructed. Figure 3 shows a strong (R-squared 0.999) linear relationship between ΔE* and AV in that when JMF AV is increased the JMF E* will decrease, leading to an increased ΔE*.

This preliminary analysis provided some level of indication that sensitivity of changes to certain mix volumetrics or construction parameters can be captured ahead of production with the featured software. However, there are several areas for additional study efforts. A very high R-squared value can present some over emphasis in that five data points are featured, but since only one variable was significant, the normalization technique utilized was not fully functional. There was no comparison variable to assess the degree of significance. These shortcomings would suggest that additional testing is needed.

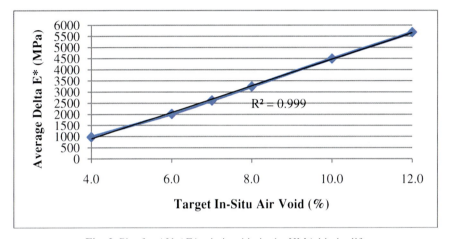

Fig. 3. Plot for AV-ΔE* relationship in the HMA binder lift

6 Conclusions and Recommendations

Based on the test results, AV is a statistically significant parameter impacting traditional fatigue cracking predictions in the quality-related specification software. These preliminary results show a strong level of statistical significance with regard to this parameter, providing users with knowledge towards expected pay factor values. The results also show a linear relationship between ΔE* and JMF parameter. Owner agencies can use this information to plan accordingly for specification development, with a focus on AV, due to its influence on stiffness and service life predictions for traditional fatigue cracking applications. However, given the complexity of dynamic modulus predictions, further analysis is warranted to obtain a more comprehensive understanding of the software's interactions and, ultimately, pay factor assignment. The results of this study are valuable in that they present opportunities for future investigations, as this preliminary analysis shows that particular JMF values relay a direct and notable change in stiffness values, and hence PLD/pay factor. Recommendations for further study include: 1) analysis of multiple HMA paving projects, 2) a factorial analysis performed to vary major input categories, instead of specific volumetric values, 3) nonlinear regression techniques to formulate alternate prediction models which can then be used in a more simplistic form by contractors and transportation agencies, and 4) a well-developed study on pay factor variance as a result of changes in as-built characteristics which will benefit the transportation and contracting industries for implementation. Judging by the level of importance associated with stiffness values throughout the program, a procedure seeking to relate stiffness directly to service life with a small number of variables appears to be an optimal choice for future endeavors. By relating stiffness, major input categories (structural design, traffic, etc.), and service life, software users can predict a service life parameter, without embarking on a full mixture design process or E* testing sweep. The future enhancements of the quality-related specification process could present cost benefits, and further the successful implementation of a quality-related specification concept for asphalt pavements.

References

[1] Moulthrop, J., Witczak, M.W., et al.: National Cooperative Highway Research Program. NCHRP Report 704: Beta Testing and Validation of HMA PRS (2011)
[2] Chamberlin, W.P.: National Cooperative Highway Research Program. NCHRP Synthesis of Highway Practice 212: Performance-Related Specifications for Highway Construction and Rehabilitation (1995)
[3] Jeong, M.G.: Implementation of a Simple Performance Test Procedure in a Hot Mix Asphalt Quality Assurance Program. Ph.D. Dissertation, Arizona State University (2010)
[4] Epps, J.A., Hand, A., Seeds, S., et al.: National Cooperative Highway Research Program. NCHRP Report 455: Recommended Performance-Related Specification for Hot Mix Asphalt Construction: Results of the Westrack Project (2002)
[5] Hand, A.J., Martin, A.E., Sebaaly, P.E., Weitzel, D.: Evaluating Field Performance: Case Study Including Hot Mix Asphalt Performance-Related Specifications. American Society of Civil Engineers Journal of Transportation Engineering 130(2), 251–260 (2004)
[6] Buttlar, W.G., Harrell, M.: Development of End-Result and Performance-Related Specifications for Asphalt Pavement Construction in Illinois, Crossroads 2000 Proceedings, pp. 195–202. Iowa State University and Iowa Department of Transportation, Ames (1998)
[7] Witczak, M.W.: Development of Performance-Related Specifications for Asphalt Pavements in the State of Arizona, Report FHWA-SPR-08-402-2, Arizona Department of Transportation (2008)
[8] Rhode Island Department of Transportation Materials Section, Full-Depth Reclamation of Rt. 102. Contract-Specific Specification (2010)
[9] Reigle, J.A.: Development of an Integrated Project-Level Pavement Management Model Using Risk Analysis. Ph.D. Dissertation, West Virginia University (2000)

Evaluating Root Resistance of Asphaltic Pavement Focusing on Woody Plants' Root Growth

Saori Ishihara[1], Kyoji Tanaka[2], and Yasuji Shinohara[3]

[1] Researcher, Tokyo Institute of Technology, Dr. Eng.
[2] Prof. Emeritus, Tokyo Institute of Technology, Dr. Eng.
[3] Associate Prof., Tokyo Institute of Technology, Dr. Eng.

Abstract. Pavement failures like cracking and rising, caused by the growth of plant roots are often observed in asphaltic pavement around roadside trees. To avoid trouble, the resistance of pavements to root growth should be estimated using a suitable test before installation. The aim of this study is to develop a test method for evaluating pavements' resistance to the thickening of roots as they grow. The method uses a simulated root developed to mimic the mechanical power of a growing root for evaluating the performance of pavements more easily and quickly.

First, we measured the force exerted by a growing root using a cherry tree over a period of four months from April to July. The enlargement force reached approximately 440 N/cm. Next, we developed an apparatus to reproduce the thickening growth of the root based on the earlier measurement. The test was carried out using an asphaltic pavement consisting of a 30 mm thick asphalt layer placed over various thicknesses of sand beds.

Cracking was observed during the test, and it was found that increasing the thickness of the sand bed reduced the damage to asphaltic pavement. The simulated root was useful for evaluating pavements' root resistance. Finally, we confirmed the appropriateness of this test by comparing its results with a numerical model.

1 Introduction

Damage to asphaltic pavement, such as cracking along roots and uplifting caused by the growth of roadside trees' roots is seen often. Fig. 1 shows examples of pavement damage caused by root thickening. The damage can progress until the pavement is completely separated as shown on the right in Fig. 1. This is a serious issue that not only hinders walking, but also reduces the walkway's aesthetic value. As the damage progresses, the pavement is usually replaced or repaired, but it requires time and money and is not a permanent solution. The same damage may occur again several years later even if the pavement is replaced or repaired consistently. If the resistance of a pavement to such root damage could be known before it is installed, the extent of damage could be reduced.

Fig. 1. Damage to asphaltic pavement caused by the growth and thickening of plant roots

Several studies have investigated damage to pavement due to root enlargement [1]-[7]. Although some studies focus on site investigation, including tree species, descriptions of pavement damage and the distribution of root systems, other studies have proposed countermeasures such as increasing the planting area [1], choosing suitable tree species [1], installing root barrier sheets [6], and controlling the soil under sidewalks [7]. Using real trees to evaluate whether pavements are root-resistant would be the most suitable way to predict the success of these proposed countermeasures. This, however, would require a long time to produce results, and would not help us determine whether root resistance would last beyond the experimental period. To avoid these problems, it would be very helpful to have a method to quickly and easily evaluate pavements' root resistance before installation.

The aim of this study is to develop a quicker, easier test method for evaluating pavements' root resistance. To ensure rapid results and simplified evaluation, simulated roots, which reproduce the behaviour of actual roots, should be used.

2 Measuring the Force Exerted by Thickening Roots

2.1 A System to Measure the Force Exerted by Enlarging Roots

Root growth has been studied, but no studies have yet to measure the force exerted by thickening roots. Thus, we initiated the measurement of the force in this study.

Roots grow along their radial axes and length wise. Roots that are thickening often cause pavement damage. Therefore, we developed a system to measure the force exerts by the growing root in the radial direction. The system is shown in Fig. 2. The top and bottom of the root was firmly held with square aluminum square bars, and the force exerted by the growing root against the aluminium bars was measured using load cells attached symmetrically on top of the root. The square bars and load cells were fixed with bolts to ensure that root growth is measured under constrained conditions.

2.2 Measurement Procedure

Cherry trees are often planted along roadsides in Japan; thus the measurement system was affixed to a cherry tree of approximately 21 years old with a circumference at breast height of about 138 cm. The apparatus was set approximately 2 m away from the tree. To avoid the effects of direct sunlight, rain, and wind and to protect the roots, a metal plate cover was placed over the device. Measurements were obtained from March through early August of 2009 during which root enlargement was assumed to occur.

2.3 Measurement Results

The results of our measurements are shown in Fig. 3. Although no increase in load was observed in March, the load increased gradually in repeating cycles in the beginning of April, decreasing from dawn to daytime and increasing from the evening to midnight of each day due to transpiration. From the middle of May, the load increased rapidly and began to slow down in late July. The maximum value was obtained on July 29, 2009 when the force exerted by root enlargement was 440 N/cm. Subsequently, the force levelled off; thus the measurements were stopped on August 2, 2009.

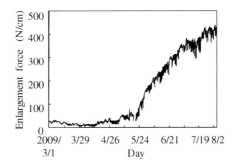

Fig. 2. Device for measuring the force exerted by root enlargement

Fig. 3. Variation of force exerted by the growth of a cherry root

3 Preparation of Simulated Roots

3.1 Basic Concept

Root growth occurs in the circumferential direction, but, as shown in Fig. 1, upward forces cause most of the damage. Therefore, root resistance can be evaluated simply by exerting an upward uniaxial compressive force to simulate the root growth mechanism.

3.2 Mechanism and Construction of the Simulated Root

We simulated root enlargement by pushing on the upper and lower semicircular aluminum columns that act like the cross section of a root, using a small hydraulic jack as shown in Fig. 4. The hydraulic jack had a maximum stroke of 10 mm; thus the simulated root could be enlarged up to 10 mm. Load cells were attached inside the simulated root, allowing measurement of the pressure generated by the root during the test. The diameter of the simulated root was set to 100 mm. This was based on the diameter of the cherry tree root observed earlier. The length of the simulated root was also set to 100 mm.

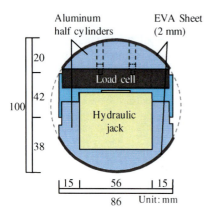

Fig. 4. Cross-section of simulated root

Table 1. Thickness and hardness of surface materials for the simulated root

Material	Thickness	Hardness[*]
Cork sheet	2 mm	51.5°
Rubber sheet	3 mm	63.2°
EVA sheet (ethylene-vinyl acetate)	2 mm	27.2°
	3 mm	28.3°
	5 mm	29.4°
Sponge sheet	5 mm	7.4°
EPDM-reinforced sheet	1.5 mm	66.1°

[*]measured by durometer (A)

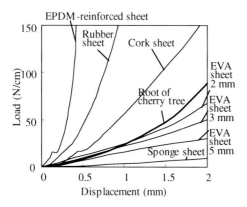

Fig. 5. Relationship between displacement and load measured in compression test of simulated roots covered with various surface materials and the root of a cherry tree (loading rate 1 mm/min, 20°C)

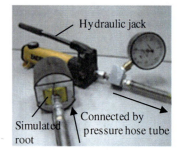

Fig. 6. Simulated root and compression system

Aluminium, however, is harder than cherry root. To approximate the mechanical properties of the cherry root, the surface of the aluminum columns was covered with an elastic material. To select the ideal surface material, aluminum columns with various elastic sheets as shown in Table 1 and the actual root of a cherry tree were subjected to compression tests at a rate of 1 mm/min. The results of these tests are shown in Fig. 5. A sheet of ethylene-vinyl acetate (EVA) 2 mm in thickness behaved the most like the cherry root; thus EVA was used to cover the columns. The simulated root and the compression system are shown in Fig. 6.

4 Evaluation of Asphaltic Pavement's Root Resistance Using a Simulated Root

4.1 Specimen

We studied the usefulness of using simulated roots to evaluate root resistance using asphaltic pavement specimens obtained from a sidewalk. Asphaltic pavements vary widely from full-fledged pavements supported by sufficiently thick layers of roadbed materials such as crushed stone, gravel, or slag to simple pavements with a layer of sand laid as the roadbed material. In this study, a relatively simple type of pavement, shown in Fig. 7, was used to confirm the usefulness of simulated roots. This pavement specimen comprised a layer of sand placed level as the roadbed material with asphalt applied over it. The thickness of the roadbed above the root affects how much cracking and lifting of the asphaltic pavement occurs; thus we prepared specimens with 4 different roadbed thicknesses (0 mm, 10 mm, 20 mm, and 50 mm) to be placed above the simulated root.

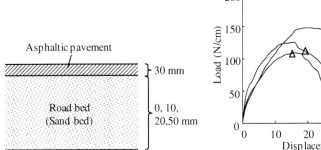

Fig. 7. Cross-section of the asphaltic pavement specimen

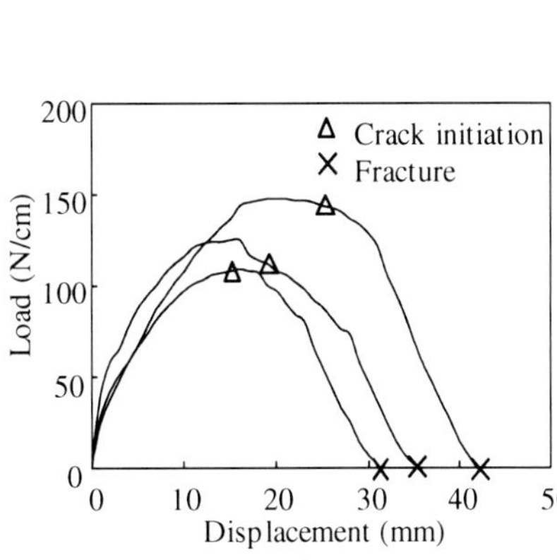

Fig. 8. Relationship between load and displacement measured by three-point bending test of asphaltic pavement

Asphaltic pavements are usually somewhat flexible just after construction. Many kinds of damage cannot be observed immediately, and are visible only after the pavement has deteriorated somewhat over time. Therefore, asphaltic pavement that had been used as a sidewalk for about 10 years was used in the test. The relation between load and displacement in the asphalt, measured with a three-point bending test, is shown in Fig. 8. The Young's modulus calculated using these results was 304.6 N/mm^2. To obtain specimens from this pavement, we cut out test pieces that are 680 mm long, 100 mm wide, and 30 mm thick.

4.2 Test Method

The following test procedures were employed: First, the simulated root was installed in a container with inside dimensions of 700 mm by 110 mm by 330 mm. Then, the container was filled with sand to simulate a roadbed with a specified thickness, making sure the sand was packed sufficiently tightly. The asphaltic pavement was then laid, and both edges were fixed with supporting bars. Next, oil was transferred to the simulated root from the hydraulic jack to pressurize the roadbed and asphaltic pavement from underneath. In our previous trial using a cherry root (section 2.3), the enlargement force increased to 440 N/cm very slowly with an average increase of 5 N/cm per day over 3 months from May to July, as shown in Fig. 3. However, in our indoor test, we manually exerted pressure on the pavement with a target pressurization rate of 5 N/cm per minute to obtain results quickly.

The test conditions are shown in Fig. 9. Cracks on the asphaltic pavement surface were monitored visually and we measured the height to which the asphaltic pavement center was lifted with a displacement meter. The test was considered complete when cracks appeared in the asphaltic pavement or when the maximum deformation (10 mm) of the simulated root was achieved.

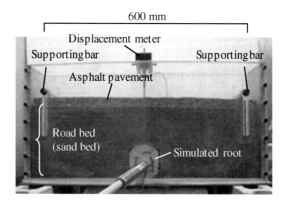

Fig. 9. Root-resistance test

4.3 Results and Discussion

Fig. 10 shows the height to which the asphaltic pavement surface was lifted when the simulated root was pressurized. An increase in the pressurization force by the simulated root caused a gradual elevation in the asphaltic pavement surface. Cracks were generated on the asphalt surface when the roadbed thickness above the simulated root was 0 mm or 10mm. The pressurization force at which cracking occurred was about 300 N when the roadbed thickness was 10 mm, but only 100 N when the roadbed thickness was 0 mm. The conditions under which cracks formed on the asphaltic pavement surface are shown in Fig. 11. Irrespective of the roadbed thickness, cracking usually occurred at points directly above the center of the simulated root.

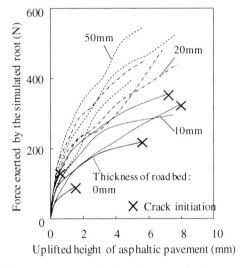

Fig.10. Relationship between pressurization of the simulated root and the height to which the asphalt pavement was lifted

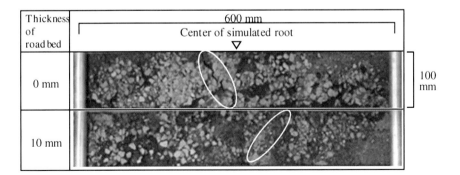

Fig. 11. Cracks occurred on the surface of the asphaltic pavement (No cracks were observed with sand bed thickness of 20mm and 50mm)

We also observed that the height to which the pavement was lifted tended to decrease as the thickness of the roadbed increased. We surmise that this is because the roadbed layer acts as a buffer and disperses the force from the simulated root; conversely, in the case of thin pavements, the force from the simulated root is applied directly to the asphaltic pavement. In either case, the distance between the asphalt and the root was considered to be an important factor in determining the damage to the asphaltic pavement.

The nature of this damage was similar to that observed in actual asphaltic pavement sidewalks, leading us to believe that the simulated root developed in this study will be useful in evaluating root resistance.

4.4 Comparison with Results of Numerical Modelling

Modelling for numerical calculation The height to which the asphaltic pavement was lifted measured in the root resistance test using the simulated root was compared with numerical simulations conducted using a finite element method. Table 2 shows the physical constants used in the numerical calculations.

The conditions of the analysis were as follows:

(1) Physical characteristics

The tensile strength and ultimate strain of the asphalt were calculated using inverse analysis of the tension softening property, and Young's modulus was measured using the previous three-point bending test. We used published values for the Young's modulus and Poisson ratio for the sand. We assumed that cracks were generated when the load approached the tensile strength of the asphalt.

(2) Interface element

We set interface elements between the asphalt and the sand bed, and between the simulated root and the sand bed. The interface elements had a defined stiffness in tension, but none for compression.

(3) Model of simulated root

The simulated root was modelled as the only moving part, and was constrained not to undergo any deformation. In the model, displacement of the simulated root was set at 0.05 mm increments up to a maximum of 10mm of displacement achieved in 200 steps. An element breakdown and the constraint conditions of the numerical simulation are shown in Fig. 12.

Table 2. Physical constants used in the numerical simulation

	Asphalt	Sand bed
Yang's module (N/mm^2)	304.6	200.0*
Poisson ration	0.35*	0.40*
Tensile strength (N/mm^2)	0.61	-
Ultimate strain	0.26	-

*Published Values

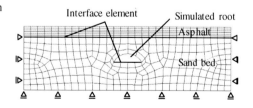

Fig. 12. Element breakdown and constraint conditions for the numerical simulation

4.5 Results of Calculation

The results of the simulated root test and the numerical simulation with the thickness of sand bed set at 50 mm are compared in Fig. 13. The uplift of the asphaltic pavement calculated by the numerical simulation is greater than that measured by the resistance test using the simulated root. For modelling purposes, the sand is defined as an isotropic material with a Young's modulus and a Poisson ratio as shown in Table 2; thus, movement between the sand particles was not taken into consideration. Such movements in the physical experiment may account for the difference. The two simulations, however, are generally consistent and agree well with each other.

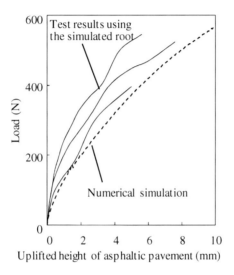

Fig. 13. Comparison of test results with numerical simulation (Thickness of sand bed : 50mm)

5 Conclusion

In this research we developed a simulated root that makes it possible to evaluate root resistance quickly and easily.

First, we developed a system to measure the enlargement force exerted by root growth. We used this system to measure the force of a cherry root and determined that the maximum force exerted by root thickening between mid-April and late July was approximately 440 N/cm.

Then, we developed a simulated root, consisting of a pair of semicircular aluminum columns, a hydraulic jack, a load cell and a 2 mm-thick EVA sheet based on the measured root shape and mechanical properties of a cherry root.

Finally, we conducted a root resistance test on a simple asphaltic pavement as an example. We found the simulated root developed in our study to be useful in evaluating root resistance.

References

[1] Okawa, H., Kurihara, S.: Damages of Sidewalk Pavement due to Plant. Pavement 42(7), 21–24 (2007) (Japanese)
[2] Day, R.W.: Damage of structures due to tree roots. Journal of Performance of Constructed Facilities 5(3), 200–207 (1991)
[3] Nicoll, B.C., Armstrong, A.: Street tree root architecture and pavement damage. Arboriculture Research and Information Note - Arboricultural Advisory and Information Service, No.138/97 (1997)
[4] Nicoll, B.C., Armstrong: Development of Prunus root systems in a city street: pavement damage and root architecture. Arboricultural Journal 22(3), 259–270 (1998)
[5] Iwata, A., Kida, Y., Kouno, T., Karizumi, N.: Study on the Influence on the Road Structures by the Root Growth of Japanese Zelkova Street Trees. Journal of the Japanese Institute of Landscape Architecture 59, 49–52 (1996) (Japanese)
[6] Thomas Smiley, E.: Comparison of methods to reduce sidewalk damage from tree roots. Arboriculture & Urban Forestry 34(3), 179–183 (2008)
[7] Grabosky, J., Bassuk, N.: A new rooting volumes under sidewalks, urban tree soil to safelyh increase. Journal of Arboriculture 21(4), 187–201 (1995)
[8] Huck, M.G., Klepper, B., Taylor, H.M.: Diurnal Variations in Root Diameter. Plant Physiology 45, 529–530 (1970)
[9] Genard, M., Fishman, S., Vercambre, G., Huguet, J.-G., Bussi, C., Besset, J., Habib, R.: A Biophysical Analysis of Stem and Root Diameter Variations in Woody Plants. Plant Physiology 126, 188–202 (2001)

20 Years of Research on Asphalt Reinforcement – Achievements and Future Needs

Arian H. De Bondt[1]

Ooms Civiel bv, Scharwoude 9, 1634 EA Scharwoude, The Netherlands
adebondt@ooms.nl

Abstract. A proper road network is crucial for the economy. For the primary system, the motorways, this implies that its usage should be safe, the arteries should be spread around the economic centres and there should be an undisrupted traffic flow. For the secondary system, the rural roads, this implies that the width and bearing capacity should be such that there are no limitations for the distribution of goods.

The basis of the primary road network has been built. The focus is now on maintenance. Due to the increased cargo weight per truck axle, the introduction of super-single tyres, the growing traffic and the political difficulties to achieve widening of roads, the individual layers of pavement structures are degrading at a faster rate, while there is no time frame left to do structural maintenance. Also there is the desire that pavements should last longer and when a treatment on a specific jobsite is necessary, it should be carried out in a short(er) period of time. Furthermore, maintenance intervals should become more accurately predictable, so they can be planned better in combination with other works on for instance bridges, safety barriers, traffic management infrastructure, etc. With respect to the rural roads, it has become clear that because of the reduced budgets, more cost-effective ways to upgrade pavements are needed.

From the foregoing it can be concluded that there is a clear need for reliable, robust and cost-effective (structural) maintenance technologies, such as the (right) use of grids in asphalt layers.

1 History of Grids in Bituminous Layers

During the 1960's the first applications on the use of grids in asphaltic layers have been reported in the USA. The experience with steel welded fabrics was positive from the effectiveness point of view. However, this type of reinforcement was more or less impossible to place and remove. This was the reason that it died.

In the early nineteen-seventies a synthetic grid was trialled in the Netherlands. Problems were reported with respect to the installation, the lack of bond (adhesion) with the surrounding asphalt and the anchorage. Market penetration was also not successful, because of insufficient know-how about the mechanism behind these products. Road authorities were for instance trying to check the degree of improvement by means of deflection measurements or product rolls were too narrow (creating pullout problems). Last but not least, at that time the focus in the sector was on new roads.

Around 1985 different types of grid products successfully entered the market in the Netherlands. These consisted of polyester, polypropylene, glass and steel. All products in fact originated from other fields within the construction industry. This is also the reason that in the early days failures took place, because some products were found not to be suited for the quite special demands in case of the application in an asphalt overlay. After some improvements, reinforcement by means of grids has become an accepted method to tackle reflective cracking in the Netherlands since the 1990's.

2 Phenomenon of Reflective Cracking in Asphalt Pavements

Many pavements, the life of which are thought to be extended by means of a new surface course, reveal soon after construction of this overlay a crack pattern similar to that which was visible in the old existing surface [1]. This propagation of cracks or joints from the old pavement into and through the overlay is commonly known as "reflective cracking", Fig. 1. It occurs in all types of pavement structures (flexible and semi-rigid/composite) and imposes heavy strains on national and local road authorities. This is because cracks in pavement surfaces allow water penetration into the structure (weakening its foundation), cause ravelling at the edges of the crack (thus breaking windshields of cars), increase the roughness (thereby disrupting comfort and creating dynamic loadings) and also generate noise and vibrations. Since filling cracks with bitumen is not a durable option (must be repeated after each winter), is unsafe for motorists and creates poor esthetics, a better alternative is required. More than twenty years of experience has shown that in general (but not always) grid reinforced overlays can be a good and cost-effective solution.

Fig. 1. Typical reflective cracking on a cement treated base (left) and concrete slabs (right)

Depending on a number of factors, reflective cracking can be caused by the following mechanisms [1]:

- Traffic (especially during cold periods);
- Daily temperature cycles (extreme);
- Seasonal temperature variations (summer/winter);
- Shrinkage of soils in dry periods;
- Downward subsoil movements (uneven settlements);
- Upward subsoil movements (frost heave).

The first three mechanisms are the major ones worldwide. In the case of traffic, the crack initiation phase as well as the crack propagation phase within the overlay mixture are both of importance for the overall life, whereas in case of temperature cycles the crack initiation phase is the dominant factor [2, 3].

Based on proper investigations and analyses (also in-situ), it is possible for an experienced pavement engineer to deduce which cracking mechanism has been or will be active on a specific site [1]. It is important to perform this work, because for example depending on the type of loading, an asphalt mixture responds differently with respect to the phenomenon cracking. Also, the effect of maintenance options and via this, their cost-effectiveness, strongly depends on the specific project circumstances/details.

3 Mechanisms of Asphalt Reinforcement

The effect of a grid reinforcement in asphalt depends on the:

- Type and severity (movement) of the cracks/joints;
- Characteristics and nature of loading of the pavement structure;
- Location within the overlay;
- Properties of the grid (mechanical, durability, etc.);
- Anchorage (bond) method/procedure;
- Quality of installation (including the paving operation).

It has been found by means of finite element computations, laboratory work and (limited) field experience that depending on the factors listed above, a lifetime increase of up to a factor 5 compared to a similar unreinforced overlay can be achieved. A factor x means that it takes x times more traffic repetitions or temperature cycles, before a reflected crack is visible at the surface of the overlay. Typical grid reinforcements, with tensile strength values from 15 up to 250 kN/m[1] (according to EN 15381:2008), are mainly activated during the crack propagation phase in case of traffic loading (in bending as well as in shearing mode). In case of thermally induced reflective cracking the reinforcement is already activated during the crack initiation phase.

The fact that grid reinforcement in asphalt is also beneficial under so-called pure mode II shearing action in a (new!) crack in an overlay has become apparent from extensive fundamental research carried out by de Bondt at Delft University of Technology [1]. To get insight into the effect of the presence of reinforcement on the load carrying capacity of cracks in asphaltic mixtures, shear tests were performed on plain as well as reinforced cracks, Fig. 2. The specimens were taken from a pavement trial section specially made for this purpose; the asphalt layers were laid down and compacted with ordinary (routine) construction equipment. First of all, aggregate interlock tests were performed under different confining pressures. The measurement data enabled the development of a theoretical saw-teeth model (including a specific crack inclination angle α) explaining and describing the observed behaviour.

Fig. 2. Newly developed 4-point shear testing device for plain cracks (left) and reinforced cracks (right)

From a series of shear tests on reinforced cracks, it could be concluded that these types of cracks can transfer shear without externally applied normal pressure. This is possible, because adequately anchored reinforcement is capable of generating a normal force $\sum n$ at the crack (via crack dilatancy), which allows friction $\sum f$ along the planes of the crack to occur, Fig. 3. The contribution of the (indirectly via the grid) generated friction along the teeth of the crack is even larger than the one generated by the reinforcement directly. It is obvious that this mechanism only occurs if asphaltic mixtures are composed with proper sized mineral aggregates (grading > 2 mm); in case of so-called sand mixes there is no shear carrying capability of a reinforced asphalt crack in pure mode II.

From measurements on several commercially available reinforcing systems, it appeared that not only the axial product stiffness EA of a reinforcement is an important factor, but also its resistance to pullout and anchorage length. The way in which the junctions between the ribs/strands of a reinforcing product are manufactured, controls if pullout restraint is developed via bearing of the mineral aggregate of the asphalt mixture in the grid apertures or via friction and adhesion along the strands. It is important to realise that the component adhesion in generating pullout restraint is a typical aspect of the application of grids in asphalt layers; this because bitumen sticks (and its shear stiffness is temperature and rate dependent).

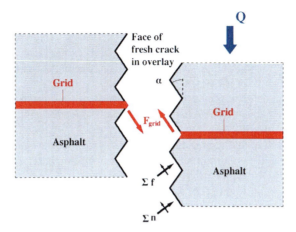

Fig. 3. Schematic illustration of forces acting in a reinforced crack under external shear loading (Q)

It also became clear that the required anchorage length in the field and in fact also during laboratory tests, depends on the type of loading (traffic or temperature cycles) and testing temperature.

This effect is often underestimated by laboratories, which do not analyse the mechanics of their test set-up before starting to work with it; this leads to unwanted biased results. Adequate mobilisation of the reinforcement is an important issue. To study this topic, dedicated research is required [4].

4 Design of Reinforced Asphalt Pavements

The current situation in practice with respect to design is that:

- Generally applicable (accepted and standardised) design methods for (reinforced and unreinforced) maintenance treatments are not available.
- Design is often based on personal experience (often hardly documented) or extrapolated laboratory simulation data; the latter is not always allowed since circumstances differ in the field from project to project.
- Tender specifications are incomplete or not representative.
- Criteria for the evaluation of alternatives are missing.
- Product characterisation is not uniform, despite CE-marking.

Since the beneficial effect of a given grid reinforcement highly depends on the type of cracking mechanism which is dominant on a particular jobsite and is also extremely case dependent (e.g. the bearing capacity of the soil can play an important role), proper design based on (extrapolated) laboratory experiments and/or accelerated load testing data is not possible. The requirements for any design model or procedure for grids in asphaltic pavements, which is meant to be used for routine purposes, can be summarised as follows [5]:

- It should tackle the right project specific cracking mechanism.
- If relevant, the traffic characteristics (number, type of vehicles, speed) specific for the jobsite need to be taken into account.
- If relevant, the temperature variations in time (day/night, season) specific for the jobsite have to be incorporated.
- The pavement and soil properties relevant for the jobsite should be used.
- In case of maintenance, the existing condition of the pavement has to be one of the input parameters.
- The mechanical and durability characteristics of the grid (the in-situ stiffness/strength including the potential effect of damage during installation) must be incorporated in sufficient detail.
- The interaction between grid and surrounding asphalt mixtures has to be taken into account.
- The computational engine (procedure), which is behind the design method should be described in such a way that it can be evaluated (judged) by third parties.
- The method should have been validated with long-term field monitoring data.
- Life-cycle costing analyses should be possible in an easy way.
- For an average jobsite the (user-friendly) design process should not take too long for an average skilled pavement engineer who is familiar with the mechanistic-empirical approach.
- The end result of the design process should be tender specifications and a sketch of the laying plan of the grid. The tender specifications should be in a generally accepted format, where product description is according to international standards (CEN, ISO).

Furthermore, it is recommended that parameters which give an indication about road user costs and driving comfort are outputted. This because it is interesting for clients to know the effect of maintaining the pavement with a grid on parameters such as the Present Serviceability Index or the International Roughness Index.

5 Current and Potential Field of Application

At the moment grid reinforced asphalt is used in overlays on flexible pavements which show alligator (fatigue) cracking, transverse (low-temperature) or block cracking, longitudinal top-down cracking, pavement edge stability cracking problems or wide (open) longitudinal construction joints. Furthermore, they are applied in asphaltic overlays on transverse and longitudinal joints in PCC-slabs or on top of continuously reinforced concrete. Also there are applications in asphalt overlays on top of cracked asphalt on cement treated bases; this is in fact on old reflected cracks. Applications along road widenings (the transition new/old) can be found in all pavement types.

Depending on the application area, the characteristics of the grid system and the quality of the installation procedure, there have been of course positive and negative experiences in the field. This also has to do with the fact that not each product is suited for each situation. Fig. 4 shows an example of the results of long-term field performance monitoring on a semi-rigid pavement structure.

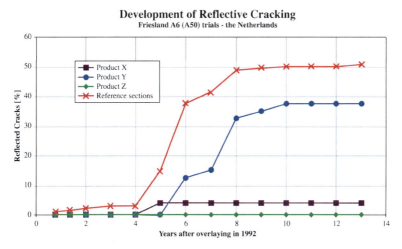

Fig. 4. Long-term performance of the motorway A6 (A50) in the Netherlands

Hardly ever it is decided to apply grid reinforcement in a new construction. This has to do with the impossibility to back-up the reinforced case with a design or it is simply not (seen as) a cost-effective option. Out-of-the-box thinking might suggest that in case of for example a 30-year PPP-project the introduction of a grid (with a certain minimum product stiffness and durable pullout resistance) at the bottom of only the slow lane (lane 3) of a new or reconstructed motorway, creates a perpetual pavement without the need for an entire cross-section of thick asphalt across all lanes, Fig. 5.

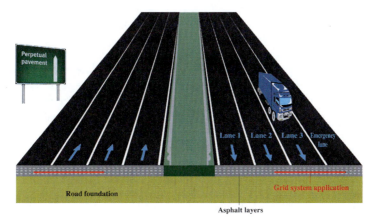

Fig. 5. Potential grid application in large scale new construction

6 Innovative Examples of Application

Areas in asphalt pavement construction where grids are successfully applied outside the standard way of using them, are for instance invisible joint systems, slab rocking details and asphalt solar collector systems. In case of invisible joint systems [6], specific grids are used in different asphalt layers on top of each other (up to even 6 layers). This to make a jointless and maintenance free transition between a bridge and the road possible; it is done in such a way that an asphalt plug joint (which has a short life) is not necessary. In case of slab rocking details a combination of a special grid, polymer modified bitumen and an optimized asphalt mixture is composed in such a way, that it is not needed to take out a PCC-slab which shows rocking (large differential movement) at the joint. In asphalt solar collector systems, grids are used to enable first of all the installation of the pipes; during the service life of the road they make sure that the structural integrity of the pavement is kept.

7 Conclusions

During the past 20 years it has become clear that grid reinforcement applications in asphaltic pavements are clearly solving the needs in our society. There is even more potential. To materialise this potential, improve the cost-effectiveness and avoid the risk of bad (non-suited) applications, research needs have come up in recent time. The main issues are design and adequate product characterisation.

8 Future Research Needs

Based on twenty years of experience in fundamental research (including laboratory testing), design, practical application/installation, long-term field performance, product development, standardisation and strategic market overviews, the following research needs can be listed:

- Design procedures for the major areas of application (linked to unreinforced design tradition/methods).
- Development/improvement of a series of laboratory test methods to fully characterise the mechanical and durability properties of new (unknown) high-quality and low-quality (surrogate) grids entering the market (to become part of some form of CEN type testing); part of this research is prenormative work.
- Perform (and report!) long-term field performance studies to validate (future) design procedures.
- Clarify the issue of the optimum adhesion between (cracked) pavement layers; this currently creates unnecessary confusion in practice.
- Development of an adapted (dedicated) 4-point bending fatigue test for reinforced asphalt samples, which can be carried out in each asphalt laboratory capable of doing the standard European Union 4-point bending fatigue test for asphalt mixes.
- Whole-life costing on grid applications in new asphaltic pavements.

References

[1] de Bondt, A.H.: Anti-Reflective Cracking Design of (Reinforced) Asphaltic Overlays. Ph.D.-Thesis, Delft University of Technology (1999)
[2] de Bondt, A.H.: Effect of Reinforcement Properties. In: Proceedings of the 4th RILEM Conference on Reflective Cracking, Ottawa, pp. 13–22 (2000)
[3] Brooker, T., Foulkes, M.D., Kennedy, C.K.: Influence of Mix Design on Reflection Cracking Growth Rates through Asphalt Surfacings. In: Proceedings of 6th International Conference on the Structural Design of Asphalt Pavements, pp. 107–120 (1987)
[4] de Bondt, A.H.: Development of a laboratory pullout test set-up for asphalt reinforcement, Internal report, Ooms R&D Laboratory, Bayex-Ooms research project (1997)
[5] de Bondt, A.H.: COST Action 348 – Reinforcement of pavements with steel meshes and geosynthetics, Report Work Package 4: Selection of Design Models and Design Procedures (2006)
[6] de Bondt, A.H., Schrader, J.: Jointless asphalt pavements at bridge ends. In: 3D Finite Element Modeling of Pavement Structures, Amsterdam, The Netherlands, pp. 459–473 (2002)

Concrete Pavement Strength Investigations at the FAA National Airport Pavement Test Facility

Edward H. Guo[1], David R. Brill[2], and Hao Yin[3]

[1] Consultant
[2] Federal Aviation Administration, USA
[3] Gemini Technologies, Inc., USA

Abstract. The Federal Aviation Administration (FAA) conducted airplane gear load tests on three new concrete pavement test items at the National Airport Pavement Test Facility (NAPTF) to determine the in-situ concrete slab strength and estimate the actual stress ratios to be expected under full-scale traffic loads. Full-scale static loads were applied incrementally to designated slabs within the test items to identify the cracking loads for both bottom-up and top-down cracks. Crack initiation was determined by monitoring real-time strain gage measurements. Some of the strength tests were supplemented by rolling-wheel traffic loads to try to propagate the already initiated cracks to full depth. A limited set of beam fatigue tests on standard lab-cured beams cast at the time of construction provided laboratory data for comparison. The immediate practical result of the strength tests was to allow the FAA to set the wheel loads for the CC6 full-scale trafficking phase at approximately 80 percent of the slab cracking strength measured for the low-strength test item. The long-term goal of these investigations is to relate pavement cracking strength to flexural beam strength from standard ASTM C 78 tests, and to relate the pavement stress ratio to the stress ratio in laboratory fatigue tests performed on standard concrete specimens. Both tests are needed to satisfy the requirements of fatigue theory.

1 Introduction and Background of CC6 Tests

To understand the fatigue behaviour of concrete pavements at the structural level, it is necessary to consider the pavement strength. Most published fatigue results for concrete pavements were obtained at the material level by testing beam specimens. Thus, almost all "fatigue" models for design fail to completely satisfy the fundamental requirements of fatigue theory – the two key variables, stress and strength, are obtained from two different structures. The results of full-scale tests in this paper can be used to derive the actual stress ratio, avoiding problems of correspondence between the test specimen and the full structure.

The NAPTF, located at the FAA William J. Hughes Technical Center, Atlantic City International Airport, New Jersey, USA, is a unique facility for full-scale testing of airport pavements. The current cycle of rigid pavement full-scale tests at the NAPTF has been designated Construction Cycle 6 (CC6). The three test items in CC6 have been constructed with identical cross-sections, but with three different concrete mixes designed to give different values of flexural strength R. In addition, two different subbase materials (econocrete and hot-mix asphalt) provide a total of six combinations of concrete strength and subbase type (Figure 1). As shown in Figure 1, gear loads are applied by the north and south carriages, each of which is configured to simulate a 4-wheel (2D) gear with lateral wander. In Figure 1, the gears are shown in the center wander position, designated "track 0." Details of the wander pattern used for trafficking may be found on the NAPTF web site [1]. The material designations in Figure 1, e.g., item P-501, Portland cement concrete (PCC), refer to FAA specifications found in [2].

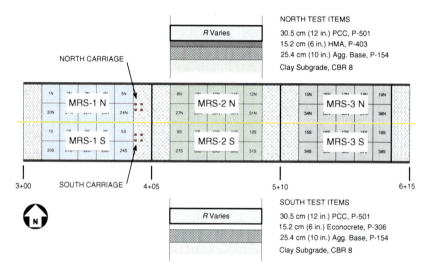

Fig. 1. General layout of CC6 test items. Stationing is shown in hundreds of feet (100 ft. = 30.5 m). All slabs are 4.6 × 4.6 m (15 × 15 ft.).

Three different concrete mixes were used to obtain a separation in concrete strength. Table 1 shows that the actual 28-day flexural strength values, as determined by ASTM C78 4-point beam tests [3], were close to the target values, except for the low-strength mix (MRS-1), which was higher. Nevertheless, a statistically significant separation in mean test item concrete strengths was achieved. Each reported value in Table 1 is an average of 12 samples, which includes both laboratory- and field-cured specimens.

Concrete Pavement Strength Investigations

Table 1. PCC Strength Values for CC6 Test Items

Test Item	Target R, MPa (psi)	28-day Flexural Strength R, MPa		28-day Compressive Strength F'_c, MPa	
		Mean	Std. Dev.	Mean	Std. Dev.
MRS-1	3.45 (500)	4.56	0.31	25.8	2.51
MRS-2	5.17 (750)	5.26	0.74	30.3	3.95
MRS-3	6.89 (1000)	6.70	1.24	41.0	1.59

2 Slab Strength Tests

Full-scale slab strength tests were performed using the NAPTF vehicle to provide the load while monitoring the response using strain gages in real time. The position of the gear depended on whether the test was for bottom-up or top-down cracking.

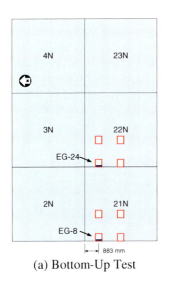
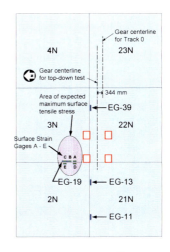

(a) Bottom-Up Test (b) Top-Down Test

Fig. 2. Position of 2D gear for slab strength tests on test item MRS-1N

2.1 Bottom-Up Strength Tests

Bottom-up strength tests were performed on two slabs (21N and 22N) in test item MRS-1N (low-strength) on June 30, 2011. The 2D gear was positioned as shown in Figure 2(a) so as to produce the maximum tensile stress at the bottom of the slab near the embedded strain gages EG-8 and EG-24. A small initial wheel load (2268 kg per wheel) was applied, held for approximately 10 s, then released. During this interval, strain gage data was acquired at a rate of 20 Hz. The wheel load was increased in 2268 kg (5000 lb.) increments and the procedure repeated until a bottom-up crack formed (Figure 3). The test result shown in Figure 3 is typical of the case where the bottom-up macro crack develops near, but not through the strain gage. Under the constant load condition, the strain increment may (a) reach a stable value (linear elastic condition), (b) increase, or (c) decrease. Increasing strain response under a sustained load may be explained by microcracking in the concrete matrix in the vicinity of the strain gage. This is seen in Figure 3, for example in the response to the 24,948 kg (55,000 lb.) wheel load. When the macro crack finally develops at some distance from the gage itself, this is indicated by decreasing strain under the load, as seen in the strain record for the 29,484 kg (65,000 lb.) wheel load.

At each discrete load level, the tensile strain increment corresponding to the applied load was computed and plotted against the wheel load (Figure 4). As expected, the relationship observed between the recorded strain and the applied load was approximately linear at low loads but turned highly nonlinear after a crack formed. The wheel load and strain increment corresponding to crack formation were determined with reference to Figure 4. Based on Figure 4, the strain increment related to the pavement strength for both slab 21N and slab 22N was estimated to be 123 microstrains. This level of strain corresponded to a wheel load of 29,484 kg (65,000 lbs.) for slab 21N but only 24,948 kg (55,000 lbs) for slab 22N. Although the strain response recorded by EG-08 was significantly lower than EG-24 under the same load (Figure 4), the measured pavement strengths are still similar (136 vs. 128 microstrains). In other words, the pavement strength has been verified experimentally as an indicator of concrete pavement resistance to crack initiation.

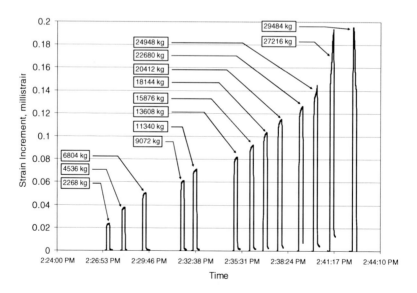

Fig. 3. Record of bottom-up strength test on MRS-1N, gage EG-24 (slab 22N)

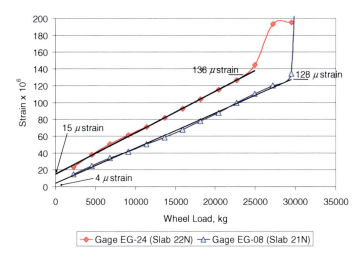

Fig. 4. Strain versus wheel load for bottom-up strength tests on MRS-1N

2.2 Top-Down Strength Test

A single top-down strength test was performed on test item MRS-1N on August 10, 2011. The concept behind this test is that the static gear load should be positioned in such a way as to maximize the ratio of tensile stress on the top surface of the slab to the tensile stress on the bottom of the slab under the wheel, thereby maximizing the opportunity for an initial crack to form at the surface. A previous study [4] involving finite element analysis of surface strain gage measurements at the NAPTF found that a gear alignment with one set of wheels along the longitudinal joint and the other on the adjacent slab produces high surface tensile stress at a short lateral distance from the gear. Thus, for the top-down test, the gear was positioned straddling two slabs (3N and 22N) as shown in Figure 2(b). The shaded oval on the transverse joint between slabs 2N and 3N indicates where the maximum surface stress was expected to occur. Along this joint, embedded gage EG-19, located 2.54 cm (1 in.) from the surface, was suitable for recording load-related strains. In addition, the five surface strain gages shown in Figure 2(b), designated A though E, were installed along the transverse joint. Gage C was installed directly over EG-19, and gages B and A were, respectively, 0.305 m (12 in.) and 0.610 m (24 in.) south of gage C. Gages D and E were located opposite A and C, respectively, on slab 2N.

Prior to performing the strength test, a series of load positioning tests was conducted at a relatively low wheel load. The purpose of these tests was to determine the optimal gear offset position to induce top-down cracking at the transverse joint (EG-19) location, while also minimizing the risk that a top-down crack would occur first along the longitudinal joint. A sequence of six load positions spaced laterally at 0.127 m (5 in.) was used. At each lateral offset, the gear was loaded to 11,340 kg (25,000 lbs.) per wheel and held for 10 s. Strain gage readings were taken at EG-19, longitudinal gages EG-11, EG-13 and EG-39, and surface gages A through E. From analysis of these data, it was determined that the best choice was to place the gear centerline 344 mm north of Track 0, as shown in Figure 2(b). This was the offset used for the subsequent strength test, as well as for the second phase zero-wander trafficking discussed in a later section.

Figure 5 shows surface strain gage readings from the top-down strength test. As in the previous bottom-up test, the procedure was to apply a load, hold for 10 s, release, then apply the next higher load, increasing the wheel load in 2268 kg (5000 lb.) increments until a crack formed. In Figure 5, it is clear from the sudden change in the gage B reading for the 24,950 kg (55,000 lb.) load level that a crack has formed at the top surface. The load-related strain increment at the surface was approximately 135 microstrains. This strain can be related to a load-related maximum fiber stress provided the modulus of elasticity E of the concrete is known. However, it must also be considered that the total stress operating on the concrete slab, and which leads to the rupture, is the sum of the load-related stress and the built-in, or residual, stress:

$$\sigma_{tot} = E\varepsilon_{load} + \sigma_{res} \qquad (1)$$

where ε_{load} is the load-related strain increment measured at the strain gage. If the in-situ residual stress can be measured then the total stress driving the crack formation can be estimated. An innovative means of measuring the residual stress in the slab was developed at the NAPTF [5] and built on by work at the University of Illinois under the FAA Center of Excellence for Airport Technology (CEAT) program [6]. In this procedure, a surface strain gage is applied to the slab. Then sawcuts are made on both sides of the gage to a sufficient depth to release the residual stress. After allowing sufficient time for the heat generated by the sawcut

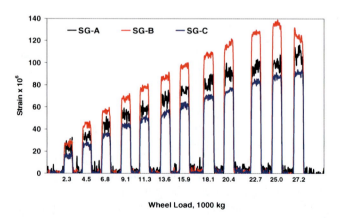

Fig. 5. Surface strain gage readings vs. wheel load for top-down strength test

operation to dissipate, the difference between the initial and final strain gage readings is proportional to the built-in stress. A series of such tests was performed on slab 22N, which gave inconclusive results. While an accurate estimate of residual stress is needed to compare the total slab top-down stress at rupture to the rupture stress from C78 tests, it is important to recognize that the wheel load based on strain ratio is not affected by the built-in stress.

3 Concrete Beam Tests

Prior to running the traffic test, an appropriate wheel load needed to be determined. Ideally, the wheel load should provide a reasonable number of passes to failure without exceeding the cracking strength for the low-strength slabs. A total of twelve $150 \times 150 \times 550$ mm concrete beams were tested for both flexural strength (following ASTM C78 [3]), and fatigue resistance using the same test equipment. These beams were cast from four slabs in MRS-1N (low-strength) during concrete placement (April 2010) and had been cured under laboratory conditions since then.

One of the key considerations in the fatigue test is the loading frequency. Kesler [7] found that loading frequencies between 1 and 7 Hz did not significantly affect the fatigue resistance of the concrete. In low frequency fatigue tests, time-dependent characteristics such as creep and shrinkage influence the results. Awad and Hilsdorf [8] found the fatigue resistance of concrete in the low-cycle (high stress ratio) regime to be highly load rate sensitive, but for lower stress ratios the loading rate did not have an appreciable affect on fatigue life. In the present study, the loading frequency was set at 2 Hz with no unloading between pulses. A similar loading rate was previously used in large-scale concrete slab fatigue tests by Roesler et al. [9]. The ratio of minimum to maximum cyclic load was maintained at 10 percent during fatigue testing. Fatigue tests were conducted at two stress ratios, 0.7 and 0.8. The flexural strength values for the four slabs are given in Table 2. The values are fairly consistent, except for slab 2N, which was somewhat higher than the others. The average beam flexural strength was 4.82 MPa (699 psi). On the other hand, fatigue test results demonstrated very high scatter. As given in Table 2, the number of cycles to failure for an 80% stress ratio ranged from 363 (slab 2N) to over 5000. Overall, the mean fatigue strength results proved to be consistent with the 5% probability curve proposed by Hilsdorf and Kesler [10]. Based on the cycles to failure in the beam test for the 80% stress ratio, it was decided to use a load ratio (i.e., a ratio of traffic load to measured in situ slab cracking load) of 0.8 in the subsequent traffic tests. The beam data in Table 2 represent part of a more extensive program of beam fatigue tests (in progress), involving up to 240 beam samples from all three CC6 test item placements. The complete data set will be analyzed to determine how well the number of fatigue cycles corresponds to observed coverages to failure in the CC6 tests.

Table 2. Summary of Concrete Beam Test Results

Slab ID	Air, %	Unit Wt., kg/m^3	w/c ratio	Max. Load, kN	Flexural Strength, MPa	Stress Ratio	Cycles to Failure
2N	7.0	2320	0.48	41.2	5.5	-	-
				30.5	-	0.8	363
				26.7	-	0.7	865
4N	7.0	2275	0.54	37.2	4.5	-	-
				30.5	-	0.8	1384
				26.7	-	0.7	2945
22N	6.0	2278	0.48	36.7	4.6	-	-
				30.5	-	0.8	776
				26.7	-	0.7	4637
24N	6.5	2272	0.59	37.1	4.6	-	-
				30.5	-	0.8	5112
				26.7	-	0.7	4141

4 Zero-Wander Traffic Tests

The objectives of the zero-wander traffic tests included the following:

(1) Comparisons between distresses developed under zero-wander traffic and under normal wander will help to evaluate the accuracy of the existing pass-to-coverage model used in FAARFIELD.
(2) The rigid pavement failure can best be understood as a progression through three stages [11]. Stage one is from the new pavement condition to the initiation of a crack. In this case, the crack was initiated by the static load, and additional traffic was applied to complete stage two, characterized by the development of a full-length and full-depth crack. The stage two failure mechanism can be quantified more easily in a zero-wander test.

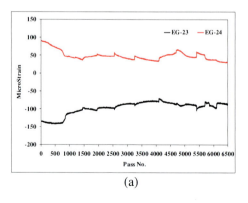

(a)   (b)

Fig. 6. (a) Strain gage readings during bottom-up crack propagation test. (b) Core from slab 22N indicating progress of bottom-up crack after 6500 passes.

4.1 Bottom-Up Crack Propagation Test

After the initial bottom-up crack was created by static loading of slab 22N, a rolling gear load at 1.34 m/s (3 mph) was applied to the slab at 80% of the bottom-up cracking load (195.7 kN per wheel) to propagate the crack to the surface. The lateral gear position of Figure 2(a) was used with no wander. Peak strain histories from gages EG-23 (top) and EG-24 (bottom) are presented in Figure 6(a). Up to the first 1000 passes, peak readings from both gages decreased dramatically, indicating that, initially, the bottom-up crack propagated quickly. This initial phase was followed by a period of stabilization attributed to the change in structure. After 6500 passes, the test was terminated without observing the crack at the surface. Cores (Figure 6(b)) verified the presence of the incomplete bottom-up crack. The reason that the bottom-up crack stabilized at this point is not definitely known, but it is probably linked to a redistribution of bending stresses in the post-cracked structure. In any case, it is noted that the longitudinal crack in question eventually did reach the surface after application of an additional 1650 passes of the 2D gear during the traffic test phase (i.e., with full wander) at a slightly higher per-wheel load of 200.1 kN (45,000 lbs.).

4.2 Top-Down Crack Propagation Test

Although strain gage analysis (Fig. 5) identified the beginning of a surface crack in the vicinity of SG-B, it was not visible to the naked eye. A rolling gear load at 80% of the surface cracking load (195.7 kN per wheel) was applied to slabs 3N and 22N to propagate the surface crack to full depth. The lateral gear position used was the same as for the top-down static load test, Fig. 2(b). The top-down crack propagated much more quickly than the bottom-up crack under the same wheel load. A top-down crack was visible after 18 passes, and traffic was stopped after 190 passes, when strain gage analysis indicated that the crack had progressed through the slab.

5 Conclusions

A series of tests was performed to obtain the fracture load for full-scale rigid test items at the NAPTF, for both bottom-up and top-down load configurations. As a result of these investigations, an initial wheel load of approximately 80% of the rupture load was established for trafficking all CC6 test items. Traffic tests are underway at the time of writing. Concrete flexural strength obtained from beam samples (ASTM C78) is an essential input parameter for FAA rigid pavement design. However, the resistance to failure of a beam is significantly different from that of a pavement. The full-scale tests described in this paper provide reliable data for relating the strength and fatigue strength of pavement slabs to standard beams. Finally, the test allowed comparison of the failure mechanism for bottom-up and top-down cracks. Under similar traffic, the top-down crack progressed into the full-depth and full-length condition much more quickly than did the bottom-up crack.

Acknowledgments/Disclaimer. The work described in this paper was supported by the FAA Airport Technology R&D Branch, Dr. Satish K. Agrawal, Manager. Special thanks are due to Dr. Gordon F. Hayhoe, NAPTF Manager, for technical leadership in test planning and organization, Mr. Chuck Teubert of SRA International, Inc. for construction and test management, and Dr. Shelley Stoffels of the Pennsylvania State University for concrete sample testing support. The contents of the paper reflect the views of the authors, who are responsible for the facts and accuracy of the data presented within. The contents do not necessarily reflect the official views and policies of the FAA. The paper does not constitute a standard, specification, or regulation.

References

[1] Web page of the FAA Airport Technology R&D Team, Construction Cycle Two (CC-2) Test Items,
http://www.airporttech.tc.faa.gov/naptf/cc2d.asp
[2] Federal Aviation Administration: Standards for Specifying Construction. Advisory Circular 150/5370-10E, FAA, Washington, DC, USA (2009)
[3] ASTM Standard C78: Standard Test Method for Flexural Strength of Concrete (Using Simple Beam with Third-Point Loading). ASTM International, West Conshohocken, PA, USA (2009), http://www.astm.org
[4] Brill, D.R., Wang, Q., Guo, E.H.: Finite Element Simulation of Surface Strain Gage Measurements in Rigid Airport Pavements. In: Proceedings of the 2009 European Airport Pavement Workshop, Amsterdam, May 13-14 (2009)
[5] Guo, E.H., Pecht, F., Ricalde, L.: Pavement Cracking, Mechanisms, Modeling, Detection, Testing and Case Historie. In: Al-Qadi, I.L., Scarpas, T., Loizos, A. (eds.) Proc. of the 6th RILEM Conference on Cracking in Pavements, pp. 25–34. CRC Press (2008)
[6] Marks, D.G., Lange, D.A.: Development of Residual Stress Measurement for Concrete Pavements Through Cantilevered Beam Testing. In: Proc. 2010 FAA Airport Tech. Transfer Conf., Atlantic City, NJ, USA (April 2010)
[7] Kesler, C.E.: Effect of Speed of Testing on Flexural Fatigue Strength of Plain Concrete. In: Proc. Highway Research Board, vol. 32, pp. 251–258 (1953)
[8] Awad, M.E., Hilsdorf, H.K.: Strength and Deformation Characteristics of Plain Concrete Subjected to High Repeated and Sustained Loads. In: Abeles Symposium, Fatigue of Concrete, vol. SP-41, pp. 1–13. ACI Pub. (1974)
[9] Roesler, J.R., Hiller, J.E., Littleton, P.C.: Large-Scale Airfield Concrete Slab Fatigue Tests. In: Proc. of the 9th Intl. Conference on Concrete Pavement, Colorado Springs, Colorado, USA, August 13-18, pp. 1247–1268 (2005)
[10] Hilsdorf, H.K., Kesler, C.E.: Fatigue Strength of Concrete Under Varying Flexural Stresses. ACI Journal and Proceedings 63(10) (1966)
[11] Guo, E., Hayhoe, G.: Three-Stage Failure Mechanisms of Concrete Pavement – Failure Stage One: Initiation of Cracks. In: Proceedings of 7th DUT Workshop on Concrete Pavements, Seville, Spain, October 10-11 (2010)

The Effects Non-uniform Contact Pressure Distribution Has on Surface Distress of Flexible Pavements Using a Finite Element Method

Dermot B. Casey[1], Andrew C. Collop[2], Gordon D. Airey[1], and James R. Grenfell[1]

[1] Nottingham Transportation Engineering Centre (NTEC), University of Nottingham, University Park, Nottingham NG7 2RD, UK
dermot.casey@nottingham.ac.uk
[2] Faculty of Technology, De Montfort University, Leicester,
acollop@dmu.ac.uk

Abstract. The current practice in pavement design is to use a circular uniformly distributed load as the input to ascertain the maximum stresses in the pavement. This is not the reality; tyre-pavement contact stress distributions are very complex. The distress on the surface of the pavement in the form of rutting and surface initiated cracking is very much dependent on these complex stresses. This study investigates the effects that non-uniform contact pressure distributions have in comparison to the traditional circular loading on the initiation and rate of accumulation of this distress. The problem has been modelled using the CAPA-3D finite element software. The traditional circular load was modelled for two different asphalt materials with different moduli. The stresses in a number of key locations were recorded and measured. Then the non-uniform loading was modelled using the same procedure. What was of particular interest was the difference in the peak surface stresses and positions between the two different methods of loading. The non-uniform loading created significantly larger stresses on the surface in comparison to the circular loading. The non-uniform loading also created significant shearing forces close to the surface. This leads to a greater propensity for the surface to develop rutting and cracking to develop at the surface. The differences started to become less evident with depth and/or distance from the loading area for the principal stresses. It is recommended that for the design of surface layers non-uniform contact pressures should be used.

1 Introduction

Tyre inflation pressure has increased significantly since the AASHO road tests of the 1960s (1). There has also been an increase in the permitted axle loading. This has led to an increase in both the inflation pressure and axle load and a change in tyre type from bias ply tyres to radial ply tyres. This creates increased loads with higher inflation pressures on tyres with greater non-uniformity of pressure (2).

This added to the development of wide base tyres with smaller contact areas increases the problem further. It is believed that this leads to increased damage to the pavement and premature pavement damage, especially at the surface. The trend in increased inflation pressures has been seen in a study by Morton (3). The survey showed that from 1974-1995 the average inflation increased from 620kPa to 733kPa.

The current practice in pavement engineering is to use a circular uniform contact patch to represent a tyre load. This is used in conjunction with a Layered elastic program like BISAR to estimate the maximum tensile strain at the bottom of the asphalt package and the maximum compressive strain at the top of the subgrade. However, this method was shown to overestimate the tensile strain at the bottom of the asphalt layers and the compressive strain at the top of subgrade (4). This study also studied the effects of tyre pressures on pavement response; it was shown that increased tyre pressure can increase the propensity of the pavement to fatigue and rutting damage. De Beer has done a number of studies using the Vehicle-Road Surface Pressure Transducer Array (VRSPTA) to quantify the magnitude and range of contact pressure geometries in the vertical, lateral and longitudinal directions (5-8). This has given a great insight in the true nature of contact pressures that pavements are subjected to. It has been shown that the reality of contact pressure is far removed from the idealised scenario of a uniform circular contact patch. The pressure is highly non-uniform with peaks of vertical pressure 1-2 times the inflation pressure (the inflation pressure is usually used to represent the contact pressure). The shape is predominately rectangular and the width is relatively constant over a range of inflation pressures and axle loads. The shape of the contact pressure is dependent on the combination of tyre type, inflation pressure and axle load. This makes the increase in the inflation pressure, axle load and change in tyre designs rather worrying.

The non-uniformity of contact pressure leads to high stresses and strains on and near the surface of the pavement. These stresses and strains are higher than those created at the bottom of the asphalt and the top of the subgrade. This then gives rise to premature distress and maintenance interventions and causes an increased economic cost to the infrastructure stakeholders. The phenomenon of top down cracking has been highlighted (9-11). The cause of this cracking has not been conclusively proven but tyre contact pressures are believed to be a leading factor in this behaviour (12). The distress mode of surface rutting is also linked to the contact pressure being imposed on the pavement (13) . These two modes of distress are influenced by the type and magnitude of stresses and strains on and near the surface of the asphalt layer. The nature of the contact pressure influences both the size of the peak stresses and strains but also where they are observed e.g. inside or outside the contact area (14).

In this paper, the effects of non-uniform vertical contact pressure on pavement response were investigated using a linear elastic constitutive model in the CAPA-3D Finite element package. The pavement response was predicted for the near surface stresses and strains under the loading area and out from it. A uniform circular contact patch was applied and a rectangular contact patch with simplified non-uniform contact pressure from De Beer's VRSPTA was applied to a Finite Element mesh. These two loads and two different asphalt moduli (one high, one low) were used to investigate the effect of non-uniform contact pressure.

2 Objectives

- To establish the variation in the stresses induced on the surface and the near surface by non-uniform contact pressure in comparison with uniform contact pressure.
- To illustrate the affect that the Young's modulus has on the observed stresses on the surface and the near surface.
- To illustrate how the effects of the non-uniform contact pressure reduce rapidly with depth and/or distance from the contact area.

3 Procedure

3.1 Mesh Validation

Choosing the correct mesh geometry with proper boundary conditions is a critical element of the analysis. It is an area that has the largest input from the user and as such is prone to the most mistakes. The refinement of the mesh was a time consuming activity but is essential to balance the refinement of the mesh with the conflicting requirement of minimising the computational resources required. It was decided to create a complete model with no axes of symmetry present due to the nature of non-uniform contact pressures. The method of validating the mesh was to compare the stress and strain outputs at specific locations with that of BISAR for the uniform circular loading. BISAR is a well used layered elastic program that has been validated numerous times and represents an excellent benchmarking tool for the Finite Element model. The output from the near surface and under the loading was of particular interest and was used for the validating procedure.

The mesh that is presented in Figure 1 is the mesh that was chosen from the analysis of various mesh geometries and boundary conditions. It represents a good mix between accuracy of stresses in the analysis area and reasonable running times. The base of the mesh is fixed in the x, y and z planes, and the four sides of the model are restrained in the horizontal direction. The overall dimension of the model of 4m by 4m in the x and z planes and 2.45m in the y plane was chosen to make the end effects of the boundary conditions negligible to the analysis. The layers are the subgrade 2m in y-axis, base 0.3m in y-axis, and bituminous layers 0.15m in y-axis

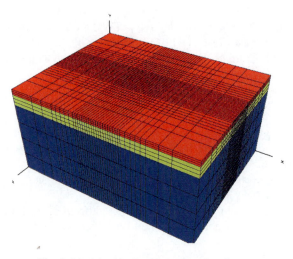

Fig. 1. Model with the selected mesh refinement

Table 1. Model Material Properties

	Young's Modulus (MPa)	*Poisson Ratio*
Asphalt	4500, 2000	0.35
Base	700	0.35
Subgrade	100	0.35

Stress and strain outputs at a depth of 2.5mm and 27.5mm compared well with that of BISAR. The Finite Elements solution slightly underestimated the stresses and strains which is common for Finite Elements. This is an acceptable error as it is expected and only represents a small variation and will be constant in both the uniform and non-uniform analysis. The material properties of the model are shown in Table 1. The subgrade and base properties are constant but there are two moduli used for the asphalt layer. The base modulus is on the high side as it is the purpose of this study to investigate the stresses and strains in the asphalt layer.

3.2 Selection of Loading

The loading of the circular contact patch was chosen as 150mm radius with a uniform vertical loading of 666kPa. This is a common representation of a tyre loading both for the contact pressure and the area of loading. This also is approximately the area of the non-uniform contact patch which lends itself to a good comparison. The loading was chosen as it is a common contact pressure and will show the stresses created by this loading scenario. The non-uniform contact pressure is a rectangle of 240mm wide by 285mm long for a 315/80R22.5 tyre that is commonly used on the steering axle. The measurements of the tyre were obtained from De Beer's VRSPTA. The author has divided the pressure into 9 regions of constant contact pressure based on the distribution of the readings. The width is divided into 20% 60% 20% as is the length. The readings in these areas are summed and averaged to obtain the reading for that region. This gives a good estimation of the nature of the contact pressure of the tyre. The magnitude of the loading on these regions can be seen in Figure 2.

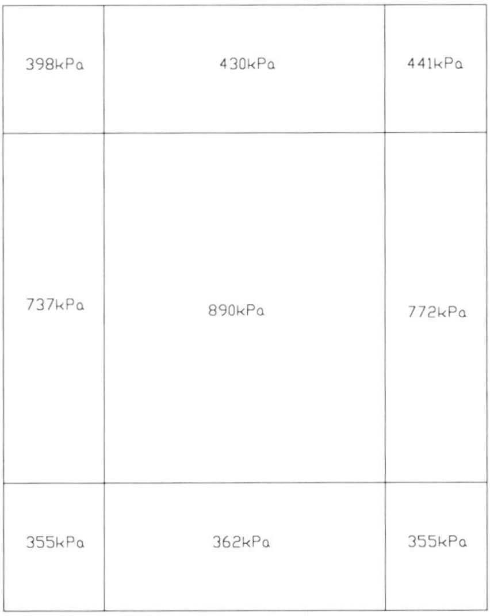

398kPa	430kPa	441kPa
737kPa	890kPa	772kPa
355kPa	362kPa	355kPa

Fig. 2. Map of the non-uniform contact patch

3.3 Output Positions

CAPA-3D gives text stress and strain output at the integration points in the elements. These points are the positions where these outputs are most accurate and as such these are the points that were used for visualising the output from each analysis. The purpose of this paper is to compare the near surface stresses; therefore, the upper most elements were used. These points are at a depth of 2.5mm, then to observe the effect depth has on the output, points at 27.5mm were also represented. The cross section has been taken underneath the loaded area moving out along the x plane to a distance of 500mm from the centre of loading. A close up of this area of the mesh can be seen in Figure 3.

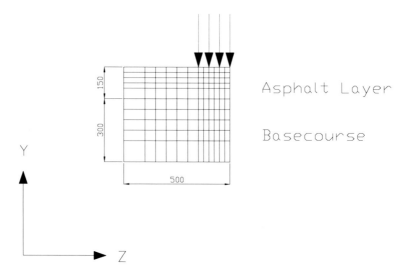

Fig. 3. Close-up of the area of interest for analysis

This area will be used to show the differences in all the loading scenarios and the change in the two moduli for the asphalt layer.

4 Results and Discussion

4.1 Near Surface Stresses

In this section the graph from the 2.5mm deep points will be presented. The graphs of the principal stresses and the shearing stresses will be presented and discussed. The variation in stresses can be easily seen from the graphs, especially in the shearing stresses which are very interesting to observe. The legend of the graphs refers to the type of loading uniform/non-uniform and the two asphalt moduli of 2000MPa (low) and 5000MPa (high).

In Figure 4 the variation in the principal stresses can be seen from the different loading schemes and asphalt layer moduli. The stress levels and shape are shared by the non-uniform high and low analysis. The uniform loading shows that the low modulus has less stress in comparison to the high modulus for the x and z stress. The nature of the uniform and non-uniform pressure is completely different. The uniform stays constant over the majority of the loaded area. On the other hand, the non-uniform varies widely across the loading area and only settles down outside the area of loading like the uniform pressure. Another interesting observation is that outside the loading area the stress caused by the non-uniform loading is less that the high modulus uniform analysis for the x and z stress.

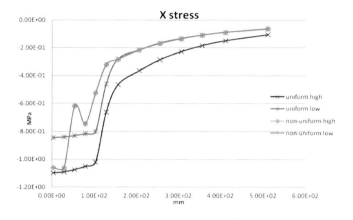

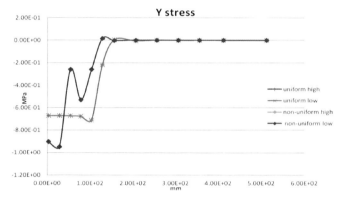

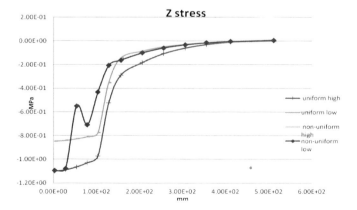

Fig. 4. Principal Stresses for 2.5mm Depth

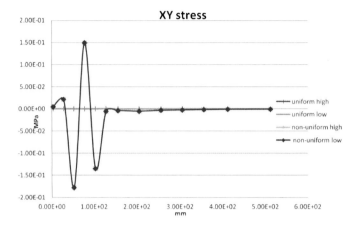

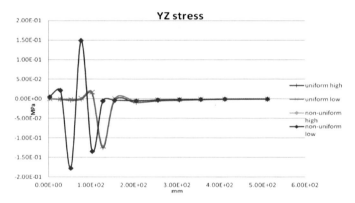

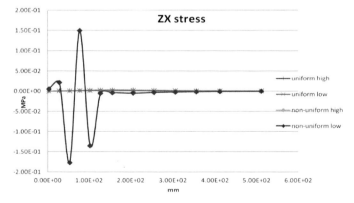

Fig. 5. Shear Stresses for 2.5mm Depth

The difference in shearing stresses is starkest when contrasting the uniform and non-uniform loading. In all three components of shearing the non-uniform stresses vary widely from – to + sign. Whereas the uniform contact it only exhibits shearing in the ZY direction and then only at the edge of the loading area. In the XY and ZX the uniform pressure develops no shear stress. The non-uniform demonstrates shearing across the loading area. This is quite a difference and is something that is not accounted for in design of pavements, materials and testing standards.

4.2 27.5mm Deep Principal Stresses

These graphs (Figure 6) of the principal stresses show the reduction in stress with depth and also the normalisation of these stresses. The non-uniform stress still has variation under the loading area, but the magnitude of this is less and the variation is less. There is also now a change in the values for the high and low modulus non-uniform analysis. The loads are spreading out becoming reduced and the non-uniform load is becoming less important for these stress components.

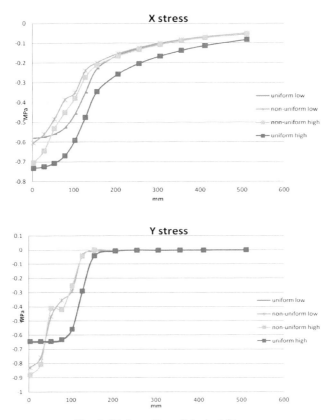

Fig. 6. 27. 5mm Deep Principal Stresses

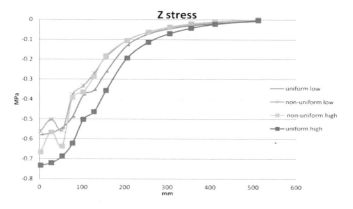

Fig. 6. *(continued)*

5 Conclusions

The analysis that was undertaken in this paper has led to the following conclusions:

- The non-uniform contact pressure leads to varying principal stresses under the contact area at near surface depths (2.5mm).
- The shearing stresses are significant close to the surface for the non-uniform contact, whereas the uniform contact pressure produces approximately zero shearing for the XY and ZX. There is shearing produced outside the contact area for the uniform pressure in the YZ, however, this is less than the non-uniform shear.
- The variation of the principal stresses for the non-uniform contact pressure is less varied with depth.
- At the 2.5mm depth the modulus has no effect on the output for the Y stress for the uniform and non-uniform contact pressures. There is a variation for the X and Z stresses between the low and high modulus for the uniform contact pressure.

Overall, it is observed that the non-uniform contact pressure leads to much more variation of the stresses under the contact area and close to the surface (2.5mm depth). This leads to stresses of a different nature than are predicted by the uniform contact pressure used in the conventional design procedure. There are also significant shearing forces for the non-uniform contact pressure at the 2.5mm depth leading to stresses at this depth of a different nature to the uniform contact pressures. This can lead to cracking initiating in the surface through the shearing forces and the increase in variation of the principal stresses at the 2.5mm depth under the contact area

Acknowledgements. The authors would like to acknowledge the support of the European Commission under the Marie Curie Intra-European Fellowship Programme. The authors would like to acknowledgement Morris DeBeer of CSIR in South Africa for supplying contact pressure measurements. The authors would also like to thank the members of the CAPA-3D group at TU Delft under the leadership of Prof. Tom Scarpas for their help.

References

[1] American Association of State Highway & Transportation Office, AASHTO interim guide for design of pavement structures, ch. III revised (1981)
[2] Greene, J., Toros, U., Kim, S., Byron, T., Choubane, B.: Impact of Wide-Base Single Tires on Pavement Damage. Transportation Research Record (2155), 82–90 (2010)
[3] Morton, B.S., Luttig, E., Horak, E., Visser, A.T.: The Effects of Axle Load Spectra and Tyre Inflation Pressures on Standard Pavement Design Methods. In: 8th Conference on Asphalt Pavements for Southern Africa, CAPSA 2004 (2004)
[4] Wang, F., Machemehl, R.B.: Southwest Region University Transportation C, University of Texas at Austin. Center for Transportation R, & University Transportation Centers P, Predicting truck tire pressure effects upon pavement performance (Southwest Region University Transportation Center, Center for Transportation Research, University of Texas at Austin, Austin, Tex.) (2006)
[5] De Beer, M., Fisher, C., Jooste, F.J.: Determination of pneumatic tyre/pavement interface contact stresses under moving loads and some effects on pavements with thin asphalt surfacing layers. In: Eight (8th) International Conference on Asphalt Pavements (8th ICAP 1997), pp. 179–227 (1997)
[6] Blab, R.: Introducing Improved Loading Assumptions into Analytical Pavement Models Based on Measured Contact Stresses of Tires. In: International Conference on Accelerated Pavement Testing, Reno, NV (1999)
[7] Fernando, E.G., Musani, D., Park, D.-W., Liu, W.: Evalution of Effects of Tire Size and Inflation Pressure on Tire Contact Stresses and Pavement Response, p. 288 (2006)
[8] De Beer, M., Fisher, C., Kannemeyer, L.: Tyre-pavement interface contact stresses on flexible pavements – quo vadis? In: 8th Conference on Asphalt Pavements for Southern Africa (CAPSA 2004), pp. 1–22 (2004)
[9] Uhlmeyer, J.S., Willoughby, K., Pierce, L.M., Mahoney, J.P.: Top-Down Cracking in Washington State Asphalt Concrete Wearing Courses. Transportation Research Record (1730), 110 (2000)
[10] Jacobs, M.M.J., Hopman, P.C., Molenaar, A.A.A.: The crack growth mechanism in asphaltic mixes. Heron-English Edition 40(3), 181–200 (1995)
[11] Matsuno, S., Nishizawa, T.: Mechanism of Longitudinal Surface Cracking in Asphalt Pavement. In: Proceedings of the 7th International Conference on Asphalt Pavements, Nottingham, pp. 277–291 (1992)
[12] Baladi, G.Y., Schorsch, M.R., Svasdisant, T.: Determining the causes of top-down cracks in bituminous pavements (Michigan Dept. of Transportation, Construction & Technology Division, Testing and Research Section, Lansing, MI) (2003)
[13] Novak, M., Birgisson, B., Roque, R.: Tire contact stresses and their effects on instability rutting of asphalt mixture pavements - Three-dimensional finite element analysis. Pavement Management and Rigid and Flexible Pavement Design (1853), 150–156 (2003)
[14] Perret, J.: The effect of loading conditions on pavement responses calculated using a linear-elastic model (2003)

Finite Element Analysis of a New Test Specimen for Investigating Mixed Mode Cracks in Asphalt Overlays

M.R.M. Aliha[1], M. Ameri[1], A. Mansourian[2], and M.R. Ayatollahi[1]

[1] Iran University of Science and Technology, Narmak, 16846-13114, Tehran, Iran
[2] Transportation Research Institute, Iran Ministry of Road and Transportation, Tehran, Iran

Abstract. Cracking is a common mode of deterioration in asphalt pavements. In general, cracks in the asphalt pavements experience a combination of opening and sliding deformation due to thermal and traffic loads. In this research, a new test specimen called ASCB is proposed for mixed mode I/II fracture toughness study of asphalt materials. The ASCB specimen is a semi-circular specimen containing a crack normal to the specimen edge and subjected to asymmetric three-point bend loading. Simple geometry and convenience of testing set up are two primary advantages of the ASCB specimen. In addition, the disc shape of specimen facilitates its preparation using the conventional gyratory compactor machines or using the asphalt field coring devices. The stress intensity factors (K_I and K_{II}) are fundamental parameters in order to characterize the load bearing capacity of asphalt failure due to brittle fracture or fatigue crack growth. Hence, in this paper the stress intensity factors of the ASCB specimen are calculated from several finite element analyses and for different mixed mode loading conditions. The numerical results show that the complete mode mixities ranging from pure mode I (opening mode) to pure mode II (in-plane sliding) can be achieved from the ASCB specimen by changing the loading support positions relative to the crack plane. It is also shown that the suggested laboratory specimen is also very suitable for simulating the stress and deformation fields of real cracked pavements which are subjected to the loads induced by the wheels of the moving vehicles.

1 Introduction

Service life of asphalt pavements is an important issue in most countries having long networks of roads and highways. Annually huge amount of money is spent for maintenance of asphalt pavements [1,2]. Cracking is a primary and common mode of deterioration and one of the main causes for overall failure of asphalt pavement of roads and highways especially in cold regions [3-5]. Fig. 1 shows three basic modes of deformations namely: mode I (opening), mode II (in-plane sliding) and mode III (out of plane tearing) in a typical cracked asphalt pavement which can be induced by thermal/mechanical loads. Those cracks which are found across the pavement surface can be subjected to complex states of stress and deformation induced by cyclic thermal loads or mechanical traffic loads. In general, such cracks in the asphalt pavements experience a combination of opening and sliding deformation; often called mixed mode I/II loading.

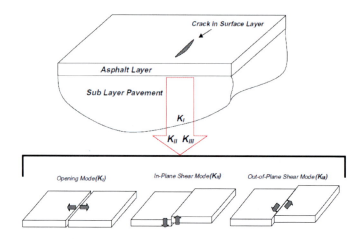

Fig. 1. Three modes of deformations for a cracked asphalt overlay

Since the formation of cracks in asphalt layers is almost inevitable, the investigation of mixed mode crack growth in asphalt pavements is important for estimating the suitable rehabilitation time of pavements and the service capability of the roads and highways. For any given cracked material, mixed mode fracture is usually studied experimentally using appropriate test methods. Some test specimens have been used in the past by researchers in order to investigate experimentally the mixed mode fracture resistance of asphalts. The rectangular beam specimen subjected to asymmetric three or four-point bend loading [6,7] and the inclined edge crack semi-circular bend specimen are two well-known examples [8].

A suitable test specimen should have simple configuration, inexpensive preparation procedure, convenience of testing set up and also the ability of introducing complete mode mixities ranging from pure mode I to pure mode II. In this research, a new test specimen called ASCB is proposed for mixed mode fracture toughness study of asphalt materials. First the specimen is described and then its capabilities and advantages are investigated by means of finite element analyses.

2 New Test Configuration

Figure 2 shows the geometry and loading conditions for the asymmetric semi-circular bend (ASCB) specimen. In this test configuration, a semi-circular specimen of radius R that contains an edge crack of length a emanating normal to the flat edge of the specimen is loaded asymmetrically by a three-point bend fixture. Simple geometry and convenience of testing set up are two primary advantages of the ASCB specimen. In addition, the circular shape of specimen facilitates its preparation using the conventional gyratory compactor machines or using the asphalt field coring devices, without any additional machining procedure. Moreover, the crack is always along the symmetry line of the semi-circle. The state of mode mixity in the ASCB specimen can be easily altered by changing the locations of two bottom supports (*S1* and *S2*). When the bottom loads are applied symmetric to the crack line (i.e. when *S1* = *S2*) the specimen is subjected to pure mode I. But for asymmetric loading (i.e. *S1* ≠ *S2*), mode II appears in the crack deformation in addition to mode I. The mode I and mode II contributions can be controlled simply by choosing appropriate values for *S1* and *S2*. Hence different mode mixities can be obtained in the proposed specimen. The specimen has been frequently used in the past but only for the simple case of symmetric loading conditions in order to obtain pure mode I fracture toughness for several engineering materials including, concrete and asphalt [5, 9-11].

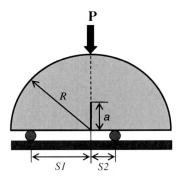

Fig. 2. Suggested ASCB specimen for mixed mode fracture toughness studies on asphalt materials

The stress intensity factors (K_I and K_{II}) are fundamental parameters in order to characterize the load bearing capacity of asphalt pavement due to brittle fracture or fatigue crack growth. Hence, in this paper the stress intensity factors of the ASCB specimen are calculated using several finite element analyses and for different mixed mode loading conditions (i.e. loading positions and crack lengths). More details of calculations will be given in the next section.

3 Numerical Analysis

The stress intensity factors K_I and K_{II} for the ASCB specimen are functions of the crack length (*a*) and the locations of loading supports defined by *S1* and *S2* and can be written as:

$$K_I = Y_I \frac{P}{2Rt}\sqrt{\pi a} \qquad (1)$$

$$K_{II} = Y_{II} \frac{P}{2Rt}\sqrt{\pi a} \qquad (2)$$

where t is the specimen thickness and Y_I and Y_{II} are the geometry factors corresponding to mode I and mode II, respectively. These geometry factors are functions of a/R, $S1/R$ and $S2/R$. For calculating Y_I and Y_{II}, different models of the ASCB specimen were analyzed. Fig. 3 shows a typical mesh pattern generated for simulating the ASCB specimen. In the models, the following geometry and loading conditions were considered: $R = 60$ mm, $t = 20$ mm, $P = 1000$ N and different values for crack lengths. $S1$ was set at a fixed value of 40 mm and $S2$ varied from zero to 40 mm to change the state of mode mixity. Although, asphalt is a composite material, it is often modeled by an equivalent isotropic and homogenous material. Moreover, at subzero temperatures, asphalt of pavements usually behaves as a linear elastic and brittle material (see e.g. [12–14]). Hence, most of the researchers have used the hypothesis of linear elastic fracture mechanics for modeling the cracked asphalt pavement under low temperature conditions [12–14].

Therefore in this research, the construction material is assumed to be isotropic, homogenous and linearly elastic for obtaining the fracture parameters of the ASCB specimen. The elastic material properties of a typical asphalt mixture as $E = 2760$ MPa and $v = 0.35$ were also considered in the finite element models.

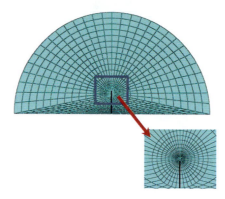

Fig. 3. A typical finite element mesh used for simulating ASCB specimen

A total number of 2162 eight-noded plane stress elements were used for each model. The singular elements were considered in the first ring of elements surrounding the crack tip for producing the square root singularity of stress/strain field. A J-integral based method which is readily available in ABAQUS finite element code was used for obtaining directely the stress intensity factors of the sugusted ASCB specimen. Figs. 4 and 5 show the values of Y_I and Y_{II} calculated from several finite element analyses performed for different loading conditions in the ASCB specimen. It is seen from these Figures that for the symmetric loading conditions (i.e. $S1 = S2$), Y_{II} equals zero and thus the specimen is subjected to pure mode I loading. By changing the location of the second loading support $S2$, the mode II component also appears in the ASCB specimen. It is seen from Figs. 4 and 5 that by moving $S2$ towards the crack plane, the mode I geometry factor Y_I decreases and the mode II geometry factor Y_{II} increases. According to Figs. 4 and 5, the mode I geometry factor increases by increasing a/R, $S1/R$ and $S2/R$. But Y_{II} decreases when $S2/R$ becomes greater. For each value of crack length ratio (a/R), there is a specific value for $S2$ where Y_I becomes zero while Y_{II} is non-zero. This loading situation corresponds to pure mode II conditions. Fig. 6 shows the loading and geometry conditions that correspond to pure mode II deformation in the ASCB specimen for some combinations of a/R, $S1/R$ and $S2/R$.

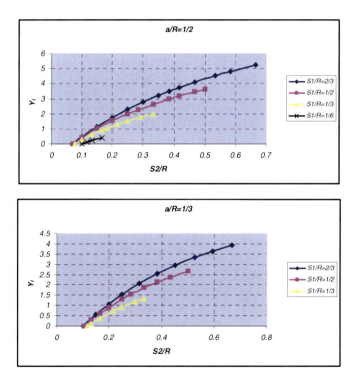

Fig. 4. Mode I geometry factor Y_I for different values of a/R, $S1/R$ and $S2/R$ in the analyzed ASCB specimen

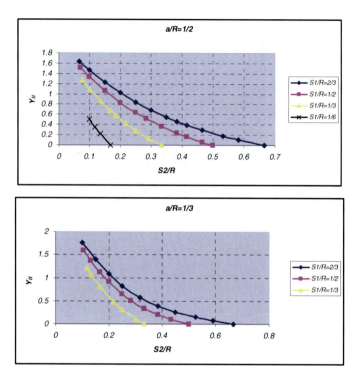

Fig. 5. Mode II geometry factor Y_{II} for different values of *a/R*, *S1/R* and *S2/R* in the analyzed ASCB specimen

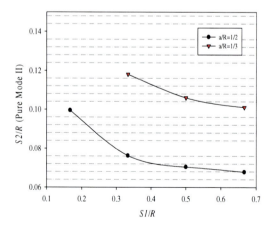

Fig. 6. Positions of bottom supports for pure mode II conditions in the ASCB specimen

4 Disscussion

A review of the test specimens suggested in the past for conducting mixed mode fracture experiments shows that some of them can not be used as a favorite test specimen for asphalt materials. For example, the compact tension-shear (CTS) specimen suggested by Richard [15] and Arcan et al. [16] consists of a cracked specimen loaded through a complicated loading fixture. The auxiliary fixture not only makes the experiments more expensive but also may sometimes be a source of error in the test results due to possible manufacturing inaccuracies. Moreover, such samples require direct tensile loading which is not suitable for asphalt materials. The ASCB specimen suggested in this paper does not need any additional loading fixture since it can directly be tested by an ordinary three-point bend fixture which is normally available in standard fracture testing machines. Furthermore, the application of compressive loads in the ASCB experiments makes it more suitable for conducting fracture tests on asphalt specimens which are weak against tensile loads. Although almost all of the mixed mode test specimens can be used for pure mode I and mixed mode fracture tests, some of them are not able to provide pure mode II. For instance, the edge cracked rectangular beam specimen under asymmetric three-point bend loading [6,7] can produce only limited combinations of mode I and mode II. In particular, they cannot be used for pure mode II tests. Another advantage for the ASCB specimen is its ability for providing complete combinations of mode I and mode II , since the numerical finite element results show that the complete mode mixities ranging from pure mode I (opening mode) to pure mode II (in-plane sliding) and various intermediate mode mixities can be achieved from the ASCB specimen by changing the loading support positions relative to the crack plane. It is worth mentioning that the cracked test specimens which contain only one crack tip (like the ASCB specimen), are often preferred to centrally cracked specimens that contain two crack tips (e.g. the centrally cracked Brazilian disk specimen). This is because the crack extension does not necessarily initiate from the two crack tips simultaneously. The delay between the fracture initiations at the two crack tips is not controllable and can sometimes be a likely source of error in the test results. The laboratory specimen suggested in this paper is also very suitable for simulating the stress and deformation fields of real cracked pavements which are subjected to the loads induced by the wheels of the moving vehicles. As shown shematically in Fig. 7, the position of moving vehichles relative to a crack in the surface of an overlay may change the state of crack flank deformation from opening (mode I) to sliding (mode II) [17]. Thus the numerical results obtained in this research for ASCB specimen can be used for investigating the real cracked overlays subjected to traffic loading.

Because of the advantages elaborated above, the ASCB specimen can be recommended as a favorite test sample for conducting mixed mode fracture experiments on asphalts. However, the practical and experimental ability of the ASCB specimen for providing reliable predictions for asphat fracture behaviour is another key issue for the proposed specimen. This subject is currently being studied in a complementary experimental research work by the authors.

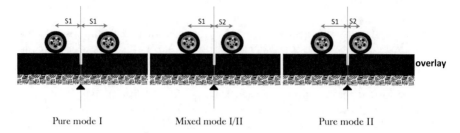

Fig. 7. Analogy between the loading conditions of the suggested ASCB specimen with those conditions of real cracked pavement induced by wheel loading.

5 Conclusions

1- A new test configuration called the asymmetric semi-circular bend (ASCB) specimen was suggested for mixed mode I/II fracture experiments on asphalt materials.
2- Stress intensity factors were calculated numerically for the ASCB specimen for different values of crack length ratios and bottom loading support positions.
3- The simple geometry and loading set up, the convenience of specimen preparation using the conventional gyratory compactor machines or asphalt field coring devices, the ease of generating a crack in the specimen, the application of compressive loads rather than the tensile loads, the ability of introducing full combinations of mode I and mode II and the ability of simulating the stress/strain field for the real pavements are the main advantages of the ASCB specimen for asphalt mixtures.

References

[1] Lugmayr, R., Jamek, M., Tschegg, E.: Adv. Test Charac. Bituminous Mater. II, 807 (2009)
[2] Kim, H., Buttlar, W.: Cold Reg. Sci. Tech. 57, 123 (2009)
[3] Labuz, J., Dai, S.: Cracking of asphalt concrete at low temperatures. Research report. Center for transportation studies. University of Minnesota (1994)
[4] Anderson, D., Lapalu, L., Marasteanu, M., Le Hir, Y.M., Martin, D., Planche, J.P., et al.: J. Transport Res. Board 1766, 1 (2001)
[5] Li, X., Marasteanu, M.O.: Exp. Mech. 50, 867 (2010)
[6] Kim, H., Wagoner, M., Buttlar, W.G.: Mater. Struct. 42, 677 (2009)
[7] Braham, A., Peterson, C., Buttlar, W.: Adv. Testing Charac. Bitum. Mater. I, 785 (2009)
[8] Artamendi, I., Khalid, H.: Int. J. Road Mater. Pave. Des. 7, 163 (2006)
[9] Molenaar, A.A.A., Scarpas, A., Liu, X., Erkens, G.: J. Assoc. Asphalt Technol. 71, 794 (2002)
[10] Chen, X., Li, W., Li, H.: J. Southeast Univ. 25(4), 527 (2009)

[11] Molenaar, J.M.M., Molenaar, A.A.A., Liu, X.: In: 6th International RILEM Symp. Perf. Testing Evalu. Bitum. Mater., p. 618 (2003)
[12] Akbulut, H., Aslantas, K.: Mater. Des. 26(4), 383 (2004)
[13] Li, X., Marasteanu, M.: Eng. Frac. Mech. 77(7), 1175 (2010)
[14] Kim, H., Wagoner, M., Buttlar, W.: Constr. Build. Mater. 23(5), 2112 (2009)
[15] Richard, H.A., Benitz, K.: Int. J. Fract. 22, 55 (1983)
[16] Arcan, M., Hashin, Z., Volosnin, A.: Exp. Mech. 18, 141 (1978)
[17] Ameri, M., Mansourian, A., Heidary Khavas, M., Aliha, M.R.M., Ayatollahi, M.R.: Eng. Fract. Mech. 78(8), 1817 (2011)

Modelling of the Initiation and Development of Transverse Cracks in Jointed Plain Concrete Pavements for Dutch Conditions

Mauricio Pradena[1] and Lambert Houben[2]

[1] Assistant Professor, Civil Engineering Department, University of Concepción, Chile – PhD candidate,
Section Road and Railway Engineering,
Delft University of Technology, the Netherlands
[2] Associate Professor, Section Road and Railway Engineering,
Delft University of Technology, the Netherlands

Abstract. In a previous study concerning the cracking at transverse joints in jointed plain concrete pavements (JPCP), the authors uses equations from the standard Eurocode 2 for the time-dependent concrete properties and considers the thermal deformation and the shrinkage for different design and construction conditions. For properties that are not available in standards, the authors made assumptions based on engineering judgment.

In this present paper the starting point is that previous modelling, and now the assumptions made in that work are studied more in depth. The paper also includes an improvement of the development of the concrete elastic modulus and strength. According with that, the objective of this paper is to improve the theoretical background of the assumptions and the modelling of the initiation and development of cracks at joints in JPCP for Dutch conditions. Concerning one of the most important influencing factors, the relaxation, a new equation is proposed. From both the theoretical and the practical point of view these improvements describe in a better way the process of cracking in JPCP for Dutch conditions.

1 Introduction

The starting point of the present work is a previous modelling made for the authors concerning the transverse cracking in JPCP [1]. Figure 1 shows the factorial of this modelling with the independent variables. The dependent variables were the time of occurrence and locations of the cracks, the crack widths and the tensile stress in the middle of the slabs after the cracking process is completed.

2 Alternative Approach to the Modelling Process

The assumptions of the original modelling are based on engineering judgment and they are related with required properties that are not available in standards. In the present work the assumptions made in his work are studied more in depth. Also it includes an improvement of the development of the concrete elastic modulus and strength through the maturity method.

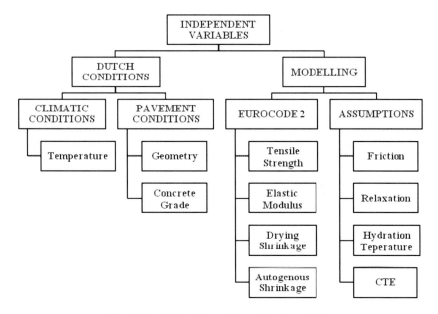

Fig. 1. Factorial design of the original modelling

2.1 The Maturity Method

A better modelling of the development of the concrete elastic modulus and strength at early-age is obtained when the degree of hydration concept or the maturity method is applied. The advantage of the maturity method is that it has also very practical application at the construction process that helps to reach the objectives of the concrete pavement design.

The rate of hydration of cement increases with increasing temperature. Consequently, the mechanical properties of concrete and their development in time are also influenced by the temperature. In fact every combination of temperature and time that yields the same maturity will have the same value of the property.

The Nurse-Saul maturity relationship is the accumulated product of time and temperature. But the Nurse-Saul approach is limited in that it assumes that the rate of strength gain is a linear function. On the other hand, the Arrhenius method takes into account the nonlinearity in the rate of cement hydration. This method expresses a maturity index in terms of an equivalent age, which represents the equivalent duration of curing at the reference temperature that would result in the same value of maturity as the curing period for the given average temperature [2].

$$te = \sum \exp\{-(E/R)*[1/273+T_a)-1/(273+T_r)]\}*\Delta t \qquad (1)$$

Te = equivalent age at reference curing temperature (hours)
E = activation energy (J/mol)
R = universal gas constant, 8.3144 J/(mol K)
T_a = average concrete temperature during time interval Δt (°C)
T_r = reference temperature (°C)
Δt = time interval (hours)

2.2 The Development of the Heat of Hydration

The hydration of the cement releases heat resulting in temperature increase. This hydration temperature is described by.

$$T_h = \frac{m_c * Q}{d_c * c_c} \qquad (2)$$

T_h = hydration temperature (°C)
m_c = mass of the cement per m³ of concrete (kg)
Q = hydration heat released till time t (kc/kg)
d_c = density of concrete (kg/m³)
c_c = specific heat of the concrete (kc/kg/°C)

The hydration heat released can be calculated using *HYMOSTRUC* software [3].

2.3 The Relaxation Factor

The experimental data on stress relaxation at early ages are very limited. Theoretical expressions and relationships with the development of creep are commonly used. But Morimoto and Koyanagi made laboratory tests, at room temperature, and they found the following empirical equation of the relaxation as a function of time [4].

$$R = \frac{0.32+0.85*t}{0.32+t} \qquad (3)$$

2.4 Development of the Coefficient of Thermal Expansion

In general the standards and recommendations include a constant value for the coefficient of thermal expansion (CTE) of hardened concrete. For instance, in the Eurocode 2 the CTE-value $10*10^{-6}$ °C^{-1} is advised for concrete [5]. However, the CTE is not necessarily constant in fresh concrete. The CTE is high in the first hours and it drops rapidly to a minimum value at t_0 = 12-14 hours after mixing. Beyond this minimum point the CTE increases gradually.

The time "zero" (t_0) is the time when strength and stiffness of concrete is defined to be zero, i.e. that the deformations occurring before this "Time Zero" can be ignored for stress calculation purposes, since they do not result in stresses [6].

The development of the CTE can be expressed by the equation 4 [7].

$$CTE(t_e) = CTE(0) + [CTE(28) - CTE(0)] \left\{ \exp \left[s_{CTE} \left(1 - \sqrt{\frac{28}{t_e - t_0}} \right) \right] \right\}^{n_{CTE}} \quad (4)$$

$CTE(0)$	=	the start-value at $t0$ (°C^{-1})
$CTE(28)$	=	the CTE-value at 28 days (°C^{-1})
s_{CTE} and n_{CTE}	=	curve-fitting parameters
te	=	the equivalent time (hours)

2.5 Friction with the Base

Zhang and Li present an analytical model for prediction of shrinkage-induced stresses and displacements in concrete pavements due to the restraint of the supporting base. The following is the governing equilibrium equation [8].

$$\frac{d^2u}{dx^2} - \frac{\tau}{E * h} = 0 \quad (5)$$

Where τ = slab-base interfacial friction (MPa); u=average displacement through the JPCP slab thickness (mm); E=elastic modulus of concrete (MPa) and h=height of the slab (mm).

3 Application to the JPCP in Dutch conditions

The typical Dutch condition for JPCP includes enough length to consider the pavement fixed at the beginning and at the end, i.e. no deformations are allowed there. The original modelling of the friction with the base considers this situation through the so called breathing length [9], but the Zhang and Li modelling introduces the friction from the beginning and from the end [8]. This is representative for pavements of small length, such as streets. Accordingly, in the present work the original model remains used.

Dutch climatic conditions mean that the amplitude of the average daily temperature of the concrete pavement ($T_{ampyear}$) has been taken as 10°C, where $T_{ampyear}$ is described by means of a sine-function with its maximum at August 1. The amplitude of the daily temperature of the concrete pavement (T_{ampday}) has been taken as 5°C, and also T_{ampday} is described by means of a sine-function that reaches its maximum at 4 PM. According with that the climate-dependent temperature T_3 of the plain concrete pavement is [9].

$$T_3 = T_2(T_1, t_2) + T_{ampyear} * \sin[(t/24) + t_1] - T_{ampyear} * \sin(15 * t) \qquad (6)$$

t   = time (number of hours) after construction
t_1 = time of construction (number of days after May 1)
t_2 = clock hour (from 0 to 24 hours) of construction at day of construction
T_1 = temperature at the day of construction of the JPCP (°C)
T_2 = temperature at the hour of construction of the JPCP (°C)

4 Results of the Modelling Process

4.1 Scenario of Evaluation

The objective of this present work is to improve the modelling of the transverse cracks in JPCP for Dutch conditions. For showing the improvements in a better way, focusing in the comparison between the original modelling and the alternative approach, a typical scenario of evaluation was chosen, i.e. transverse joint spacing (slab length) 4.5 m, concrete grade C28/35, joint depth 30% and pavement constructed at November 1st. An extensive work with different concrete grades, joint depths and slab lengths can be found in [1].

4.2 Comparison of Models

Table 1 shows a summary of the variables in the original modelling [1] and the present work. E_c is the elastic modulus of concrete.

Table 1. Comparison of models

Independent variables	Original modelling	Alternative approach
Strength and elastic modulus	Eurocode 2	Eurocode 2 + Eqn. 1
Hydration temperature	$T_h(t) = t^2 * e^{-0.27*t}$	Eqn. 2
Stress Relaxation	$R(t) = e^{-0.0003*t}$	Eqn. 3
Coefficient of thermal expansion	$CTE(t) = 3.095 * 10^{-10} * E_c(t)$	Eqn. 4

The alternative approach for the modelling of the independent variables improves the theoretical background, but the most significant changes for the behaviour of the pavement result from the stress relaxation and the introduction of the maturity in the calculation of the concrete properties. For that reason these two improvements are considered in a sensitive analysis.

In figure 2 the hourly cyclic variation of the tensile strength can be observed, because the maturity introduces the concrete temperature during the time interval Δt for the calculation of the strength.

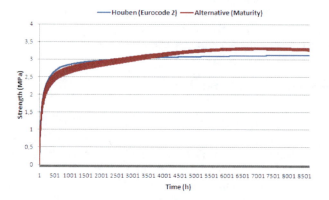

Fig. 2. Tensile strength according the two models

4.3 Sensitive Analysis

The modelling was made first for the maturity method and second for the stress relaxation, both under *ceteris paribus* condition.

Maturity method: It can be seen in table 2 that the introduction of the maturity in the modelling does not change significantly the trends of the results.

Table 2. Comparison of the results with the maturity method

Dependent variables	Original modelling	Alternative approach
Joints cracked (%)	100	100
Risk of cracks in slab (ratio)	0.51	0.71
Maximum crack width (mm)	1.31	1.19
Average width of 1st series of cracks after 1 year (mm)	0.70	0.65

Stress Relaxation: Table 3 shows the result of the alternative approach for the stress relaxation. According to experience and preliminary field measurements in Belgium and Chile, in similar climatic conditions to the Netherlands, the crack width seems to be greater than the original model indicates but smaller than the result of the alternative approach.

A similar analysis also has been made for a JPCP constructed at August 1, 4 PM, i.e. at the hottest moment of the year. The alternative approach then leads to shrinkage cracks in the slabs, so in between the transverse joints, and this is not observed in practice. This means that in the alternative approach the relaxation is too small. On the other hand, there are relaxation models with a fast fall at the beginning, until 50% of relaxation [10], however the application of this model results in a low percentage of joints cracked and a very small crack width and that is also not observed on JPCP in practice.

Furthermore, one has also to realize that a JPCP is opened to traffic 2 weeks to some months after construction. The structural design of JPCP is done considering traffic load stresses and temperature gradient stresses, but omitting shrinkage stresses. This indicates that the shrinkage stresses after some months cannot be very large.

4.4 New Approach

Taking into account all the theoretical and practical reasons mentioned previously, a new equation for the relaxation factor is proposed (figure 3).

The 1st and 3rd graph of figure 4 show the maximum shrinkage tensile stress in the JPCP during a period of 1 year according to the original model and the new approach, respectively. The 2nd and 4th graph show the width of the subsequent series of transverse joints that crack through according to the original model and the new approach, respectively. Due to the smaller stress relaxation the new approach yields higher maximum stresses, larger seasonal and daily stress amplitudes and larger widths of the cracked transverse joints.

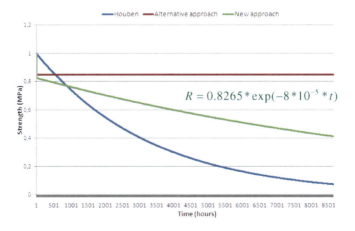

Fig. 3. Different stress relaxation factors

As is showed in table 3 the trend of the results according to the new approach is between the two previous models. It is expected that this equation can be validated through ongoing field measurements on JPCPs in Belgium and Chile.

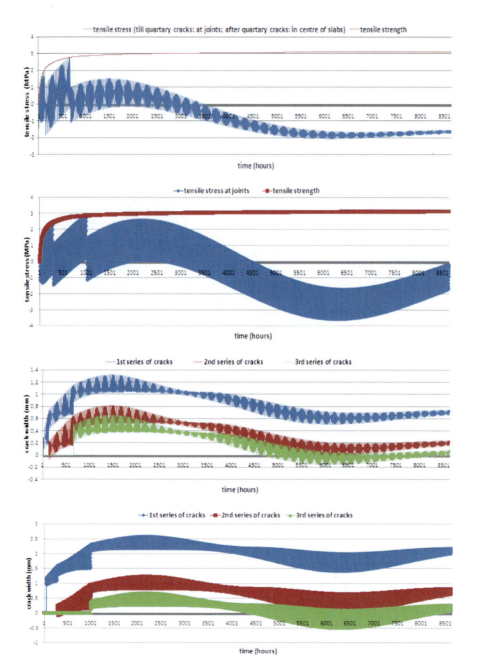

Fig. 4. Tensile stresses and crack width of the original modelling (2 upper graphs) and the new approach (2 lower graphs)

Table 3. Results of the different models

Dependent variables	Original modelling	Alternative approach	New approach
Joints cracked (%)	100	100	55
Risk of cracks in the slab (ratio)	0.51	0.58	very small
Maximum crack width (mm)	1.31	2.90	2.56
Average width of 1st series of cracks after 1 year (mm)	0.70	2.52	2.16

5 Conclusion

In this work the theoretical background of the assumptions made in the original modelling for the independent variables is improved. The trends of the dependent variables did not change significantly, with the exception of the alternative stress relaxation. In this case high values of crack widths and the possibility of slabs cracked at medium term were found, because of the almost constant small relaxation and the addition of the effect of the traffic loads and temperature gradients. Both, the width of the cracks and the possibility of cracked slabs, are not in agreement with the reality of JPCP for Dutch conditions. According to experience and preliminary field measurements in Belgium and Chile, in similar climatic conditions as the Netherlands, an equation for the relaxation factor is proposed. The trend of the results seems in agreement with reality but needs to be validated.

References

[1] Houben, L.J.M.: Transversal cracking in jointed plain concrete pavements for Dutch climatic conditions. In: Proceedings 7th International DUT-Workshop on Design and Performance of Sustainable and Durable Concrete Pavements, Carmona, Spain (2010)
[2] FHWA, Tech Brief, Fed. High. Adm., IF- 06-004, p. 6 (2005)
[3] Van Breugel, K.: Simulation of hydration and formation of structure in hardening cement-based material, Doctoral Thesis, Delft University of Technology, Delft, the Netherlands (1991)
[4] Morimoto, H., Koyanagi, W.: Estimation of stress relaxation in concrete at early ages. In: Springenschnidt, R. (ed.) Proceedings RILEM International Symposium on Thermal Cracking in Early Ages, Munich, pp. 111–116. Chapman & Hall, London (1995)
[5] Eurocode 2, Design and Calculation of concrete structures – Part 1-1: General rules and rules for buildings (in Dutch). Netherlands standard NEN-EN 1992-1-1 (en), NNI, Delft (2005)
[6] Cusson, D., Hoogeveen, T.: Cem. & Concr. Res. 37, 200–209 (2007)
[7] Atrushi, D.: Tensile and Compressive Creep of Early Age Concrete: Testing and Modelling, Doctoral Thesis, The Norwegian University of Science and Technology, Trondheim, Norway (2003)

[8] Zhang, J., Li, V.: Jour. Transp. Eng. 127(6), 455–462 (2001)
[9] Houben, L.J.M.: Model for transversal cracking in non-jointed plain concrete pavements as a function of the temperature variations and the time of construction. In: Proceedings 7th International DUT-Workshop on Design and Performance of Sustainable and Durable Concrete Pavements, Carmona, Spain (2010)
[10] Van der Ham, H.W.M., Koenders, E.A.B., van Breugel, K.: Creep, Shrink. Durab. Mech. Concr. and Concr. Struct., 431–436 (2009)

Pavement Response Excited by Road Unevennesses Using the Boundary Element Method

Arminda Almeida[1] and Luís Picado Santos[2]

[1] Department of Civil Engineering, University of Coimbra
[2] Department of Civil Engineering, Architecture and Georesources,
  Instituto Superior Técnico, Technical University of Lisbon

Abstract. Roughness is one of the major surface distresses of a pavement namely because it induces a major amplification of the loading patterns having important consequences on the structural response. This has even more significance to the pavement resistance if some cracking is associated with it. This paper examines pavement surfaces with different roughness levels from smooth surfaces to rough surfaces. First, the road profile is generated and then used as an input in a truck load simulator in order to get the dynamic amplification along the profile. Different travel speeds for trucks are also considered once speed has influence on the amplification. The pavement response is calculated using the BEM (Boundary Element Method). Only one pavement structure, with a thin asphalt concrete layer, is considered. The BEM is used because it is a powerful alternative to Finite Element Method (FEM) for problems with semi-infinite/infinite domains. In addition to that, it only requires discretization of the surface rather than the volume. The aforementioned features reduce significantly the number of nodes and elements of mesh and consequently the computational time. The results achieved until now show that the dynamic loads reduced the life of a pavement in a manner that can not be neglected. It also was referred the advantage of using BEM instead of FEM.

1 Introduction

Flexible pavements are a complex system where the different parts interact between themselves and some of their characteristics evolve during the pavement's life cycle. In order to deal with this variability an incremental design procedure should be used, meaning that it uses different design stages to represent the life cycle taking into account the characteristics of each part at each stage. The response model plays a vital role in this approach. The type of response model not only influences the assumptions of the analysis, but also the type of analysis and other functional aspects of the process. Analytical analysis can only be applied to simplified systems. For complex systems numerical analysis is used. The FEM is most probably the numerical method best known and mostly used to address pavement computations. However, there are other possibilities as for example the BEM. Each of the numerical methods has its own range of applications where they are most efficient.

This paper, using the BEM,has the objective of making the evaluationof pavement response due to dynamic loadinginduced by the roughness of the pavement surface. For that, the procedure to generate road surface profiles is first described and then these surface profiles are used as input in avehicle simulation software. After that, the DLC (Dynamic Load Coefficient) of each obtained load profile is calculated. Finally the pavement response is establishedfor the maximum value of load presented in the load profile and the pavement life is estimatedusing Shell transfers functions.

2 Generation of Road Surface Profiles

Vehicle simulations in a specific software requires road surface profiles as input. The one-dimensional random profiles were generated to have a specific spectral density. Cebon and Newland [1, 2] describe the generation procedure.

Different road surface profiles were generated corresponding to different levels of roughness. The International Standard ISO 8608 [3] defines different road class in terms of spectral density (Figure 1). Since the last three classes (F to G) correspond to unrealistic paved road surfaces, only the classes A to E were taken into account. Figure 2 depicts the generated road surface profiles.

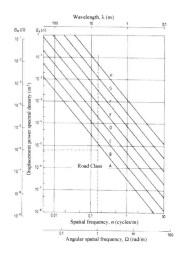

Fig. 1. Classification of roads [3]

The International Roughness Index (IRI) of the generated road surface profiles was calculated using an engineering software application called ProVAL (Profile Viewing and AnaLysis) [4]. It was developed by the Engineering Research

Division (ERD) of the University of Michigan Transportation Research Institute (UMTRI). Table 1 shows the calculated IRI values.

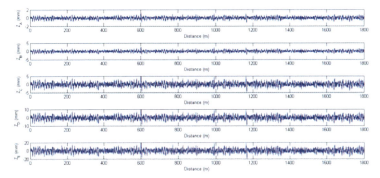

Fig. 2. Generated road surface profiles

Table 1. IRI values

Profile	IRI
Class A	0.38
Class B	0.77
Class C	1.53
Class D	3.06
Class E	6.13

3 Dynamic Loads

The TruckSim® of Mechanical Simulation [5] was the software used to perform the vehicle dynamic simulation. This tool was produced using the VehicleSim® technology which is the successor of the AutoSim® developed by Sayers [6] and can be used to generate comprehensive vehicle dynamics models for real-time simulation (RTS) applications using an ordinary PC [7]. It can perform the simulation and the analysis of the dynamic behaviour of medium to heavy trucks, buses and articulated vehicles.

Some of TruckSim' aspects or capabilities related closer with this work (for a full description, see [8]) are: road profiles are possible touse and the software allowsefficient use of high-frequency measured road roughness data; a wide range of tractor-trailer(s) combinations are available in a specific library of the software; multiple axle configurations, dual, single and wide-base tyres are also available; it supports many different suspension designs, using data that can be obtained from real or simulated kinematics and compliance tests; it includes several tire models as a table-based basic model, an extended model (more tables for camber effects), the Pacejka 5.2 version of the Magic Formula, and the MF-Tyre from TNO. With extra licenses, it is also ready to run with MF-Swift from TNO and FTire from COSIN. Pacejka [9] describe all these tire models.

Figure 3 depicts the vehicle used in the simulation. The tyres considered are 315/85R22.5 single on steer axle, 315/85R22.5 dual on drive axle and 315/85R22.5 single on semitrailer tridem axle.

Fig. 3. Simulation vehicle (TruckSim®)

The weight distribution between axles depends on the vehicle's unladen weight, dimensions, carried load and position of the load [10].Newton and Ramdas [10] present regression lines of weight distribution per axle on a 5-axle articulated vehicle, obtained from weight-in-motion (WIM) measurements on a UK motorway (Table 2).

Table 2. Regression lines for 5-axle articulated vehicles on a UK motorway [10]

Axle	Regression line	R^2
Steer	y=0.0646x+4276.9	0.4234
Drive	y=0.284x-70.856	0.7739
Semitrailer	y=0.6514x-4206	0.9242

y: Axle weight (kg); x: Gross vehicle weight (kg)

When using regression lines, the correlation concept is crucial. The better correlation (R^2=0.9242) occur in the axle of the semitrailer which is the axle considered in this research work for the pavement design.

The report of Lima and Quaresma [11], that evaluates for Portuguese circumstances the aggressiveness of heavy-vehicle traffic from data collected in a WIM station, presents statistical parameters (weight per axle and distance between axles) that matchthe articulated vehicle used in this work. The application of the regression lines to the total weight of thosePortuguese circumstances shownsmall differences, thus it is possible to define a payload that is in agreement with the result obtained with the abovementioned regression lines.

In addition, from the report of Lima and Quaresma [11] it is possible to verify that the GVW (Gross Vehicle Weight) is usually 37.5% above the legal maximum vehicle weight.

Therefore, in this work different percentages of overloading were considered, namely 10%, 20%, 30%, 40% and 50%. Different travel speeds were also considered, specifically 40 km/h, 60 km/h, 80 km/h and 100 km/h. From TruckSim® several dynamic load profiles were obtained for different road surface profiles (road class A to E), different values of overloading and different values of travel speeds.The DLC (Dynamic Load Coefficient) was the parameter used to quantify the magnitude of the induced dynamic tyre forces. It is defined as the ratio of the standard deviation of the perturbation and average or nominal static wheel load [12]. Figure 4 shows the DLC of each load profile from TruckSim.

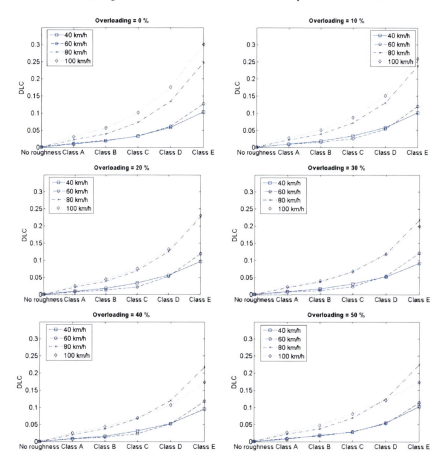

Fig. 4. Dynamic Load Coefficients

4 Pavement Response

4.1 Boundary Element Method

The basic idea of the Boundary Element Method (BEM) comes from Trefftz [13], who suggested that in contrast to the method of Ritz, only functions satisfying the differential equations exactly should be used to approximate the solution inside the domain and therefore we only need to approximate the presentboundary conditions. Thus is no need to subdivide the domain into elements. These solutions also satisfy conditions at infinity, and consequently there is no problem when dealing with the infinite domain. For a brief description of the BEM formulation and implementation see [14].In this paper only the advantages and disadvantages of BEM are summarized (Table 3).

Table 3. Advantages and disadvantages of BEM [12]

Advantages	Disadvantages
- Requires the discretization of the surface only (the dimension of the problem is effectively reduced by one) - There is no need to use elements on the planes of symmetry - It's easier to changethe structural meshes to meet different conditions - Gives a better accuracy in problems with stress concentration - To solveproblems where boundary stresses are of primary importance - It is simple and accurate in problems with infinite and semi-infinite domains (high ratio of volume to surface area) - Provides a complete solution in terms of boundary values only	- More susceptible to error when the appropriate numerical techniques are not used - The system of equations is non-symmetric and fully populated - Requires the knowledge of suitable fundamental solutions - When solvingproblems with high ratio of surface area to volume (thin plate or shell structure)

Almeida and Picado-Santos [14] compared the results obtained using the BEM(strains at the bottom of AC layer and strains at the top of subgrade in depth) with results from FEM (ADINA®,[15]) and BISAR® [16], demonstrating that the BEM results on the boundaries match the BISAR results, which is a "exact solution" tool. Regarding the mesh dimension, representing in some extent the computational effort, the boundary element mesh had 1539 nodes and 816 elements while the finite element had 8931 nodes and 1872 elements, so the BEM analyse needed 17% of FEM nodes and 44% of FEM elements.

4.2 Model Definition

Only a structure with thin asphalt concrete layer (12 cm) was considered. The granular layer has a thickness of 0.20 m and the subgrade was considered withinfinitethickness. All the materials were modelled as linear-elastic. The properties are shown in Table 4.

Table 4. Materials' properties

Layer	Modulus (MPa)	Poisson's ratio
Asphalt Concrete	4658	0.35
Granular	200	0.30
Subgrade	100	0.35

Contact area is a crucial aspect from the point of view of pavement design. As it depends on the tyre load, contact areas with different lenghts were considered. The report of action COST 334 [17] presents figures for the consideration of tyre-pavement contact area. The figures used are shown in Table 5.

Table 5. Tyre footprint size for driven and towed axles (adapted from [17])

Axle load in tonnes →		7		8		9	
Contact area width (mm)	Diameter (mm)	Contact area (cm^2)	Tyre Pressure (kPa)	Contact area (cm^2)	Tyre Pressure (kPa)	Contact area (cm^2)	Tyre Pressure (kPa)
285	1071	555	775	564	900	578	1000

The pavement response was calculated for the maximum value of each dynamic load profile. This option is related with the concept of spatial repeatability which says that the peak forces applied by a heavy vehicle fleet are concentrated at specific locations along the pavement and consequently these locations incur greater damage [2]. Table 6 presents the resultsof the maximum axle load in tonnes.

Table 6. Axle load in tonnes

			Overloading				
		0%	10	20	30	40	50
40 km/h	No roughness	7.4	8.3	9.1	10.	10.	11.
	Class A	7.6	8.5	9.4	10.	11.	12.
	Class B	7.8	8.8	9.6	10.	11.	12.
	Class C	8.1	9.2	10.	11.	11.	13.
	Class D	8.8	9.8	10.	11.	12.	13.
	Class E	10.	11.	12.	12.	14.	15.
60 km/h	No roughness	7.4	8.3	9.1	10.	10.	11.
	Class A	7.7	8.6	9.4	10.	11.	12.
	Class B	7.8	8.7	9.6	10.	11.	12.
	Class C	8.2	8.9	9.8	10.	11.	12.
	Class D	8.9	9.8	10.	12.	13.	13.
	Class E	10.	11.	13.	14.	16.	17.
80 km/h	No roughness	7.4	8.3	9.1	10.	10.	11.
	Class A	8.0	9.0	9.8	10.	11.	12.
	Class B	8.6	9.5	10.	11.	12.	13.
	Class C	9.5	10.	11.	12.	13.	14.
	Class D	11.	11.	13.	13.	14.	16.
	Class E	13.	15.	16.	16.	18.	19.
100 km/h	No roughness	7.4	8.3	9.1	10.	10.	11.
	Class A	8.1	9.0	9.8	10.	11.	12.
	Class B	8.7	9.6	10.	11.	12.	13.
	Class C	9.6	10.	11.	12.	12.	14.
	Class D	11.	12.	12.	13.	14.	15.
	Class E	14.	14.	15.	15.	16.	18.

From the data of Table 5 were made extrapolations in order to obtain a contact area for each value of Table 6. Taking the width of the concat area constant and equal to 285 mm, a contact area's length for each value of Table 6 was determined. However, althought 144 values of vertical contact stress have been considered, only 7 contact areas' lengths were considered instead of 144, which correspond to the average valueswithin eachrange of Table 7.

Table 7. Contact area's length

	1	2	3	4	5	6	7
Range (cm)	19 - 20	20 – 21	21 – 22	22 – 23	23 – 24	24 - 25	25 - 26
Average (cm)	19.92	20.57	21.39	22.39	23.44	24.30	25.18

The structures were discretised using quadratic isoparametric elements which imply 8 nodes per quadrilateral element (Figure 5).

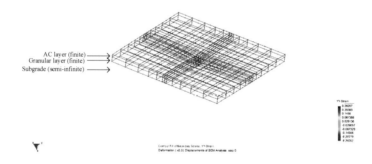

Fig. 5. BEM Mesh

4.3 Results

In order to estimate the pavement life, the Shell transfer functions were used. Figure 5 shows the number of loading cycles (fatigue life and rutting life) for different roughness levels, different travel speeds and different overloading percentages.

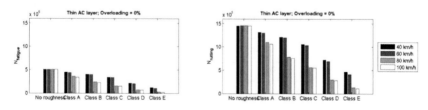

Fig. 5. Pavement life

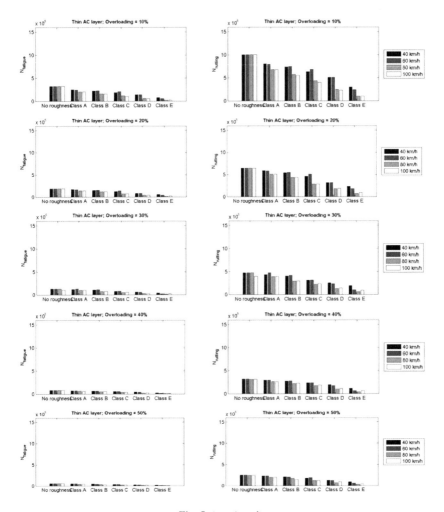

Fig. 5. *(continued)*

5 Discussion and Conclusion

In general, the DLC increases with roughness level and travel speed. As expected, this result is more pronounced for higher travel speeds (less recover time for suspensions) and level of roughness. The percentages of overloading of 20%, 30%, 40% and 50% havesmaller influence on DLC than for 0% of overloading because higher load level implies less suspension amplitude so less difference for the static load. This, of course, doesn't means that any level of overloading is a less aggressive than the legal load.

In relation to the pavement life, the number of possible loading cycles before end of pavement life decreases with increasing of roughness level and travel speed. As expected, the decreasing is higher for higher percentages of overloading. Just as an example of a regular situation, for overloading equal to 0% and travel speed equal to 80 km/h, the fatigue pavement life reduces about 45% when the roughness level passes from A to C. With higher overloading, as expected at some extent in real life, the result becomes worst for the pavement.

More applications, namely considering structures with different thickness of AC and including cases with stress concentration in order to deal with surface cracking, will be made to underline the importance of DLC on pavement life and to strength the conclusion that the BEM is more suitable than the FEM regarding problems where the boundary stresses are of primary importance, as cited by Beer e Watson [18].

References

[1] Cebon, D., Newland, D.E.: Vehicle System Dynamics 12(1), 160 (1983)
[2] Cebon, D.: Handbook of Vehicle-Road Interaction.Taylor & Francis (1999)
[3] ISO 8608, Mechanical vibration - Road surface profiles - Reporting of measured data, International Organization for Standardization (1995)
[4] ProVAL@ (2011), http://www.roadprofile.com/
[5] TruckSim@ Mechanical Simulation (2011),
http://www.carsim.com/products/trucksim/index.php
[6] Sayers, M.W.: Symbolic Computer Methods to Automatically Formulate Vehicle Simulation Codes. PhD, University of Michigan Transportation Research Institute (1990)
[7] Sayers, M.W.: Vehicle System Dynamics 32(4-5), 421 (1999)
[8] MSC@, TruckSim: Math Models (2012),
http://www.carsim.com/downloads/pdf/Math_Models_T81.pdf
[9] Pacejka, H.B.: Tyre and Vehicle Dynamics, 2nd edn. Elsevier (2006)
[10] Newton, W.H., Ramdas, V.: Road User Charges Review - Engineering Advice, TRL (Transport Research Laboratory) (2009)
[11] Lima, H., Quaresma, L.: Caracterização do factor de agressividade do tráfego de veículos pesados em Portugal. Lisboa, JAE and LNEC (1999)
[12] Siddharthan, R.V., Yao, J., Sebaaly, P.E.: Journal of Transportation Engineering 124(6), 557 (1998)
[13] Beer, G., Smith, I., Duenser, C.: The Boundary Element Method with Programming - For Engineers and Scientists. Springer, New York (2008)
[14] Almeida, A., Picado-Santos, L.: In: Proceedings of the 2nd International Conference on Transport Infrastructures, paper 160, on CD. S. Paulo-Brasil (2010)
[15] ADINA@ Automatic Dynamic Incremental NonLinear Analysis (2008),
http://www.adina.com/
[16] BISAR, Shell pavement design method, BISAR PC user Manual. Shell International Petroleum Company Limited, London, England (1988)
[17] COST 334. Effects of Wide Single Tyres and Dual Tyres (Final report of the Action - version November 29 2001, Taskgroup 3 Final Report), European Commission, Directorate General Transport (November 2001)
[18] Beer, G., Watson, J.O.: Introduction to Finite and Boundary Element Methods for Engineers. John Wiley & Sons, England (1992)

Discrete Particle Element Analysis of Aggregate Interaction in Granular Mixes for Asphalt: Combined DEM and Experimental Study

Giulio Dondi, Andrea Simone, Valeria Vignali, and Giulia Manganelli

DICAM Department, Faculty of Engineering, University of Bologna

Abstract. The conventional approach to modeling asphaltic materials is to treat them at macro-scale using continuum-based methods. Numerous research works, however, show that for these mixtures it's very important to take into account their micromechanical behaviour, at the scale of aggregate particles, because it is a primary factor in terms of overall system performance. In this way the Distinct Particle Element Method (DEM) represents a very useful tool.

In previous research works the authors have performed a DEM analysis of the fatigue performances of a road pavement and they have observed a great influence on the materials response of shape and interlocking of aggregates. In order to investigate this influence, a series of triaxial tests have been conducted and numerical results have been compared with the lab ones.

The samples, in particular, are composed of different types of steel elements (spheres and angular grains), because this ideal granular material allows an accurate geometrical representation of physical test specimens to be made in DEM simulation.

1 Introduction

The greater part of asphalt mixtures is composed of aggregates. Their structure and characteristics, particularly angularity and shape, have been considered as primary factors that affect the development of the aggregate skeleton and the mechanical performance of asphalt pavements. Aggregate contact and interlocking, in fact, control the load-bearing capacity and load-transferring capability of asphalt mixes.

In reality, it is very difficult to measure and quantify the degree of aggregate interlocking directly, because the fundamental theories of packing for particles are still not entirely clear and most of existing methods are confined in two dimensional assemblies or have difficulties in distinguishing between contacts and near-contacts [1].

Because Distinct Particle Elements Method (DEM) considers particles as distinct interacting bodies, it is an excellent tool to investigate the micro-mechanical behaviour of granular materials. Interactions between particles are described by contact laws that define forces and moments created by relative motions of the particles.

A commercially available three-dimensional DEM code called Particle Flow Code (PFC) [2], developed by Itasca Consulting Group, was used in this study. In PFC3D, particles are spheres (balls) that move independently of each other and only interact at the contact points.

The role of the aggregate's shape and angularity in controlling the performance of asphalt mixtures has been highlighted by many researchers. Cheung and Dawson (2002) [3] concluded that roundness and angularity are the main factors affecting the ultimate shear strength and permanent deformation. Aho et al. (2001) [4] indicated that aggregate shape and angularity are the second most important parameters, after gradation, that affect the asphalt mixture's performance.

Dondi et al. (2007) [5] presented a DEM model to simulate the fatigue performance of an asphalt pavement under traffic loading. The materials were modelled with clumps of discrete elements with different shapes. The results showed that the introduction of parameters, such as the shape and angularity of aggregates, greatly influenced the system's response.

Mahmoud et al. (2010) [6] introduced an approach that combined the discrete element method with image processing techniques in order to analyze the combined effects of aggregate gradation, shape, stiffness and strength on hot-mix asphalt's resistance to fracture. The model was used to quantify the internal forces in asphalt mixtures and determine their relationship to aggregate fracture, something that cannot be obtained by conventional experimental methods. The results showed that the required aggregate strength depends strongly upon the aggregate's characteristics.

Shen and Yu (2011) [1, 7] have studied aggregate packing, which, as it affects the way aggregate particles form a skeleton to transmit and distribute traffic loads, influences the stability and mechanical performance of the mix. The authors have developed a two-step procedure, using a discrete element modelling simulation method. The first step involved evaluating the effect of size distribution, while the second step investigated the combined effect of size distribution and shape impact. The study demonstrated that aggregate size distribution plays a significant role, affecting both the volumetric and contact characteristics of a packed structure, such as an asphalt mixture.

In summary, aggregate shape and angularity were found to be amongst the most important parameters that affect asphalt performance, as they have a strong influence on the way grains make contact and interlock.

2 Research Approach

2.1 Introduction

To understand how the packing characteristics of the aggregate particles in asphalt mix can be affected by shape and interlocking of grains, a two-step procedure has been developed:

- in the first step, the combined effect of grain shape and angularity on packing and stability of an aggregate assembly has been investigated;
- the second one, instead, will provide a model that, including the effect of the binder, can successfully simulate the packing characteristics of asphalt mixture and contribute to the improvement of its mix design.

This paper, in particular, presents the results only of the first step, which involves three major phases:

- selection of materials, mixes and tests;
- estimation of material micro-scale parameters, which have been calibrated based from experimental data. In this step only spherical particles have been used;
- evaluation of the sensitivity of grain shape and interlocking to the response of an aggregate system. This step has been performed on specimens of spheres and angular grains.

2.2 Materials, Mixes and Tests

An ideal granular material, 420C stainless steel balls, has been used in this research. There are obvious differences between these steel elements and the real aggregates of asphalt mixtures; however, by coupling DEM simulations with physical tests on this "ideal material", it is possible to replicate the geometry of the DEM model accurately [8, 9]. The use of ideal granular material with regular and simple geometry, in fact, allows an accurate geometrical representation of physical test specimens to be made in DEM simulations. Physical tests on this material can then be used to validate DEM models and these DEM models can be confidently used to develop into the micro-scale interactions driving the macro-scale response observed in the laboratory.

As measured by the manufacturer, the spheres material density is 7800 kg/m^3.

For steel mixture a discontinuous gradation has been selected. It has been obtained from a typical Superpave gradation with nominal maximum aggregate size of 12.5 mm, removing fractions passing at small sieves for computational reasons.

Two specimen types have been considered: one uniform, containing spheres (with diameters of 2.77 mm, 11 mm and 18 mm) (figure 1), and the other non-uniform specimen, containing a mixture of spheres and angular grains.

The specimens have been subjected to triaxial test, which has been recognized as a useful experimental tool for evaluating shearing resistance, stress-strain characteristics and strength properties of a granular assembly [8].

The prepared specimens were 100 mm in diameter and 200 mm high.

All the tests have been strain controlled and the strain rate has been set 1 mm/min.

Fig. 1. Steel balls

2.3 Evaluation of Material Micro-scale Parameters

A series of preliminary validation simulations have been performed to estimate the steel micro-scale parameters. They have been obtained comparing experimental and numerical results, and all the tests have been conducted on specimens of three-dimensional assemblies of steel spheres.

In this step only spherical particles have been used, because, even if they differ from real aggregates, they can provide a close coupling between numerical simulations and physical tests. More information about the micromechanics of a real material can be achieved by incorporating more realistic particle geometries in DEM model. However, prior to incorporate the complexity of a real aggregate in DEM simulations, it is important to demonstrate the accuracy of the numerical models using simple granular material such as that used in this study.

One type of sample has been used: a mixture of 32886 spheres with diameters of 2.77 mm (32204 spheres), 11 mm (605 spheres) and 18 mm (77 spheres).

In the laboratory tests, specimen has been created using dry pluviation with a funnel to minimize the drop height (Figure 2). A filter paper has been introduced inside the framework in order to confine the sample during its preparation, avoiding the collapse caused by steel particles weight. Once wet, it doesn't increase the strength resistance of the system during the test.

Fig. 2. Preparation procedure of the lab specimens

The confining pressure (σ_3) has been set to 300 kPa (test 1S_300), 400 kPa (test 1S_400) and 500 kPa (test 1S_500).

For the DEM simulations, a triaxial cell has been modeled with a cylindrical wall, closed at the top and the bottom boundaries by planes which simulate the loading plates (Figure 3). During the tests, its velocity is controlled automatically by a function that maintains a constant confining stress in the specimen [5].

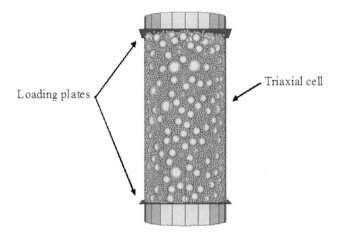

Fig. 3. DEM model for triaxial test

Since steel balls are an unbound material, bond models have been avoided. The system behavior has been defined only by a contact-stiffness model and a slip-separation model. Normal stiffness, shear stiffness and wall stiffness have been set equal to 10^7 N/m; interparticle friction coefficient and particle-boundary friction coefficient have been set equal to 0.42 and 0 respectively.

For each test have been monitored:

- confining pressure (σ_3), axial stress (σ_1) and deviator stress ($\sigma_d = \sigma_1 - \sigma_3$);
- axial strain (ε) and volumetric strain ($\Delta V/V$).

Figure 4 illustrates the variation in deviator stress and volumetric strain with axial strain for the laboratory tests (LAB) and the DEM simulations (DEM) under different confining pressures. It can be observed that the numerical and experimental results were very similar. In all tests, the deviator stress increases progressively with axial strain, until a maximum value is attained. What is more, residual and peak strength increases as the confining pressure increases. Volumetric strain, on the contrary, is negative early in the test (initial compaction of material) and subsequently increases and the material dilates. When the confining pressure increases, there is an increase in the change in volume of the specimen.

Therefore it can be concluded that results validate the steel micro-scale parameters selected.

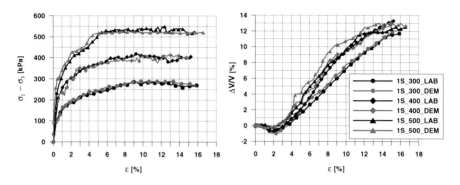

Fig. 4. Comparison of DEM simulation results and physical tests

2.4 Evaluation of the Sensitivity of the Aggregate System to Grain shape and Interlocking

Using the steel micro-scale parameters established above, in this step aggregate properties, such as shape and angularity, have been included, to develop more sophisticated models. In this way, in this research step, assemblies consisting of spheres and angular grains have been used.

For better comparison of results, samples dimensions and triaxial tests procedures are been assumed equal to what has been described in section 2.3

For the DEM simulations, angular grains have been modeled by clumps [9, 10]. A clump behaves as a rigid body because the particles comprising it remain at a fixed distance from each other [2].

The specimens, according to grain size distribution explained in section 2.2, are formed of:

- 32204 spheres with diameter of 2.77 mm, as mixtures of preliminary research step;
- clumps, that replace the medium and large spheres of the mixtures of preliminary research step.

According to table 4, three types of clumps, comprising two or three or four spheres, have been chosen. In order to have a quantitative comparison between the shapes of the four series of grains, clumps and spheres have the same external diameter and the same total particle volume for each size ranges of the gradation curve (figure 5).

A series of triaxial tests have been conducted on the assemblies created, as described in section 2.3 (table 4).

Table 4. Description of elements of specimens and of tests of analysis

Composition	Test code						
	1S	2C		3C		4C	
Small spheres	32204	32204		32204		32204	
Medium angular grains	605	2423		1615		1211	
Large angular grains	77	308		205		154	
Test name		2C_400	2C_500	3C_400	3C_500	4C_400	4C_500
σ_3 [kPa]		400	500	400	500	400	500

In the laboratory tests, angular grains have been obtained sticking together the steel spheres by a cold-weld compound (figure 5).

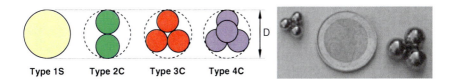

Fig. 5. Types of mix of spheres and angular grains and 4C grains for lab tests

3 Results

The results of triaxial tests are presented in figure 6 in terms of deviator stress versus axial strain, for the laboratory tests (LAB) and the DEM simulations (DEM), under confining pressure of 400 and 500 kPa. It can be observed a good agreement between numerical and experimental results, for each assembly, independently of the type of clump. For assemblies of grain with equal angularity, residual and peak strength increase with increasing confining pressure. Under any specified confining pressure, shear strength (residual and peak) increases significantly with increasing in the angularity of grains, because shear resistance arises from friction and interlocking between particles (table 5).

The shear strength, moreover, increases with decreasing of void ratio of the assemblies and with increasing of inter-particle friction coefficient. These results could be attributed to less interlocking and fewer contacts between grains (table 6). Increasing initial void ratio, in fact, leads to less dilation and lower mobilized friction angle.

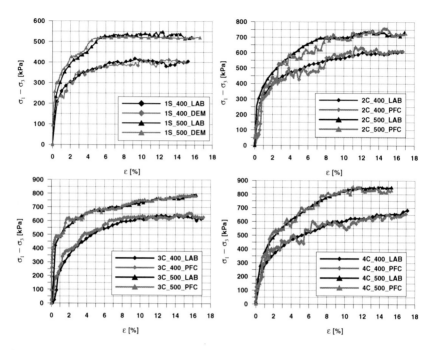

Fig. 6. Comparison of DEM simulation results and physical tests, in terms of deviator stress versus axial strain

Table 5. Peak strength for assemblies object of study [kPa]

Test code	Physical test (LAB)	DEM model (DEM)
1S_400	419	418
1S_500	549	525
2C_400	623	637
2C_500	737	762
3C_400	640	658
3C_500	795	799
4C_400	687	668
4C_500	855	857

Table 6. Void ratio and inter-particle friction coefficient for assemblies

Mix type	1S	2C	3C	4C
Void ratio	0.59	0.58	0.57	0.56
Inter-particle friction angle [°]	23	28	29	30

Because the contact force is strongly dependent on the arrangement of the particles, it is difficult to relate the total contact force to the performance of an assembly. Mean contact force, that is the total normal contact force divided by total contact numbers, has been used in this study to quantify the stability of the aggregate's structure [1]. In general, if the same external load is applied to two models, the one with lower mean contact force will have more contact points, as less stresses being transmitted through individual aggregate skeleton. Mean contact force is considered to be related to the aggregate structure's resistance to permanent deformation. Table 7 shows that adding more angular grains into the assembly, more aggregate contacts and better particles interconnection have been achieved, and thus the mean contact force decreases with improvement of the stability and load distribution capability of the structure as well as of the resistance to permanent deformation.

Table 7. Total number of contacts and mean contact force for each assembly

		Mean contact force [N]	
Confining pressure [kPa]		400	500
Mix type	1S	1.47	1.46
	2C	1.45	1.42
	3C	1.37	1.36
	4C	1.35	1.32

4 Conclusions

This paper conducts a 3D DEM analysis of aggregate packing, to characterize the roles of aggregates with different shapes in an assembly, and evaluate the aggregate contact and contact forces as an indication of the stability of grain skeleton.

It was found that the 3D DEM model developed in this study can effectively capture the effect of particle shape and angularity on the assembly performance.

The degree of aggregate contact and interlocking are found to be a function of grain shape and angularity. By adding angular grains, the quantity and magnitude of aggregate contacts increase.

Mean contact force is calculated to describe the aggregate system's resistance to permanent deformation. Given the same external load, a low mean contact force indicates more contact points, less forces transmitted through individual aggregate contact and thus a stable structure.

The understanding of particle packing, combined with particle shape and angularity, will help to develop meaningful gradation parameters that can be fundamentally related to the performance of overall aggregate structures, thereby to provide guidance on the improvement of aggregate gradation design for the asphalt mixtures. In this way, in fact, including the effect of binder, the combination of analytical and numerical approaches will able to develop a virtual testing environment. In this, once the model is calibrated, it can be used to run as many simulations are required. Thus, the virtual testing environment would be an inexpensive tool to evaluate the influence of changing different materials and design factors on the mixture response, as it provides precise control over almost every single factor.

Acknowledgements. The authors would like to take this opportunity to thank Dott. Fila and all the staff of Tecnotest (Dott. Mambrini, Eng. Viola, Eng. Nardelli) for the significant help given in carrying out laboratory tests.

References

[1] Shen, S., Yu, H.: Construction and Building Materials 25, 1362–1368 (2011)
[2] PFC 3D manual. Version 4.0. Itasca Consulting Group Inc. Minneapolis
[3] Cheung, L.W., Dawson, A.R.: Transportation Research Record (1787), 90–98 (2002)
[4] Aho, B.D., Vavrik, W.R., Carpenter, S.H.: Transportation Research Record (1761), 26–31 (2001)
[5] Dondi, G., Bragaglia, M., Vignali, V.: Advances in transport infrastructures and stakeholders expectations. In: Proceedings of the 4th International SIIV Congress, CD-ROM (2007)
[6] Mahmoud, E., Masad, E., Nazarian, S.: Journal of Materials in Civil Engineering 22, 10–20 (2010)
[7] Yu, H., Shen, S.: Construction and Building Materials 26, 302–309 (2012)
[8] Cui, L., O'Sullivan, C., O'Neill, S.: Géotechnique 57(10), 831–844 (2007)
[9] O'Sullivan, C., Cui, L.: Powder Technology 193, 289–302 (2009)
[10] Lee, Y.: PhD dissertation. School of Civil Engineering. University of Nottingham (2006)
[11] Abedi, S., Mirghasemi, A.A.: Particuology 9(4), 387–397 (2011)
[12] Cho, N., Martin, C.D., Sego, D.C.: International Journal of Rock Mechanics & Mining Sciences 44

Recent Developments and Applications of Pavement Analysis Using Nonlinear Damage (PANDA) Model

Eyad Masad[1], Rashid Abu Al-Rub[2], and Dallas N. Little[3]

[1] Professor, Texas A&M at Qatar, Doha, Qatar
[2] Assistant Professor, Texas A&M University, College Station, Texas, USA
[3] Professor, Texas A&M University, College Station, Texas, USA

Abstract. This paper presents an overview of the development and applications of the PANDA (Pavement Analysis using Nonlinear Damage Approach) model that has been under development at Texas A&M University for the past few years. In addition to the basics of the constitutive relationships used in PANDA, this paper presents examples of calibration and validation of the model using experimental laboratory data. The results demonstrate clearly the ability of the model to describe the mechanical behaviour of asphalt mixtures in terms of resistance to damage and permanent deformation. Finally, the capabilities of the model to simulate the mesoscale response of asphalt mixtures are presented and their implications in the design of asphalt mixtures are discussed.

1 PANDA Constitutive Models

This section summarizes the various constitutive models employed in PANDA to simulate damage, healing and permanent deformation. These models are all formulated to be temperature, loading rate and time dependent. The reader is referred to several papers of the authors and their co-workers for more details about these models and their implementation in finite element [1-6].

1.1 Total Strain Additive Decomposition

The total deformation of an asphalt mixture subjected to an applied stress can be decomposed into recoverable and irrecoverable components, where the extent of each is mainly affected by time, temperature, and loading rate. In this analysis, small deformations are assumed such that the total strain is additively decomposed into a viscoelastic component and a viscoplastic component:

$$\varepsilon_{ij} = \varepsilon_{ij}^{nve} + \varepsilon_{ij}^{vp} \qquad (1)$$

where ε_{ij} is the total strain tensor, ε_{ij}^{nve} is the nonlinear viscoelastic strain tensor, and ε_{ij}^{vp} is the viscoplastic strain tensor.

1.2 Effective (Undamaged) Stress Concept

Kachanov [7] has pioneered the concept of continuum damage mechanics (CDM), where he introduced a scalar measure called continuity, ζ, which is physically defined by Rabotnov [8] as:

$$\zeta = \frac{\overline{A}}{A} \qquad (2)$$

where A is the damaged (apparent) area and $\overline{A}$ is the real area (intact or undamaged area) carrying the load. In other words, $\overline{A}$ is the resulted *effective* area after micro-damages (micro-cracks and micro-voids) are removed from the damaged area A. The continuity parameter has, thus, values ranging from $\zeta = 1$ for intact (undamaged) material to $\zeta = 0$ indicating total rupture.

Odqvist and Hult [9] introduced another variable, ϕ, defining the reduction of area due to micro-damages:

$$\phi = 1 - \zeta = \frac{A - \overline{A}}{A} = \frac{A^D}{A} \qquad (3)$$

where A^D is the area of micro-damages such that $A^D = A - \overline{A}$. ϕ is the so-called damage variable or damage density which starts from $\phi = 0$ and ends with $\phi = \phi^c$ for complete rupture, where ϕ^c is the critical damage density [10].

Based on CDM definition of an effective area and the work of Abu Al-Rub and Voyiadjis [11], the relationship between the stresses in the undamaged (effective) material and the damaged material is defined as [see Chaboche [12] for a concise review of effective stress in CDM]:

$$\overline{\sigma}_{ij} = \frac{\sigma_{ij}}{(1-\phi)^2} \qquad (4)$$

where $\overline{\sigma}_{ij}$ is the effective stress tensor in the effective (undamaged) configuration, and σ_{ij} is the nominal Cauchy stress tensor in the nominal (damaged) configuration.

1.3 Nonlinear Thermo-Viscoelastic Model

In this study, the Schapery's nonlinear viscoelasticity theory is employed to model the viscoelastic response of asphalt mixtures [13]. The Schapery's viscoelastic one-dimensional single integral model is expressed here in terms of the effective stress $\bar{\sigma}$, Eq. (4), as follows:

$$\varepsilon^{nve,t} = g_0(\bar{\sigma}^t, T^t) D_0 \bar{\sigma}^t + g_1(\bar{\sigma}^t, T^t) \int_0^t \Delta D(\psi^t - \psi^\tau) \frac{d(g_2(\bar{\sigma}^\tau, T^\tau)\bar{\sigma}^\tau)}{d\tau} d\tau \qquad (5)$$

where D_0 is the instantaneous compliance, ΔD is the transient compliance, g_0, g_1, and g_2 are nonlinear parameters related to the effective stress, $\bar{\sigma}$, strain level, ε_{ij}, or temperature T at specific time τ. The parameter g_0 is the nonlinear instantaneous compliance parameter that measures the reduction or the increase in the instantaneous compliance. The transient nonlinear parameter g_1 measures the nonlinearity effect in the transient compliance. The nonlinear parameter g_2 accounts for the loading rate effect on the creep response, and ψ^t is the reduced time.

1.4 Thermo-Viscoplastic Model

Perzyna-type viscoplasticity constitutive equations as outlined in Masad et al. [14] are modified here and expressed in terms of the effective stress tensor $\bar{\sigma}_{ij}$, Eq. (5), instead of the nominal stress tenor σ_{ij}. The viscoplastic strain rate is defined through the following classical viscoplastic flow rule:

$$\dot{\varepsilon}_{ij}^{vp} = \dot{\gamma}^{vp} \frac{\partial g}{\partial \bar{\sigma}_{ij}} \qquad (6)$$

where $\dot{\gamma}^{vp}$ and g are the viscoplastic multiplier and the viscoplastic potential function, respectively. Physically, $\dot{\gamma}^{vp}$ is a positive scalar which determines the magnitude of $\dot{\varepsilon}_{ij}^{vp}$, whereas $\partial g / \partial \bar{\sigma}_{ij}$ determines the direction of $\dot{\varepsilon}_{ij}^{vp}$.

In this study, a modified Drucker-Prager yield function that distinguishes between the distinct behavior of asphalt mixture in contraction and extension and the sensitivity to confining pressures is employed as presented in Masad et al. [14]. However, this modified Drucker-Prager yield function is expressed here as a function of the effective (undamaged) stresses, $\bar{\sigma}_{ij}$, as follows:

$$f = F\left(\bar{\sigma}_{ij}\right) - \kappa\left(\varepsilon_e^{vp}\right) = \bar{\tau} - \alpha\bar{I}_1 - \kappa\left(\varepsilon_e^{vp}\right) \quad (7)$$

where α is a material parameter related to the material's internal friction, $\kappa\left(\varepsilon_e^{vp}\right)$ is an isotropic hardening function associated with the cohesive characteristics of the material and depends on the effective viscoplastic strain ε_e^{vp}, $\bar{I}_1 = \bar{\sigma}_{kk}$ is the first stress invariant, and $\bar{\tau}$ is the deviatoric effective shear stress modified to distinguish between the behavior under contraction and extension loading conditions.

1.5 Thermo-Viscodamage Model

Initially, Darabi et al. [5] proposed the following form of the viscodamage evolution law as an exponential form of the total effective strain:

$$\dot{\phi} = \Gamma^{\varphi} \exp(k\varepsilon_{eff}^{Tot}) \quad (8)$$

where Γ^{φ} is a damage viscousity parameter, ε_{eff}^{Tot} is the effective total strain, $\varepsilon_{eff}^{Tot} = \sqrt{\varepsilon_{ij}\varepsilon_{ij}}$, where ε_{ij} is given by Eq. (1) including both viscoelastic and viscoplastic parts, and k is a material parameter. The dependence of the damage density evolution equation on the total strain makes damage coupled to viscoelasticity and viscoplasticity, and to include implicitly the effects of time, rate, and temperature dependency. However, time of rupture in creep test and peak point in the stress-strain diagram for the constant strain rate test are highly stress dependent. As a result, one may assume that the damage viscousity variable in Eq. (8) is a function of stress, and has the following power law:

$$\Gamma^{\varphi} = \Gamma_0^{\varphi}(\frac{Y}{Y_0})^q \quad (9)$$

where q is the stress dependency parameter, Γ_0^{φ} and Y_0 are the reference damage viscousity parameter and reference damage force obtained at a reference stress for a creep test, and Y is the damage driving force in the nominal (damaged) configuration, which can be assumed to have a form similar to the Drucker-Prager-type function, $F(\bar{\sigma}_{ij})$, in Eq. (7), scuch that [15]:

$$Y = \tau - \alpha I_1 \quad (10)$$

where τ is as introduced in Eq. (20), but is a function of σ_{ij} and not $\bar{\sigma}_{ij}$, and $I_1 = \sigma_{kk}$. In continuum damage mechanics, Y is interpreted as the energy release rate necessary for damage nucleation and growth [10]. Assuming the damage force to have a Drucker-Prager-like form allows the damage evolution to be dependent on confining pressures, and taskes into consideration the distinct response of asphalt mixtures under extention and compression loading conditions. Also, assuming the damage viscousity parameter to be a function of the damage force, Y, in the nominal (damaged) configuration instead of the effective (undamaged) configuration allows one to include damage history effects, such that by using the effective stress concept in Eq. (4) one can rewrite Y as follows:

$$Y = \overline{Y}(1-\phi)^2 \tag{11}$$

Moreover, the damage density evolution highly depends on temperature. In this work, the proposed damage evolution law is coupled with temperature through a damage temperature function $G(T)$, which is identified based on experimental observations, such that one can write the following thermo-viscodamage evolution law [5]:

$$\dot{\phi} = \Gamma_0^\varphi [\frac{\overline{Y}(1-\phi)^2}{Y_0}]^q \exp(k\varepsilon_{eff}^{Tot})G(T) \tag{12}$$

1.6 Healing Model

It is shown in Abu Al-Rub et al. [1] that coupled viscoelastic, viscoplastic, and viscodamage constitutive models significantly underestimate the number of loading cycles up to failure of asphalt mixtures in case of repeated creep-recovery tests, especially, when relatively long rest periods (or unloading times) are introduced between the loading cycles. The reason for this underestimation is related to micro-damage healing occurring during the rest periods. Asphaltic materials have inherent micro-damage self-healing capacity that is more evident during unloading times and increasing temperatures. To remedy this issue, Abu Al-Rub et al. [1] proposed a phenomenological-based micro-damage healing model based on continuum damage mechanics such that the density of healed micro-cracks, h, are calculated based on the following evolution law:

$$\dot{h} = \Gamma^h \left(1-\phi\right)^{m_1} \left(1-h\right)^{m_2} \tag{13}$$

where Γ^h is the healing viscosity parameter controlling the rate of the micro-damage healing, and m_1 and m_2 are model material parameters. The effective stress concept as presented in Eq. (4) is then modified as follows:

$$\overline{\sigma}_{ij} = \frac{\sigma_{ij}}{1-\phi(1-h)} \tag{14}$$

The healing internal state variable ranges from $0 \leq h \leq 1$; $h = 0$ for no healing and $h = 1$ when all micro-cracks are healed.

2 Examples of PANDA Validation

2.1 Nottingham Database

The model was calibratted and validated using experimental data on asphalt mixtures tested using different stress levels, strain rates, and temperatures as outlined in Grenfell et al. [16]. The asphalt mixture is described as 10 mm Dense Bitumen Macadam (DBM) which is a continuously graded mixture.

The comparisons between experiments and model predictions for different temperatures and stress levels for creep tests are shown in Figures 1 and 2. The corresponding damage density versus total strain are also plotted in Figure 3. Figure 3 shows that the damage density is close to zero or at least insignificant at low strain levels, and increases as strain and applied stress is increased. Figure 3 also shows that the damage density grows almost with a constant slope for a while, where in this region the steady creep or secondary creep occurs. After this region damage grows with a higher rate until the rupture point. This region corresponds to tertiary creep. It is interesting to note that the damage density evolution follows an S-like curve, which is physically sound.

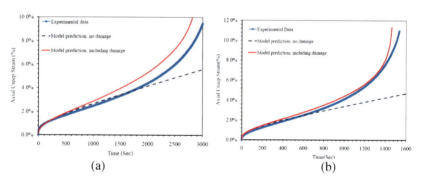

Fig. 1. The comparison of the creep response between experimental measurements and model predictions at $T = 10^{o}C$ and stress levels of (a) $\sigma = 2000\text{kPa}$ and (b) $\sigma = 2500\text{kPa}$

2.2 ALF Database

The experimental data presented in this section were obtained from North Carolina State University based on testing asphalt mixtures that were used in the Accelerated Loading Facility (ALF) of the Federal Highway Administration (FHWA). Details about these experimental measurements are available in Kim et al. [17]

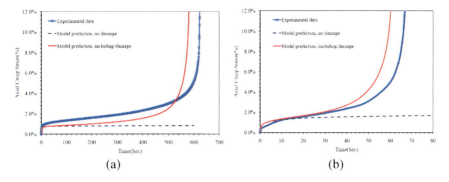

Fig. 2. The comparison of the creep response between experimental measurements and model predictions at $T = 40^\circ C$ and stress levels of (a) $\sigma = 500\text{kPa}$ and (b) $\sigma = 750\text{kPa}$

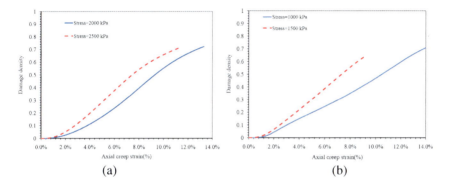

Fig. 3. Model prediction results for damage density versus total strain at different stress levels and different temperatures: (a) $T = 10^\circ C$, (b) $T = 20^\circ C$

Examples of the experimental and model results for tensile creep-recovery tests are shown in Figure 4. The results clearly show that the model is able to capture the accumulated damage of asphalt mixtures especially when the healing behavior of the mixtures is included in the model.

2.3 Mescoscale Results

X-ray computed tomography (CT) was used to capture the three dimensional microstructure of an asphalt mixture with a diameter of 50 mm and a height of 75 mm.

The modulus of elasticity and Poisson's ratio for the aggregate were assumed to be 25 GPa and 0.25, respectively. The matrix (or mastic) was modeled using the PANDA constitutive laws. Compressive repeated creep-recovery tests were simulated, and examples of the damage density distributions, viscoelastic strain distribution, and effective viscoplastic distribution at different times are shown in Figure 5. The mesoscale simulations are very useful to determine the effect of the matrix properties on the overall mixture performance. In addition, these simulations can be used in order to determine the optimum mixture design (aggregate gradation and volumetrics) that minimize localized damage and enhance performance.

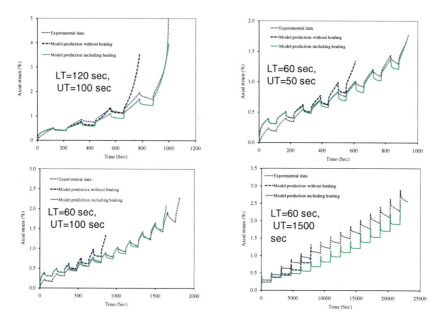

Fig. 4. Model prediction results for tensile creep-recovery tests of the control mixture in the ALF experiment for different loading time (LT) and unloading time (UT) at T =20°C

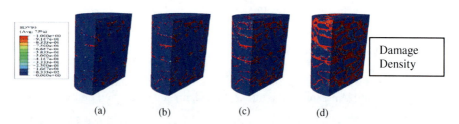

Fig. 5. Mesoscale response under repeated creep-recovery compressive test at: (a) 50 seconds (4 cycles), (b) 100 seconds (9 cycles), (c) 150 seconds (13 cycles), and (d) failure (16 cycles)

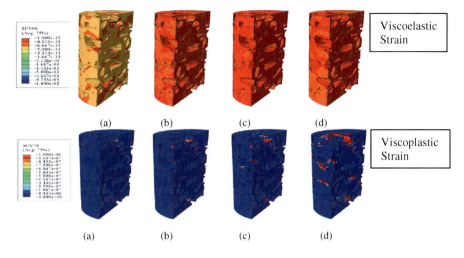

Fig. 5. *(continued)*

3 Conclusions

This paper presents an overview of the PANDA model. This is a temperature, rate-, and time-dependent continuum damage model coupled to temperature-dependent viscoelasticity and viscoplasticity models for accurately predicting the nonlinear behavior of asphalt mixes.

The model has been used to model the response of asphalt mixtures under creep and repeated creep-recovery tests. The results validate the model since its predictions of the numerical results compare well with the experimental measurements validating the model capabilities.

The PANDA model was used to simulate the mesoscale response of asphalt mixtures. The outcomes of these simulations are the macroscopic response as well as the damage, viscoelastic strain, and viscoplastic strain distribution within the asphalt mixture microstructure. The mesoscale simulations can be used in guiding virtual testing of asphalt mixtures.

Acknowledgements. The authors acknowledge the support of the US Federal Highway Administration through the Asphalt Research Consortium (ARC). The ARC funding supported the development of the constitutive model presented in this study. In addition, the authors would like to acknowledge the financial support provided by Qatar Nation Research Fund (QNRF) through the National Priority Research Program project 08-310-2-110. The QNRF funding supported the developed meso-scale model presented in this study.

References

[1] Abu Al-Rub, R.K., Darabi, M.K., Little, D., Masad, E.A.: Int. J. of Eng. Sci. 48, 966–990 (2010)
[2] Huang, C.W., Abu Al-Rub, R.K., Masad, E.A., Little, D., Airey, G.: Int. J. Pavement Engineering 12, 433–447 (2011)
[3] Darabi, M.K., Abu Al-Rub, R.K., Masad, E.A., Little, D.: Int. J. Num. and Anal. Methods in Geomechanics (2011) (in press)
[4] Huang, C.W., Abu Al-Rub, R., Masad, E., Little, D.: J. of Mat. in Civil Eng., ASCE 23, 56–68 (2011)
[5] Darabi, M., Abu Al-Rub, R., Masad, E., Huang, C.W., Little, D.: Int. J. of Solids and Structures 48, 191–207 (2011)
[6] Abu Al-Rub, R.K., Darabi, M.K., You, T., Masad, E.A., Little, D.N.: Int. J. of Roads and Airports 1, 68–84 (2011)
[7] Kachanov, L.M.: On time to rupture in creep conditions. Izviestia Akademii Nauk SSSR, Otdelenie Tekhnicheskikh Nauk 8, 26–31 (1958) (in Russian)
[8] Yu Rabotnov, N.: North-Holland, Amsterdam (1969)
[9] Odqvist, F.K.G., Hult, J.: Some aspects of creep rupture. Arkiv foK r Fysik 19, 379–382 (1961)
[10] Abu Al-Rub, R.K., Voyiadjis, G.Z.: Int. J. Sol. Struc. 40, 2611–2643 (2003)
[11] Abu Al-Rub, R.K., Voyiadjis, G.Z.: Int. J. of Dam. Mech. 18(2), 115–154 (2009)
[12] Chaboche, J.L.: Chapter 2: Damage Mechanics. Comprehensive Structural Integrity 2, 213–284 (2003)
[13] Schapery, R.A.: Polymer Engineering and Science 9, 295–310 (1969)
[14] Masad, E., Tashman, L., Little, D., Zbib, H.: J. Mech. Mat. 37, 1242–1256 (2005)
[15] Graham, M.: Damaged Viscoelastic-Viscoplastic Model for Asphalt Concrete. M.S. Thesis, Texas A&M University, College Station, Texas (2009)
[16] Grenfell, J., Collop, A., Airey, G., Taherkhani, H., Scarpas, A.T.: J. Asso. Asph. Pav. Tech. 77, 49–516 (2008)
[17] Kim, Y.R., Guddati, M.N., Underwood, B.S., Yun, T.Y., Subramanian, S., Savadatti, S.: Report No. FHWA-HRT-08-073, U.S. Federal Highway Administration (2009)

Laboratory and Computational Evaluation of Compact Tension Fracture Test and Texas Overlay Tester for Asphalt Concrete

Eshan V. Dave[1], Sarfraz Ahmed[2], and William G. Buttlar[3]

[1] University of Minnesota Duluth
[2] Pakistan University of Science and Technology
[3] University of Illinois at Urbana-Champaign

Abstract. Reflective cracking is the primary mode of failure for pavements rehabilitated with asphalt overlays in many instances. The Texas Overlay Tester (OLT) has been utilized by several researchers and practitioners to evaluate the reflective cracking resistance of asphalt overlays. The OLT is a simulative test procedure that emulates the portion of asphalt overlay located directly on top of the crack or discontinuity in the underlying pavement. The testing involves cyclic horizontal displacement of the underlying layer to initiate and propagate the crack. The number of cycles required to form the crack through asphalt overlay is typically utilized as a performance parameter indicative of cracking resistance of the asphalt mixture. This paper describes a comprehensive analysis of the OLT through comparative laboratory fracture testing and computational modelling. The compact tension (CT) test geometry has been recently adapted to characterize the fracture properties of asphalt concrete, and can be used to extract useful mode I (tensile) local fracture properties such as material strength and fracture energy.

Laboratory creep and fracture testing was conducted for two hot-mix asphalt samples. Both OLT and CT tests were conducted for each mixture, and both tests were simulated using the finite element technique. The simulation results and the laboratory findings demonstrate the relative pros and cons of each approach (fracture test versus simulative test). Reasons for the significantly higher variability found in the OLT as compared to the CT test are hypothesized and discussed. The development and implementation of a phenomenological cohesive zone fatigue (CZF) model specifically tailored for this study is presented. The CZF model utilizes fracture properties obtained from CT test along with a functional degradation of those properties under cyclic straining as calibrated using OLT results. The calibrated model was shown to be in favourable agreement with laboratory testing results. Extensions and limitations of model are also discussed.

1 Introduction and Background

The use of asphalt overlays to rehabilitated distressed pavements is quite extensive. The most prominent failure mode for asphalt overlays is through reflective cracking, whereby cracks and joints from underlying distressed pavements causes stress concentrations in the overlay leading to formation of cracks. Significant research efforts have been made on development of laboratory characterization methods to evaluate the reflective cracking resistance of overlay mixtures. The Texas Overlay Tester (OLT) has gained significant popularity in recent years [1-3]; this method has been refined by Zhou et al. [4] for use in standard material specifications. The OLT test procedure simulates the straining of asphalt overlay placed over jointed or cracking pavement that undergoes horizontal movement in direction of traffic. The compact tension (CT) fracture test procedure for asphalt concrete has been formalized by Ahmed et al. [5] and has been utilized for evaluation of thin overlays [6-8]. Previous studies on evaluation of OLT focussed primarily on lab testing or computer simulations using the linear elastic fracture mechanics approaches [9]. In this study, cohesive zone fracture approach was utilized to study the material failure mechanism in OLT and it is compared with the same for CT test. Two asphalt mixes were tested using both tests for comparative purposes. Finally, a cohesive zone fatigue model is proposed that utilizes fracture properties from CT tests and can be calibrated using OLT data to simulate fatigue induced damage and cracking in asphalt concrete.

2 Research Approach

The technical efforts undertaken in this work were divided into two major components, namely: laboratory testing efforts and computational modelling efforts. The laboratory tests were conducted on two hot-mix asphalts, one plant produced and other lab produced. Both mix types represented identical volumetric mix design, aggregate gradations and same asphalt binder grades. The asphalt mixtures were provided to the researchers from the Texas Department of Transportation (TXDOT). The nominal maximum aggregate size of the mixes was 9.5 mm with 5.1% asphalt content. Both mixes were produced using the Superpave PG 64-22 grade binder and consisted of 20% fractionated recycled asphalt pavement (FRAP). Laboratory characterization of mixes included OLT, CT and indirect tensile creep tests. The OLT tests were conducted in accordance with the Tex-248-F [1] procedure by TXDOT and the results were shared with the authors of this paper. The CT tests were conducted for both mixes at -12, 0, and +12 °C. The specimen preparation and laboratory testing procedures for CT test are discussed elsewhere [2,3,4]. The indirect tensile creep tests were conducted in accordance with the AASHTO T-322 [5] test procedure at test temperatures of -12, 0, +12 °C. As discussed before, the primary objective for conducting OLT and CT tests was to make quantitative and qualitative comparisons between the results obtained from these tests and explore the suitability of OLT in predicting cracking potential of asphalt mixtures.

A series of computational modelling was conducted to simulate the OLT and CT tests. The computational modelling was conducted using finite element method. The crack initiation and propagation in the CT and OLT tests were simulated through use of cohesive zone fracture model. A variety of results were extracted from computer simulations to make comparisons between OLT and CT tests including stress distribution and determination of local fracture properties. A phenomenological fatigue damage model was also developed and calibrated using the OLT results. The details of computational models and data analysis are discussed later in the paper.

3 Laboratory Test Results

The laboratory test results are discussed in this section. The OLT test procedure has been described in detail by Zhou et al. [4] and the TXDOT TEX-248-F test specifications [9]. The test procedure involves repeated loading and unloading of asphalt concrete sample at 25 °C through a controlled displacement test. Each displacement cycle for this test is of 0.63 mm magnitude applied in a triangular waveform. The reported performance properties of the mixture from OLT include the starting load and the number of cycles to failure. The starting load is defined as the amount of peak force needed during the first displacement cycle. The number of cycles to failure represents the number of repetitions of triangular displacement to reach 93% reduction in the load as compared to starting load. The results for number of replicate samples from both mixes are tabulated in Table 1. The test results indicate a relatively high variability in both parameters with values of coefficient of variance (CoV) ranging from 4.6% to 49.9%. In light of the CoV values the relative performance differences between the two mixtures is not significant. In absolute sense the lab mixtures demonstrated marginally superior cracking performance.

Table 1. Texas overlay tester results

LAB MIX			PLANT MIX		
Sample Number	Starting Load (N)	Cycles to Failure	Sample Number	Starting Load (N)	Cycles to Failure
1	2655.6	17	1	3180.5	43
2	3202.7	39	2	2953.6	43
3	4461.6	69	3	3136.0	48
4	3727.6	58	4	2926.9	18
Average	3511.9	45.8	5	3251.7	46
CoV	21.9%	49.9%	6	3140.4	22
			7	3322.8	22
			Average	3130.3	34.6
			CoV	4.6%	38.2%

The indirect tensile creep test results were obtained through testing of three replicate samples for each mix. The creep compliance measurements were shifted using time-temperature superposition principle to obtain the master-curves.

The CT tests were conducted with three replicate samples for each mix tested at -12, 0 and +12 °C. The tests were conducted to yield a constant rate of crack mouth opening displacement (CMOD) of 0.0167 mm/s. The load-displacement data for the lab mix at +12 °C are shown in Figure 1(a). Contrary to strength tests the fracture tests, such as CT test, focus on measurement of the necessary amount of energy that is required to propagate a crack through the material rather than focus on the amount of stress necessary to initiate a crack. This energy measure is commonly referred to as fracture energy of the material. In the case of materials that exhibit quasi-brittle and ductile failure behaviour this property is of particular interest. This is primarily due to the fact that the material has significant capacity to carry load once the peak capacity, as commonly indicated by tensile strength, is reached. Fracture energy can be determined by normalizing the fracture work against the newly formed area by process of fracture. The fracture work is the area under the load-displacement curve. The fracture energies of both plant and lab mixtures at three test temperature is presented in Figure 1(b). The average of three measurements and corresponding CoV for each set is also shown on the plot. The fracture energies of two mixtures are almost identical at -12 and 0 °C. At +12 °C the plant mixture has marginally higher fracture resistance, the OLT results showed similar distinction. Overall, the CoV for lab mixture was observed to be higher than lab mixture, this observation is also consistent with results obtained from OLT. It is anticipated that the mixing and aging procedures attributed to the greater CoV for lab specimens.

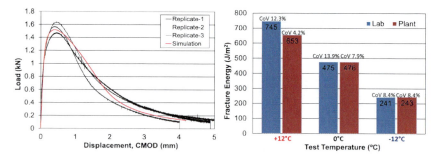

Fig. 1. (a) Load-displacement curves from CT test for lab mixture at temperature of +12 °C; (b) Fracture energy measurements for lab and plant mixtures (number on the bar is average fracture energy from three test replicates).

4 Modelling Efforts

The computer simulations reported in this paper were conducted for twin-fold objectives:

- Evaluation and comparison of OLT and CT test procedures from perspective of stress distribution in specimen; and
- Development of a phenomenological cohesive zone fatigue model.

The computer simulations were conducted using the commercially available finite element (FE) program ABAQUS. However, due to specialized nature of this research several features were programed and utilized within the framework of

ABAQUS. These include user defined materials, user defined elements, and subroutine for application of repeated loading. The FE models for both OLT and CT tests were developed with two-dimensional plane-strain approximation using four-node quadrilateral elements. The FE meshes for OLT and CT tests along with the dimensions and boundary conditions are shown in Figure 2.

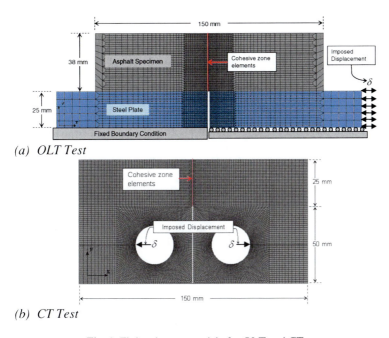

(a) OLT Test

(b) CT Test

Fig. 2. Finite element models for OLT and CT tests

The asphalt concrete materials for both OLT and CT tests were simulated as linear viscoelastic. Cohesive zone fracture elements were utilized for simulation of damage and cracking along the potential crack path. Bilinear model customized for asphalt concrete by Song et al. [6] was utilized for providing the relationship between the normal traction and corresponding displacement jump. The approach undertaken for this paper was to utilize CT test for extracting local material fracture properties, namely, fracture energy and cohesive strength and utilize those properties for simulation of OLT. Due to lack of fracture property measurements at 25°C, which is the test temperature for OLT, the local fracture properties were extrapolated from -12, 0 and +12°C as shown in Figure 3. The cohesive (tensile) strength of the material did not appear to change between -12, 0, and +12°C; this was kept constant at 2.30 MPa for plant mix and 2.60 MPa for lab mix. The fracture property at warmer temperatures is one topic that was identified through this work as a future research area. In this work a linear extrapolation assumption was made as the authors of this work did not feel comfortable with use of higher order function with only three available data points for each mixture. However in future, better suited extrapolation models are expected to be available. Detailed descriptions of local property extraction procedure are presented elsewhere [7], in summary the procedure involves simulation of fracture test and adjusting local property inputs to the cohesive zone model until the global responses from experiment matches the simulation. For example, Figure 2 shows the experimental measurements of load and crack mouth opening displacements for three replicates as well as the simulated response used to extracted local fracture properties.

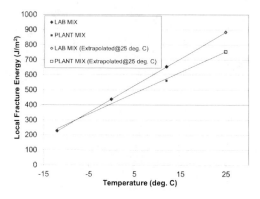

Fig. 3. Local fracture energy measurements extracted from CT test at -12, 0, and +12 °C and extrapolations to +25 °C

4.1 Stress Distribution in OLT and CT Tests

The distribution of stresses in a material fracture characterization tests is an important consideration. For example, strength tests require uniform stress distribution along the potential failure plane versus fracture tests require localization of stresses at a discontinuity or pre-crack. This section briefly presents stress distribution in OLT and CT tests along the potential failure plane.

The stress distribution in the loading direction (x-direction) for OLT and CT is shown in Figure 4. The stresses for OLT are shown for the first displacement cycle. The stress plots are limited to tensile magnitudes. From the stress distribution of OLT it can be seen that significant amount of the potential failure region has uniform tensile stress distribution. Near the very bottom of asphalt concrete the stresses have reduced to close to zero indicating formation of small macro-crack. The uniform tensile stressing of material over the complete thickness often results in difficulty to obtain high repeatability of test results, such as in case of direct tension tests. This is mainly due to heterogeneity of asphalt mixture, which also results in quite non-repeatable aggregate and mastic distribution in the region situated directly above the discontinuity in steel plates. The rigidity of the OLT system in terms of only horizontal movement (x-direction) of steel plates with constraint in vertical direction (y-direction) restricts the sample to fail through stress localization and through propagation of crack; it forces material to have uniform strain thus causing stress states comparable to a direct tension type loading conditions. Furthermore the stress distribution in x-direction for the whole specimen indicated that peak stresses are limited in narrow band along middle $1/4^{th}$ of the specimen. The contour plots that shown this behaviour are not presented in the paper for brevity. In these simulations, the OLT test does not satisfy the requirements of fracture test due to lack of stress localization and corresponding crack propagation. Due to presence of cohesive zone model it can be seen that as displacement increases the amount of stresses decreases due to fracture dissipation (shown by points indicated on plot as 2, 3, 4, and 5).

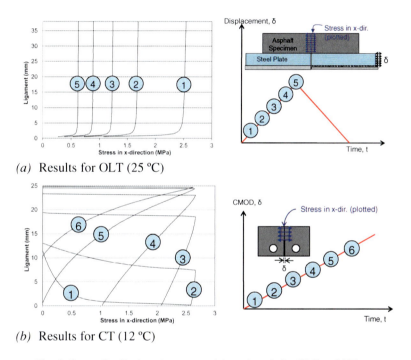

(a) Results for OLT (25 °C)

(b) Results for CT (12 °C)

Fig. 4. Stress distribution along potential crack path for OLT and CT tests

The stress distribution for CT test indicates that the stresses localize at the crack tip and thereafter the stress concentration gradually moves with the movement of crack tip. The point 1 shown on the plot is right at the moment when peak stress at crack tip approaches material strength. The point 2 corresponds to moment when the material begins to undergo softening and hence the stress at the actual crack tip begins to drop. This effect continues from point 2 through 5 and in the meanwhile the location of damage threshold gradually moves along the potential crack path. The actual crack tip moves along the potential crack path from point 5 to point 6. At the point 6 about 40% of ligament has fully cracking and remaining portion has undergone softening or damage.

Thus, by comparison of stress distributions in OLT and CT it can be deduced that the OLT procedure simulates the failure of overlay due to repeated uniform straining whereas the CT test procedure conducts evaluation of overlay to resist the movement of crack through it. Thus, each test serves very different purpose and also the corresponding outcomes from the tests could potentially be very different. Also, we get some insight into lower repeatability of OLT test procedure which is hypothesized due to very uniform stress distribution along potential failure plane.

4.2 Cohesive Zone Fatigue Model

The OLT simulates repetitive straining of asphalt material. By using CT and OLT test data a phenomenological cohesive zone fatigue model was developed and implemented. The traditional cohesive fracture model does not account for degradation of material capacity due to repeated loading. Thus, the traditional cohesive zone approach was extended to account for reduction in material capacity with increasing load repetitions. A common approach utilized in modelling of fatigue damage in asphalt concrete is through stiffness degradation, whereby, with each repetition of load the stiffness of material is reduced to account for damage.

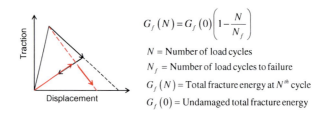

$$G_f(N) = G_f(0)\left(1 - \frac{N}{N_f}\right)$$

N = Number of load cycles
N_f = Number of load cycles to failure
$G_f(N)$ = Total fracture energy at N^{th} cycle
$G_f(0)$ = Undamaged total fracture energy

Fig. 5. Cohesive zone fatigue model (graphical representation and functional form)

This approach presents significant difficulty from perspective of pavement modelling due to non-discrete representation of crack. In the present approach the fatigue damage is accounted through dissipation of fracture energy in each load cycle. The dissipation only occurs after the material has started to undergo softening. In terms of the local cohesive zone law, Figure 5 and corresponding equation describes the model. As evident from the equation describing the model, in the present form the model assumes a constant dissipation of fracture energy in each load repetition. Furthermore, the reloading stiffness is also assumed to be same as unloading stiffness. Both of these assumptions are needed to be validated and/or modified through future research efforts. In its present form the model requires only one additional model input as compared to traditional cohesive zone approach, that is the number of cycles to failure (N_f).

Using the aforementioned model and results from CT tests, the OLT was simulated for both mixes previously described in this paper. The results from OLT were used as calibration data set to determine the N_f parameter for the model. The simulation results for both mixes in terms of the load histories as function of time are presented in Figure 6. It can be seen from the plot that with this model a significant amount of energy dissipation is observed in the first loading cycle as evident by the significant drop in the load. The table in Figure 8 shows the simulation results as well as laboratory test results. Comparisons shows that with the cohesive zone fatigue model it was possible to match the lab measured OLT results in terms of number of cycles to failure; however the peak loads did not match well. The lack of match between peak load from model and testing can be attributed to assumption that the constant cohesive strength for -12, 0 and +12 °C is also applicable to 25 °C.

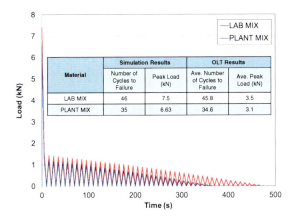

Fig. 6. OLT simulation results using cohesive zone fatigue model. Insert: simulation and laboratory measure results

5 Summary, Conclusions and Future Extensions

The laboratory evaluation and computer simulations provided great insight into the mechanisms of material failure for the OLT and CT tests. The laboratory tests showed relatively high variability amongst test replicates for OLT and relatively lower variability for CT tests at multiple temperatures. The computer simulations showed that OLT test imposes very uniform tensile stresses near the center of the specimen. The uniform stress distribution in tensile mode in heterogeneous materials such as asphalt concrete often causes high test variability; the high variability on OLT is anticipated from this effect. The reflective cracking resistance of material in OLT is measured by its endurance to carry load in horizontal direction under repeated displacement cycling. Thus, a material that has high strain tolerance is expected to yield superior performance in the OLT. On the other hand, a material that has high strength characteristics and moderate to low strain tolerance will give inferior results. From the perspective of fracture, both strength and strain tolerance are necessary features as strength determines the necessary stress conditions for onset of damage, while strain-tolerance controls the propagation characteristics of crack. Use of fracture test, such as CT, allows measurement of both of these quantities and also makes it possible to extract local material properties which in-turn can be used for modelling purposes. Thus OLT allows for quantifying asphalt concrete's reflective cracking resistance by simulative means and in-turn can have higher test variability. The CT test determines material's cracking resistance from fracture perspective and has relatively lower test variability. On the other hand, CT test is limited to monotonic conditions, whereas OLT allows for accounting of repeated loading behaviour. In this limited study two mixes were evaluated, more testing efforts are needed to expand upon the findings from this study.

Combination the fatigue damage characteristics from OLT with the fundamental fracture properties from CT test in form of a phenomenological cohesive zone fatigue model was explored in this paper. The preliminary results show promising outcome. Significant future research efforts are needed to further qualify the proposed model and build additional material effects, such as effect of strain/stress amplitudes, temperature, strength degradation and effect of rest periods.

References

[1] Zhou, F., Sheng, H., Xiaodi, H., Scullion, T., Magdy, M., Walubita, L.F.: J. Trans. Eng. 136(4), 353 (2010)
[2] Zhou, F., Sheng, H., Scullion, T.: Asphalt Pavement Technology. In: Proceedings of the Technical Sessions: Journal of the Association of Asphalt Paving Technologists (AAPT), vol. 79, pp. 597–634. AAPT, Sacramento (2010)
[3] Zhou, F., Sheng, H., Scullion, T.: Road Pavement Material Characterization and Rehabilitation. In: Proceedings of the 2009 GeoHunan International Conference, Geotechnical Special Publication, vol. (191), pp. 65–73. American Society of Civil Engineers, Reston (2009)
[4] Zhou, F., Scullion, T.: Overlay Tester: A Rapid Performance Related Crack Resistance Test. Texas Transportation Institute, College Station (2005)
[5] Ahmed, S.: Fracture Characterization of Thin Bonded Asphalt Concrete Overlay Systems, Ph.D. Dissertation, University of Illinois at Urbana-Champaign, Urbana, IL (2011)
[6] Ahmed, S., Dave, E., Behnia, B., Buttlar, W.: Materials and Structures (2011) article in press, available online
[7] Ahmed, S., Dave, E., Buttlar, W., Exline, M.: Asphalt Pavement Technology. In: Proceedings of the Technical Sessions: Journal of the Association of Asphalt Paving Technologists (AAPT), vol. 79, pp. 443–472. AAPT, Sacramento (2010)
[8] Ahmed, S., Dave, E., Buttlar, W., Exline, M.: Int. J. Pav. Eng. (2011) (article in review)
[9] Zhou, F., Sheng, H., Scullion, T., Chen, D., Qi, X., Claros, G.: Asphalt Pavement Technology. In: Proceedings of the Technical Sessions: Journal of the Association of Asphalt Paving Technologists (AAPT), vol. 76, pp. 627–662. AAPT, San Antonio (2007)

Crack Fundamental Element (CFE) for Multi-scale Crack Classification

Yuchun Huang[1] and Yichang (James) Tsai[2]

[1] School of Civil and Environmental Engineering, Georgia Institute of Technology,
790 Atlantic Dr., Atlanta, GA, USA, 30332
[2] School of Civil and Environmental Engineering, Georgia Institute of Technology,
210 Technology Circle, Savannah, GA, USA, 31407

Abstract. With the advance of sensor and information technology, high-resolution 2D image and 3D range data are available to support crack classification. However, crack classification still remains a challenge because state Departments of Transportation (DOTs) engineers often use multi-scale crack characteristics (e.g. crack width/length, intersection, pattern, etc) to classify the crack types, and these characteristics are not fully modelled for a reliable crack classification. Based on the new 3D range data, this paper proposes a Crack Fundamental Element (CFE) to characterize cracks at different scales. After an analysis of the fundamental and multi-scale crack characteristics, CFE is proposed for the fundamental line segment approximation of the crack characteristics on multi-scale grid cell analysis, and it is characterized by its density, relative area, bounding box, length, width, center, and orientation. Based on the low-level CFEs, a topological crack graphical representation is, for the first time, built by extending the CFEs into significant crack curves, intersecting crack curves, and approximating polygons of closed crack pieces/spalls at multiple scales. The crack can then be classified using the characteristics of CFEs and their density measures on multi-scale levels. An experimental test using actual 3D data taken in Savannah, Georgia, demonstrates the feasibility of the proposed CFE for multi-scale crack classification. Future research is also discussed.

1 Introduction

Cracks in pavement come from constant overloading, asphalt aging, environmental impacts, improper structural design, etc. Progressive pavement cracking can weaken pavement because it allows water to penetrate and exposes the base to aging. Cracking can then cause accelerated deterioration of pavements. The proper treatment of pavement cracks in an identified crack segment at the right moment is important for cost-effective pavement maintenance. Many transportation agencies, including the Georgia Department of Transportation (GDOT), the Texas Department of Transportation (TxDOT), etc., have invested major resources in their pavement condition survey and evaluation procedures to enhance their decision-making capabilities for determining the best pavement treatments. Traditionally, the collection of pavement crack data is done by visual and manual inspection and analysis. It is dangerous, subjective, costly, time-consuming, and labor-intensive. 2D video log pavement images have been used for automatic pavement distress evaluation, which includes crack segmentation and classification. Crack segmentation is the preliminary process of differentiating the pavement image pixels containing cracks from pixels without cracks and then connecting the crack pixels together. Crack classification evaluates the type, extent, and severity level of pavement cracks, or deducts value of pavement performance for Pavement Management System (PMS) according to crack identification protocols of different state DOTs.

Automatic crack classification in the literature can be divided into two categories: index-based and intelligence-based crack classification. The index-based crack classification focuses on the extraction and utilization of indicators to classify pavement cracks. The method to extract the indicators could be wavelet, statistical analysis, or other image processing methods. Zhou et al. [13] and Nejad and Zakeri [10-11] utilized wavelet transform to extract the indicators for crack classification. Cheng, et al. [4] and Cheng and Miyojim [3] classified the cracks based on the statistical features of graphic properties, such as orientations. For the index-based classification methods, it is difficult to intepret the indecators into crack types. The intelligence-based crack classification, in contrast, learns decision rules from the training samples, which mimics the engineering practices and is more accurate and robust than simple thresholding on the crack indices/indicators. One of the most representative neural network solutions is provided by Lee, H., et al. [7-8]. Three different image-based, histogram-based, and proximity-based neural networks were designed to classify cracks into longitudinal, transverse, block, and alligator cracking with an accurate classification rate of over 90%. The issue with the intelligence-based crack classification is the lack of direct physical meaning of the decision making that can guide the uniform pavement treatment. Numerous large-scale studies [1-2, 5, 12] have shown that current automatic crack classification and quantification results usually have poor correlation with manual survey results. The main challenges of transiting automatic classification results to current survey manuals lie in

1) Different protocols of crack classification
   Outputs of automatic crack classification algorithms are limited to several typical crack types: longitudinal, transverse, block, and alligator, etc. However, the identification of crack type and severity levels in transportation agencies is far more complicated and diverse. None of the existing studies is able to accommodate directly the current Long Term Pavement Protocol (LTPP) or state DOTs' manuals, nor can they be directly implemented by transportation agencies.
2) Different characteristics of crack classification
   Pavement survey manuals of transportation agencies usually have a detailed identification of crack types and severity levels by many crack characteristics, such as crack location, width, and more advanced crack

patterns (e.g., predominant curves, polygonal crack pieces/spalls, crack networks, etc). However, it is difficult to comprehensively extract equivalent crack characteristics from the 2D images to achieve the same accuracy with engineers' field judgments due to lighting non-uniformity, depth absence, and resolution of data acquisition [9].

3) Different scales of crack classification
Pavement cracks are evaluated at different scales in different protocols for the segment, project, and network level survey. Crack characteristics vary at different scales. A dominant crack curve at the segment level could be ignored at the project and network level. While engineers' judgments are adaptable to this multi-scale characterization of cracks, this also challenges the adaptive characterization of cracks under a unified framework based on the fixed-rate video logging of 2D pavement images.

With the advances in sensing and information technology, especially the Global Positioning System (GPS) devices and 3D laser techniques, high-resolution 2D image and 3D range data are now available to support automatic pavement crack detection and classification. Based on the new 3D range data, this paper proposes a Crack Fundamental Element (CFE) to characterize cracks at different scales in support of multi-scale crack classification. After an analysis of the fundamental and multi-scale crack characteristics, CFE is proposed for the fundamental line segment approximation of the crack characteristics on multi-scale grid cell analysis of a pavement image, and it is characterized by its density, relative area, bounding box, length, width, center, and orientation. A topological graphical representation of the crack pattern is built by extending the CFEs into significant crack curves, intersecting crack curves, and approximating polygons of closed crack pieces/spall on multiple scales. The crack can then be classified using the characteristics of CFEs and their density measures on multi-scale levels. An experimental test using the actual 3D data taken in Savannah, Georgia, will be used to demonstrate the feasibility of the proposed CFE for multi-scale crack classification. To the best of the author's knowledge, the proposed CFE is the first to accommodate different protocols of crack classification in state DOTs due to its multi-scale modeling of the physical and fundamental characterization of cracks by CFEs.

2 Crack Fundamental Elements

In this section, crack fundamental elements are presented to characterize the cracks in the identification of pavement cracks among different state DOT practices. As shown in Fig. 1, the comprehensive and accurate charatieristics of a pavement crack are now available in both 2D and 3D data. In the left-hand photo, the 2D intensity data enable us to find the pavement marking and locate the relative position of cracking occurrence. In the right-hand photo of Fig. 1, the 3D range data provide a stronger potential to characterize crack details. The new 3D data can provide a transverse resolution of one millimeter besides the depth measured by the triangulation principle. Crack location, depth, and width at one point in the 3D date are, hereafter, called the fundamental crack characteristics for crack classification. Crack pattern, together with these crack characteristics, are crucial for differentiating crack types and severity levels in transportation agencies' pavement survey practice. However, the complexity of crack pattern makes it difficult to describe and measure accurately and automatically.

Fig. 1. Combined 2D and 3D crack data by 3D laser (left: 2D intensity data; right: 3D range data)

With the advance of high resolution 2D and 3D pavement data, obviously, the next step is to provide a fundamental analysis of multi-scale crack characteristics, in a way that more advanced measures for crack classification could be extracted. Crack Fundamental Element (CFE) is proposed to topologically represent crack characteristics, including crack pattern, from multiple scales. Crack Fundamental Element (CFE) is a group of crack segments, which are clustered together due to their similar or relevant graphic properties or crack characteristics regarding crack type and severity level classification, together with their bounding box. Crack Fundamental Element is the basic component of the proposed model, and the crack pattern is characterized based on the evolution and analysis of CFEs at different levels.

The primary cues for pavement crack classification are that 1) crack pixels can usually be distinguished from their surroundings by their elevation change in the 3D range data or intensity change in the 2D pavement image; 2) crack pixels are geometrically connected in the pattern of a thin strip to formulate the crack curves; 3) crack pixels look more like linear segments in a small window than non-crack pixels. Instead of looking at cracks in the 3D range data pixel by pixel, it is simple and intuitive to take the small linear segments in windows of multiple scales as the fundamental elements for multi-scale crack classification. To classify cracks, the exact segmentation of each crack pixel is unnecessary. The simple but effective characterization of a crack curve is the approximation of the curve by many line segments. Each of these small straight crack segments, with uniform crack width, is considered as an initial CFE. Fig. 2 shows an illustration of initial CFE results.

Crack Fundamental Element (CFE) for Multi-scale Crack Classification

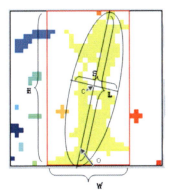

Fig. 2. Results of initial Crack Fundamental Element [6]

When it comes to the higher scales, the CFEs are no more single crack line segments, but a group of crack curves. These crack curves usually share the similar or relevant graphic properties and crack characteristics, such as adjacent locations or similar orientations. CFEs at higher scales are clustered from the lower scale CFEs. A bounding box is introduced to combine these crack curves together and provide uniform graphic properties of this CFE. In this study, ellipses are used as the shape of bounding boxes for CFEs to fit the group of crack curves.

Besides CFE's density (D), relative area (A), bounding box (W, H), and line similarity that were developed in [6] for crack segmentation, four more indicators, length (L), width (S), center (C), and orientation (O), which are shown in Fig. 2, are proposed to characterize the graphic properties of the crack fundamental element at different scales to support crack classification.

(1) Length
Length (L) is defined as the length of the major axis of the ellipse that has the same normalized second central moments as the region of CFE-like pixels in the grid cell:

$$L = 2 * \sqrt{2\left((\mu_{xx} + \mu_{yy}) + \Delta\right)} \tag{1}$$

where
$\mu_{xx} = \frac{\sum_{i=1}^{N} x_i^2}{N}, \mu_{yy} = \frac{\sum_{i=1}^{N} y_i^2}{N}, \mu_{xy} = \frac{\sum_{i=1}^{N} x_i y_i}{N}$, and $\Delta = \sqrt{\left(\mu_{xx} - \mu_{yy}\right)^2 + 4 * \mu_{xy}^2}$ are the second moments of the region.

(2) Width
Width (S) is defined as the length of the minor axis of the ellipse that has the same normalized second central moments as the region of CFE-like pixels in the grid cell:

$$L = 2 * \sqrt{2\left((\mu_{xx} + \mu_{yy}) - \Delta\right)} \tag{2}$$

(3) Center
Center (C) is defined as the centroid of CFE-like pixels in the grid cell:

$$C = (\mu_x, \mu_y) = \left(\frac{\sum_{i=1}^{N} x_i}{N}, \frac{\sum_{i=1}^{N} y_i}{N} \right) \tag{3}$$

(4) Orientation
Orientation (O) is a measure of the CFE direction relative to the horizontal axis of a grid cell:

$$O = \begin{cases} \frac{180}{\pi} * \tan^{-1}\left(\frac{\mu_{xx} + \mu_{yy} + \Delta}{2\mu_{xy}} \right), & \mu_{xx} < \mu_{yy} \\ \frac{180}{\pi} * \tan^{-1}\left(\frac{2\mu_{xy}}{\mu_{xx} + \mu_{yy} + \Delta} \right), & \mu_{xx} \geq \mu_{yy} \end{cases} \tag{4}$$

The main purpose of calculating CFE graphical properties is that the graphical properties at lower scales will be used to further connect and cluster CFEs into higher scales through crack clusteing. It can be seen that all the above indicators are based on one specific CFE and can be extended to a larger scale. Also, CFE represents more significant crack pattern if it runs on the larger scale. This kind of multi-scale characterization of CFE is exactly the property that is needed for multi-scale crack classification. To find the CFE on a larger scale, we can extend CFEs in the four or eight adjacent smaller CFEs by recalculating the length, width, center, and orientation from those of smaller CFEs. The criteria of extension depend on the distance of the center and the deviation of the orientation of two adjacent CFEs. If the distance of two CFEs is greater than a threshold, or if two CFEs deviate from each other greatly, no extension is needed, since larger distance or deviation means the discontinuity of two adjacent CFEs or the appearance of a new separate CFE. Obviously, more advanced distance measures of the four indicators of CFEs can be employed with little extra computation cost.

3 Experimental Results

In this section, the proposed CFE-based graphical representation for multi-scale crack classification is tested using the actual pavement data collected on the State Route 275 (SR-275) in Georgia.

1) Ground truth

The actual pavement data that were collected from SR-275 in Georgia have the resolution of 4 mm at driving direction and 1 mm at transverse direction. GDOT (the Georgia Department of Transportation) pavement engineers helped to establish the ground truth. Fig. 3 (a), as an example, shows the ground truth crack map of the range data in Fig. 1 (b).

2) CFE-based graphical representation (CFE clustering)

Fig. 3 (b) ~ (d) are the multi-scale analysis of cracks in Fig. 3 (a) using CFE-based multi-scale analysis, where Fig. 3 (b) is the lowest scale and Fig. 3 (d) is the highest scale analysis. The lowest scale CFE analysis starts with grouping of continuous crack pixels of crack width variation less than 2 mm for two adjacent pixels on multiple grid cell analysis [6]. The ellipses in Fig. 3 (b) are the CFEs; the red points are the centers and the two green lines inside each ellipse are the length and width of the CFE. The index and orientation of each CFE are marked in a string of the format of "index -- orientation" beside its center.

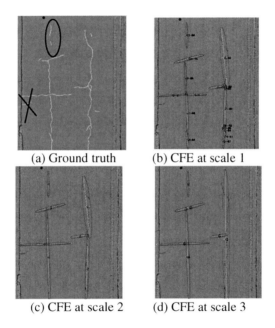

(a) Ground truth   (b) CFE at scale 1

(c) CFE at scale 2   (d) CFE at scale 3

Fig. 3. Results of CFE-based crack graphical representation for crack classification

Compared to the ground truth in Fig. 3 (a), we can see that most of the cracks are well reflected in the lowest CFE analysis. The two crack segments within the ellipse of location 1 in Fig. 3 (a) are connected due to their neighboring relationship and similar cracking direction (less than 15 degree of deviation in terms of their orientations), while most small, isolated crack segments are ignored if they are not too wide and deviate a lot horizontally or vertically. The small segment of location 2 in Fig. 3 (a) locates on the road edge, which is not considered in our analysis for simplicity.

Based on the lowest scale CFE analysis, we further perform the extending, intersecting, and/or polygon-approximating operation of CFEs on two higher scales in Fig. 3 (c) and (d). It can be seen that less significant CFEs are merged or abandoned at a higher scale CFE analysis. There are a total of 17, 7, and 6 CFEs, respectively, from CFEs at the first scale to the third scale. The multi-scale CFE analysis stops if no operation is necessary in terms of the overlapping, deviation, and/or width of CFEs.

3) Results of CFE-based graphical representation for multi-scale crack classification

Table 1 lists the result of CFE-based graphical representation of the highest scale in Fig. 3 (d), where the unit of the position coordinates and length/width is millimeter, and the unit of orientation is degree relative to the horizontal axis from left to right. There are 6 dominant crack curves and 3 intersections of transverse and longitudinal curves. No crack pieces/spalls are detected in Fig. 3 (a). It can be seen that the significant crack characteristics is well extracted to classify cracks according to different protocols. Further, the applications of the data at a lower scale (e.g scale 1 or 2) of CFE analysis can also be explored in the future, such as automated crack sealing.

Table 1. Result of the highest scale CFE analysis for crack classification

CFE #	Center (mm)	Length (mm)	Width (mm)	Orientation (°)
1	(1285, 696)	1016	65	89
2	(1366, 1585)	1261	105	15
3	(1123, 2957)	2103	70	2
4	(1281, 3447)	3712	102	88
5	(2485, 2654)	729	100	8
6	(2772, 2790)	5293	277	90

4 Conclusions and Recommendations

With the advance of sensor and information technology, high-resolution 2D image and 3D range data are available to support crack classification. However, crack classification still remains a challenge because state Departments of Transportation (DOTs) engineers often use multi-scale crack characteristics (e.g. crack width/length, intersection, and pattern) to classify the crack types, and these characteristics are not modelled adequately for a reliable crack classification. Therefore, there is a need to consistently classify pavement cracks from the physical and fundamental perspectives according to different state DOTs' protocols by taking advantage of the new 3D data.

Based on the new 3D range data, this paper proposes a Crack Fundamental Element (CFE) to characterize cracks at different scales in support of multi-scale crack classification. Crack Fundamental Element (CFE) is a group of crack segments, which are clustered together based on to their similar or relevant graphic properties or crack characteristics regarding crack type and severity level classification, together with their bounding box. The graphic properties of CFEs are characterized by their density, relative area, bounding box, length, width, center, and orientation. The fundamental straight line segment approximation is defined as the initial level of CFE. Based on the low-scale CFEs, a topological CFE clustering is conducted to build the high-scale CFEs. The crack can then be classified using the characteristics of CFEs and their measures at multi-scale levels. An experimental test using the actual 3D data taken in Savannah, Georgia, demonstrates the concept of the proposed CFE for multi-scale crack classification, with the focus on CFE clustering. To the best of the author's knowledge, the proposed CFE is the first to accommodate different protocols of crack classification in state DOTs due to its multi-scale modeling of the physical and fundamental characterization of cracks by CFEs.

Although the proposed CFE concept demonstrates its capability for multi-scale crack classification, we recommend the following:

1) This paper presents the concept of Crack Fundamental Element (CFE), with the focus on CFE graphic properties and clustering process at different scales. A complete CFE-based multi-scale crack analysis model, including detailed crack characteristics at different CFE scales along with the topological crack representation, needs to be developed.
2) More comprehensive tests with the proposed CFE be conducted for different types, severity levels, and extent of cracks on the project- and network- level;
3) More sophisticated algorithms be developed for the extending, intersecting, and approximating operations in the CFE;
4) An innovative and consistent decision tree for crack treatment be devised based on the results of the topological crack graphical representation of CFEs.

Acknowledgements. The authors would like to thank the US DOT RITA program for its support. This paper was sponsored by US DOT RITA program (RITARS-11-H-GAT). The authors would like to thank the assistance provided by Mr. Caesar Singh, the program manager of US DOT. The authors also would like to thank the Georgia Department of Transportation for establishing ground truth for this study. We also would like to thank Chenglong Jiang for his support during this paper's development. The views, opinions, findings and conclusions reflected in this paper are the responsibility of the authors only and do not represent the official policy or position of the USDOT, RITA, or any State or other entity.

References

[1] Albitres, C.M.C., Smith, R.E., et al.: Comparison of automated pavement distress data collection procedures for local agencies in San Francisco Bay Area, California. Transportation Research Record (1990), 119–126 (2007)

[2] Capuruço, R.A.C., Tighe, S.L., et al.: Performance evaluation of sensor- and image-based technologies for automated pavement condition surveys. Transportation Research Record (1968), 47–52 (2006)

[3] Cheng, H.D., Chen, J.-R., et al.: Novel approach to pavement cracking detection based on fuzzy set theory. Journal of Computing in Civil Engineering 13(4), 270–280 (1999)
[4] Cheng, H.D., Miyojim, M.: Automatic pavement distress detection system. Information Sciences 108(1-4), 219–240 (1998)
[5] Fu, P., Harvey, J., et al.: New Method for Classifying and Quantifying Flexible Pavement Cracking in Automated Pavement Condition Survey. Transportation Research Board Annual Meeting (2011)
[6] Huang, Y., Tsai, J.: Enhanced Pavement Distress Segmentation Algorithm using Dynamic Programming and Connected Component Analysis. Transportation ResearchBoard Annual Meeting (2011)
[7] Lee, B.J., Lee, H.D.: Position-invariant neural network for digital pavement crack analysis. Computer-Aided Civil and Infrastructure Engineering 19(2), 105–118 (2004)
[8] Lee, H., Kim, J.: Development of a crack type index. Transportation Research Record (1940), 99–109 (2005)
[9] Kaul, V., Tsai, Y.: A Quantitative Performance Evaluation of Pavement Distress Segmentation Algorithms. Transportation Research Record: Journal of the Transportation Research Board 2153, 106–113 (2010)
[10] Nejad, F.M., Zakeri, H.: An expert system based on wavelet transform and radon neural network for pavement distress classification. Expert Systems with Applications 38(6), 7088–7101 (2011)
[11] Nejad, F.M., Zakeri, H.: An optimum feature extraction method based on Wavelet-Radon Transform and Dynamic Neural Network for pavement distress classification. Expert Systems with Applications 38(8), 9442–9460 (2011)
[12] Tighe, S.L., Ningyuan, L., et al.: Evaluation of semiautomated and automated pavement distress collection for network-level pavement management. Transportation Research Record (2084), 11–17 (2008)
[13] Zhou, J., Huang, P.S., et al.: Wavelet-based pavement distress detection and evaluation. Optical Engineering 45(2) (2006)

Cracking Models for Use in Pavement Maintenance Management

Adelino Ferreira[1], Rui Micaelo[2], and Ricardo Souza[1]

[1] Department of Civil Engineering, University of Coimbra, Portugal
{adelino,ricardosouza}@dec.uc.pt
[2] Department of Civil Engineering, Universidade Nova de Lisboa, Portugal
ruilbm@fct.unl.pt

Abstract. With the recent approval of the Portuguese Law No. 110/2009 of 18 May, within the scope of road concession contracts, the concessionaires need to submit to the Portuguese Road Infrastructures Institute a Quality Control Plan (QCP) and a Maintenance and Operation Manual (MOM). These documents require the revision of current Pavement Management Systems to consider pavement performance prediction models for each pavement state parameter so that it permits time definition of maintenance and rehabilitation (M&R) interventions for the fulfilment of the values defined in the QCP in each year of the concession period. The QCP presents the admissible values for each pavement state parameter (cracking, rutting, roughness, etc.) that a concessionaire of highways needs to verify.

Nevertheless, a concessionaire, beyond the annual pavement inspections to demonstrate the fulfilment of the QCP, wants to predict the proper time to apply M&R preventive interventions at a minimum cost for the complete concession period.

This paper describes the state-of-the-art in terms of cracking models. The selected models evaluate the cracking area evolution for a set of representative Portuguese pavements structures and traffic conditions. The Indian and HDM-4 deterioration models were considered to be the most promising to implement in a new Portuguese Maintenance Optimisation System, i.e. to provide a good solution to the pavement maintenance management problem involving not only periodic maintenance but also routine maintenance (crack sealing, rut levelling, patching, etc.).

1 Introduction

A Pavement Management System (PMS) can be defined as a set of tools which helps a road network administration to optimize maintenance and rehabilitation (M&R) actions for keeping the pavements in good service condition. One of the modules of a PMS is the Pavement Performance Model (PPM), which is a mathematical representation that can be used to predict the future state of pavements, based on current state, deterioration factors and effects resulting from M&R actions [1]. Currently the PMS of Estradas de Portugal S.A., the Portuguese

Road Administration, uses for PPM the AASHTO pavement performance model that computes a global pavement condition index, the present serviceability index (PSI), based on several factors like the traffic, the material properties and the drainage and environmental conditions [2]. The extent and severity of distresses at the time that the pavement condition index reaches the warning level restricts the implementation of more cost-effective techniques [3]. In 2007 and 2009 the Portuguese Government published legislation [4, 5] established EP – Estradas de Portugal, S.A., as the global road network concessionaire and the basis of the concession contract. Within this contract it was established that concessionaires have to submit to the Portuguese Road Infrastructures Institute (InIR), the supervisor institution, on a regular basis, a Quality Control Plan (QCP) and a Maintenance and operation Manual (MOM). The QCP defines the limits of pavement condition parameters (rutting, cracking, roughness, etc.) than can be found at any time of the concession period. Therefore, these two documents, specially the first one, require knowing the pavement condition ahead, not only the general pavement service condition, but quantifying the extent and the magnitude of pavement distresses and the actions to be implemented in every situation. When a concessionaire does not fulfil the QCP, InIR can apply a contractual infraction, in which the global sum varies, according to its gravity, between €5,000 and €100,000, or daily values that can vary between €500 and €5,000 [4]. A concessionaire, beyond the annual pavement inspections to demonstrate to the InIR and the concessor (the Portuguese State or represented by the EP - Estradas de Portugal, S.A.) the fulfilment of the QCP, wants to predict the year when their pavements do not fulfil the admissible values for some state parameter. A concessionaire knowing this information can apply M&R preventive interventions at a minimum cost in order to effectively fulfil the QCP in all the remaining years of the concession period. This paper describes the state-of-the-art in terms of cracking models, and evaluates the performance of the selected models considering the cracking area evolution for a set of representative Portuguese pavements structures and traffic conditions.

2 Cracking Prediction Models

In this study, the methodology used was to analyse several pavement cracking deterioration models available in the literature with the objective of its integration in the Portuguese Pavement Management Systems. The models selected are the following ones: the Brazilian model; the PAVENET-R model; the HDM-4 model; the Ker Lee Wu (KLW) model; the Indian model; and the Austroads model.

2.1 Brazilian Model (1994)

This model was developed by Visser, Queiroz and Caroca [6] based on a long-term pavement monitoring program carried out in Brazil between 1975 and 1985. The road sections were unbound granular base flexible pavements in areas with tropical to subtropical climate with an average annual precipitation between 1200 and 1700 mm/year. Cracked area over time is predicted with Eqn. (1), developed with multiple regression and probabilistic time failure analysis, which depends on traffic volume, the pavement bearing capacity (load deflection) and age. For existing pavements with asphalt surfacing the model only applies to cracking progression prediction, while for asphalt overlays and slurry seals the model comprises two-phases, initiation time and progression prediction. Two-phase models give an extra opportunity to calibrate the model for cracking prediction over time. This model considers the Benkelman beam to measure pavement deflections, but this equipment is not used in Europe any more. However, some regressions relating the modified structural number with the Benkelman beam maximum deflection have been proposed by several authors such as Eqn. (2) by Paterson [7, 8]. The modified structural number is the evolution of the AASHTO structural number by considering the subgrade contribution.

$$C_t = (B \times 10^{-2}) \times \log(N80c_t) \times (0.0456 + 0.00501 \times Y_t) - 18.53 - C_0 \qquad (1)$$

$$SNC = 3.2 \times B^{-0.63} \qquad (2)$$

Where C_t is the pavement cracked area (class 2 cracking or worse, i.e. crack width larger than 1 mm) in year t (m^2/100m^2); Y_t the age of pavement since original construction or since subsequent AC overlay (years); $N80c_t$ is the cumulative 80 kN equivalent single axle load (ESAL) at age t (ESAL/lane); B is the Benkelman beam maximum deflection for the existing pavement (mm); C_0 is the cracking offset term calculated to ensure that predicted cracking conforms with the initial value at the start of analysis; SNC is the modified structural number.

2.2 PAVENET-R Model (1996)

The cracking prediction model defined by Eqn. (3) is used in the computer model PAVENET-R [9] aiming at the optimization of the maintenance-rehabilitation problem at the network level. The cracked area over time is predicted based on traffic and the pavement AASHTO structural number calculated using Eqn. (4). As for the previous model it is an only one phase model (just dealing with progression) and it does not include a variable that accounts for the existing cracked area at the beginning of the analysis.

$$C_t = 617.14 \times N80c_t \times SN^{-SN} \qquad (3)$$

$$SN = \sum_{n=1}^{N} H_n \times C_n^e \times C_n^d \qquad (4)$$

Where C_t is the total cracked area in year t (m^2/100m^2); $N80c_t$ is the cumulative equivalent standard axle load (ESAL) at age t (million ESAL/lane); SN is a structural number; C_n^e is the structural coefficient of layer n; C_n^d is the drainage coefficient of layer n; and H_n is the thickness of layer n (mm).

2.3 INDIAN Model (1994)

The Indian model was derived from a pavement performance study carried out during the 90's with extensive monitorisation of pavement sections (145) along national and state highways in four Indian states [10]. The cracking prediction model is a two-phase model, considering the time to cracking initiation calculated using Eqn. (5), and the cracking progression calculated using Eqn. (6). As for the previous models, just two variables were included (traffic and the pavement structural number). The model is applicable to pavements with asphalt surfacing (excluding surface dressing and slurry seal). The climate where the pavement data was gathered varies from arid to humid subtropical, being far from the Portuguese Mediterranean climate.

$$T_{ci} = 4.00 \times e^{-1.09 \times \frac{n \times N80c_1}{SNC^2}} \tag{5}$$

$$C_t = C_i + 4.26 \times \left(\frac{n \times (N80c_t - N80c_{ti})}{SNC} \right)^{0.65} \times SC_i^{0.32} \times (t - t_i) \tag{6}$$

Where T_{ci} is the time to structural cracking initiation (years) - $C_t = 2.0\%$; C_t is the total cracked pavement area in year t (m²/100m²); $N80c_t$ is the cumulative 80 kN equivalent single axle load (ESAL) at age t (million ESAL/lane); $N80c_{Tci}$ is the cumulative 80 kN equivalent single axle load (ESAL) at age of cracking initiation (million ESAL/lane); SNC is the modified structural number for the pavement; n is the number of lanes in the road section; C_i is the total cracked pavement area at the beginning of the analysis period (m²/100m²); t_i is the time at the beginning of the analysis period (years); SC_i is the minimum of $\{C_i; 100 - C_i\}$ (m²/100m²).

2.4 HDM-4 Model (2000)

The Highway Development and Management (HDM-4) system uses a cracking pavement performance model applied in two phases [6, 10-14]: the time to structural crack initiation and the structural crack progression. The HDM-4 is the successor of the World Bank Highway Design and Maintenance Standards Model HDM-III, which has been used by various road agencies all over the world for the last 20 years. Eqn. (7) is used to calculate the time to structural cracking initiation (years). Eqn. (8), (9), (10) and (11) are used to calculate the percentage of cracking over time (class 2 or worse), designed as the "all cracking" model. This formulation is valid for flexible pavements with asphalt or surface treatment as surface course and granular or asphalt base course.

$$T_{ci} = CDS^2 \times 4.21 \cdot e^{0.14 \times SNC - 17.1 \cdot N80c_g / (8 \times SNC^2)} + CRT \tag{7}$$

$$C_t = C_i + dC \tag{8}$$

$$Y = 0.828 \times z \times (t - t_i) + [\min \{\max(C_i; 0.5); 100 - \max (C_i; 0.5)\}]^{0.45} \qquad (9)$$

$$dC = \frac{CRP}{CDS} \times z \times \left(Y^{0.45} - \min \{\max(C_i; 0.5); 100 - \max (C_i; 0.5)\}\right) \quad \text{if } Y \geq 0 \qquad (10)$$

$$dC = \frac{CRP}{CDS} \times (100 - C_t) \quad \text{if } Y < 0 \qquad (11)$$

Where T_{ci} is the time to structural cracking initiation (years) - $C_t = 0.5\%$; C_t is the total cracked pavement area in year t (m²/100m²); C_i is the total cracked pavement area at the beginning of the analysis period (m²/100m²); $N80c_t$ is the cumulative 80 kN equivalent single axle load (ESAL) at age t (million ESAL/lane); SNC is the modified structural number for the pavement; Y is an auxiliary variable; CDS is the construction defects indicator in asphalt layers (0.5 to 1.5 according to real binder content – dry, normal and rich); CRT is the crack retardation time due to maintenance (years) (default value 0); CRP is the crack propagation indicator (1 - 0.12×CRT); z is an auxiliary variable (+1 if $C_i \leq 50$ and -1 otherwise).

The AASHTO structural number (SN), calculated using Eqn. (4), does not include the subgrade contribution as it is considered in the pavement design procedure through the resilient modulus. In opposition, HDM models (version III and 4) consider a different version of the structural number, the modified structural number (SNC), calculated using Eqn. (12), which takes into account the subgrade strength which is calculated using Eqn. (13) [7].

$$SNC = 0.0396 \sum_{n=1}^{N} (H_n / 25.4) \times C_n^e \times C_n^d + SNSG \qquad (12)$$

$$SNSG = 3.51 \times \log(CBR) - 0.85 \times [\log(CBR)]^2 - 1.43 \quad \text{if } CBR \geq 3 \qquad (13)$$

Where SNC is the modified structural number; C_n^e is the structural coefficient of layer n; C_n^d is the drainage coefficient of layer n; and H_n is the thickness of layer n (mm).

Thermal cracking is not considered to be an important source of cracking in Portuguese road pavements due to small number of days with subfreezing temperatures (Mediterranean climate). In a different position, reflexion cracking, which is the progression of cracks upwards to the surface from previously cracked asphalt pavements or cement stabilized materials, is important in every cracking deterioration model. It allows prediction of this pavement distress evolution after M&R actions have been taken, considering the common large concession periods. The HDM-4 model for reflexion cracking prediction, Eqn. (14) and (15), is based on Gulden and Malaysian studies, considering a two-phase model as for the other cracking models.

$$T_{ri} = \frac{685}{ADH} \times B^{-0.5} \times \left(1 - \frac{\min\{H_{OV}; 199\}}{200}\right)^{-2} \qquad (14)$$

$$Cr_t = \min\left\{Cr_i + 0.0182 \times ADH \times B^{0.5} \times \left(\max\left\{0; 1 - \frac{H_{OV}}{200}\right\}\right)^2 \times (t - tr_i); C_{OV}\right\} \quad (15)$$

Where Tr_i is the time to reflexion cracking initiation (years); H_{OV} is the thickness of the new asphalt concrete layer (mm) (< 200 mm); Cr_t is the total cracked pavement area in year t, limited to the amount of cracked area before overlay C_{OV} (m^2/100m^2); ADH is the average daily heavy traffic; tr_i is the time at the beginning of the analysis period; B is the average Benkelman beam deflection on both wheel-paths (mm).

2.5 Ker Lee Wu (KLW) Model (2008)

Ker, Lee and Wu [15] developed a fatigue cracking prediction model (Eqn. (16)) using LTPP data, for the pavements with asphalt surface course on granular or bound base. The model determines cracking area over time based on traffic, pavement age, climate (precipitation, air temperature and freeze-thaw cycles) and load pavement response (tensile strain). The model was developed to improve pavement design but it can be applied as a PPM.

$$C_t = \exp\begin{pmatrix} -18.08 + 0.943 \times \sqrt{Y_t} + 0.832 \times \log(1000 \times N80_t) \\ + 0.121 \times \sqrt{precip} + 0.869 \times \sqrt{temp} \\ + 31.489 \times (\varepsilon_t \times 1000)^2 + 3.242 \times \log(ft) \end{pmatrix} \quad (16)$$

Where C_t is the total cracked pavement area in year t (m^2/100m^2); $N80_t$ is the number of 80 kN equivalent single axle load (ESAL) applications in year t (million ESAL/lane); *precip* is the average annual precipitation (mm); *temp* is the mean annual temperature (°C); ε_t is the tensile strain at the bottom of the AC layer; *ft* is the yearly freeze-thaw cycles; Y_t is the time since the pavement's construction or its last rehabilitation (years).

2.6 Austroads Model (2010)

Austroads has recently developed road deterioration models for the roughness, rutting and cracking prediction of sealed granular pavements, which represents 85% of sealed pavements in Australia [16]. Cracking deterioration models were determined based on data collected with the RTA/NSW Road Crack equipment on arterial roads in South Australia between 1999 and 2004. As most of collected cracking data (1384 of 1675 samples) was on asphalt pavements, a cracking model was determined for this pavement type (Eqn. (17) and (18)). The two-phase model (cracking initiation and progression) was initially developed for sealed granular pavements and then adapted to asphalt pavements. Therefore, initiation of cracking is estimated using the same Eqn. to sealed granular pavements (seal life) that depends on climate (air temperatures), bitumen (ARRB test result) and maximum aggregate dimension. Cracked area over time is predicted based on pavement's age, time since cracking initiation and climate (with Thornthwaite Moisture Index). It is referred that traffic and pavement bearing capacity were not considered statistically significant for the models because these variables could not be reliable assessed in the data set.

$$T_{ci} = \left(\frac{0.158 \times TMIN - 0.107 \times R + 0.84}{0.0498 \times T - 0.0216 \times D - 0.000381 \times S^2} \right)^2 \quad (17)$$

$$C_t = K \times \left[100 - 200 \times \left(1 + e^{\left(\frac{0.682 \times (t - T_{ci})}{\left(\frac{200 - TI_t}{25} \right)^{3.5}} \right)} \right)^{-1} \right] \quad (18)$$

Where T_{ci} is the seal life (years); TI_i is the Thornthwaite Moisture Index for climate pavement conditions at year t; $TMIN$ is the yearly average of the daily minimum air temperature (°C); $TMAX$ is the yearly average of the daily maximum air temperature (°C); T is the average of $TMAX$ and $TMIN$ values; D is the ARRB Durability Test result; S is the nominal size of seal (nominal stone size, mm); R is the risk factor with a scale from 1 (very low risk) to 10 (very high risk); C_t is the total cracked pavement area in year t (m^2/100m^2); K is the calibration factor.

This cracking prediction model has several drawbacks for the implementation in the Portuguese PMS, namely for having been developed from sealed granular pavements, which is not a common pavement type (at least on main roads), for including a variable from a lab test not used in Europe and for not including a traffic related variable that it is considered to induce most of pavements cracking (fatigue).

3 Cracking Prediction Models Testing

The selected cracking models were tested by comparing the evolution of cracked area over time (design period, taken usually as 20 years) for different pavement structures. The set of representative pavement structures were selected based on the structures proposed by the Portuguese Pavements Design Manual [17] as function of traffic level and foundation capacity. Table 1 presents the levels of the selected daily traffic (T1, T3 and T5) and the corresponding proposed pavement structures for a subgrade F3 (CBR ratio of 20%). In the analysis it was additionally considered a subgrade F2 (CBR ratio of 10%), which requires extra 40 mm of asphalt thickness in each case. The pavements were considered to be situated in central area of Portugal (Coimbra district). The temperature and precipitation values were determined with the weather data collected over the period 1971-2000 by the Portuguese Meteorology Institute [18].

Table 1. Pavement data

	Parameter		Pavements		
			P4	P9	P14
Traffic	$AADT_h$ (per way and lane)		300 (T5)	800 (T3)	2000 (T1)
	Traffic growth rate (%)		3	4	5
	Heavy vehicles damage factor		3	4.5	5.5
Structure	Asphalt surface layer	H_n (mm)	40	50	60
		E (MPa)	4000	4000	4000
	Asphalt base layer	H_n (mm)	140	190	220
		E (MPa)	4000	4000	4000
	Granular sub-base	H_n (mm)	200	200	200
		E (MPa)	200	200	200
	Foundation	CBR (%)	20	20	20
Climate	Average daily temperature (°C)		15.5	15.5	15.5
	Average yearly precipitation (mm)		905.1	905.1	905.1

4 Results

Figure 1 shows pavement cracked area evolution predicted by all PPM for a pavement P9 (traffic level T3) and two subgrade levels (F2 and F3) during 20 years that is usually considered for the pavement design. It is considered that no M&R actions are implemented during the 20 years. HDM-4 predicts considerably larger values of cracked area during the analysis. At the end of the analysis period HDM-4 predicted cracked area is around the double of the second largest value and cracking is spread all over the road pavement (100%). In a lowest to largest predicted cracked area in year 20, the PPM sequence is the following: KLW; Austroads; Brazilian; Indian; PAVENET-R and HDM-4. HDM-4 predicts cracking initiation around years 5 to 6 and after that cracking spreads faster than predicted by any other method. In opposition, cracked area predicted by KLW and Austroads PPM is very low, less than 15% in year 20. When a lower bearing capacity foundation is considered (F2 subgrade instead of F3), with extra 40 mm of asphalt, cracked area evolution difference is almost imperceptible with the exception of using the PAVENET-R PPM. This PPM uses the AASHTO structural number, which does not consider the subgrade effect. As the pavement thickness is increased to compensate the subgrade strength, the SN value increases and predicted cracking is lower at any time. Most PPM results show that the extra 40 mm of asphalt thickness indicated in the design manual is adequate for this foundation bearing capacity variation. Figure 2 shows the cracked area evolution predicted by all models for two pavement structures (P4 and P14) with different traffic levels (T1 and T5) and F3 subgrade.

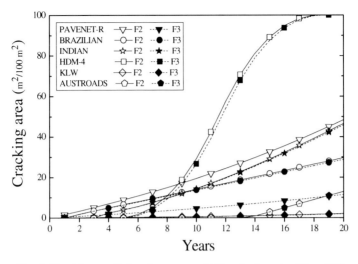

Fig. 1. Cracking prediction for pavement P9 and subgrades F2 and F3

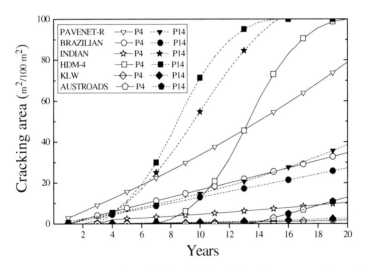

Fig. 2. Cracking prediction for traffic levels T5 and T1 and pavements P4 and P14

The KLW, Austroads and Brazilian models predict similar values of cracked area during the analysis period. The Austroads method predicts exactly the same values as it does not consider traffic and subgrade in the model. The HDM-4 and Indian models predict larger cracked area values for pavement P14 (highest traffic level) and PAVENET-R predicts less cracked area. The largest difference between predictions is obtained with the Indian model. The cracking growing rate is very similar for both situations in the HDM-4 prediction and the extent of cracking area at any moment is solely dependent on time for cracking initiation. The PAVENET-R model predicts a 50% reduction of cracked area in year 20 as the pavement changes from P4 to P14, which allows concluding that the model is much more dependent on pavement bearing capacity than on traffic.

5 Conclusions

The Austroads model is not adequate to include into Portuguese PMS since it does not distinguish different pavement structures used in Portugal as well as traffic levels. The number of years to intervention is considered too optimistic. The KLW model is not satisfactory since the predicted cracked areas are extremely low during the design period, i.e. not in agreement to current Portuguese roads condition. The PAVENET-R model is considered not to be adequate because it amplifies too much the pavement bearing capacity (with exclusion of the foundation strength) and lessens too much the traffic influence. In the Brazilian model, the values obtained are logical though pavements sections used for the model development have different conditions (climate, soils) than can be found in Portugal. The Indian and HDM-4 models were considered to produce acceptable results and therefore it is recommended that a full verification and validation of both models should be conducted using Portuguese pavement condition time series data.

References

[1] Ferreira, A., Picado-Santos, L., Antunes, A.: Pavement performance modelling: state of the art. In: Proceedings of Seventh International Conference on Civil and Structural Engineering Computing, pp. 157–264. Civil-Comp Press, Oxford (1999)
[2] Ferreira, A., et al.: Selection of pavement performance models for use in the Portuguese PMS. International Journal of Pavement Engineering 12(1), 87–97 (2011)
[3] Lou, Z., Lu, J., Gunaratne, M.: Road surface crack condition forecast using neural network models, p. 93. University of South Florida, Florida (2003)
[4] MOPTC, Portuguese Law No. 110/2009 of May 18. Ministry of Public Works, Transports and Communications, Daily of the Republic, 1ª Série - No. pp. 3061-3099 (May 18, 2009) (in Portuguese)
[5] MOPTC, Portuguese Law No. 380/2007 of November 13. Ministry of Public Works, Transports and Communications, Daily of the Republic 1st Series - No. pp. 8403-8437 (2007) (in Portuguese)
[6] Visser, A., Queiroz, C., Caroca, A.: Total cost rehabilitation method for use in pavement management. In: Proceedings of 3rd International Conference on Managing Pavements, Texas, USA, pp. 37–44 (1994)
[7] NDLI: Modelling road deterioration and maintenance effects in HDM-4, pp.351, International Study of Highway Development and Management Tools, ND Lea International Ltd., Vancouver, British Columbia, Canada (1995)
[8] Bennett, C.R.: Comparison of loadman and Benkelman beam deflection Measurements, p. 38. Four States Pavement Management Project, India (1994)

[9] Fwa, T.F., Chan, W.T., Tan, C.Y.: Genetic-Algorithm Programming of Road Maintenance and Rehabilitation. Journal of Transportation Engineering 122(3), 246–253 (1996)
[10] Jain, S.S., Aggarwal, S., Parida, M.: HDM-4 Pavement Deterioration Models for Indian National Highway Network. Journal of Transportation Engineering 131(8), 623–631 (2005)
[11] Mrawira, D., et al.: Sensitivity Analysis of Computer Models: World Bank HDM-III Model. Journal of Transportation Engineering 125(5), 421–428 (1999)
[12] PIARC: Highway development and management, volume one – Overview of HDM-4, pp.43, World Road Association, Paris, France (2000)
[13] Odoki, J., Akena, R.: Energy balance framework for appraising road projects. Proceedings of the Institution of Civil Engineers - Transport 161(1), 23–35 (2008)
[14] Ihs, A., Sjögren, L.: An overview of HDM-4 and the Swedish pavement management system. In: VTI - Infrastructure Maintenance, p. 31. Linköping, Sweden (2003)
[15] Ker, H.-W., Lee, Y.-H., Wu, P.-H.: Development of Fatigue Cracking Prediction Models Using Long-Term Pavement Performance Database. Journal of Transportation Engineering 134(11), 477–482 (2008)
[16] Austroads: Interim network level functional road deterioration models, p. 45. Austroads Ltd., Australia (2010)
[17] JAE, Manual of pavement structures for the Portuguese road network, pp. 1–54. Junta Autónoma de Estradas, Portugal (1995) (in Portuguese)
[18] IM: Climate Normal 71-00. Institute of Meteorology IP Portugal (2011), http://www.meteo.pt/en/oclima/clima.normais/006/ (cited September1, 2011]

Multi-cracks Modeling in Reflective Cracking

Jorge Pais[1], Manuel Minhoto[2], and Shakir Shatnawi[3]

[1] University of Minho, Portugal
[2] Polytechnic Institute of Bragança, Portugal
[3] Shatec Engineering Consultants, LLC, California, USA

Abstract. Reflective cracking is a major concern for engineers facing the problem of road maintenance and rehabilitation. The problem appears due to the presence of cracks in the old pavement layers that propagate into the pavement overlay layer when traffic load passes over the cracks and due to the temperature variation. The stress concentration in the overlay just above the existing cracks is responsible for the appearance and crack propagation throughout the overlay. The analysis of the reflective cracking phenomenon is usually made by numerical modeling simulating the presence of cracks in the existing pavement and the stress concentration in the crack tip is assessed to predict either the cracking propagation rate or the expected fatigue life of the overlay. Numerical modeling to study reflective cracking is made by simulating one crack in the existing pavement and the loading is usually applied considering the shear mode of crack opening. Sometimes the simulation considers the mode I of crack opening, mainly when temperature effects are predominant. Thus, this paper presents a study where multiple cracks are modeled to assess the reflective cracking phenomenon and to compare to the case of only one crack. The modeling with only one crack was made simulating both mode I and mode II of crack opening taking into account the traffic effects. The influence of multiple cracks was expressed in terms of stress and strain in the zone above existing cracks. One of the conclusions from the current study is that the presence of multiple cracks can lead to a state of stress/strain higher than those obtained with only one crack. Also the position of the crack modeled in the finite elements analysis have a significant influence in the state of stress/strain obtained. However, the consideration of only one crack is sufficient to obtain significant results in the reflective cracking modeling.

1 Introduction

Cracks in the pavements tend to reflect through an overlay placed on the cracked pavement due to the traffic and temperature effects, depending on the magnitude of the stress concentrations at the tip of the crack, the resistance of the overlay material to crack propagation and the characteristics of the interface between the overlay and the existing pavement. The stress concentration at the crack tip results of the bending, shearing and tearing actions of traffic loads and tensile and bending actions caused by temperature and moisture movements as well as temperature and moisture gradients [1].

To assess the cracking in pavement overlays, many investigations have been conducted in terms of experimental and numerical modeling. The first studies started by Majidzadeh et al [2] with the application of the fracture mechanics in the analysis of pavement fatigue where the fatigue life of paving mixtures in terms of material constants, geometry, boundary conditions, and the state of stress is predicted. In that work, fatigue is defined in terms of crack initiation, influence on crack growth, and critical stress intensity at the critical failure point. The laboratory tests to support that study utilized notched and unnotched beams supported in an elastic foundation to predict the fracture parameters. The determination of the fracture parameters (i.e the stress intensity factors) were made based on experimental assumptions from fracture tests.

Later on, Van Gurp and Molenaar [3] developed a procedure to predict the reflective cracking in asphalt overlays using linear elastic finite element models, by analyzing the crack propagation form the old cracked asphalt layers through the new overlay. The models used only considered the traffic influence simulating mode I and II of crack opening by applying the load above the crack and adjacent to the crack, respectively, as represented in Figure 1.

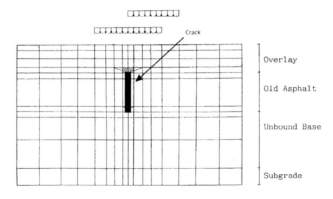

Fig. 1. Single crack modeling in mode I and II of crack opening [3]

Since then, the numerical modeling has been used in the assessment of reflective cracking by different methods. Paulino et al [4] applied a cohesive zone fracture model to simulate crack initiation and propagation in asphalt concrete using intrinsic constitutive laws to connect traditional finite elements to simulate localized damage and softening behavior. Nesnas and Nunn [5] used a finite element model with multi-cracks, as indicated in Figure 2, to investigate the top-down cracking in cement treated base pavements.

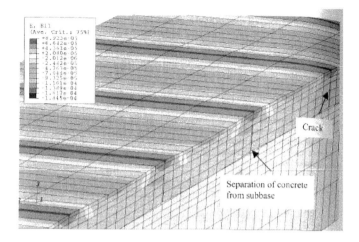

Fig. 2. Multi-cracks simulation for top-down cracking [5]

More recently, Wu and Harvey [6] developed a finite element model to evaluate the performance of several asphalt mixes that contain binders with recycled rubber. The mechanistic model was based on non-local continuum damage mechanics and the finite element method and the damage evolution law parameters were identified using laboratory fatigue test data. The finite element model (Figure 3) was the first model used in the reflective cracking analysis that was created with multiple cracks, simulating the alligator cracking in the existing pavements before the placement of a pavement overlay.

However, these studies did not investigate the influence of the existence of multiple cracks or the influence of the spacing between cracks in the evaluation of the reflective cracking. This subject takes a significant importance because cracked pavements, mainly the flexible pavements, usually present multiple cracks (alligator cracks) before the pavement overlaying.

Thus, this paper aims to study the influence of the existence of multiple cracks in finite element models on reflective cracking. This influence is assessed by the state of stress and strain in the pavement overlay, just above the existing cracks in the old pavement.

This effort consisted of developing a 2D finite element model which was created, using the plain strain mode, in which 10 cracks were modeled in the cracked layer and spaced 10 cm from each other. The model has the ability to easily change the crack spacing, the elimination of some cracks to create any configuration of cracking with any cracking spacing, from a pavement with only one crack to a pavement up to 10 cracks.

The application of this model resulted in creating different cracking configurations to study the influence of crack spacing on the reflective cracking phenomenon. The study investigated three different overlay configurations: 10, 20 and 30 cm overlay thicknesses over an existing cracked pavement.

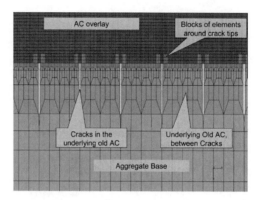

Fig. 3. Multi-cracks modeling to simulate HVS effect [6]

2 Finite Elements Model

A 2D finite elements model was created for a pavement configuration consisting of an overlay layer, an existing asphalt cracked layer, a granular base layer and a subgrade layer, as schematically represented in Figure 4. The model considers the existence of full friction as interface between old and new pavement layers. The existing asphalt layer was modeled with 10 cracks, numbered from 1 to 10, starting from the left side of the model. The distance between cracks was set to 10 cm and the crack width was set to 3 mm. The model has the ability to easily allow the elimination of some cracks to create any configuration of cracking with any cracking spacing, from a pavement with only one crack to a pavement up to 10 cracks.

The mesh of the model was designed by using quadrilateral, two-dimensional structural-solid elements, with eight nodes, with two degrees of freedom at each node. The mesh was designed to apply a load with a dual wheel configuration representing a standard axle wheel of 80 kN (Figure 5), applied on the pavement surface in a representative area of the tire-pavement contact. The finite element model used in the numerical analysis was developed in a general finite elements code, ANSYS(R) Academic Teaching Introductory, V12.1.

The finite elements model was configured to create 25 different pavements configurations, varying the cracking configuration. The first 10 models present only one crack, from crack #1 to crack #10, representing all possibilities of isolated cracks. The following pavements represent the configurations of multi-cracks varying the cracking space from 10 cm up to 50 cm, as indicated in Figure 6.

These models were applied to three different pavement structures, varying the overlay thickness and keeping the same thickness for the layers of the existing asphalt layer and for the granular base. The thickness and stiffness of the layers for the three structures are shown in Table 1. The materials were modeled assuming a linear elastic behavior.

The finite elements model was designed as a plain strain problem, using plane structural solid elements, defined by eight nodes and having two degrees of freedom at each node.

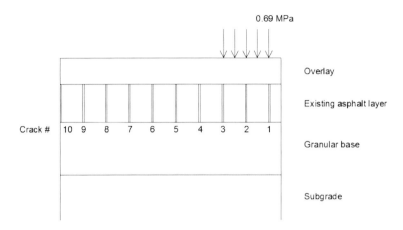

Fig. 4. Schematic representation of the finite elements model

The state of stress and strain in this type of problems is governed by a stress concentration at the crack tip and needs to use the fracture mechanics for a correct assessment. To avoid the use of the fracture mechanics, the state of stress and strain in the crack tip was calculated just above the modeled cracks, 0.3 and 0.25 mm from the crack edge, as indicated in Figure 7. For each case, the horizontal, vertical, shear and Von Mises strain was calculated.

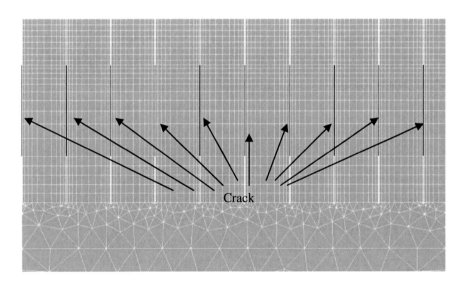

Fig. 5. Finite elements model

Table 1. Thickness and stiffness of pavement layers

Layers		Structure 1	Structure 2	Structure 3
Overlay	thickness	0.10 m	0.20 m	0.30 m
	stiffness	5000 MPa	5000 MPa	5000 MPa
Existing asphalt layer	thickness	0.20 m		
	stiffness	2000 MPa		
Granular base	thickness	0.20 m		
	stiffness	160 MPa		
Subgrade	stiffness	80 MPa		

3 Modeling Results

One of the objectives of this work was to evaluate the difference between modeling of multiples cracks as compared to a single crack. This can be observed by the representation of the Von Mises strain presented in Figure 8 and 9, respectively for a pavement with only one crack modeled (crack #3) and the pavement with 10 cm spaced cracks (case 11). The analysis of these figures show the difference between the state of strain in the overlay associated with the presence of either a single or multiple cracks. The difference is also visible in the state of strain above the existing cracks which is responsible for the reflective cracking.

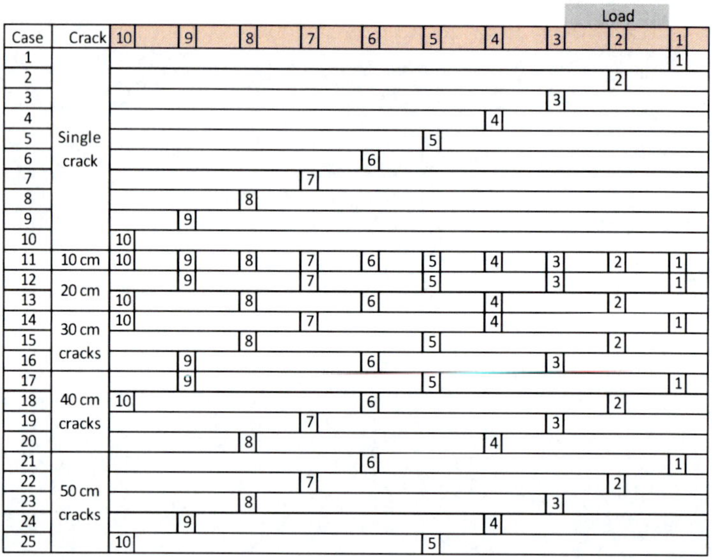

Fig. 6. Representation of cracking configuration

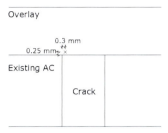

Fig. 7. Location for stress and strain calculation

The first analysis of this study of carried out for the cases of single cracks, i.e. the cases 1 to 10 where for each model only one crack was modeled in each pavement. For these cases, the strain level in the pavement with 10 cm thickness overlay is indicated in Figure 10, where Ex represents the horizontal strain, Ey represents the vertical strain, Exy represents the shear strain and the Evm represents the Von Mises shear strain.

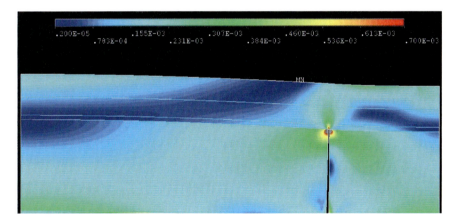

Fig. 8. Von Mises strain in a pavement with a single crack (#3)

The analysis of single crack modeling indicates that the vertical strain is almost constant when cracks below the load (1, 2 and 3) are modeled. Also, for the other cracks, the vertical strain is constant. In terms of horizontal, shear and Von Mises strain, they increase as the cracks moves away from the load but after crack 4 the strain level reduces significantly, except for the horizontal strain.

This analysis shows that when only one crack is modeled, that should be the crack 4, which is 10 cm away from the load. Usually, the modeling of a single crack simulating the mode II of crack opening is simulated by crack 3, which is around 20% less them the strain level above crack 4.

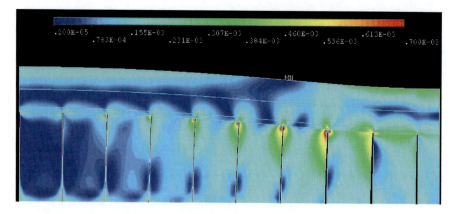

Fig. 9. Von Mises strain in a pavement with multiple cracks

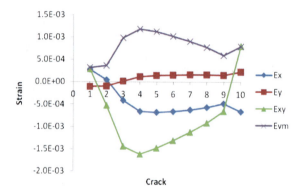

Fig. 10. Strain in the overlay for models with only one crack

For the other pavements, with 20 and 30 cm overlay thickness, the conclusion is identical, ie, the maximum state of strain appears not for the crack adjacent to the load but away from the load, as it can be observed in Figure 11 where the Von Mises strain is represented as function of the overlay thickness and crack number.

As the overlay thickness increases, the Von Mises strain (Figure 11) decreases and the maximum strain appears for crack 6, ie, 30 cm away from the load.

The modeling of multiple cracks with a spacing of 10 cm produces the results presented in Figure 12, for the overlay with 10 cm thickness, which are similar to the ones obtained for single cracks modeling. However, the comparison between Figure 10 (single cracks) and Figure 12 (10 cm cracks) indicates that the presence of 10 cm multiple cracks reduces the strain in the overlay compared to the single cracks modeling.

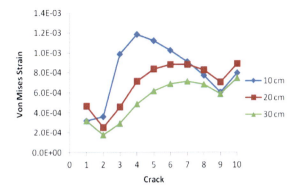

Fig. 11. Influence of overlay thickness of Von Mises strain for single crack

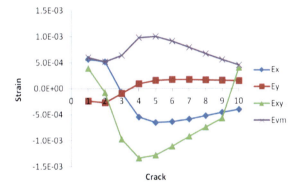

Fig. 12. Strain in the overlay for 10 cm crack spacing

For the other pavements (20 and 30 cm overlay thickness) the behavior of the Von Mises strain is identical to the observed for single cracks modeling where maximum strain appears for crack number 6 but the strain level is lesser than the one observed for single cracks. Identical results were obtained for the other strain components as well as for stress components. The analysis of 10 cm spaced cracks shows that, for this crack spacing, the consideration of multiple cracks is unfavorable.

Identical conclusion to those obtained for 10 cm spaced cracks can be obtained for 20 and 30 cm spaced cracks, as it can be observed in Figures 14 and 15. For these cases, the Von Mises strain presents the highest value for crack #5. However, the maximum Von Mises strain for 20 and 30 cm spaced cracks is identical to the one observed for 10 cm spaced cracks, which is less than the one observed for single cracks (#4 and #5).

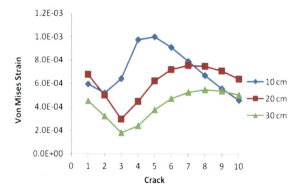

Fig. 13. Influence of overlay thickness of Von Mises strain for 10 cm crack spacing.

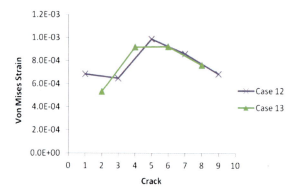

Fig. 14. Von Mises strain in the overlay for 20 cm crack spacing

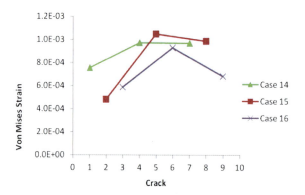

Fig. 15. Von Mises strain in the overlay for 30 cm crack spacing

Different conclusions can be obtained from the results for 40 and 50 cm spaced cracks where a Von Mises strain level greater or identical to the one obtained for single cracks (mainly for crack #4) was achieved. This appears mainly when a 50 cm crack spacing exists and there is a crack below the load and the other crack is away from the load. If the first crack is not below the load, then the strain level in that crack and in the other cracks is reduced compared to the maximum observed for a single crack #4. However, the maximum value observed for these cases is almost identical to the one observed for a single crack #4 and thus it is enough to consider the existence of only one crack, not adjacent to the load but some centimeters away from the load, depending of the overlay thickness.

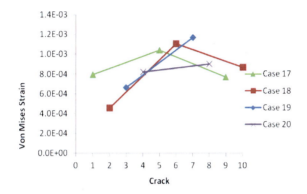

Fig. 16. Von Mises strain in the overlay for 40 cm crack spacing

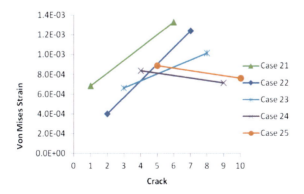

Fig. 17. Von Mises strain in the overlay for 50 cm crack spacing

4 Conclusions

This paper presented a study about the influence of multiple cracks in finite element models on the reflective cracking. This influence was assessed by evaluating the Von Mises strain in the pavement overlay, just above the existing cracks in the old pavement. In the 2D finite elements models, single and multiple cracks were simulated considering a distance between cracks ranging from 10 cm to 50 cm. The following conclusions can be made from the analysis of the results:

- There are significant differences in the state of stress in the pavement overlays due the presence of multiple cracks compared to the modeling of single cracks;
- The modeling of a single crack must be made by a crack considerable away from the load. That distance depends on the overlay thickness and for the cases studied it should be greater than 10 cm;
- The multiple cracks modeling do not increase the strain level in the overlay. However, the larger the distance between cracks, the higher and more significant the strain level can be when compared to the one obtained for a single crack.

References

[1] Molenaar, A.A.A., Potter, J.: Prevention of Reflective Cracking in Pavements. In: Vanelstraete, A., Francken, L. (eds.) RILEM Report 18, pp. 2–6. E&FN Spon, Boundary Row, London SE1 8HN (1997) ISBN: 0 419 22950 7

[2] Majidzadeh, K., Kauffmann, E.M., Ramsamooj, D.V.: Application of fracture mechanics in the analysis of pavement fatigue. In: Proceeding of the Association of Asphalt Paving Technologists, vol. 40, pp. 227–246 (1971)

[3] Van Gurp, C., Molenaar, A.A.A.: Simplified method to predict reflective cracking in asphalt overlays. In: Proceeding of the Reflective Cracking in Pavements – Assessment and Control Conference, Liege (1989)

[4] Paulino, G.H., Song, S.H., Buttlar, W.: Cohesive zone modeling of fracture in asphalt concrete. In: Proceeding of the Cracking in Pavements – Mitigation, Risk Assessment and Prevention Conference, Limoges (2004)

[5] Nesnas, K., Nunn, M.: A model for top-down reflection cracking in composite pavements. In: Proceeding of the Cracking in Pavements – Mitigation, Risk Assessment and Prevention Conference, Limoges (2004)

[6] Wu, R.Z., Harvey, J.T.: Evaluation of reflective cracking performance of asphalt mixes with asphalt rubber binder using HVS tests and non-local continuum damage mechanics. In: Scarpas, Loizos (eds.) Pavement Cracking – Al-Qadi, p. 978. Taylor & Francis Group, London (2008) ISBN: 978-0-415-47575-4

Using Black Space Diagrams to Predict Age-Induced Cracking

Gayle King[1], Mike Anderson[2], Doug Hanson[3], and Phil Blankenship[4]

[1] GHK, Inc.
[2] Asphalt Institute
[3] AMEC
[4] Asphalt Institute

Abstract. Asphalt aging is typically monitored through rheological changes at high pavement temperatures. Asphalt quality in laboratory aging experiments is then ranked using classic measures such as absolute viscosity ratios or changes in ring and ball softening point. However, this study [1,2] suggests that location on Black Space Diagrams at lower pavement temparatures is a better predictor for block cracking and related failure mechanisms associated with highly oxidized asphalt.

Three unmodified asphalts were PAV aged for 20, 40, and 80 hours. Modulus and phase angle were determined using DSR for intermediate temperatures, while stiffness and m-values using BBR were measured at low temperatures. Proposed aging functions such as Glover-Rowe's Damage Parameter and Anderson's R-value were compared to lab results on Black Space Diagrams. The key finding from the binder phase of this study is that location in Black Space is an important performance measure for cracking. However, the initial quality of the asphalt as determined in Black Space is just as important to performance as rheological changes in modulus and phase angle occurring during aging.

Aged mixtures were then tested in the BBR to determine whether binder aging trends would translate to mixture properties. One surprising finding was that microdamage forms in unconfined, highly aged mixture specimens as they cool. Findings from this study suggest that environmental effects models and timing strategies for pavement preservation should be revised to consider binder properties in Black Space, where both the initial asphalt quality and the effects of oxidation on key physical properties can be monitored. A new approach links initial binder quality with oxidative aging to propose a non-load induced cracking parameter for performance-related specifications.

1 Introduction

As pavements age, they crack and ravel. Although stresses induced by traffic exacerbate distress, the evolving rheological properties of aging asphalt are sufficiently damaging to cause a pavement to crack from thermal stresses alone.

The source of thermal stress could have several origins, including bending stresses caused by thermal or stiffness gradients with depth, or tensile stresses within the material itself induced by cooling when there is a difference in thermal expansion coefficients between asphalt and aggregate. Unfortunately, the damage mechanisms for block cracking, surface raveling, or other age-related cracking are not well understood, nor are reliable tools available to predict imminent damage so that pavement preservation strategies can be implemented proactively.

1.1 Black Space Diagrams

Although recent research has primarily focused on fracture energy as the failure property most directly tied crack prediction models, there is much information to be gained by considering simple rheological plots of G* versus Phase Angle, commonly referred to as Black Space Diagrams. Black Space analysis is particularly convenient because rheological data needed for a mastercurve is directly measured in the dynamic shear rheometer (DSR), with no mathematical shifts required to account for time-temperature superposition (TTS). Black space also captures phase changes in the binder, such as wax crystallization known to cause low temperature physical hardening. Such effects can be missed if TTS software incorrectly over-shifts data to align individual frequency sweeps. Because crack growth accelerates as pavements cool, this same low temperature portion of the Black Space mastercurve is most relevant to age-related damage mechanisms. It is also important to remember that asphalt has a very low phase angle at low pavement temperature. Under such cold conditions, Bending Beam Rheometer (BBR) measurements of stiffness and m-value are reasonable surrogates for G* and Phase Angle, and plots of stiffness versus m-value serve as reasonable surrogates for Black Space Diagrams.

2 Aging in Black Space

Domke [3] evaluated the evolution of low temperature binder properties with aging as part of his PhD thesis. He correctly but misleadingly concluded that stiffness and m-value change by similar percentages during aging. Glover [4] re-evaluated Domke's data and showed that when m-value changes by a given percentage, the change in PG binder grade is much greater than observed for an equivalent percentage change in binder stiffness. As shown in Figure 1, Glover plotted the continuous grade for stiffness (Temperature where S=300 MPa) versus the continuous grade for m-value (Temperature where m-value = 0.30) for two asphalts as they aged.

As Glover noted, the binder becomes much more m-controlled as it ages. The temperature at which m-value reaches 0.30 increases by 3 to 5°C for each 1°C increase in temperature where binder stiffness equals 300 MPa.

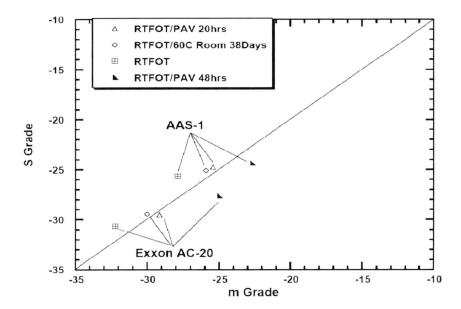

Fig. 1. Comparison of BBR m-Value & S Grades During Aging (°C)[5]

Glover [6] also derived a fatigue cracking parameter, G'/(η'/G'), from a Maxwell model. He correlated this parameter to binder ductility (15°C, 1 cm/mn), which Kandhal [7] had previously tied to age-induced cracking during field pavement studies. Anderson [8] confirmed the correlation between Kandhal's ductility and Glover's fatigue parameter with a laboratory aging study, and further found a very high correlation between Glover's parameter and the evolving m-control as exemplified by Figure #1. Anderson quantified this change in m-control by introducing a new parameter, ΔT_c, defined as the difference between BBR critical temperatures for stiffness and m-value as determined for 300 MPa and 0.30 respectively. The extraordinary correlation between ΔT_c as determined at very low temperatures using the BBR, Glover's fatigue parameter as measured in the DSR at 15°C and 0.005 rad/sec, and Kandhal's ductility recommendations (15°C, 1 cm/min) was unexpected.

One concludes from these findings that oxidation-induced embrittlement of the binder is a primary cause for block cracking and surface ravelling. Although some hardening occurs, the loss in phase angle at low temperatures is dramatic, and clearly contributes to age-induced damage. However, if a location in Black Space can be tied to damage, then the initial quality of the asphalt (S or m-control) should be just as important as any changes that occur during aging.

2.1 Glover-Rowe Damage Parameter

In an AAPT prepared discussion to the Anderson paper cited earlier, Rowe [9] agreed with the authors that Black Space Diagrams could be useful for comparing the various proposed damage parameters. He derived a new form of the Glover parameter as represented by equation (1). So long as the test frequency (ω) is known, variables G* and phase angle (δ) can be plotted to create a damage curve in Black Space.

$$G'/(\frac{\eta'}{G'}) = G \times ((\cos \delta)^2/\sin \delta) \times \omega \quad (1)$$

Given a Black Space function as defined by this new Glover-Rowe (G-R) parameter, an aged binder can be tested for degree of damage without imposing a rigid single test temperature and frequency. For example, Kandhal suggested that damage from binder aging was initiated when ductility fell to 5 cm, and cracking was serious when ductility reached 3 cm. Glover used his correlations with Kandhal's ductility to predict damage onset when G'/(η'/G') equals 900 Pa*s. Dividing this limit by Glover's test frequency (0.005 rad/s), Rowe suggested a failure curve in Black Space to represent the onset of cracking as shown by equation (2). Beginning with Kandhal's second observation that surface cracking is apparent when ductility falls to 3 cm, the corresponding value of the Glover-Rowe parameter is represented by equation (3).

Damage onset (5 cm ductility): $\quad G \times ((\cos \delta)^2/\sin \delta) = 180\ kPa \quad (2)$

Significant cracking (3 cm ductility): $\quad G \times ((\cos \delta)^2/\sin \delta) = 450\ kPa \quad (3)$

These two equations provide a damage zone in Black Space that correlates well with Kandhal's pavement damage studies, which were specific to use of unmodified asphalts within the local climate conditions in Pennsylvania.

As reported by Anderson in a previous paper from this study [10], three asphalts (West Texas, Gulf-Southeast, and Western Canadian) were RTFO aged, and then PAV aged for 20, 40, and 80 hours respectively. As shown on the Black Space Diagram in figure 2, the Western Canadian asphalt has much better initial properties than the other two asphalts, but appears to deteriorate more rapidly with aging. Thanks to its better initial quality, the Western Canadian has not reached the damage zone after 40 hours in the PAV, whereas the West Texas asphalt has passed through the damage zone into the failure region after 20 hours PAV aging. Also note the difference in shape of the current SuperPave fatigue parameter (G* x sin δ = 5 MPa), versus the Glover-Rowe curve. Given the binder aging data, the curve shape for G* x sin δ does not seem a logical damage indicator. At best, the failure limit would need to change if Glover's test conditions (15°C, 005 rad/s) were used for DSR frequency sweeps.

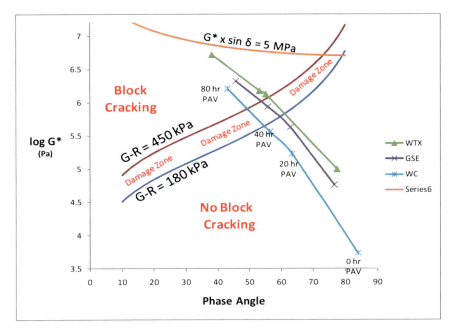

Fig. 2. PAV-aged Binders Passing through the Glover-Rowe Damage Zone

2.2 R-Value as a Damage Parameter

During SHRP research, Anderson and his team at Penn State developed equations to fit traditional rheological mastercurves of G* or phase angle versus temperature or frequency. An important parameter from those mastercurves is the R-value, which was shown to increase with asphalt aging.

$$R = \frac{(\log 2) * \log \dfrac{G^*(\omega)}{G_g}}{\log\left(1 - \dfrac{\delta(\omega)}{90}\right)}$$

where: $G^*(\omega)$ = complex shear modulus at frequency ω (rad/s), Pa
G_g = glassy modulus, Pa (assumed to be 1E+09 Pa)
$\delta(\omega)$ = phase angle at frequency ω (rad/s), degrees (valid between 10 and 70°)

It is interesting to similarly view the PAV aged results from this study on a Black Space Diagram by replacing the Glover-Rowe damage zone with R-values as potential damage parameters (see figure 3).

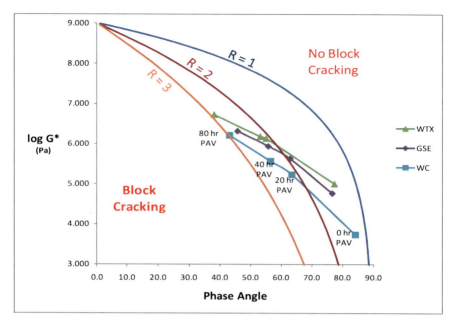

Fig. 3. Using R Values to Track PAV Aging in Black Space

Similar to Glover-Rowe, damage zones could be based upon ranges in R-value. For example, a range of R values between 2.3 and 2.7 might predict the onset and propagation of damage in a comparable range to that shown in figure 2 above. However, it is clear from the two figures that the damage curves using R-value are very different in shape from those for Glover-Rowe. R-value necessarily predicts that the WC asphalt has a higher R value than WTX after 40 hours of PAV aging, and is therefore more likely to crack. This ranking is very different from that predicted by the Glover-Rowe parameter, for which the WC 40 hr PAV sample has not yet reached the damage zone, whereas the WTX 40 hr PAV sample has totally passed through the damage zone into the failure region for block cracking. Although the Glover-Rowe parameter appears to better predict damage in line with Kandhal's ductility observations, a rigorous field cracking study is needed to establish damage zones against which these various theoretical approaches can be compared.

3   Bending Beam Rheometer Tests on Mixtures

The rate of asphalt oxidation varies significantly with pavement depth. The most pronounced rheological changes occur near the surface, so any mixture test predicting the onset of surface cracking for in-place pavements should use thin specimens of ½" or less. Furthermore, test capabilities must include low pavement temperatures where cracking is most likely to initiate. In a study evaluating the performance of RAP in asphalt mixtures, Marasteanu [11] found that the Bending Beam Rheometer can test asphalt mixtures in thin beams to derive creep compliance curves. He then applied the Hirsch model to back-calculate the binder stiffness and obtain critical cracking temperatures. The BBR Mixture Bending Test (Sliver Test) was included in the laboratory mixture aging phase of this study.

4 Testing Protocol

Mixture Aging: Loose, uncovered mixtures were aged in a force-draft oven for 4hr, 24hr, and 48hr at 135°C. After aging, mixtures were compacted in a SuperPave Gyratory Compactor (SGC) to target air voids that are representative of in-pace airfield pavements. Following compaction, eight BBR-sized beams were cut from each specimen. *Note: During phase 1 of this study, compacted specimens were aged using typical SuperPave mixture aging protocols. This initial test series raised concerns that aging of compacted specimens does not result in uniform oxidation throughout.*

BBR Testing: The aged mixture beams were tested in the BBR using Marasteanu's method which applies a 500 g load. Four beams were tested at each of two temperatures selected from the standard PG grading temperatures immediately above and below the continuous low temperature PG grade of the binder itself. For West Texas Sour asphalt, the BBR test temperatures were -6°C and -12°C. For Gulf-Southeast and Western Canadian, the BBR test temperatures were -12°C and -18°C.

4.1 Analysis of BBR Rheological Properties Stiffness (S) and m-value

The working hypothesis for this study assumes that, as asphalt ages, low temperature relaxation properties deteriorate more quickly than modulus. As the phase angle falls, the binder can no longer flow fast enough to heal damage that might accumulate in the mix. The binder testing phase of this study validated the hypothesis by demonstrating that the BBR critical temperature for m-value deteriorates much faster than the critical temperature for stiffness during PAV aging. Marasteanu's BBR method was selected to determine whether these same trends exist in mixes. Results for S and m-value are shown graphically in Figure 4. Detailed data with replicates and statistics can be found in the project report.

The graphs in Figure 4 contradict findings from the binder study that BBR stiffness continues to increase and m-value continues to decrease with additional aging. For all three mixes tested at the temperature immediately above its low temperature PG grade, the mixture stiffens and the m-value continues to drop with longer aging times as expected. However, at test temperatures below the recommended PG binder grade, stiffness reaches a maximum (~20,000 Mpa) and m-value exhibits a minimum (~0.13±0.01) after approximately 24 hours of aging, and then both S and m-value reverse direction in a manner that is inconsistent with the comparable binder rheology after extended PAV aging. The only plausible explanation is that damage (micro cracking) has occurred in the highly aged mixture specimen, either as it was cooled below its critical cracking temperature or during the first sixty seconds of loading in the BBR.

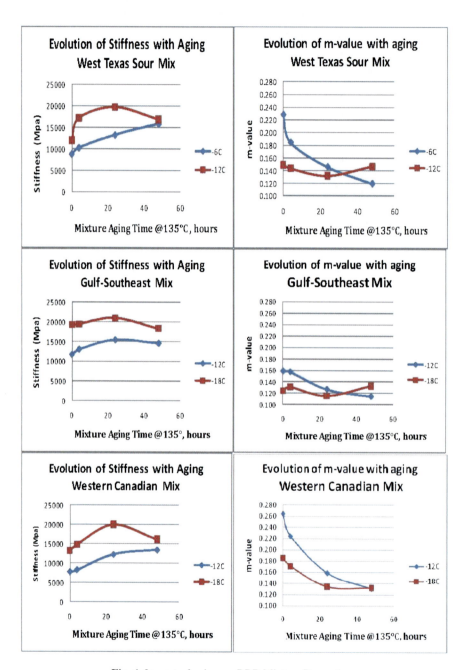

Fig. 4. Impact of aging on BBR Mixture Properties

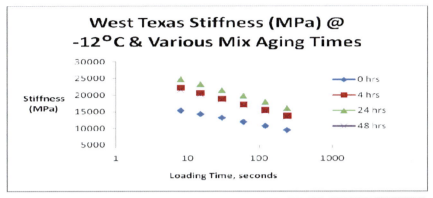

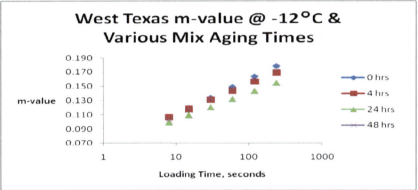

Fig. 5. BBR loading curves for oven-aged WTS mix @-12°C

The BBR loading curves for undamaged and damaged specimens were then compared to determine whether the load itself is responsible for the microdamage. Results for the West Texas Sour mix (see figure 5) reflect typical behavior for all three mixes when tested below the critical cracking temperature. Notice that the BBR loading curves for the 4 hr and 48 hr aged specimens overlay almost perfectly, but both are considerably less stiff than the 24 hr aged specimen. Likewise, results for m-value reflect the same behavior, with the 4 hr and 48 hr aged specimens overlaping, but both at higher m-values than the intermediate 24 hr aged sample. Because the shape of the BBR loading curve does not change for the damaged 48 hr aged specimens, the microcracks must have developed during cooling, not during BBR loading. Furthermore, those microcracks are stable, such that they do not appear to propogate further as the BBR beam is loaded.

Damage hypothesis: Traditional thermal cracking theory assumes that asphalt concrete shrinks upon cooling. Since the pavement is restrained in the the longitudinal direction, mixture shrinkage creates tensile stresses that result in transverse thermal cracks. The Thermal Stress Restrained Specimen Test (TSRST) mimics this failure in the laboratory. However, small BBR specimens are not restrained as they cool in the bath. Tensile stresses must develop between aggregate and binder even when the specimen is not restrained. The coefficient of thermal expansion for the asphalt mixture is no longer important. One must consider the coefficients of thermal expansion for binder and aggregate as two separate functions. Micromechanics can then be used to analyze stress development caused by differential shrinkage between the two phases. This hypothesis fits well with field observations that small, unconfined blocks of aged asphalt concrete continue to break into smaller chuncks, even on airport runway pavements that have been abandoned for twenty years.

5 Conclusions

Black Space Diagrams offer a means to combine the rheological properties of an unaged bitumen with oxidation-induced changes to predict when binder embrittlement might lead to block cracking. This approach offers a convenient method to compare age-related damage parameters, such as Glover-Rowe, R-value, and $G^* \times \sin\delta$, to lab and field data. Identified ties between failure strain in tension (ductility) and rheology are also intriguing.

The mixture phase of this study asked the question, "Can BBR tests on mixture specimens predict the onset of cracking?" For all three unmodified binders used in the single mix design tested here, microdamage thought to be associated with block cracking became significant as the stiffness of the aged mixture approached 20,000 MPa and m-value approached 0.13, regardless of the binder grade or the test temperature. Because aggregate has such a strong influence on mixture stiffness, this study must be expanded to other mixes and modified binders before significant conclusions can be drawn regarding failure limits. However, by analogy to the Glover/Rowe parameter evaluated in the binder phase of this study, a preferred solution would combine stiffness and m-value to create a damage region in Black Space (or a plot of S vs m-value). A field validation study evaluating pavements with varying cracking severity is now underway to refine conclusions that damage zones plotted on Black Diagrams can predict approaching pavement damage before visible cracks form.

Acknowledgements. The authors gratefully acknowedge the financial support from FAA through AAPTP project 6-1, as well as the work of Dr. Mihai Marasteanu and his team at U. Minn that provided the bending beam data for mixtures, and the theortical contributions of Dr. Geoff Rowe that enabled rheological comparisons to be made in Black Space.

References

[1] Hanson, D.I., Blankenship, P.B., King, G.N., Anderson, R.M.: Techniques for Prevention and Remediation of Non-Load-Related Distresses on HMA Airport Pavements – Phase II, Final Report, Airfield Asphalt Pavement Technology Program, Project 06-01 (2010)
[2] Anderson, R.M., King, G.N., Hanson, D.I., Blankenship, P.B.: AAPT 80, 615 (2011)
[3] Domke, C.H.: Asphalt Compositional Effects on Physical and Chemical Properties, Dissertation submitted to Texas A&M University (1999)

[4] Glover, C.J., Davison, R.R., Domke, C.H., Ruan, Y., Juristyarini, P., Knorr, D.B., Jung, S.H.: Development of a New Method for Assessing Asphalt Binder Durability with Field Evaluation, Federal Highway Administration and Texas Department of Transportation, Report # FHWA/TX-05/1872-2 (2005)
[5] Knorr Jr., D.B., Davison, R.R., Glover, C.J.: Transp. Res. Rec. 1810, 9–16 (2002); reprinted from this publication with permission from TRB
[6] See citation #4
[7] Kandhal, P.S.: ASTM STP 628: Low-Temperature Properties of Bituminous Materials and Compacted Bituminous Paving Mixtures. In: Marek, C.R. (ed.), American Society for Testing and Materials, Philadelphia, PA (1977)
[8] See citation #2
[9] Rowe, G.M.: Prepared Discussion following the Anderson AAPT paper cited previously. AAPT 80, 649–662 (2011)
[10] See citation #2
[11] Zofka, A., Marasteanu, M., Clyne, T., Li, X., Hoffmann, O.: Development of a Simple Asphalt Test for Determination of Asphalt Blending Charts. MNDOT report MN/RC 2004-44 (2004)

Top-Down Cracking Prediction Tool for Hot Mix Asphalt Pavements

Cheolmin Baek[1], Senganal Thirunavukkarasu[2], B. Shane Underwood[2], Murthy N. Guddati[2], and Y. Richard Kim[2]

[1] Korea Institute of ConstructionTechnology, Korea
[2] North CarolinaStateUniversity, Raleigh, North Carolina, USA

Abstract. This paper presents an analysis tool for predicting top-down cracking (TDC) of hot-mix asphalt (HMA) pavements. TDC is known to involve a complicated set of interactive mechanisms, perhaps more so than other HMA distresses. Such complexity makes it difficult to predict TDC reliably using conventional material models and analysis tools. Over the years, the viscoelastoplastic continuum damage (VEPCD) model has been improved to better understand and predict the behavior of asphalt concrete materials. The ability of the VEPCD model to accurately capture various critical phenomena has been demonstrated. For fatigue cracking evaluation of pavement structures, the viscoelastic continuum damage (VECD) model has been incorporated into a finite element code as VECD-FEP++. To use this code in the prediction of TDC requires the enhancement and incorporation of additional sub-models to account for the effects of aging, healing, thermal stress, viscoplasticity and mode of loading. The Enhanced Integrated Climatic Model (EICM) is also integrated into the framework.The flexible nature of the VECD-FEP++ modeling technique allows cracks to initiate and propagate wherever the fundamental material law suggests. As a result, much more realistic and accurate cracking simulations can be accomplished using the VECD-FEP++.To demonstrate the full capabilities of the VECD-FEP++, two example simulations were carried out, and the results indicate that the interactions among the sub-models and overall trends in terms of pavement behavior were reasonably captured.After proper calibration, this tool could provide quantitative predictions of the extent and severity of TDC.

1 Introduction

The top-down cracking (TDC) is known to involve a complicated set of interactive mechanisms, perhaps more so than other hot-mix asphalt (HMA) distresses. Such complexity makes it difficult to predict TDC reliably using conventional material models and analysis tools [1]. Over the years, the viscoelastoplastic continuum damage (VEPCD) model has been improved to better understand and predict the behavior of asphalt concrete materials [2]. The ability of the VEPCD model to accurately capture various critical phenomena has been demonstrated. For fatigue cracking evaluation of pavement structures, the viscoelastic continuum damage

(VECD) model has been incorporated into a finite element code as VECD-FEP++,which is formulated for the axisymmetric analysis [3, 4]. To use this code in the prediction of TDC requires the enhancement and incorporation of additional sub-models to account for the effects of aging, healing, thermal stress, viscoplasticity and mode of loading. The Enhanced Integrated Climatic Model (EICM, [5]) is also integrated into the framework.In the VECD-FEP++, the damage is calculated for each element based on its state of stress, temperature, loading rate, and boundary conditions. Therefore, it is not necessary to assume a priori the location of distress initiation, nor the path of distress evolution. Not having to make such assumptions is a feature of the VECD-FEP++ that is essential in modeling TDC in various HMA pavements. The flexible nature of the VECD-FEP++ modeling technique allows cracks to initiate and propagate wherever the fundamental material law suggests. As a result, much more realistic and accurate cracking simulations can be accomplished using the VECD-FEP++. In this paper, the advanced tool to predict the TDC will be presented with a brief description ofthe key components and by running the example simulations, the capability of VECD-FEP++ will be demonstrated.

2 Model Framework

The overall framework guiding the VECD-FEP++ analysis is shown in Figure 1. The analysis is divided into five sub-modules: the input module, the material properties sub-models, the analytical sub-models, the performance prediction module and the output module. These modules provide an analytical/computational method for identifying the location and time of crack initiation in the pavement structure.A complete review of the all modelsand modules is beyond the scope of this paper; however, interested readers are directed to [6, 7] for a more thorough review and for citations to additional resources.

2.1 Input Module

Preprocessors have been developed to facilitate easy and rapid analysis of pavement systems using the FEP++. Specifically, the preprocessor helps in the rapid development of input models for analysis and also helps in making consistent changes for repeated analysis. This tool is ANSI-compliant and developed with portable libraries, thus making it easy to transfer to other platforms.

2.2 Material Property Sub-Models

Linear viscoelastic (LVE) model. Viscoelastic materials exhibit time and temperature dependence, meaning that the material response is not only a function of the current input, but the entire input history. By contrast, the response of an elastic material is dependent only on the current input. For the uniaxial loading considered in this research, the non-aging, LVE constitutive relationships are expressed in the convolution integral form, as shown in Equations [1] and [2]:

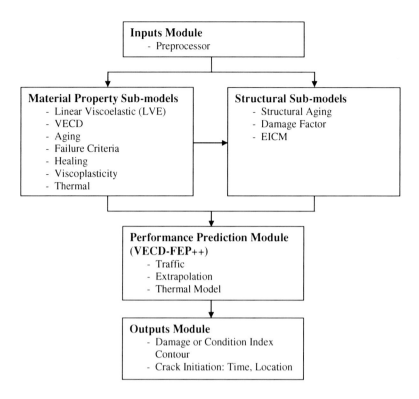

Fig. 1. VECD-FEP++ model framework

$$\sigma = \int_0^t E(t-\tau)\frac{d\varepsilon}{d\tau}d\tau \text{ and} \quad (1)$$

$$\varepsilon = \int_0^t D(t-\tau)\frac{d\sigma}{d\tau}d\tau, \quad (2)$$

where $E(t)$ and $D(t)$ are the relaxation modulus and creep compliance, respectively (the τ term is the integration variable). The relaxation modulus and creep compliance are important material properties, along with the complex modulus, in LVE theory. Because these two properties are the responses for respective unit inputs, they are called unit response functions. These unit response functions can be obtained either by experimental tests performed in the LVE range or by converting another unit response function, as suggested by Park and Schapery[8].

Viscoelastci continuum damage (VECD) model. On the simplest level, continuum damage mechanics considers a damaged body with some stiffness as an undamaged body with a reduced stiffness. Continuum damage theories thus attempt to quantify two values: damage and effective stiffness. Further, these theories ignore specific microscale behaviors and, instead, characterize a material using macroscale observations, i.e., the net effect of microstructural changes on observable properties. In the macroscale, the most convenient method to assess the effective stiffness is to use the instantaneous secant modulus. As discussed in the subsequent sections, direct use of the stress-strain secant modulus in asphalt concrete (AC) is complicated by time dependence. Damage is oftentimes more difficult to quantify and generally relies on macroscale measurements combined with rigorous theoretical considerations. For the VECD model, Schapery's work potential theory [9], which is based on thermodynamic principles, is appropriate for the purpose of quantifying damage. Within Schapery's theory, damage is quantified by an internal state variable (S) that accounts for microstructural changes in the material.

Aging model. The VECD constitutive model is based on the assumption that the material is a non-aging system.However, the aging of binder is well recognized as a contributing factor to the TDC of asphalt pavement. To incorporate aging effects into the current VECD model, significant experimental and analytical work has been done. A complete discussion of such work is given elsewhere [10].The approach adopted in this study is to subject the asphalt mixtures to various aging conditions and then measure the physical properties of the aged mixtures.SHRP methods [11]for the laboratory aging of asphalt concrete specimen was utilized and the dynamic modulus test, the direct tension monotonic test, fatigue test were performed to characterize the aged asphalt mixture. To incorporate the effects of aging into the LVE and VECD model, all coefficients of LVE and VECD model for aged mixtures wererelated to the aging time suggested by SHRP for the lab-to-field aging times [11].Figure 2 presents the relationship between the sigmoidal coefficients of LVE model and aging time as an example. From the equations in the figure, it can be observed that the models are formulated using the ratio of the aged values to the original (un-aged) values. This normalized formulation is chosen so that the final function can be applied universally to other mixtures to simulate aging effects.

2.3 *Structural Sub-Models*

To predict the pavement performance, the material sub-models must be converted to, or implemented into, structural models to consider the different structures, boundary conditions, climate conditions, etc.

The variation of temperature in a pavement has two effects: a change in stiffness of the AC and a change in the thermal stress due to thermal expansion of the material. Thermal stress is generated in the pavement depending on the boundary conditions. These two effects of temperature have been implemented in the FEP++. The actual temperature variation that is used for the pavement performance prediction is generated from the EICM. Temperature profiles generated from the EICM are input directly into the FEP++ preprocessor.

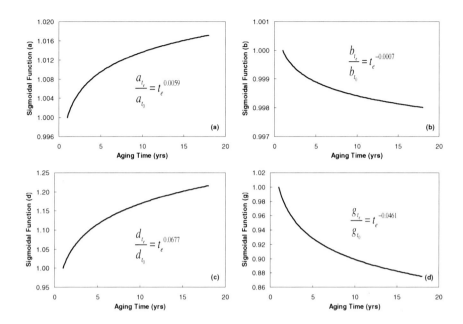

Fig. 2. Relationship of sigmoidal function coefficients to aging time

Structural Aging model. The time scale used for the material level aging model corresponds physically only to that used for the top layer of a real pavement cross-section. To apply this model to other depths, the age of each sublayer relative to that of the surface must be found. This goal is achieved by coupling the principles of the Global Aging System (GAS), first proposed by Mirza and Witczak [12], with an effective time concept.The GAS model predicts the viscosity of the asphalt binder as a function of depth, mean annual air temperature (MAAT) representing the effect of geographical location, and rolling thin film oven (RTFO) binder viscosity.The effective time is determined by finding, for some physical time and depth, the time that gives the same viscosity for the binder at the pavement surface. A flow chart of the structural aging model, including the equiviscosity concept as well as the plot of effective time versus depth, is shown in Figure 3.A ten-year-old pavement can serve as an example of the effective time concept whereby after ten years of service, the surface layer has aged ten years, but the material at a depth of three inches may behave as the surface layer behaved when the pavement was only four years old. In this example then, the effective time of the sub-layer three inches from the surface ten years after construction is four years. To compute the material properties of this sub-layer at year ten, material aging models (described in the previous section) can be used to find the value of the coefficients at four years.

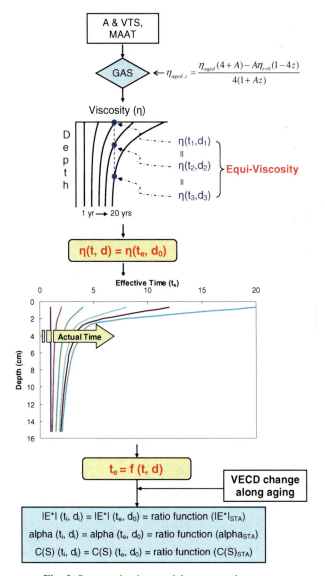

Fig. 3. Structural aging model conceptual map

2.4 Performance Prediction Module

The performance of a pavement can be characterized by predicting the damage accumulation in a pavement under the realistic conditions. In this research, the VECD-FEP++ was utilized as a mean of predicting the pavement performance. To integrate the effects of variables that are important in TDC behavior, the submodels presented in Figure 1 have been incorporated into the existing VECD-FEP++ with the extrapolation method.

The FEP++ uses an extrapolation scheme that can significantly reduce the running time while still capturing the essential characteristics. The current scheme computes the damage caused by load and thermal variations at representative times in a day. These data are then extrapolated using a nonlinear scheme to obtain the total damage accumulation in a given month. This damage is then applied to the pavement as the initial condition for the next month's simulation. This process is continued for the entire simulation period.

2.5 Outputs Module

The output module consists of the tools and techniques necessary to view and interpret the VECD-FEP++ performance predictions.It creates a single file that can be opened, processed, and manipulated to view visual interpretations (contours) of the predicted damage, stress distribution, or other quantities of interest. This file can also be processed to extract different indices to quantify the visual observations. The example of contour plots is shown in Table 1.

3 Pavement Simulation

An example simulation of the FEP++ using the VECD model and all accompanying analytical sub-models was carried out to demonstrate the capabilities of the modeling approach. Two pavement structures were investigatedas shown in Figure 4: thin (127 mm or 5 in.) and thick (304.8 mm or 12 in.).The Control asphalt concrete mixture used in the FHWA ALF research was selected for the simulation because all the necessary model data for both mixes were available. The base layer for the thin pavement structure was 203 mm (8 in.) thick, and the subgrade for both structures was considered to be semi-infinite.The unbound material layers were assumed to be linear elastic. Base layer modulus was 276 MPa (40 ksi) and subgrade modulus was 83 MPa (12 ksi). The pavement temperature was generated from the EICM for WashingtonD.C.A moving load was simulated by applying a 0.1 second haversine loading pulse with a magnitude of 40 kN (9 kip) and contact pressure of 689 kPa (100 psi) on the pavement surface, followed by 62.2 seconds or 622 seconds of rest for thin and thick pavement respectively.

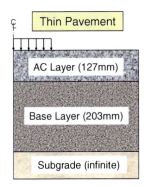
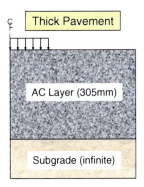

Fig. 4. Pavement structures for FEP++ simulation

Results from the VECD-FEP++ simulations are shown in Table 1 for a thin asphalt concrete pavement and for a thick asphalt concrete pavement. It is seen that the output takes the form of contours for normalized pseudo stiffness. The normalization process transforms the pseudo stiffness such that it is equal to zero at failure and is equal to one in the fully intact condition. In these contours, the areas with a gray-scale value closest to white correspond to heavily damaged areas, i.e., where the normalized pseudo stiffness values are close to zero. Note that only the asphalt concrete layers are shown, but the substructure for the simulations represents typical pavements.

Simulations began in October and were performed until any element reached a normalized pseudo stiffness of 0. As a result, the thin pavement was failed after 14 months while the thick pavement was failed after 75 months. The thin pavement was failed at the bottom and center of the AC layer while the thick pavement was failed at the top and wheel load edge of the AC layer. Such results follow the field observations, i.e., thin pavement tends to show bottom-up cracking, whereas thick pavement tends to show TDC [13]. It was also observed from examining the damage evolution between the months of Mar and July that pavement healing may constitute a major component of a pavement's total fatigue performance.

Table 1. Contours for Example Simulations

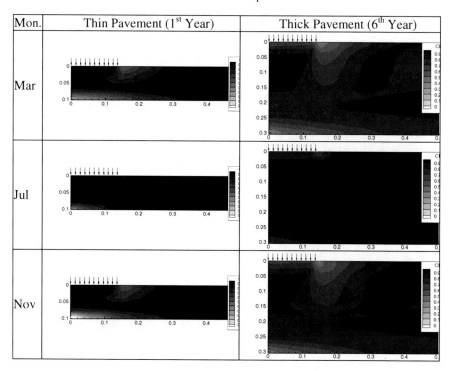

4 Summary

An enhanced VECD-FEP++ model for predicting the TDC in HMA layers has been established in this research. This effort was accomplished by developing, modifying, and/or investigating several important material property models, such as an aging model, healing model, failure criteria, viscoplasticity, and thermal stress, and then finally incorporating these sub-models into the existing VECD model. The material models were converted to and/or combined with the structural models. These sub-models were implemented into the VECD-FEP++, and an extrapolation method for TDC prediction was developed.

To demonstrate the full capabilities of the VECD-FEP++, two example simulations were carried out, and the results indicate that the interactions among the sub-models and overall trends in terms of pavement behavior were reasonably captured.After proper calibration, this tool could provide quantitative predictions of the extent and severity of TDC.

References

[1] Nesnas, K., Nunn, M.: In: Proceedings of 5th International RILEM Conference (2004)
[2] Underwood, B.S., Kim, Y.R., Guddati, M.N.: Journal of the Association of Asphalt Paving Technologists. AAPT 75, 577–636 (2006)
[3] Kim, Y.R., Baek, C., Underwood, B.S., Subramanian, V., Guddati, M.N., Lee, K.: KSCE. Journal of Civil Engineering 12(2), 109–120 (2008)
[4] Guddati, M.N., Savadatti, S., Thirunavukkarasu, S.: FEP++: An Object Oriented Finite Element Program in C++ for NonlinearDynamic Analysis. North CarolinaStateUniversity / University of Texas, Austin (1995-current)
[5] Guide for Mechanistic-Empirical Design of New and Rehabilitated Structures. NCHRP ReportI-37A. TRB, Part 2, ch.3, National Research Council, Washington, DC., (2004)
[6] Roque, R., Zou, J., Kim, Y.R., Baek, C.M., Thirunavukkarasu, S., Underwood, B.S., Guddati, M.N.: Top-Down Cracking of Hot Mix Asphalt Layers: Models for Initiation and Propagation. Final Report, NCHRP 1-42A, National Cooperative Highway Research Program, Washington, DC (2010)
[7] Baek, C.: Investigation of Top-Down Cracking Mechanisms Using the Viscoelastic Continuum Damage Finite Element Program. Ph.D. Dissertation. North Carolina State University, Raleigh, NC (2010)
[8] Park, S.W., Schapery, R.A.: International Journal of Solids and Structures 36, 1653–1657 (1999)
[9] Schapery, R.A.: J. Mech. Phys. Solids 38, 215–253 (1990)
[10] Baek, C., Underwood, B.S., Kim, Y.R.: Effects of Oxidative Aging on Asphalt Mixture Properties.CD-ROM. Transportation Research Board of the National Academies, Washington, DC (2012)
[11] Bell, C.A., Wieder, A.J., Fellin, M.J.: Laboratory Aging of Asphalt-Aggregate Mixtures: Field Validation. SHRP-A-390. Strategic Highway Research Program. National Research Council, Washington, DC (1994)

[12] Mirza, M.W., Witczak, M.W.: Journal of the Association of Asphalt Paving Technologists, AAPT 64, 393–430 (1995)
[13] Uhlmeyer, J.S., Willoughby, K., Pierce, L.M., Mahoney, J.P.: Transportation Research Record: Journal of the Transportation Research Board, 1730, 110–116 (2000)

A Theoretical Investigation into the 4 Point Bending Test

M. Huurman[1,2], R. Gelpke[1], and Maarten M.J. Jacobs[1]

[1] BAM wegen, The Netherlands
[2] Delft University of Technology, The Netherlands

Abstract. In the Netherlands use is made of a mechanistic asphalt pavement design strategy. Asphalt stiffness and fatigue properties are important inputs for this design approach. In the Netherlands these properties are traditionally determined by 4 point bending tests (4pb).

In this paper the accuracy of that test is discussed on the basis of Finite Element modelling. It is shown that application of the EN-standard results in an underestimation of the stiffness by 3.2% (at 7500 MPa) and an underestimation of the phase lag by 0.4° (lag=20° and 20 Hz.). It is shown that a thorough analysis of the 4pb may reduce these errors to a stiffness that is underestimated by 0.88% and a phase lag which is overestimated by 0.1°.

1 Introduction

The Netherlands have a long history in the mechanistic design of asphalt pavements. The chosen design route amongst others demand mechanical properties of asphalt concrete (AC) as input. Especially stiffness and fatigue properties are essential. Traditionally these properties are determined by use of the 4 point bending test (4pb). Today the 4pb is standardised by European norms [1, 2]. The norms describe that the 4pb demands that the specimen is clamped by the test device at four locations. All clamps should allow for free rotation and translation in the longitudinal direction. The outer clamps should be fixed to prevent vertical movement; the inner clamps are excited by vertical loading, see Figure 1.

Interpretation of test results is done on basis of the Euler-Bernoulli bending beam theory. In this theory a 3Dimensional bending beam is represented by a beam without height and width, i.e. a line or 1D beam. Resistance of the 1D beam to bending is dictated by the well known letter combination EI. Here E stands for the stiffness modulus and I reflects the moment of inertia of the beam cross section.

$$I = \frac{b.h^3}{12} \qquad (1)$$

Where. I: moment of inertia [mm⁴], b: beam width [mm], h: beam height [mm].

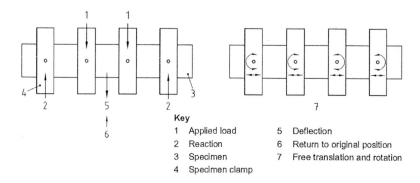

Fig. 1. Schematic representation of the 4pb as described in EN standards [2]

The previous suggests that the 4pb is a simple test that can be analysed on basis of a simple theory while results are of great value for mechanistic design purposes. However such a conclusion cannot be justified because of the following reasons.

First of all the true beam has to be grabbed by clamps. This locally introduces stress and strain into the beam material. These are not considered during test interpretation on basis of a 1D beam. However, these extra stresses and strains may well introduce (fatigue) damage or introduce local non-linear effects.

Secondly it is not possible to construct a 4pb test device that truly meets the conditions as depicted in Figure 1. One has to consider things like friction and play which may both vary through time due to wear and tear. Also deformation of set-up parts may affect test results.

Thirdly in the 4pb shear forces act on the beam in the area between the inner and the outer clamps. These forces act to deform the beam whereas these deformations are not considered in the prescribed test interpretation.

Fourthly the clamps limit the freedom of cross-section deformation. This implies that beam deformation close to the clamps cannot be described by the Euler-Bernoulli bending beam theory.

The Dutch history in mechanistic design combined with the obligation to type test led to a strong increase in availability of 4pb equipment in the Netherlands. At the moment at least 21 machines are available in the Netherlands of which 18 are produced by Zwick Roell.

From the above the following is concluded: 4pb is an important test for Dutch pavement design, The test appears simple but is complex in reality, The test is highly appreciated in the Netherlands and data is produced on daily basis, at least 18 machines are of the same type.

These conclusions triggered the authors to thoroughly investigate the accuracy of the 4pb device as a first step towards the possibilities of utilisation of the 4pb in scientific research. Hereto the most commonly used machine in the Netherlands is modelled in detail. Results of this work are discussed hereafter.

2 Four Point Bending Machine

The Zwick Roell machine is the starting point of this study. Figures 2 & 3 give some pictures of the device in combination with the FE model that was made.

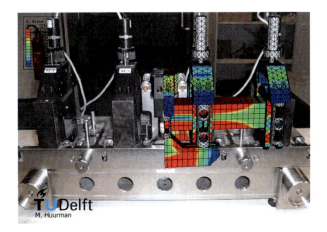

Fig. 2. Overview of the 4pb device and its model representation

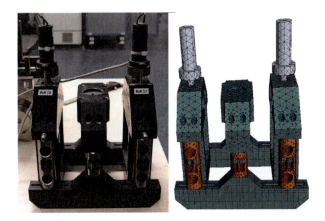

Fig. 3. Detail of model and set-up. By application of six elastic hinges the inner clamps are connected to the hydraulic actuator of which the seating is formed by a depression in the central bridge.

The machine measures the deflection in the middle of the beam. As shown in Figure 2 & 3 elastic hinges are applied to try and meet the boundary conditions as depicted in Figure 1. These hinges are not susceptible to play, slip and stick or wear and tear. The most important characteristics of the 4pb considered here are the following:

Effective beam length, L= 420 mm; Distance between inner clamps, l=140 mm; Beam height, h=50 mm; Beam width, b=50 mm; Location of deflection measurement, x =210 mm; Width of the clamps =10 mm; Clamping force to follow viscous deformation =130 N.

For FE modelling steel and aluminium are assigned the following properties.

Table 1. Properties of steel and aluminium

	E [MPa]	ν [-]	Specific mass [kg/m3]
Aluminium	70000	0.25	2700
Steel	210000	0.2	7800

It is stated explicitly that the mesh refinement and time step size are based on a study into 4pb FE modelling issues discussed elsewhere [3].

3 Elastic Response

3.1 Application of the Standard

Since AC is a visco-elastic material its stiffness is represented by a complex modulus, E*. Due to its visco-elastic properties a phase lag, ϕ, exists between stress and strain. The stiffness and phase lag of the tested material is determined by application of the following equations [1].

$$|E^*| = \sqrt{E_1^2 + E_2^2} \quad \text{and} \quad \phi = \arctan\left(E_2/E_1\right) \tag{2}$$

$$E_1 = \gamma \cdot \left(\frac{F}{Z} \cdot \cos(\phi) \cdot \frac{\mu}{10^3} \cdot \omega^2\right) \tag{3}$$

$$E_2 = \gamma \cdot \left(\frac{F}{Z} \cdot \sin(\phi)\right) \tag{4}$$

Where. E*: Complex modulus, E_1: Real component of complex modulus [MPa], E_2: Imaginary component of complex modulus [MPa], ϕ: Phase lag between force and displacement [degr.], γ: Geometrical factor [mm-1], F: Force applied to the beam [N], Z: Beam deflection at mid span [mm], μ: Mass factor [gr], ω: Fequency of applied force and displacement [Hz]

The factors γ and μ are determined as follows [1]:

$$R(x) = \frac{12L}{A} \cdot \left[\frac{1}{3x/L - 3x^2/L^2 - A^2/L^2}\right] \quad \text{and} \quad A = \frac{L-l}{2} \tag{5}$$

$$\gamma = \frac{L^2 A}{bh^3} \cdot \left(\frac{3}{4} - \frac{A^2}{L^2}\right) \tag{6}$$

$$\mu = R(x)\left(\frac{M}{\pi^4} + \frac{m}{R(A)}\right) \tag{7}$$

Where. L: effective length of the beam [mm], l: distance between inner clamps [mm], x: location where deflection is measured [mm], b: beam width [mm], h: beam height [mm], M: mass of the tested beam [gr], m: mass of moving machine parts [gr].

To check whether the listed equations lead to accurate test interpretation first geometrical non-linear static elastic simulations are done. In these calculations the phase lag, ϕ, is nil per definition. As a result E_2 remains nil, also effects of inertia remain absent. As a result equations 2, 3 and 4 reduce to the following.

$$|E^*| = \sqrt{E_1^2 + 0^2} \quad \text{or} \quad |E^*| = E_1 \quad \text{with} \quad E_1 = \gamma \cdot \left(\frac{F}{Z}\right) \tag{8}$$

In the simulations the beam material is assigned a stiffness of 7500 MPa and the Poisson's ratio is set to 0.35. Application of the above equations on simulation results leads to an AC stiffness that varies from 7260.75 MPa at very small beam excitement to 7260.72 MPa at 100 µm/m strain. From these results it is concluded that geometrical non-linear effects play a very limited role only. Therefore these effects are neglected from hereon. It is furthermore concluded that the back calculated 4pb stiffness includes an error of -3.19% when the standard is applied to a perfectly well functioning machine without further correction.

In the sections hereafter effort is made to explain the indicated error in 4pb back calculated stiffness. The most obvious reasons for the introduction of errors are addressed.

3.2 Shear Effect

The beam in the 4pb is subjected to shear in the area between the inner and the outer clamps. The shear force to which the beam is subjected in this area equals F/2, i.e. half the applied total load. Shear deformation is calculated by the following equation.

$$Z_s = \frac{F/2 \cdot A}{G \times 0.85 \cdot b \cdot h} \quad \text{with} \quad G = \frac{E}{2 \cdot (1+v)} \tag{9}$$

The shear deformation effect can be worked into the geometrical factor γ which will then become γ^*.

$$\gamma^* = \gamma + \frac{(1+v) \cdot A}{0.85 b \cdot h} \quad \text{or} \quad \gamma^* = \frac{L^2 A}{bh^3} \cdot \left(\frac{3}{4} - \frac{A^2}{L^2}\right) + \frac{(1+v) \cdot A}{0.85 b \cdot h} \tag{10}$$

Where. Z_s: Shear deflection of central part of the beam [mm], G: shear modulus [Mpa], v: Poisson's ratio [-] (mostly assumed 0.35 for AC mixtures). γ^*: Corrected geometrical factor [mm-1]

By correction for shear the 4pb back calculated stiffness increases to 7516.54 MPa at a Poisson's ratio of 0.35. This means that the error now becomes +0.22%.

3.3 Clamp Effects

The 4pb specimen is clamped between 10 mm wide metal clamps. These clamps locally restrain the beam's freedom of cross-section deformation. This effect was earlier discussed in [3], the magnitude of this effect is investigated using a model of an idealised 4pb. In this idealised 4pb the boundary conditions as depicted in Figure 1 are fully met. The specimen is excited by forces introduced via rigid 10 mm wide clamps. Figure 4 gives a visual impression of the model.

By comparison of the results of the idealised 4pb model with results of the bending beam theory it was found that the restraints at the clamps lead to an increase of 4pb bending stiffness. This increase is dependant on the Poisson's ratio. Table 2 gives the overestimation of the stiffness by application of the bending beam theory. Correction with the determined overestimation compensates for the clamp effect.

Table 2. Due to clamp restrained cross-section deformation the 4pb stiffness should be reduced with a factor depending on the Poisson's ratio

Poisson's ratio	0.0	0.15	0.25	0.35	0.45
Clamp correction, cc, to work on shear corrected stiffness	1.29%	1.78%	2.33%	3.07%	4.02%

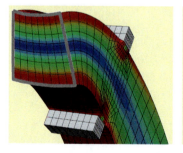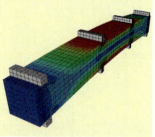

Fig. 4. Left: Due to bending and the effect of the Poisson's ratio the cross-section of the beam deforms, at the inner clamps this deformation is restrained (deformation: 250x). Right: Model overview (deformation: 10x).

As stated the correction for the clamp effect should work on the bending part of the beam stiffness only. This leads to the following modification of γ^*.

$$\gamma^* = \left(\frac{L^2 A}{bh^3} \cdot \left(\frac{3}{4} - \frac{A^2}{L^2} \right) + \frac{(1+\upsilon).A}{0.85 b \cdot h} \right) \cdot (1 - cc) \tag{11}$$

Where. Cc: correction factor for clamp effect [-]

Due to the clamp effect at ν=0.35 the 4pb beam stiffness is reduced to 7285.79 MPa, implying that the error in the 4pb stiffness is increased to -2.86%.

3.4 Clamp Movements

During the analysis of 4pb results it is assumed that the outer clamps do not allow for vertical movement. However, the model indicates that the centre line of the AC beam moves up and down at the outer clamps. In the case considered here these movements in equal 7.86581-06 mm/N applied force. These deformations effectively reduce the central deflection, Z, and should thus not be included in the 4pb back calculation procedure. For this reason Z needs to be corrected for clamp deformation.

$$Z_c = Z_m - S_s . F \tag{12}$$

Where. Z_c: Corrected beam deflection [mm], Z_m: Measured beam deflection [mm], S_s: support stiffness [mm/N], i.e. 7.86581e-06 mm/N.

Taking the end support deformation into account the 4pb back calculated stiffness becomes 7454.43 MPa, so reducing the mistake in the back calculation to -0.61%.

3.5 Hinge stiffness

Figure 1 depicts the boundary conditions that need to be matched by 4pb machines. It should be clear that it is very hard, if not impossible, to meet these boundary conditions. In the machine considered here effort is made to meet the prescribed conditions by use of elastic hinges. These hinges allow for limited clamp rotations and limited horizontal clamp translations. The nature of the elastic hinges is such that their behaviour is constant, i.e. not affected by wear & tear, slip-stick and maintenance such as regular cleaning or lubrication. Furthermore the nature of the hinges guarantees a set-up which is absolutely free of play. A disadvantage of the hinges is that they have a very small resistance to deformation. Figure 5 gives an impression of the elastic hinges in action.

As the clamping force is applied material is squeezed away from the clamped area. This slightly lengthens the beam resulting in clamp translation, see Figure 5 left. However, it should be clear that the elastic hinges mainly need to allow for rotation. By an analysis of the elastic hinges that is beyond the scope of this paper it was determined that the hinges have a rotational stiffness of 14400 Nmm.

From the bending beam theory it is known that rotations at the clamps are dependant on the deflection due to bending in the centre of the beam.

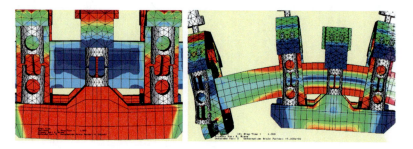

Fig. 5. Left: Clamps allowing for clamp translation (deformation factor: 1250x). Right Elastic hinges at 100 μm/m imposed strain (deformation factor: 500x).

$$R_i = c_i \cdot Z_b = \text{and} \quad R_o = c_o \cdot Z_b \qquad (13)$$

Where. Z_b: Bending deflection at the centre of the beam [mm], R_i: Rotation of inner clamps [-], R_o: Rotation of outer clamps [-], c_i: constant for determination of inner clamp rotation= 0.003727 [mm-1], c_o: constant for determination of outer clamp rotation 0.025455 [mm-1]

Knowing the rotational stiffness of the elastic hinges equation (13) translates into the following equation for moments applied at the clamps. Please note that two sets of two flexible elements in series are applied to hold the outer clamps.

$$M_i = c_i.s_i.Z_b \quad \text{and} \quad M_o = c_o.s_o.Z_b \quad \text{so that} \quad M_t = (c_i.s_i + c_o.s_o).Z_b \qquad (14)$$

Where. M_i and M_o: moment acting on inner and outer clamps respectively [Nmm], M_t: moment acting on the central part of the beam [Nmm], s_i: rotational stiffness of inner clamps= 28800 [Nmm], s_o: rotational stiffness of outer clamps= 14400 [Nmm]

The moment acting in the centre of the beam, ½.F.A, is reduced with M_t due to the rotational stiffness of the clamps. This acts to reduce the 4pb back calculated stiffness. When applying the equations from the standard this may be obtained by a correction of F.

$$\tfrac{1}{2}.F_b.A = \tfrac{1}{2}.F_m.A - (c_i.s_i + c_o.s_o).Z_b \quad \text{or} \quad F_b = F_m - 2.(c_i.s_i + c_o.s_o).\tfrac{Z_b}{A} \qquad (15)$$

Where. F_b: Force applied to bend the beam [N], F_m: Force measured by the 4pb machine [N]

Taking the effects of clamp stiffness into account 4pb back calculated stiffness becomes 7437.28 MPa, so reducing the mistake in the back calculation to -0.84%.

3.6 Conclusions

Non-linear geometrical effects in the 4pb considered here may be neglected.

When applying the EN standard the 4pb determines a stiffness that is 3.19% too low. The main reasons for this error are the following.

Neglecting shear deformation: underestimation of E by 3.52%
Neglecting clamp effects on cross section: overestimation of E by 3.07%
Neglecting clamp movements: underestimation of E by 2.31%
Neglecting hinge stiffness: overestimation of E by 0.23%

Taking the above into account the accuracy of the back calculated stiffness will increase dramatically, theoretically leading to an unexplained error of 0.84%.

4 Visco-Elastic Response

4.1 Simulation

To investigate the accuracy of the 4pb with respect to phase lag determination a visco-elastic simulation was done on basis of an asphalt with a stiffness of 7500 MPa and a lag of 20° at 20 Hz subjected to 20 Hz loading. In the simulation the clamping force of 130 N per clamp was applied in 10 seconds. After a 1 sec rest period the 20 Hz displacement controlled loading was applied. Figure 6 gives an impression of the results obtained after the clamping procedure was completed. The figure indicates that sinusoidal functions are fitted to the response signals from the last load cycle applied in the simulated test. The input amplitude of actuator displacement was 0.03532 mm. This analysis leads to the following results.

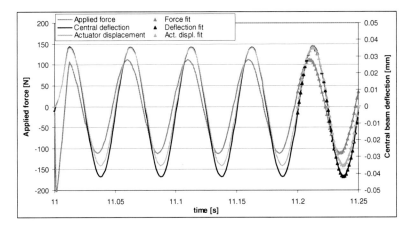

Fig. 6. Obtained force and beam deflection signals

Table 3. Results of a simulated test at 20 Hz on asphalt with a stiffness of 7500 MPa and a 20° lag

	Actuator displacement	Central deflection	Applied force
Amplitude	0.03531 mm	0.03883 mm	111.77 N
Phase lag	18.67 degr.	19.61 degr.	0.00 degr.

When applying the EN standard the listed results translate into a material with a complex modulus of 7266.6 MPa and a phase lag of 19.6°. The error in obtained stiffness is thus – 3.11% and the error in phase lag is -0.4°.

Hereafter effort is made to correct these results for the shear effect, clamp effect, clamp movements and hinge stiffness.

4.2 Shear Effect and Clamp Effects

As discussed earlier neglecting the effects of shear results in an underestimation of stiffness by 3.52%. Opposite to this the clamps act to restrain cross section deformation resulting in an overestimation of stiffness by 3.07%. Following the procedures discussed earlier the back calculated stiffness E* becomes 7291.64 MPa resulting in an error of -2.78% when correcting for these effects. The phase lag is not influenced by this correction and the error in the lag remains 0.4°.

4.3 Clamp Movements

It was determined earlier that the vertical deformation in the outer clamps equals 7.86581-06 mm per N applied force. These deformations effectively reduce the central deflection, Z, and should thus not be included in the 4pb back calculation procedure. For this reason Z needs to be corrected for clamp deformation, see equation 12. Since the clamps are made of elastic materials this correction should act on the elastic (i.e. non-delayed) deformation only. With this equation 12 translates into.

$$Z_{ce} = Z_{me} - S_s.F \qquad (16)$$

So that.

$$Z_c = \sqrt{Z_{ce}^2 + Z_{mv}^2} \quad \text{and} \quad \delta_c = atg\left(Z_v / Z_{ce}\right) \qquad (17)$$

Where. Z_c: Corrected beam deflection [mm], Z_{ce}: Corrected beam deflection at maximum F [mm], Z_{me}: Measured beam deflection at maximum F [mm], Z_{mv}: Measured beam deflection at F=0 N [mm], δc: corrected phase lag [°], S_s: support stiffness [mm/N], i.e. 7.86581e-06 mm/N.

With the above the 4pb back calculated stiffness becomes 7450.3 MPa, so reducing the mistake in the stiffness back calculation to -0.66%. The correction also affects the phase lag which now becomes 20.05° leading to an error of 0.05°.

4.4 Hinge Stiffness

As discussed the hinges have minor resistance against rotation. As explained this implies that a fraction of the applied force is used to rotate the clamps and not so much for actually bending the beam. This effectively reduces the force for bending and thus reduces the back calculated beam stiffness.

Equation 15 indicates the reduction of force applied to the beam at maximum deflection. To take this effect into account the following equations are applied.

$$F_{Z\max} = F_m \cdot \cos(\partial_c) - 2 \cdot (c_i \cdot s_i + c_o \cdot s_o) \cdot \frac{Z_c}{A} \quad \text{and} \quad F_{Z=0} = F_m \cdot \sin(\partial_c) \quad (18)$$

$$F_c = \sqrt{F_{Z\max}^2 + F_{Z=0}^2} \quad \text{and} \quad \delta_c = atg\left(\frac{F_{Z=0}}{F_{Z\max}}\right) \quad (19)$$

Where. F_{Zmax}: Force applied at maximum beam deflection corrected for clamp deformation [N], F_m: Amplitude of force applied by the 4pb machine [N], $F_{Z=0}$: Force applied at zero beam deflection corrected for clamp deformation [N], δ_c: phase lag corrected for clamp deformation [°], Z_c: amplitude of beam deflection corrected for shear, and clamps deformation [mm], F_c: Amplitude of applied force corrected for clamp rotation stiffness [N].

When this reduction is taken into account the back calculated beam stiffness becomes 7434.2 MPa resulting in an error of -0.88%. The phase lag is also affected by this correction and becomes 20.1° leading to an error of -0.1°.

4.5 Conclusions

When applying the EN standard the 4pb as discussed determines a stiffness that is 3.11% too low. The main reasons for this error are the following.
Neglecting shear deformation: underestimation of E by 3.52%
Neglecting clamp effects on cross section: overestimation of E by 3.07%
Neglecting clamp movements: underestimation of E by 2.18%
Neglecting hinge stiffness: overestimation of E by 0.22%

Taking the above into account the accuracy of the back calculated stiffness will increase dramatically, theoretically leading to an unexplained error of 0.88%.

The phase lag that is determined is 0.39° too small. The main reasons for this error are the following.
Neglecting clamp stiffness: underestimation of lag by 0.44°
Neglecting hinge stiffness: underestimation of lag by 0.05°

Taking the above into account the accuracy of the back calculated lag will increase, theoretically leading to an unexplained error of 0.1°.

5 Conclusions

From the previous, taking into account the considered material (E*=7500 MPa, lag at 20 Hz=20°), it is concluded that the 4pb test executed at 20 Hz on an ideal machine leads to two sources of error which both affect stiffness and not phase lag.

1) Neglecting shear deformation between the outer and inner clamps, resulting in a reduction of stiffness with 3.52%.
2) Neglecting the clamp restrained inner cross-sections, resulting in an increase of stiffness with 3.07%.

Other sources of error are found in the 4pb machine considered here. These sources of error are related to the built quality of the machine and are the following.

1) Neglecting the hinge stiffness will lead to an increase of stiffness with 0.22% and a reduction of lag by 0.05° (at 20° material lag)
2) Neglecting the vertical movements of the outer clams results in a reduction of stiffness with 2.18% and a reduction of lag with 0.44°.

Taking into account the above a source of unknown error remains. This source results in an underestimation of stiffness with 0.84% to 0.88% (elastic vs -elastic simulation) and an overestimation of lag by 0.1°.

It is stated explicitly that the discussed 4pb machine compensates for errors as discussed via calibration. However, it should be clear that the accuracy of any machine increases with built quality reducing the need for correction of data. The built quality of the machine discussed here is more than adequate since errors that follow from built quality remain smaller than the intrinsic errors in the 4pb test itself.

Literature

[1] NEN-EN 12697-26, Bituminous mixtures – Test methods for hot mix asphalt – Part 26: Stiffness, European Committee for Standardisation, Brussels (July 2004)
[2] NEN-EN 12697-24+A1, Bituminous mixtures – Test methods for hot mix asphalt – Part 24: Resistance to fatigue, European Committee for Standardisation, Brussels (July 2007)
[3] Huurman, M., Pronk, A.C.: Theoretical analysis of the 4 point bending test. In: Proceedings of the 7th Int. RILEM Symposium Advanced Testing and Characterization of Bituminous Materials, Rhodes, Greece (May 2009)

Multiscale Micromechanical Lattice Modeling of Cracking in Asphalt Concrete

Arash Dehghan Banadaki, Murthy N. Guddati, Y. Richard Kim[1], and Dallas N. Little[2]

[1] Department of Civil, Construction, and Environmental Engineering, North Carolina State University, Raleigh, NC 27695-7908
[2] Zachry Department of Civil Engineering, Texas A&M University, College Station, TX 77843

Abstract. A multiscale micromechanical lattice modeling technique is proposed from amongst several computational methods for predicting the performance of hot mix asphalt (HMA) under service loads. Although the lattice model has shown promise, many important details need to be addressed to ensure realistic predictions. This paper presents enhancements to the original model that have been developed over the past two years. These revisions are geared towards capturing the material behavior more accurately and efficiently than was possible with the original lattice model. Among the new enhancements that are presented in this paper are the incorporation of viscoelastic fracture with the help of the work potential-based viscoelastic continuum damage model, computationally efficient simulations under a large number of load cycles, and the incorporation of air voids to capture the reduction in stiffness and strength of the material. Efficiency of the model is improved further by incorporating novel algorithms.

1 Introduction

It is widely known that the cracking performance of hot mix asphalt (HMA) depends on the mechanical properties of its constituents, namely asphalt mastic and aggregate. Such dependence is extremely complex, and physical experimentation often is used to characterize the performance of HMA. However, considering the number and variety of possible mixtures, performing these experiments is extremely expensive and time consuming. The scope of the physical experiments can be reduced with the help of computational models that can reasonably predict the cracking behavior of a mixture using the mechanical properties of its constituent materials. With this goal in mind, Zhen et al. [1] developed a micromechanical lattice modeling technique that involves discretizing the continuum as an assembly of lattice links, with damage and cracking simulated by sequentially breaking the links [1]. The choice of this average micromechanical modeling approach is justified by the fact that asphalt concrete is inherently an imperfect material, and detailed micromechanical modeling is not warranted, given that it would be prohibitively expensive.

To further reduce computational costs, this research effort embeds the lattice modeling procedure into a multiscale framework that can link the modeling from mastic scale to mixture scale in a step-wise fashion. Additionally, a virtual fabrication technique is developed to generate a two-dimensional microstructure to reduce the need for physical fabrication and its associated costs. The first part of this paper is a general introduction to the original lattice modeling framework. The limitations of the original version are then discussed, and the new enhancements to the model are explained briefly. The effect of each enhancement is illustrated separately.

2 Methodology of Lattice Modeling

For many materials, HMA being no exception, the cracking phenomenon is driven by the nucleation, propagation, and coalescence of microcracks. This fact indicates that a micro-level simulation of this cracking phenomenon may lead to improved qualitative and quantitative understanding of cracking overall. Furthermore, because fracture mechanisms are related directly to the microstructure, a discrete micromechanical approach is an appealing method to provide valuable insights into the physical cracking process and the role of heterogeneity. One of the discrete approaches that allows a straightforward implementation of material heterogeneity at the microscopic level is lattice modeling [2].

The lattice modeling procedure analyzes either physically or virtually fabricated HMA specimens to capture the effects of the microstructure on mixture response and performance. The lattice model is comprised of three components: (1) a preprocessor that generates and converts the microstructure of a specimen into a lattice mesh, (2) a solver that analyzes the resulting lattice mesh, and (3) a postprocessor to analyze the simulation results and to average and upscale the results into a usable form at the macroscopic level or at a larger scale (figure 1).

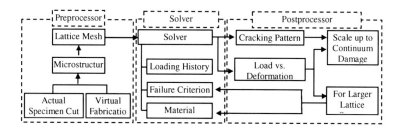

Fig. 1. Framework of lattice modeling

The elements of each of the components are described briefly in the following sections.

2.1 Microstructure and Lattice Mesh Generation

Microstructure generation involves developing detailed representative microstructures from virtually fabricated specimens using geometrical and statistical concepts, such as inverse stereology [1]. Material microstructure is generated directly using a virtual fabrication technique instead of processing physically fabricated specimens, thus resulting in a substantial reduction in experimental costs and providing more flexibility to the analyst. The obtained microstructure is then modeled as a two-phase system, i.e., homogeneous binder (or mastic) with rigid aggregate particles that do not deform or crack. In this study, a random truss lattice, which is statistically homogeneous and isotropic, is used to simulate the mastic to eliminate any computational anisotropies that are characteristic of regular lattice networks. As shown in figure 2 (a), the generation of such a lattice begins with a square mesh called the *base mesh*. Each cell of the base mesh contains a single node whose exact location (deviation from the center) is determined based on the uniform probability distribution function. Once the nodes are generated, the lattice network is constructed by linking the nodes using Delaunay's triangulation (figure 2(b)) [1]. A random mesh with regular nodes is used to simulate the aggregate particles with the purpose of emphasizing their possible configuration (figure 2(c)). The lattice link size is primarily determined by the minimum aggregate size considered in the specific scale of the microstructure.

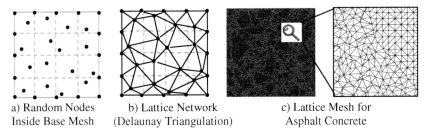

a) Random Nodes Inside Base Mesh   b) Lattice Network (Delaunay Triangulation)   c) Lattice Mesh for Asphalt Concrete

Fig. 2. Lattice representation of hot mix asphalt

2.2 Solver

The properties of each link in the resulting lattice network are chosen as the component material(s) that are underneath the link. The cross-sectional area of a lattice link for asphalt is approximated from the characteristic length (base cell size) of the lattice link [1]. The advantage of the lattice model is the convenience of introducing a failure criterion (damage parameter in the new implementation) for each link in bond-dominated materials. The choice of such a criterion determines the cracking behavior of the specimen. The brittle failure criterion (original implementation) is the simplest failure criterion by which each link fails if the stored energy in the link exceeds the surface energy of the associated microcrack. Gradual softening of the links due to the applied damage on each link is another way to introduce damage to lattice links. Based on the above parameters, the modulus and load deformation curves of the specimen can be obtained.

2.3 Methodology of Multiscale Micromechanical Lattice Modeling

Because one of the goals of lattice modeling is to relate the component material properties to the mixture properties, it is desirable to capture the effect of even the finest aggregates. Such detailed modeling, although appearing straightforward at the outset, has significant practical limitations due to the computational costs involved. Fortunately, the computational costs can be reduced significantly with the help of the multiscale modeling approach. Essentially, the multiscale approach considers the effects of different-sized aggregate particles at different length scales. Such an approach reduces the computational costs significantly while capturing the mechanical phenomena at various length scales (figure 3).

The two-dimensional apparent aggregate gradation in a cut surface can be divided into a series of subregions based on different scale lengths of observation (i.e., aggregate size). The analysis starts with the virtual fabrication of a group of representative volumetric elements (RVEs) at the last (smallest) scale, scale n. Noting that the heterogeneity at scale n is ignored at the larger scale '$n-1$', scale n RVEs can be regarded as homogeneous mastic in the analysis at scale $n-1$. The lattice analysis takes the modulus and surface energy values (replaced by damage parameters in the latest implementation) of the binder and the aggregate and calculates the effective stiffness and surface energy values of the RVEs at scale n. This calculation is accomplished by simulating the fracture test and uniaxial test for RVEs. The resulting stiffness and surface energy values (or damage parameters) from all the RVEs are statistically averaged to eliminate specimen-to-specimen variations due to the random nature of the microstructural generation. The processed mastic properties are, in turn, used for the larger scale (scale $n-1$). This procedure is applied recursively until all the subregions of the gradation are considered.

An important point to note is that, in multiscale analysis, the heterogeneity observed at the small scale and its effect on microcracking is considered in an average sense by using homogenized (averaged) mechanical properties. Considering that the main objective is to characterize the cracking behavior at the macrolevel, detailed micromechanical phenomena, such as stress singularity and the propagation of each microcrack, do not need to be simulated. Thus, in this study, the multiscale approach is sufficient and, due to its efficiency, desirable. The proposed procedure can effectively predict the response of the HMA in macroscale. However, in its original state, the framework has some limitations.

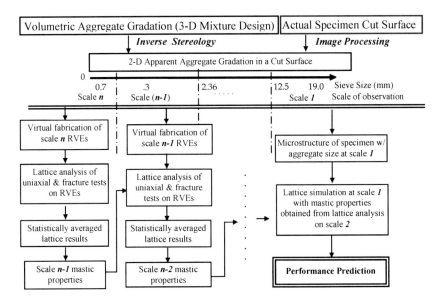

Fig. 3. Integrated lattice modeling procedure for multiscale analysis

2.4 Limitations

One of the limitations of the proposed algorithm in its original state is the prediction of post-peak behavior of the load-displacement history. Although the material shows ductile behavior after the peak of loading, using the surface energy-based fracture criterion results in brittle fracture of the specimen immediately after the peak. The desired behavior can be captured by incorporating the viscoelastic fracture.

Predicting the responses of materials under long sustained loads is the ultimate goal of the proposed model. However, applying such loads to the material in a conventional manner is practically impossible due to the excessive computational costs. Extrapolation techniques can result in an efficient algorithm while maintaining accuracy.

Another important constituent of HMA is air voids. Corners of the air voids are known to be the major source of stress concentrations that may lead to the initiation of cracks. The original version of the model does not include the effects of air voids in the microstructure. Bridging the component material properties to the performance of HMA is a computationally demanding task and requires an efficient algorithm. The efficiency of the original framework can be improved using different techniques.

The progress made in resolving the above-mentioned issues is reported below.

3 Viscoelastic Fracture

The stress-strain curves obtained from the original lattice framework typically exhibit brittle failure with a sudden drop in stress level right after the stress peak (figure 4). However, experimental results indicate that only at low temperatures or high rates of loading is such behavior noticeable. The dominant behavior of the material usually is described as the gradual breaking of the bonds due to the applied load. Damage patterns observed from the tests show that, in addition to the macrocracks that form in the specimen, other parts of the specimen experience some amount of damage. The stress-strain curves obtained from the physical experiments can show brittle and ductile behavior, depending on the temperature and rate of loading.

Careful investigation has revealed that the VECD model ([3]) appears to be suitable for this purpose. The idea is to evaluate the damage parameter (S value) of each link in each time step and to find the material integrity factor (C value, the ratio of damage stiffness to virgin stiffness) based on the amount of damage the material has experienced. Because the amount of damage correlates to the strain rate in the VECD model, the effect of the rate of loading can be captured by using the new viscoelastic fracture rule.

At any given stiffness, the pseudo strain, ε^R, for all the links in the lattice can be found using effective elastic analysis. Using Equation (1) the damage parameter can be found for each link in step $i+1$ using the damage parameter (S_i) from the previous step.

$$S_{i+1} = S_i + \Delta t \left(-\frac{1}{2}(\varepsilon^R)^2 \frac{(\partial C)_i}{\partial S} \right)^\alpha \tag{1}$$

Knowing the C vs. S relationship, Equation (2), the updated stiffness value, C_{i+1}, for all the links can be found.

$$C = e^{a(S)^b} \tag{2}$$

This procedure can be repeated until significant damage develops. The reaction history at the location of applied displacements, as well as the actual applied displacement history, can be processed appropriately to obtain the stress-strain relationships at the upper scale. This stress-strain relationship is used to develop the C vs. S relationship using the standard processing techniques proposed in [4], which, in turn, is used as the damage parameters of the lattice links at the upper scales. The introduction of viscoelastic fracture shows that this method can capture the effects of the gradual breaking of bonds as well as the rate of loading on an asphalt concrete specimen. Figure 4 shows a comparison of the stress-strain curves of three loading rates using the new fracture rule. As shown, fast loading causes a stiff response, which is consistent with experimental observations.

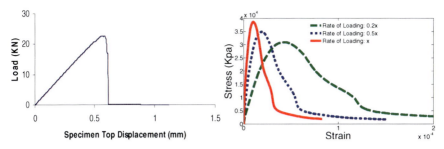

Fig. 4. Stress-strain curve a) elastic fracture criterion b) Viscoelastic fracture

4 Cyclic Loads

It is well known that the dominant type of load applied to pavements under service has a repeating nature that often is simplified as cyclic (periodic) loading. Theoretically, the same analysis procedure can be used for either cyclic or monotonic loads. However, for cyclic loading, the amplitude of the applied strains (and, hence, the amount of damage) remains quite small, and failure occurs much later compared to monotonic loading in which the applied strain keeps growing until the specimen fails. The computational costs can be reduced using the concept of extrapolation, which is based on the observation that the damage of the material does not change significantly within a given cycle. Thus, analysis for a single cycle can be performed under the assumption that the damage does not vary, in order to obtain the stress history for each link. Because damage does not vary significantly from cycle to cycle, an extrapolation technique can be used to update the damage over several cycles before another stress analysis is performed. Thus, stress analysis is performed not for each cycle, but once for a large group of cycles, which is termed an analysis *stage*.

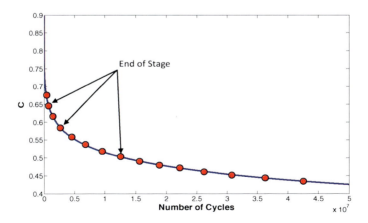

Fig. 5. C vs. n calculated for cyclic load

The above method is implemented within the lattice modeling framework whereby the damage level of each link is extrapolated for the next stage. Figure 5 shows a normalized pseudo stiffness (C) vs. number of cycles curve obtained using this procedure. Due to the fast drop in C during the first cycle of loading, multiple time steps are required to capture the correct shape of the curve [4]. A larger stage size then can be chosen after the amount of change in C shows a decrease later during the loading history.

5 Air Voids

To evaluate the effects of air voids in the model, the virtual fabrication part of the algorithm has been modified to generate cavities inside the specimen to represent air voids as a separate phase (gray spots in figure 6). The size of the voids are chosen based on the scale that is being analyzed.

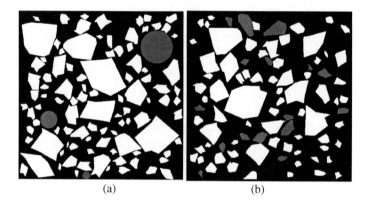

Fig. 6. Air void shapes: a) circular shapes, and b) random octagonal shapes

X-ray tomographic images show that the shapes of the air voids are not at all circular [5]. The significance of the shape of air voids becomes more important once the effect of shape on the initial stiffness of the asphalt specimens is evaluated. In this study, the initial stiffness of the specimens with circular shapes is compared with that of specimens with randomly shaped octagonal air voids (figure 6). The stiffness scale-up factors of two sets of specimens with the same air void content but different shapes have been measured and are compared in figure 7. The stiffness values are normalized against zero percent air void content. It can be concluded that, not surprisingly, the shape of the air voids plays an important role in determining the mechanical behavior of the material. As a result, it is important to include realistic air void shapes in virtually fabricated specimens. More investigation is underway to generate realistic shapes in a virtually fabricated aggregate structure.

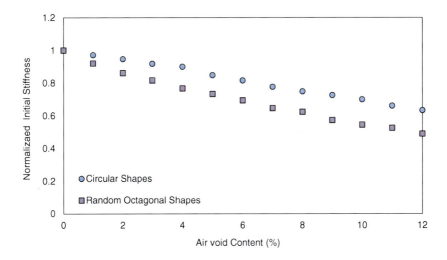

Fig. 7. Effect of shape on normalized stiffness scale-up factor

6 Efficiency

One of the most important advantages of lattice modeling compared to other techniques is its simplicity and low cost. One-dimensional elements with minimal degrees of freedom are much less expensive to create than two-dimensional elements with more degrees of freedom. However, the implementation of the algorithms demands extra attention, especially when recursive operations are performed on large matrices and vectors. Therefore, a new solver has been implemented in MATLAB with extra attention to memory efficiency and time efficiency using vectorization techniques.

In the previous implementation, conventional convolution integrals were used to find the viscoelastic response of the material. The problem with this process is that the stress in each time step for a particular link depends on the entire history of the strain in that link. However, the correspondence principle holds valid for this type of analysis and can be used to obtain the viscoelastic response of the material[6]. The interpretation of the correspondence principle for constant crosshead testing is to apply the pseudo strain to the specimen in order to obtain the entire stress history of the material by performing elastic simulations, which eventually leads to the necessary stress-strain relationship. Because the simulations are effectively elastic, significant computational effort and time can be saved. Applying the correspondence principle along with the new efficient solver makes implementing the code six times faster than for the previous version, although many additional parts were added to the algorithm that did not previously exist.

7 Summary

The original framework of multiscale lattice modeling, along with the associated limitations of the algorithm, is presented in this paper. All the limitations of the framework then are discussed individually, and the appropriate solutions are explained. It is found that viscoelastic fracture can effectively simulate both the pre- and post-peak behavior of the stress-strain curves. In addition, the effects of the rate of loading are captured using the VECD model. An efficient extrapolation method is proposed for simulating long sustaining service loads. The effects of the shape of air voids on mechanical properties also are presented. The correspondence principle is shown to be effective in increasing the efficiency of the viscoelastic solver. Further investigation is underway to resolve the remaining issues of the lattice modeling framework. These issues include: (1) incorporating realistic air void shapes in virtually fabricated microstructures, (2) capturing the effects of change in time dependency in different scales of the material, and (3) validating the results quantitatively using experimental observations.

References

[1] Feng, Z., Zhang, P., Guddati, M.N., Kim, Y.R.: The Development and evaluation of a virtual testing procedure for the prediction of the cracking performance of hot-mix asphalt. In: ASCE Conf. Proc. Characterization and Modeling Symposium at EMI 2010, vol. 385, pp. 142–158. ASCE (2010)
[2] Schlangen, E., Van Mier, J.G.M.: Cem. Concr. Compos 14, 105 (1992)
[3] Kim, Y.R., Lee, H.J., Little, D.N.: J. Assoc. Asphalt. Pav. 66, 520 (1997)
[4] Underwood, B., Kim, Y.R., Guddati, M.N.: Int. J. Pavement Eng. 11(6), 459 (2010)
[5] Kutay, M.E., Ozturk, H.I., Gibson, N.: 3D Micromechanical Simulation of Compaction of Hot Mix Asphalt Using Real Aggregate Shapes Obtained from X-ray CT. In: ASCE Conf. Proc. Characterization and Modeling Symposium at EMI 2010, vol. 385, pp. 86–98 (2010)
[6] Gross, D., Seelig, T.: Fracture Mechanics With an Introduction to Micromechanics. Springer (2006)

Accelerated Pavement Performance Modeling Using Layered Viscoelastic Analysis

Mehran Eslaminia, Senganal Thirunavukkarasu, Murthy N. Guddati, and Y. Richard Kim

Department of Civil, Construction and Environmental Engineering, North Carolina State University, Raleigh, NC, USA

Abstract. An efficient pavement performance analysis framework is developed by combining the ideas of time-scale separation and Fourier transform-based layered analysis. First, utilizing the vast difference in time scales associated with temperature and traffic load variations, the number of stress analysis runs are reduced from several million to a few dozen. Second, the computational cost of the pavement stress analysis is reduced significantly by using Fourier transform-based analysis. The resulting pavement performance prediction tool, named the layered viscoelastic continuum damage (LVECD) program, can capture the effects of viscoelasticity, temperature (thermal stresses and changes in viscoelastic properties) and the moving nature of the traffic load. The efficiency of the LVECD program is shown through 20-year pavement simulations.

1 Introduction

Reasonable stress-strain analysis is a key component in pavement design and predicting pavement life. Given the complexity of variables such as pavement life, traffic loading, and temperature variations, various approximate methods are used to predict pavement performance. Despite differences in assumptions, all of these prediction methods aim to reduce analysis that takes millions of cycles over several years to a few hundred analyses under a single cycle of loading.

The three-dimensional finite element method (3-D FEM) is a sophisticated analysis tool for pavement performance analysis that can model the response of a 3-D pavement under a moving load [1-3]. Although the 3-D FEM is capable of including the viscoelasticity and nonlinearity of pavement layers, or fatigue cracking and rutting effects, its computational cost is prohibitively expensive. Therefore, more practical approaches often are used to perform pavement performance analyses.

The most basic method is layered elastic analysis (LEA), where the pavement is idealized as a layered elastic system under a stationary axisymmetric load. In this method, the normal and radial stresses/strains often are computed using a Fourier-Bessel transform (see e.g., [4]). However, LEA leads to inaccurate responses because (1) traffic loading (i.e., tire pressure) is neither stationary nor circular in reality, and (2) asphalt concrete exhibits significant viscoelastic behaviors, especially under a moving load.

Layered viscoelastic moving load analysis (LVEMA) is an improvement over LEA in that the viscoelasticity and the moving load effects are handled efficiently with the help of Fourier transforms [5, 6]. LVEMA is more appealing than LEA for pavement stress analysis, although the stress redistribution effects due to damage still are not captured.

The goal of this study is to introduce an accelerated analysis framework based on LVEMA to predict pavement life. The basic idea of the proposed method is discussed in the first section, followed by the LVEMA formulation in the second section. Subsequently, the layered viscoelastic continuum damage (LVECD) program, developed based on the proposed framework, is introduced. Finally, the verification and efficiency of the LVECD program are demonstrated using numerical examples.

2 Outline of the Proposed Approach

The proposed framework is based on the approach found in [7] that includes various assumptions and observations regarding the pavement structure, material properties of the layers, thermal variations and traffic variations. The assumptions and the reasoning behind them are discussed below. (Note that for the sake of completeness, many arguments are repeated from [7].)

1. The pavement length (in the traffic direction) and width are both large compared to the size of the tire and pavement thickness. Thus, if the effects of fatigue/rutting on material properties/pavement structure are ignored, the pavement can be approximated as an infinite layered system where the material properties vary only in terms of depth.
2. Temperature variation is captured only in terms of pavement depth and is assumed to be constant over the entire plane that corresponds to a given depth, because the temperature variation along the length of the pavement is not significant, and the material properties are assumed to be isotropic on the plane perpendicular to the depth direction.
3. The pavement temperature profile (across the pavement depth) is assumed to be cyclic within a period of one year. Although the yearly variations can be modeled with a corresponding increase in computational cost, the variation is not significant given that the stress redistribution effects due to damage are not considered in this analysis.
4. Temperature variations are captured using hourly data. Although a fine-grained thermal variation can be captured, it is unnecessary given the approximate nature of the analysis.
5. The traffic load is idealized as a cyclic load with a constant shape (tire footprint) and speed.
6. Traffic loading varies by second(s), whereas temperature varies by hour(s). The temperature profile and the resulting effects on the material property are assumed to be fixed for the traffic analysis of a given segment.
7. Despite the nonlinear nature of the base and subgrade, they are idealized as linear elastic materials, because the effects of nonlinearity are not significant compared to the approximations inherent in the modeling of traffic and temperature variations.

The above observations and assumptions reduce the analysis from millions of load cycles to fewer than a hundred independent analyses by using a segmented analysis scheme [8]. The basic idea is to divide the pavement life into different stages, with each stage characterized by seasonal or monthly variations in temperature. The typical length of a life stage is between two weeks and a few months, depending on the desired level of accuracy. Because the yearly variations in temperature are ignored, the division into stages is restricted to the first year of pavement life, and the pavement responses during that period are assumed to repeat for the remainder of the pavement life. In addition, each life stage is divided further into analysis segments, where an analysis segment is assumed to have a constant temperature as well as a constant traffic load level and frequency. Typically, an analysis segment is a block of a few hours per day over the life stage. Note that the number of segments depends on the desired level of accuracy and the hourly variations of temperature and traffic (Figure 1).

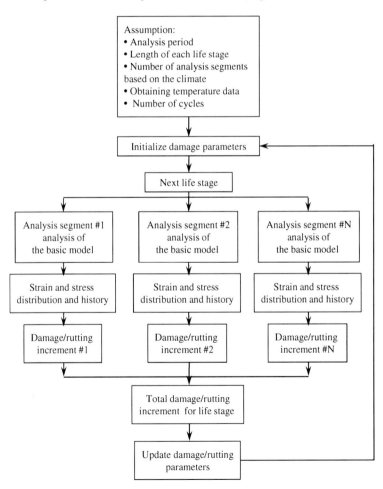

Fig. 1. Scheme of proposed framework for pavement performance analysis

Thermal and traffic stress values are computed for each analysis block during the first year of pavement life. The thermal stress values are computed using standard thermal analysis that employs the Enhanced Integrated Climate Model (EICM) temperature history. Note that because the pavement length is infinite, the plane strain conditions are appropriate for the thermal stress analysis. The traffic stress values are obtained using LVEMA, which is discussed in the next section.

3 Layered Viscoelastic Moving Load Analysis

This analysis considers an infinite pavement under a traffic load that is moving with a constant speed, V. The coordinate x is used for the transverse direction ($-\infty < x < +\infty$), y for the traffic direction ($-\infty < y < +\infty$), and z for the depth direction ($0 < z < z_{max}$; $z = 0$ is the top surface). The spatial distribution of the load at $t = 0$ is given by $\mathbf{p}(x, y)$. The precise statement of the problem is reflected in the following equations.

Strain-displacement relationship:

$$\boldsymbol{\varepsilon} = \begin{Bmatrix} \varepsilon_{xx} \\ \varepsilon_{yy} \\ \varepsilon_{zz} \\ \gamma_{yz} \\ \gamma_{zx} \\ \gamma_{xy} \end{Bmatrix} = \begin{bmatrix} \partial/\partial x & 0 & 0 \\ 0 & \partial/\partial y & 0 \\ 0 & 0 & \partial/\partial z \\ 0 & \partial/\partial z & \partial/\partial y \\ \partial/\partial z & 0 & \partial/\partial x \\ \partial/\partial y & \partial/\partial x & 0 \end{bmatrix} \begin{Bmatrix} u_x \\ u_y \\ u_z \end{Bmatrix} = \mathbf{Lu}. \quad (1)$$

Stress-strain relationship:

$$\boldsymbol{\sigma} = \begin{Bmatrix} \sigma_{xx} & \sigma_{yy} & \sigma_{zz} & \tau_{yz} & \tau_{zx} & \tau_{xy} \end{Bmatrix}^T = \int_0^t \mathbf{C}(t-\tau)\frac{d\boldsymbol{\varepsilon}}{d\tau}d\tau. \quad (2)$$

Equilibrium equations:

$$\mathbf{L}^T \boldsymbol{\sigma} = \mathbf{f}. \quad (3)$$

Bottom boundary condition:

$$u_x = u_y = u_z = 0 \text{ at } z = z_{max}. \quad (4)$$

Top boundary condition:

$$\mathbf{t}_r = \begin{Bmatrix} \tau_{xz} & \sigma_{zz} & \tau_{zy} \end{Bmatrix}^T = \mathbf{p}(x, t - y/V). \quad (5)$$

In Equations (1) to (6), $\mathbf{u}$ is the displacement vector; $\boldsymbol{\varepsilon}$ is the strain vector; $\mathbf{L}$ is the strain displacement operator; $\boldsymbol{\sigma}$ is the stress vector; $\mathbf{C}$ is the stress-strain matrix;

and **f** is the body force vector. The load **p** has the argument (t-y/V), indicating that the load is moving with a constant velocity, V. Note further that, in general, **p** has the components of tire pressure (vertical direction) and friction (horizontal direction).

Fourier transform in t, x, and y. Given that (1) the material properties and geometry do not vary with t or x or y, (2) the material properties are linear, and (3) t, x, and y are unbounded, the Fourier transform can be applied to reduce the problem dimension. The following definition of Fourier transform is employed in this paper:

$$\hat{f}(k_x, k_y, z, \omega) = \int_{-\infty}^{+\infty}\int_{-\infty}^{+\infty}\int_{-\infty}^{+\infty} f(x, y, z, t) e^{-ik_x x} e^{-ik_y y} e^{-i\omega t} dx\, dy\, dt, \quad (6)$$

where $\hat{f}$ is the Fourier transform of a generic function, f; ω is the (temporal) frequency; and k_x and k_y are the wave numbers (spatial frequencies) along the x and y axes, repectively. Thus, the problem definitions (Equations (1) to (5)) should be replaced by

$$\hat{L}(k_x, k_y) = \begin{bmatrix} ik_x & 0 & 0 \\ 0 & ik_y & 0 \\ 0 & 0 & \partial/\partial z \\ 0 & \partial/\partial z & ik_y \\ \partial/\partial z & 0 & ik_x \\ ik_y & ik_x & 0 \end{bmatrix}, \quad (7)$$

$$\hat{\varepsilon} = \hat{L}\hat{u}, \quad (8)$$

$$\hat{\sigma} = i\omega \hat{C}(\omega)\hat{\varepsilon}, \quad (9)$$

$$\hat{L}^T\hat{\sigma} = \hat{f}, \quad (10)$$

$$\hat{u}_x = \hat{u}_y = \hat{u}_z = 0 \text{ at } z = z_{max}, \quad (11)$$

$$\hat{t}_r = \{\hat{\tau}_{xz} \quad \hat{\sigma}_{zz} \quad \hat{\tau}_{zy}\}^T = \hat{p}(k_x, k_y, \omega). \quad (12)$$

The Fourier transform eliminates the need for convolution in Equation (2) and also reduces the dimension of the governing equation to one. However, the problem must be solved independently for a sweep of frequencies and wave numbers, and the resulting responses must be inversely Fourier transformed to obtain the response histories and variations in the x and y directions. Analysis of each one-dimensional problem in the frequency domain is carried out with the help of an optimized finite element algortihm.

Relation between wave number and temporal frequency. Because the load moves with a constant speed, all the responses (stress and deformation profiles) move with the same speed, i.e., the response relative to the location of the load remains constant. In other words, all the deformations and stresses take the form, $f(t-y/V)$, similar to the expression for the load in Equation (5). The Fourier transform of such a translating function takes the form,

$$\hat{f}(k_y,\omega) = \int_{-\infty}^{+\infty}\int_{-\infty}^{+\infty} f\left(t-\frac{y}{V}\right)e^{-i\omega t}e^{-ik_y z}dtdy = \int_{-\infty}^{+\infty}\int_{-\infty}^{+\infty} f(\bar{t})e^{-i\omega \bar{t}}e^{-i\left(k+\frac{\omega}{V}\right)z}d\bar{t}dz, \qquad (13)$$

where $\bar{t} = t - y/V$. Clearly, $\hat{f}$ is nonzero only when $k = -\omega/V$, and it is equal to

$$\hat{f}(\omega) = \int_{-\infty}^{+\infty} f(t)e^{-i\omega t}dt. \qquad (14)$$

The implication is that the solutions of Equations (8) to (12) do not need to be found for all sets of ω and k_y, but for all ω, with $k_y = -\omega/V$. Thus, the assumption of constant speed results in a significant reduction in the computational cost of analysis.

4 Fatigue Analysis

Fatigue cracking analysis is carried out based on the viscoelastic continuum damage (VECD) model [3]. Essentially, the damage is computed for a single cycle in each life stage using the VECD model, which is then projected using a nonlinear extrapolation scheme [8] to compute the net damage over the duration of the life stage. Note that, because the stress redistribution effect due to damage is not considered in this analysis, and the yearly temperature variation is assumed to be cyclic, the pavement response is also cyclic within the same period, and the stress analysis only needs to be performed for a single year. Thus, fatigue cracking analysis of the entire pavement life can be performed efficiently using the responses obtained for a single year.

5 LVECD Program

The proposed framework has been incorporated into a pavement performance analysis program referred to as the LVECD program (Figure 2). Some of the salient features of the LVECD program include:

1. Both pavement performance analysis and pavement response analysis can be carried out through the same user interface.
2. The program supports various material types, including elastic, viscoelastic, and transversely anisotropic materials.

3. Traffic loading is considered as a repeated application of a design vehicle (multiple axles and multiple wheels) in the current version of the software. The LVECD program supports hourly and monthly variations of truck traffic, and also the annual growth of the annual average daily truck traffic (AADTT), i.e., linear and compound growth.
4. Analysis results, i.e., the pavement responses and damage, are provided in the form of tables, plots and contours in the user interface of the program.
5. Most of the analysis parameters are determined automatically based on climate conditions and loading conditions.
6. The LVECD software has been designed to use the temperature data provided by the EICM software for performing thermal analysis.
7. Numerical algorithms in the LVECD program are optimized for pavement performance analysis.

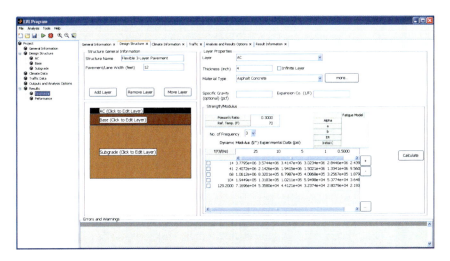

Fig. 2. LVECD program (screenshot)

6 Numerical Example

In this section, two numerical examples are used to illustrate the proposed analysis framework. The first example verifies the LVECD program by comparing it to the 3D-Move Analysis program [9]. The other example is a simulation of pavement performance under realistic conditions that demonstrates the efficiency and the reasonableness of the proposed framework.

6.1 Example 1: Verification

For the verification process, a three-layer system is considered: 10.16 cm (4 in.) for the asphalt concrete layer, 20.32 cm (8 in.) for the base, and 460 cm (15 ft) for the subgrade. The material in the asphalt concrete layer is styrene-butadiene-styrene (SBS) at 25°C (77°F). The base and subgrade layers are linearly elastic with modulus values of 276 MPa (40 ksi) and 49.8 MPa (12 ksi), respectively, and Poisson's ratios of 0.35 and 0.4, respectively. The system is subjected to a circular load with a total load of 40 KN (9,000 lb) and constant pressure (P) of 758 kPa (110 psi) on the surface (at z = 0), moving with a constant velocity (V) of 26.82 m/s (60 mph). The center of the load is assumed to be at y = 0 at t = 0. The results from the LVECD program are compared with those obtained from 3D-Move Analysis ([9]). From Figure 3, it is concluded that the new formulation is accurate for viscoelastic systems.

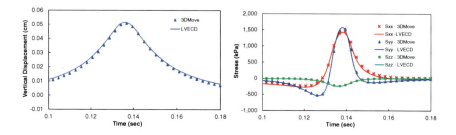

Fig. 3. Left: surface displacement history at origin. Right: stress history at the bottom of the AC layer.

6.2 Example 2: Pavement Performance

The pavement structure consists of a three-layer system: 10.16 cm (4 in.) for the asphalt concrete layer, 20.32 cm (8 in.) for the base, and 460 cm (15 ft) for the subgrade. The material in the asphalt concrete layer is SBS. The base and subgrade layers are linearly elastic with modulus values of 276 MPa (40 ksi) and 83 MPa (20 ksi), respectively, and Poisson's ratios of 0.35 and 0.4, respectively. The system is subjected to an equivalent single axle load (ESAL) (i.e., single axle and single tire) with a total axle load of 80 kN (18,000 lb), moving with a constant velocity of 26.82 m/s (60 mph). The contact pressure distribution is haversine with a diameter of 27.18 cm (10.7 in.)

For the purpose of these simulations, one month is selected as the duration of a life stage, and three analysis segments are assumed for each life stage: one for 5 AM – 2 PM, one for 2 – 9 PM, and one for 9 PM – 5 AM. In addition, the Florida temperature profile, which is computed by a weighted averaging method, is assumed for the simulations. The traffic loading consists of ten million ESALs equally distributed over the three simulation segments for a 20-year period. The total time needed to perform thermal and traffic stress simulations and to extrapolate the normalized stiffness ($C(S)$) is about 30 minutes. The distribution of C, which represents the fatigue cracking distribution, is shown in Figure 4 for the different life stages.

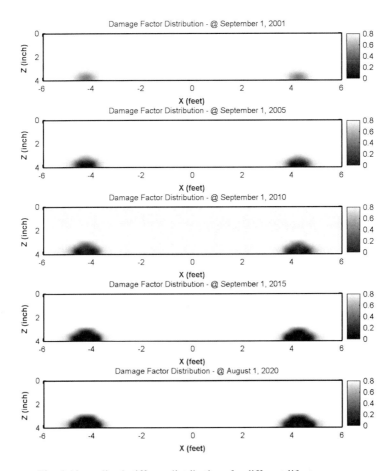

Fig. 4. Normalized stiffness distributions for different life stages

7 Conclusions

An accelerated analysis framework is proposed in this paper for pavement performance analysis. The procedure is based on several observations specific to pavement performance modeling. An efficient algorithm is proposed for thermal and traffic stress analyses by combining simplified assumptions and the Fourier transform. The framework has been implemented in LVECD software to perform fatigue cracking analysis of pavement structures under millions of cycles of traffic loading. Because of the fully optimized numerical algorithm included in the LVECD program, 20-year fatigue cracking simulations can be carried out in 30 minutes, which is a reasonable analysis time to determine pavement performance.

References

[1] Zaghloul, S., White, T.: Transp. Res. Rec. 1388, 60 (1993)
[2] Yoo, P.J., Al-Qadi, I.L., Elseifi, M.A., Janajreh, I.: Int. J. Pavement Eng. 7(1), 73 (2006)
[3] Kim, Y.R., Guddati, M.N., Underwood, B.S., Yun, T.Y., Subramanian, V., Savadatti, S.: Development of a Multiaxial VEPCD-FEP++. Department of Civil Engineering, North Carolina State University, Raleigh, NC (2008)
[4] Huang, Y.H.: Pavement Analysis and Design, 2nd edn. Prentice-Hall, Englewood Cliffs (2003)
[5] Siddharthan, R.V., Krishnamenon, N., Sebaaly, P.E.: Transp. Res. Rec. 1709, 43 (2000)
[6] Eslaminia, M., Thirunavukkarasu, S., Guddati, M.N.: Layered Viscoelastic Continuum Damage Program. Department of Civil Engineering, North Carolina State University, Raleigh, NC (2011)
[7] Eslaminia, M., Guddati, M.N.: Int. J. Pavement Eng. (2011) (accepted)
[8] Baek, C.: Top-Down Cracking Mechanisms Using the Viscoelastic Continuum Damage Finite Element Program, Ph.D. Dissertation, Department of Civil Engineering, North Carolina State University, Raleigh, NC (2010)
[9] 3D-Move Analysis Program - Version 1.2, University of Nevada, Reno, Nevada (2010)

Numerical Investigations on the Deformation Behavior of Concrete Pavements

Viktória Maláries[1] and Harald S. Müller[2]

[1] Technische Universität Darmstadt, Institute of Concrete Structures, Department of Construction Materials
[2] Karlsruhe Institute of Technology, Institute of Concrete Structures and Building Materials, Department of Building Materials

Abstract. In recent years, numerous cracks were observed in concrete pavements of federal highways (BAB) across Germany. These could be identified as the result of an alkali silica reaction (ASR). For this reason the main focus of this research project was to get a realistic view about the damaging processes in concrete pavements caused by an ASR in combination with thermal, hygric and mechanical loads by means of numerical investigations. In addition, the analyses should provide insights on how to better counteract the damage potential of ASR in the future. This paper will present selected results of comprehensive numerical investigations on the deformation behavior of concrete pavements exposed to various loads and possible failures resulting from these.

1 Introduction

Since the beginning of the German motorway system, concrete pavements are extensively used in its construction. Today, about 30 percent of the motorway network, which is the longest in Europe with a total length of 12,000 km, is made of concrete pavements. Many years of experience should make it possible to realize long-life concrete traffic areas. However, surface defects such as slab cracking often appear shortly after manufacture. These cracks not only reduce the driving comfort but also substantially affect traffic safety. Furthermore, major rehabilitation and reconstruction of pavements are expensive to accomplish. To prevent future damages, exposures of concrete slabs and the cause of incurred defects should be analyzed. Based on the results, improved concepts of road design and construction can be developed.

This paper will present the results of comprehensive numerical investigations on the deformation behavior of concrete pavements exposed to various loads and possible failures resulting from these. Concrete pavements are generally exposed to loads caused by the weather, i.e. loads due to temperature and moisture as well as loads due to traffic. In recent years, numerous cracks were observed in concrete pavements all over Germany. These could be identified as the result of an alkali silica reaction (ASR). For this reason also loads caused by strains due to an ASR were analyzed numerically.

2 Methods and Requirements

The two dimensional numerical investigations were carried out using the FE-program DIANA [3]; in total a time period of 10 years was considered. The single loads from moisture and temperature conditions, ASR as well as traffic were analyzed separately. Afterwards, the stresses resulting from these simulations were superimposed. Consequently, the influences of the single exposures as well as their combination with other load cases on the cracking could be studied.

The numerical model was set up to reflect the construction of those pavements, where crack damages occurred due to ASR. Therefore, concrete pavement systems with and without bond design according to RStO[1] 01, tab. 2, row 1.1 (see Figure 1), respectively, were modeled under different boundary conditions and using realistic nonlinear constitutive laws. In the numerical calculations for pavement designs with geotextile interface layer between the concrete slab and the cement-treated base (CTB), no bond (i.e. no bond interaction) was assumed.

The FE-mesh fineness of the numerical model takes account of the expected cracking as well as the different hygric and thermal gradients in the pavement. The material properties of the pavement concrete were assigned to isoparametrical 8 node rectangular elements. To simulate the substructure 6 node interface elements were used (see Figure 1, right).

For the simulations of the pavement a normal strength C1 (f_{cm} = 45 MPa) and a high strength C2 (f_{cm} = 70 MPa) concrete were considered (see Table 1). Additional properties, such as the modulus of elasticity or the tensile strength, were determined based on current standards (table 9 in DIN 1045-1:2001 [4] and CEB-FIP Model Code 1990 [2]). The hygrothermal properties of the pavement concrete were set according to DIN EN 12524:2000 [5].

To describe the viscoelastic deformation behavior of concrete in the numerical investigations, the approach according to CEB-FIP Model Code 1990 [2] was used. Cracking was considered by the cohesive crack model (crack band model) given in Bažant und Oh [1]. The heterogeneity of concrete was modeled according to Mechtcherine [8] by varying the tensile strength assigned to the finite elements.

The climatic boundary conditions (air temperature, humidity, solar radiation etc.) were set according to the assumptions given in Foos [7] and the results of Müller and Guse [9]. With respect to the zero-stress temperature gradient a positive distribution with a pavement temperature of approx. 30 °C was assumed for the summer and a negative vertical profile with a pavement temperature of approx. 15 °C was considered for wintertime. In both simulations (temperature and moisture) a normal seasonal cycle with a hot summer and a cold winter was assumed. The traffic load was set quasi-dynamically according to DIN-Fachbericht 101 [6].

---

[1] Guidelines for Standardization of Pavement Structures for Traffic Areas in Germany.

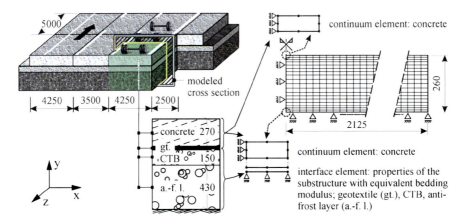

Fig. 1. Schema of a carriageway (BAB) with three lanes and the modeled cross section (left). Pavement design according to RStO 01: tab. 2, row 1.1 (middle) and FE-mesh used (right), dimensions in [mm].

Table 1. Concrete properties (mean values) of the pavements

Characteristic	Concrete C1	Concrete C2
compressive strength f_{cm} [MPa]	45	70
uniaxial tensile strength f_{ct} [MPa]	3.8	5.1
modulus of elasticity E_c [MPa]	36,000	41,000
fracture energy G_F [N/m]	86	117
Poisson's ratio v [-]	0.2	0.2

The consequences of a load resulting from ASR were considered by assuming strain rates on the upper side of the pavement of 0.2 mm/m/year for the first 5 years after production. In the period from 6 to 10 years, strain rates were set to be 0.1 mm/m/year. On the bottom side of the pavement, strain rates were set to be 0.1 mm/m/year in the first period and 0.05 mm/m/year in the second period. The strain distribution was set up to vary linearly over the cross section according to the research program supervisors.

3 Investigation Results

This paper only presents selected exemplary results, which represent the worst case for each exposure with regard to damage. Further results as well as a comprehensive discussion can be found in [10].

3.1 Moisture Induced Stresses

The drying of the upper side of the pavement due to alternating hygric loading is accompanied by the dishing of the pavement. The simulations of pavement design without bond interaction showed a deformation of about 0.5 mm (see Figure 2, above left), while those of the pavement design with bond interaction were about 0.05 mm (see Figure 2, below left). The deformations increased with higher concrete strength. Because of creeping effects, the deformations decreased after a long term period (see Figure 2, left exposure time of 5 years).

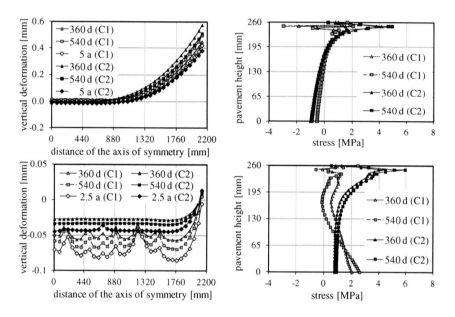

Fig. 2. Influence of concrete strength on the calculated vertical deformations at the upper side (left) and calculated stress distribution (σx) in the axis of symmetry (right) due to hygric load; assumption: design without bond, manufacturing in summer (above), design with bond, manufacturing in winter (below).

This generates tensile stresses at the top side of the pavement, which lead to cracking, if they exceed the tensile strength (see Figure 2, right). Compressive stresses could be detected at the bottom side of the pavement without bond interaction, while tensile stresses were detected for the pavement design with bond interaction. The compressive stresses result from the drying of the upper side of the pavement and are therefore more pronounced in combination with a capillary absorption of water at the bottom side. The tensile stresses are evoked due to the constraint on deformation by the cement-treated base and due to the formation of cracks at the top side of the pavement (see Figure 2, below left).

In general it is possible to reduce the risk of damage due to hygric loading by manufacturing the pavement in winter as this creates a negative vertical profile of the zero-stress temperature.

3.2 Temperature Induced Stresses

Within the thermal investigations only the pavement design without bond could be analyzed due to numerical problems that lead to an early abortion of the analyses for the pavement design with bond [10].

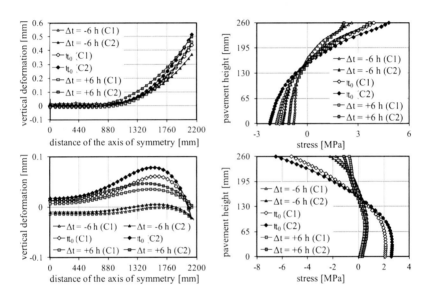

Fig. 3. Influence of concrete strength on the calculated vertical deformations at the upper side (left) and calculated stress distribution (σ_x) in the axis of symmetry (right) due to thermal load; assumption: pavement design without bond interaction; manufacturing in summer, summer day at t_0 = 3:00 am (above); manufacturing in winter, summer day at t_0 = 3:00 pm (below).

Even though a direct formation of cracks could not be detected within the scope of the thermal investigations, an increased risk of damage commonly exists due to the high tensile stresses at the top side of the pavement. Its maximum was reached on a summer day at 3:00 am by night (see Figure 3, above right). The temperature changes lead to a pavement dishing of about 0.5 mm (see Figure 3, above left). However, a manufacturing at wintertime would reduce the risk of damage due to a negative gradient of the zero-stress temperature.

Furthermore a modest risk of damage could be observed at the bottom side of the pavements made of concrete with normal strength. The maximum tensile stresses could be detected at 3:00 pm on a summer day in a pavement which was manufactured in wintertime (see Figure 3, below right). As a result of the maximum temperature, which is attained at this time of day, the pavement edge bulges upwards (see Figure 3, below left). A pavement made of concrete with high strength seems to bear less risk of cracking damages due to thermal loading.

3.3 Traffic Induced Stresses

The traffic load results in compressive stresses at the top side of the pavement and tensile stresses at its bottom side. The usage of a geotextile as an interface layer between the ground plate and the CTB (for a pavement design without bond) may cause that the transverse and longitudinal contraction joints do not break entirely. This phenomenon then provokes different deformation behavior and causes a higher stress level in the concrete slab. In this case tensile stresses with an amount of 1.4 MPa appear at the bottom side of the pavement. Therefore in combination with the stresses caused by the ASR or the stresses due to hygrothermal impact a higher risk of cracking arises.

3.4 Alkali Silica Induced Stresses

According to the specifications (see section 2) of the performed analysis the top side of the pavement expands more than its bottom side. This impact results in compressive stresses at the top side of the pavement for both pavement designs. The design with bond is subject to compressive stresses on the bottom side as well, while the pavement design without bond experiences tensile stresses (see Figures 4, 5 and 6). On the basis of the numerical results, pavements with increasing compressive strength (f_{cm} = 70 MPa) show both higher deformations and stresses. This can be traced back to the higher stiffness (modulus of elasticity) and the less creep of the high strength concrete.

Figure 4 illustrates the vertical deformations at the upper side of the pavement for the design case with bond interaction. Because of the bond between the slab and the CTB the elongation at the upper side of the pavement is restrained. With increasing exposure, the bond strength at the slab edge becomes exceeded. Due to this, the slab gets disconnected from the substructure, leading to local dishing (see Figure 4, left). The simplified consideration on the macroscale led to a linear compressive stress distribution in the pavement for a design with bond (see Figure 4, right). The superposition of these stresses with stresses of other origins can cause a "positive" effect on the total stresses. However, with respect to these results it should be noted, that changes on the mesoscale were not taken into account.

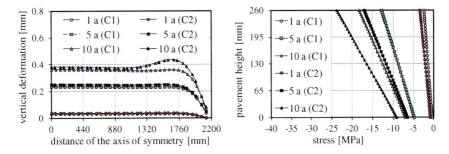

Fig. 4. Comparison of the vertical deformations at the pavement upper side (left) and stress distribution (σ_x) in pavement with f_{cm} = 45 MPa (C1) and f_{cm} = 70 MPa (C2) (right) caused by strains due to ASR; assumption: design with bond interaction, exposure duration 1, 5 and 10 years.

For the pavement design without bond interaction, the pavement center bulges upwards as a reaction to ASR (see Figures 5 and 6, left). Independent of the concrete strength, but at different points of time, maximum tensile stresses of 2.4 MPa appear at the bottom side of the pavement (see Figures 5 and 6, right). These stresses do not present a risk of damage. However, this statement is only valid as long as it is assumed that the ASR does not decrease the tensile strength.

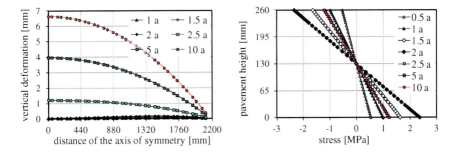

Fig. 5. Influence of exposure time on the vertical deformations at the pavement upper side (left) and on the stress distribution (σ_x) in pavement with f_{cm} = 45 MPa (C1) caused by strains due to ASR; assumption: pavement design without bond interaction.

In general two contrary aspects have to be considered regarding the evaluation of the effects of an ASR in a loaded cross section or range of a cross section. The strain augmentation due to the ASR may indeed reduce the existing tensile stresses. It has to be mentioned, however, that this will only have positive effects, lowering the risk of damage, if the tensile strength is not reduced even more at the same time as a result of the inner deterioration due to the ASR. However, this mechanism was not considered here.

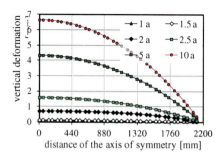 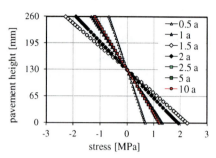

Fig. 6. Influence of exposure time on the vertical deformations at the pavement upper side (left) and on the stress distribution (σ_x) in pavement with f_{cm} = 70 MPa (C2) (below, right) caused by strains due to ASR; assumption: pavement design without bond interaction.

3.5 Superposition of the Induced Stresses

The stresses due to the single loads from moisture and temperature conditions, traffic and the ASR have to be superimposed, to enable a statement about the damage potential due to their simultaneous exposure.

The superposition of all loads led to a moderate risk of damage ($0.5 \cdot f_{ct} < \sigma_x < 0.8 \cdot f_{ct}$) for slabs which were manufactured in summer and in a pavement design without bond interaction using normal strength concrete as a consequence of the calculated tensile stresses of 2.5 MPa (see Figure 7, left). However, this only applies, if the ASR does not decrease the tensile strength of concrete, or does so insignificantly. Without the impact of ASR – which causes compressive stresses at the pavement upper side – the risk of cracking would increase, since the prevailing tensile stress of 4.2 MPa exceeds the tensile strength of 3.8 MPa.

Even though the ASR seems to have a mitigating effect on the stresses near the surface of the pavement, the load collective still constitutes a risk of damage for high strength concrete pavements (see Figure 7, right, curve "sum"). As the current model does not take a reduction of concrete tensile strength due to ASR into account, the actual situation should be even more pronounced.

If the pavement is affected by ASR-deformation in addition to the load cases described before, the risk of damage is reduced at the top side of the pavement with decreasing concrete strength as well as with a manufacturing in the winter. At the same time, the ASR-deformation may lead to the formation of cracks for the pavement design without bond due to high tensile stress at the bottom side of the pavement. A detailed analysis under the influence of temperature was not possible for the pavement design with bond. Nevertheless, theoretically, for this pavement design the strains due to ASR should reduce the risk of damage.

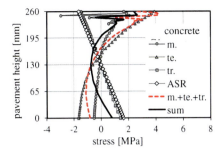 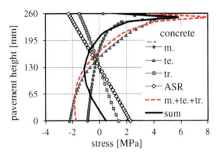

Fig. 7. Stress parts of the single exposures caused by moisture (m.), temperature (te.), traffic (tr.) and strains due to ASR (ASR) and their superposition (sum = m. + te. + tr. + ASR; m. + te. + tr.) in the axis of symmetry of the pavement; assumptions: pavement design without bond interaction, manufacturing in summer, f_{cm} = 45 MPa (C1) (left) und f_{cm} = 70 MPa (C2) (right).

4 Conclusions

First, impacts of the exposures caused by moisture, temperature, traffic and strains due to ASR on the concrete pavement were separately analyzed numerically. Afterwards, the stress results were superimposed linearly to show the critical load combinations. This could be considered a restriction of the presented research as the interaction by coupling the single effects could not be resolved. However, the loss of information resulting from this is still considered comparatively minor.

The situation is different in cases, in which the ASR seems to have a positive effect, as accompanied strains can reduce the tensile stresses resulting from other loads. In these cases, the ASR decreases the risk of cracking only, if the reduction of the effective tensile load is higher than the decrease in tensile strength, which is caused by the inner damage due to ASR. On what magnitude the ASR reduces the concrete strength and stiffness are an open research issue.

Acknowledgement. This paper is based on the experimental and numerical investigations of several research projects, which had been carried out at the Institute of Concrete Structures and Building Materials of the University of Karlsruhe under the financial support of research programs of the Federal Ministry of Transport, Building and Urban Affairs (BMVBS) ([9] and [10] among others). This financial support is gratefully acknowledged.

Literature

[1] Bažant, Z.P., Oh, B.H.: Crack band theory for fracture of concrete. Materials and Structures 16(93), 155–177 (1983)
[2] CEB: –Comite Euro-International du Beton: CEB-FIP Model Code, Bulletin D'Information, No. 213/214. Lausanne (1993)
[3] DIANA, Finite Element Analysis: User's Manuals release 9.2, TNO Building and Construction Research, Delft (2007)

[4] DIN 1045-1: Concrete, reinforced and prestressed concrete structures – Part 1: Design and construction. Beuth publishing (2001)
[5] DIN EN 12524: Building materials and products – Hygrothermal properties – Tabulated design values. Beuth publishing (2000)
[6] DIN-Fachbericht 101: Actions on bridges. Beuth publishing (2003)
[7] Foos, S.: Unbewehrte Betonfahrbahnplatten unter witterungsbeding-ten Beanspruchungen. Universität Karlsruhe (TH), Institut für Massivbau und Baustofftechnologie, Diss (2006)
[8] Mechtcherine, V.: Bruchmechanische und fraktologische Untersu-chungen zur Rissausbreitung in Beton. Universität Karlsruhe (TH), Institut für Massivbau und Baustofftechnologie, Diss (2000)
[9] Müller, H.S., Guse, U.: Untersuchungen zur Beanspruchung und Dauerhaftigkeit von Betonfahrbahnen. Abschlussbericht zum Forschungs-vorhaben der BASt 08.156/1999/ LRB, Institut für Massivbau und Baustofftechnologie der Universität Karlsruhe, TH (2005)
[10] Müller, H.S., Maláriçs, V., Soddemann, N., Guse, U.: Rechnerische Untersuchung zur Entstehung breiter Risse in Fahrbahndecken aus Beton unter Mitwirkung einer Alkali-Kieselsäure-Reaktion. Abschlussbericht zum Forschungsvorhaben der BASt 08.189/2006/ LRB, Institut für Massivbau und Baustofftechnologie der Universität Karlsruhe (TH)a (2010)

Fatigue Behaviour Modelling in the Mechanistic Empirical Pavement Design

Mofreh F. Saleh

University of Canterbury
Department of Civil and Natural Resources
Christchurch, New Zealand

Abstract. The Mechanistic empirical pavement design is based on modelling certain modes of failure for the different pavement materials. In the Australian and New Zealand guidelines, the mechanistic pavement design is based on modelling fatigue and permanent deformation as the two major modes of failures. The guidelines use the Shell fatigue performance function to model fatigue behaviour of asphalt mixes. However, there are wide range of asphalt mixes on New Zealand roads and they are all behave differently regarding their fatigue and permanent deformation performance. Therefore, the question here, can one fatigue model accurately fit the fatigue performance of all different asphalt mixes. This research examined the fatigue behaviour of two different types of dense graded hot mix asphalts, the first is made of aggregate with maximum nominal size 10 mm and the second with 14 mm maximum nominal sizes. The effects of air voids in the total mix and aggregate gradations on the fatigue behaviour were compared. It was found that the aggregate gradations have significant effect on the fatigue life with finer mixes have significantly higher fatigue lives compared to the coarser gradations. In addition, air voids in the total mix have a profound effect on the fatigue behaviour.

1 Background

An accurate modelling of the fatigue behaviour of asphalt mixes is very significant for a robust mechanistic empirical pavement design. In most mechanistic empirical design methods, fatigue of asphalt mixes is considered through a single fatigue model that is developed in the laboratory and calibrated in the field based on the actual field performance data. However, the fatigue behaviour of asphalt mixes depends on several factors such as aggregate gradation, maximum nominal size, binder type and content, air voids content and pavement temperature, in addition, to the traffic loading and traffic wander in the field [1]. There are various types of mixes with a wide range of maximum nominal size, binder content and gradations on today's roads; therefore, the fatigue performance of these mixes is expected to widely vary from one mix to the other. Thus, it is unlikely to model the fatigue performance of these widely different mixes by one single fatigue model that can address all various factors.

In the current Austroads Mechanistic Empirical pavement design, two modes of failure are considered, namely fatigue of bound materials and permanent deformation based on the subgrade compressive strain criterion [2]. The Austroads design guidelines and New Zealand supplement adopted the Shell fatigue performance function to calculate the fatigue damage of the structural asphalt. The Shell fatigue model was developed in 1978. The fatigue relationship was developed in the laboratory on a range of seven different types of asphalt mixes namely, dense asphaltic concrete, dense bitumen macadam, gravel-sand asphalt, lean bitumen macadam, rolled asphalt base course mix, French "Grave Bitume", and Bitumen-sand base course [3]. The fatigue testing was done using controlled strain (displacement) with a sinusoidal loading [1]. Different types of bitumen grades, 40/50, 80/100, 45/60, and 40/60 were used in the Shell fatigue performance developments (Shell Manual, 1978). These mixes and binder types and sources are different from those are currently used on the New Zealand highway system. Consequently, it is expected to find significant differences between the predicted and observed fatigue life when using the Shell fatigue model [4,5].

In this research, two different types of mixes that are widely used in New Zealand will be tested for fatigue behavior using four point bending fatigue. The first type is the AC10 hot mix asphalt which is made with dense graded aggregate of 10 mm maximum nominal size and a bitumen grade 80/100 penetration grade. The second type of mix is the AC14 hot mix asphalt which is made with dense graded aggregate of 14 mm maximum nominal size and bitumen of 60/70 penetration grade. The two mixes were designed by local contractors and the optimum binder content was determined. The fatigue beam specimens were designed in the University of Canterbury laboratory for a wide range of air voids from as low as 2.83% to over 13.6% for the AC 10 mixes and a range from 3.96% to 6.6% for the AC14 mixes. With this wide range of air voids, the effect of volumetrics on the fatigue behavior was examined. The fatigue behavior of the two mixes was compared. The Shell fatigue model predictions were also compared with the laboratory measured fatigue values to examine the suitability and the accuracy of the model prediction.

2 Material Properties and Sample Preparations

Two hot mix asphalts were designed by two local contractors and the optimum binder content was determined for each mix. The first mix is AC 10 hot mix asphalt which has a maximum nominal aggregate size of 10 mm and dense gradation while the second mix is AC14, which has a maximum nominal size 14 mm, and dense gradation as shown in Figure 1. Specific gravities for aggregates and maximum theoretical specific gravities for loose mixes were measured and the results are shown in table 1. In order to study the effect of the percentage of air voids on the fatigue behaviour a range of air voids content was considered. The AC10 mix was prepared in slabs with dimensions of 305x400x75 mm. The quantities of the mix was calculated using the specific gravities and percentages of the aggregates and binder to produce a range of air voids from 2.83% to about 13.6% for the AC10 mixes. The roller compactor shown in Figure 2 was used to compact the mix quantities to the required slab dimensions. The same methods was used with AC14 mixes with a range of air voids from 3.96 to 6.6%. The

Table 1. Specific gravities and optimum binder content

Property	AC10	AC14
Bulk Specific gravity of aggregates	2.62	2.83
Maximum Theoretical Specific Gravity, Gmm	2.407	2.652
Total Optimum Binder Content	6.2	5.2
Effective Optimum Binder Content	5.98	4.19

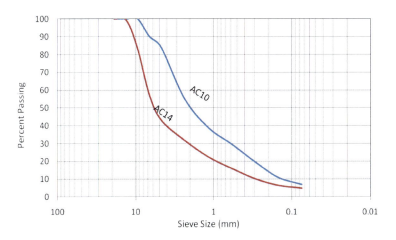

Fig. 1. Aggregate gradations for AC10 and AC14 hot mix asphalts

Fig. 2. University of Canterbury roller compactor used to preapre AC10 and AC14 fatigue beams

range of the percentages of air voids used for AC14 is much narrower than AC10 and the author is planning to address this problem with some more testing in the future using higher air voids. Twenty four beams were prepared from the AC10 mixes and 13 beams of AC 14 mix.

2.1 Fatigue Testing

The 37 beams of both AC10 and AC14 were subjected to flexural stiffness modulus testing at 20 °C using the four bending beam apparatus shown in Figure 3. The flexural stiffness modulus was measured using haversin load pulses at 10 Hz frequency and the modulus is measured after 50 cycles. The fatigue test were carried out on asphalt specimens using constant strain mode and the failure is defined as the number of cycles at which the flexural modulus will reduce to 50% of its initial value. A range of 300 to 600 με of constant strain amplitudes was used in the testing. At least two replicates were tested at each strain level.

Fig. 3. University of Canterbury bending beam fatigue apparatus

2.2 Results and Analysis

The fatigue results of the AC10 and AC14 asphalt beams were plotted and compared with the Shell fatigue model predictions. Shell fatigue model that is currently used in the Austroads mechanistic empirical design is shown in Equation 1.

$$N_f = \left[\frac{6918*(0.856*V_b + 1.08)}{S_{mix}^{0.36} * \mu\varepsilon_i} \right]^5 \tag{1}$$

N_f = Number of loading cycle to failure
V_b = percentage by volume of bitumen in the asphalt mix;
S_{mix} = asphalt mix flexural stiffness modulus in MPa; and
$\mu\varepsilon$ = Tensile strain level in microstrain units.

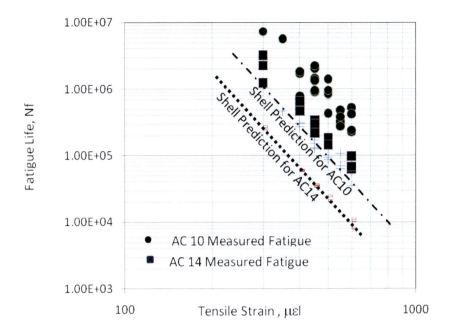

Fig. 4. Measured fatigue values for AC10 and AC14 vesus the Shell predicted values

Figure 4 shows the measured fatigue values of the AC10 and AC14 asphalt mixes and the Shell predicted fatigue values. It is clear from Figure 1 that the Shell fatigue model is significantly underestimating the laboratory measured fatigue lives for the both types of mixes with at least one order of magnitude. It is well known that the laboratory measured fatigue values are significantly lower than the actual field fatigue due to several reasons such as traffic wander, time for crack healing is much longer, type of support for the asphalt (Saleh, 2010). Therefore, it is beyond doubt that the Shell fatigue model is exceedingly underestimating fatigue life and therefore overestimating the design thickness of asphalt mixes. It is also obvious that the two asphalt mixes behave differently and the Shell model is unable to model the two different types of mixes and any calibration or adjustment to the model will make it at best predicting one type or another.

In addition, it is obviously clear that the hot mix asphalt maximum nominal size is playing a very significant role in the fatigue behaviour of the hot mix asphalts. The AC 10 hot mix asphalt showed a remarkably higher fatigue resistance compared to AC14. This indicates that finer mixes are more fatigue resistant compared to coarser mixes. The simple explanation for this behaviour is that with fine gradation the crack path is much longer and more intricate compared to the coarser gradation. This means the crack growth time will be much longer and therefore the fatigue life is much longer for finer mixes compared to coarser mixes.

2.3 Fatigue Modelling

There are several forms of the fatigue models. The most common two forms are the strain model and the stain and stiffness modulus model as shown in equation 2 and 3. The strain model shown by Equation 1 does not address the effect of mix properties such as the percentage of air voids or binder content. Although the model shown in Equation 3 addresses one of the important mix properties which is the stiffness modulus yet the model is mathematically incorrect because both strain and stiffness are dependent variables. .

$$N_f = a\varepsilon_t^{-b} \tag{2}$$

$$N_f = a\varepsilon_t^{-b} E^{-c} \tag{3}$$

Where,
- N_f = Number of cycles to fatigue failure
- ε_t = Initial tensile strain
- a,b, and c = material constants. They are based on the mix type, volumetric properties and binder type, and test conditions.

In this research the author is proposing a different form of the fatigue model that is addressing both the strain level and volumetric properties. Equation 4 shows the suggested model. In this model, the number of fatigue cycles to failure is a function of the percentage by volume of the effective binder content and the percentage of air voids content in the total mix and the tensile strain amplitude. The author believes that this model will better account for the volumetric properties of the mix which are related to the mix stiffness and the fatigue propeties. The proposed model relates the number of fatigue cycles to failure with three independent parameters, V_{be}, V_a, and strain amplitude.

$$N_f = a * \left(\frac{V_{be}}{V_a}\right)^b * \varepsilon^c \tag{4}$$

The developed fatigue model for the AC10 mix is shown in Equation 5.

$$N_f = 5.13 * 10^{18} * \left(\frac{V_{be}}{V_a}\right)^{0.467} * \varepsilon^{-4.83} \qquad R^2=0.93 \tag{5}$$

The developed fatigue model for the AC 14 mix is shown in Equation 6

$$N_f = 1.18 * 10^{19} * \left(\frac{V_{be}}{V_a}\right)^{-1.734} * \varepsilon^{-4.954} \qquad R^2=0.973 \tag{6}$$

Figures 5 and 6 show the goodness of fit of Equations 5 and 6, respectively. Both models reasonably provide a good match between the measured and predicted fatigue lives without bias.

Fatigue Behaviour Modelling in the Mechanistic Empirical Pavement Design 523

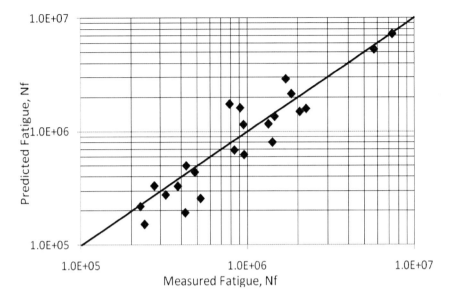

Fig. 5. The relationship between the predicted and measured fatigue life for AC10 hot mix asphalt

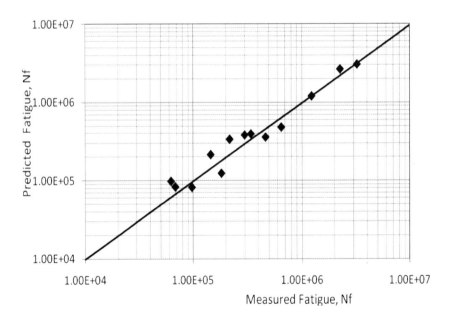

Fig. 6. The relationship between the predicted and measured fatigue life for AC14 hot mix asphalt

Decipite the high correlation of the model shown in Equation 6, the negative exponent for the ratio of (V_{be}/V_a) does not seem correct and this perhaps is a result of the narrow range of the percentage of air voids selected for the AC14. Therefore, this model is not analysed any further in this paper and the author is considering testing more samples at wider percentages of air voids to have a better understanding of the behaviour of this mix.

2.3 Effect of Air Voids on Fatigue Behaviour

In order to study the effect of mix volumetrics on the fatigue behaviour, Equation 6 can be used as shown in Figure 7. For the same strain level and binder content, increasing the air void content decreases the fatigue life. This behaviour can be attributed to the greater degree of discontinuity in the mix structure as the air voids content increases and therefore, it becomes easier for cracks to grow and propagate in the mix resulting in lower fatigue resistance.

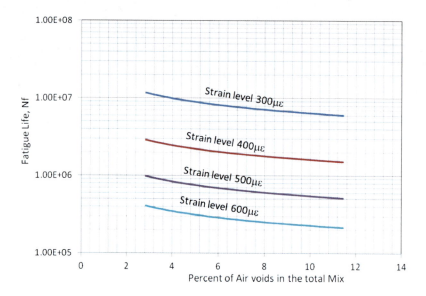

Fig. 7. Effect of the percentage of air voids on AC10 HMA fatigue Life

3 Conclusions

In this research two types of hot mix asphalts, AC10 and AC14 were used to investigate the fatigue behaviour. Hot mix asphalts were compacted at a wide range of air voids ranging from 2.83% to about 13.6% for AC10 mixes, however, AC 14 were prepared at a narrower range from 3.96 to 6.6%. Twenty-four beams of AC10 and thirteen beams of AC14 mixes were tested in four point bending beam fatigue using constant strain level ranges from 300 to 600με. The measured fatigue lives were compared with the predicted fatigue lives using the Shell fatigue model. It was clear that the Shell fatigue model is consistently underestimating the fatigue lives for both the AC10 and AC14 mixes. The fatigue behaviour of the AC 10 fine mix was significantly higher than the AC14 with at least one order of magnitude. The reason for this is the crack path is much longer and more intricate compared to the coarse graded mixes, therefore, the fatigue resistance and fatigue life for fine graded mixes are much longer. Two fatigue models were developed for the AC10 and AC14 to address the effect of volumetrics and strain amplitude on the fatigue life. The developed fatigue models provided very good match between the measured and predicted fatigue lives with a coefficient of determination R^2 of at least 0.93. The effect of air voids content for AC10 was quite notable as the higher the air voids content in the total mix, the lower the fatigue resistance.

Acknowledgement. The author would like to thank Kelly Hoara and Anthony Stubbs and John Kooloos from the Department of Civil and Natural Resources Engineering, University of Canterbury for their hard work in preparing and testing fatigue specimens. The author would like also to acknowledge the support received from Fulton Hogan contactor in particular Dr. Bryan Pidwerbesky and Martin Clay. The author is also very grateful for the help and support provided by Downer NZ contractor in particular Dr. David Hutchison and Janet Jackson.

References

1. Baburamani, P.: Asphalt Fatigue Life Prediciton Models – A Literature Review ARRB Transport Research Ltd., Vermont South, Victoria. Research Report ARR 334 (1999)
2. Austroads, Pavement Design, A Guide to Structural Design of Road Pavements (2008)
3. Shell Pavement Design Manual - Asphalt Pavements and Overlays for Road Traffic (1978)
4. Saleh, M.: Methodology for the Calibration and Validation of Shell Fatigue Performance Function Using Experimental Laboratory Data. Journal of the Road and Transport Research 19(4), 13–22 (2010)
5. Saleh, M.: Implications of Using Calibrated and Validated Performance Transfer Functions in the Mechanistic Empirical Pavement Design Procedure. International Journal of Pavement Research and Technology 4(2), 111–117 (2011)

Theoretical Analysis of Overlay Resisting Crack Propagation in Old Cement Concrete Pavement

Yang Zhong, Yuanyuan Gao, and Minglong Li

Faculty of Infrastructure Engineering, Dalian University of Technology,
Dalian 116024, Liaoning Province, PR China

Abstract. The main purpose of this study is to determine the effect of overlay on the crack propagation. In order to simplify the problem, a cement concrete pavement is modeled as an elastic plate on Winkler foundation. To drive the singular integral equations, the Fourier transform and dislocation density function are used. Lobatto—Chebyshev integration formula, as a numerical method, is used to solve the singular integral equations. The numerical solution of stress intensity factor at the crack tip is derived. In order to examine the effect of overlay for resisting crack propagation, numerical analyses are carried out on a cement concrete pavement with embedded crack and a concrete pavement with an asphalt overlay. Results show that the thickness and the shear modulus are two significant factors influencing the crack propagation.

1 Introduction

Cement concrete pavement is a typical road surface used in urban roads, highways and airports. Since more and more cars and trucks are traveling on the road each day, damages in the form of cracks could be induced in the pavement structure due to the increasing traffic load. A popular method to resist the crack propagation is to place an asphalt overlay on the original cement concrete pavement. Therefore, determination of the thickness and relevant parameters of the asphalt overlay becomes a important and urgent subject. In 1950s, the fracture mechanics, which is a branch of solid mechanics, was forwarded as a discipline studying the strength of materials and structures containing cracks [1]. The complex function and integral transform based on the fracture mechanics theory are effective methods to derive the analytical solution for describing the problem. For instance, Zak, William[2] studied the stress intensity factor of two infinite planes with infinite crack and the crack terminated on the interface by using the complex function method. F. Erdogan and G. Gupta have obtained the analytical solution of composite materials containing cracks in the interlayer by using integral transform [3]. Integral transform was also adopted by Sei UEDA and Tatsuya MUKAI [4] to solve the problem of multilayered composite with a crack perpendicular to the boundary and with the normal and symmetric uniform load distributed on the crack. However, complex function and integral transform methods can not work well in all the conditions. For example, a perfect theoretical model for cement concrete pavement with crack perpendicular to the interface has not been built. At present, the pavement with a crack can be mostly analyzed by the finite element method [5, 6]. However, compared with the approximate solution from finite element model, the theoretical approach is better in the computational complexity and accuracy. In this study, models are developed for describing the crack propagation in cement concrete pavement with and without asphalt overlay. Stress intensity factor is taken into account as a parameter to judge whether the crack will extend or not. For achieving the analytical expression of the intensity factor, Fourier transform and dislocation density function are used to drive the singular integral equations in this study. Lobatto—Chebyshev integration formula is employed to solve the singular integral equations. The numerical results of stress intensity factor at the crack tips are calculated in order to examine the usefulness of overlay for resisting crack propagation. The results from the numerical analysis are discussed and the factors that affect the crack propagation are investigated.

2 Description of the Problem

The cement concrete pavement with a crack perpendicular to the interface is considered as a plate on an elastic foundation. Figure1 and Figure2 show the concrete pavement models without and with the overlay respectively. There is an embedded crack perpendicular to the boundaries of the plate as shown in the figures. In this study, only the plane strain is concerned, thus the effect of volume force is ignored. And the analytical solution for the model in Figure2, which is more complicated, is discussed in detail.

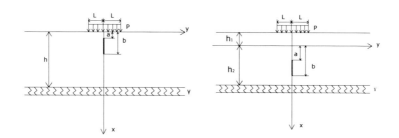

Figs. 1-2. Model of cement concrete pavement with crack

The boundary conditions of the second model (Figure 2):

$$\sigma_{xy1}(x,0) = 0, v_1(x,0) = 0, -h_1 < x < 0 \tag{1}$$

$$\sigma_{xy2}(x,0) = 0, 0 < x < h_2, v_2(x,0) = 0, 0 < x < a \text{ or } b < x < h_2 \tag{2}$$

$$\sigma_{xx1}(-h_1, y) = P, -L < y < L; \sigma_{xy1}(-h_1, y) = 0, -\infty < y < \infty \tag{3}$$

$$\sigma_{xx2}(h_2,y) = \gamma u_2(h_2,y), \sigma_{xy2}(h_2,y) = 0, -\infty < y < \infty \qquad (4)$$

$$u_1(0,y) = u_2(0,y), v_1(0,y) = v_2(0,y), -\infty < y < \infty \qquad (5)$$

$$\sigma_{xx1}(0,y) = \sigma_{xx2}(0,y), \sigma_{xy1}(0,y) = \sigma_{xy1}(0,y), -\infty < y < \infty \qquad (6)$$

$$\sigma_{yy2}(x,0) = 0, a < x < b;$$
$$\sigma_{xxn} \to 0, \sigma_{yyn} \to 0, \sigma_{xyn} \to 0, y \to \infty, n = 1,2 \qquad (7)$$

where $\sigma_{xxn}, \sigma_{yyn}$ are the x, y components of the stress vector in the n layer, respectively. σ_{xyn} is the shearing stress in the n layer. u_n and v_n are the x, y components of the displacement vector in the n layer, respectively. $n = 1,2$. γ is the stiffness of the foundation.

3 Derivation and Solution of Singular Integral Equations

The governing equations of plane elasticity are expressed as [7]:

$$(1+k)\frac{\partial^2 u}{\partial x^2} + (3-k)\frac{\partial^2 v}{\partial x \partial y} + (k-1)(\frac{\partial^2 u}{\partial y^2} + \frac{\partial^2 v}{\partial x \partial y}) = 0 \qquad (8)$$

$$(3-k)\frac{\partial^2 u}{\partial x \partial y} + (1+k)\frac{\partial^2 v}{\partial y^2} + (k-1)(\frac{\partial^2 v}{\partial x^2} + \frac{\partial^2 u}{\partial x \partial y}) = 0 \qquad (9)$$

where $k = (3-v)/(1+v)$ for plane stress, $k = 3-4v$ for plane strain and u, v are the x, y components of the displacement vector, respectively. v is Poisson's ratio. In this paper, $k = 3-4v$ is selected corresponding to the case of plane strain.

From Hooke's Law, the stress components can be expressed as:

$$\sigma_{xx} = \frac{G}{k-1}[(1+k)\frac{\partial u}{\partial x} + (3-k)\frac{\partial v}{\partial y}]; \sigma_{yy} = \frac{G}{k-1}[(3-k)\frac{\partial u}{\partial x} + (1+k)\frac{\partial v}{\partial y}]; \sigma_{xy} = G(\frac{\partial u}{\partial y} + \frac{\partial v}{\partial x}) \qquad (10)$$

where G is the shear modulus of the material.

In order to achieving the solutions of the displacement components, the displacement u_n, v_n in the n layer can be expressed with Fourier integral formulas[8]:

$$u_n(x,y) = \frac{2}{\pi}\int_0^\infty f_{n1}(x,\eta)\cos(\eta y)d\eta + \frac{1}{2\pi}\int_{-\infty}^\infty g_{n1}(\xi,y)e^{i\xi x}d\xi. \qquad (11)$$

$$v_n(x,y) = \frac{2}{\pi}\int_0^\infty f_{n2}(x,\eta)\sin(\eta y)d\eta + \frac{1}{2\pi}\int_{-\infty}^\infty g_{n2}(\xi,y)e^{i\xi x}d\xi. \quad (12)$$

Substituting Eqs. (11) and (12) into Eqs. (8) and (9), one can obtain the expressions of displacements u_n and v_n.

$$\begin{aligned}u_n(x,y) = &\frac{2}{\pi}\int_0^\infty [e^{\eta x}(A_{n1}+A_{n2}x)+e^{-\eta x}(A_{n3}+A_{n4}x)]\cos(\eta y)d\eta \\ &+\frac{1}{2\pi}\int_{-\infty}^\infty e^{i\xi x-|\xi|y}(B_{n3}+B_{n4}y)d\xi\end{aligned} \quad (13)$$

$$\begin{aligned}v_n(x,y) = &\frac{2}{\pi}\int_0^\infty \frac{1}{\eta}\{-e^{\eta x}[\eta A_{n1}+(\eta x+k)]A_{n2} \\ &+e^{-\eta x}[\eta A_{n3}+(\eta x-k)A_{n4}]\}\sin(\eta y)d\eta \\ &+\frac{1}{2\pi}\int_{-\infty}^\infty \frac{ie^{i\xi x-|\xi|y}}{\xi}[|\xi|B_{n3}+(k+|\xi|y)B_{n4}]d\xi\end{aligned} \quad (14)$$

where B_{n3} and B_{n4} are the functions of ξ and A_{n1}, A_{n2}, A_{n3} and A_{n4} are the functions of η.

Substituting Eqs. (11) and (12) into Eq. (10), the components of stress are given by:

$$\begin{aligned}\sigma_{xxn} = &\frac{2G_n}{\pi}\int_0^\infty \{e^{\eta x}[2\eta A_{n1}+(k-1+2\eta x)A_{n2}] \\ &+e^{-\eta x}[-2\eta A_{n3}+(k-1-2\eta x)A_{n4}]\}\cos(\eta y)d\eta \\ &+\frac{\mu}{2\pi}\int_{-\infty}^\infty \frac{ie^{i\xi x-|\xi|y}}{\xi}[2(k-1)\xi^2(B_{n3}+B_{n4}y) \\ &+|\xi|(k^2-4k+3)B_{n4}]d\xi\end{aligned} \quad (15)$$

$$\begin{aligned}\sigma_{yyn} = &\frac{2G_n}{\pi}\int_0^\infty \{e^{\eta x}[-2\eta A_{n1}-(k+3+2\eta x)A_{n2}] \\ &+e^{-\eta x}[2\eta A_{n3}-(k+3-2\eta x)A_{n4}]\}\cos(\eta y)d\eta \\ &+\frac{\mu}{2\pi}\int_{-\infty}^\infty \frac{ie^{i\xi x-|\xi|y}}{\xi}[2(k-1)\xi^2 B_{n3}+2(1-k)\xi^2 y B_{n4} \\ &+(1-k^2)|\xi|B_{n4}]d\xi\end{aligned} \quad (16)$$

$$\sigma_{xyn} = \frac{2G_n}{\pi}\int_0^\infty \{e^{\eta x}[-2\eta A_{n1}-(k+1+2\eta x)A_{n2}]$$
$$+e^{-\eta x}[-2\eta A_{n3}+(k+1-2\eta x)A_{n4}]\}\sin(\eta y)d\eta \qquad (17)$$
$$-\frac{\mu}{2\pi}\int_{-\infty}^{\infty} e^{i\xi x-|\xi|y}\left[(k-1)B_{n4}+2|\xi|(B_{n3}+B_{n4}y)\right]d\xi$$

For convenience, dislocation density function is introduced, which is defined as:

$$\phi(x) = \frac{\partial v(x,0)}{\partial x}; \phi(x)=0, 0<x<a \text{ or } b<x<h; \int_a^b \phi(x)dx = 0.$$

Based on Eqs. (1) and (2) and the theory of residue, B_{13}, B_{14}, B_{23} and B_{24} are expressed as:

$$B_{13}=0, B_{14}=0, B_{23}=\int_a^b \frac{(k-1)e^{i\xi x}}{(k+1)|\xi|}\phi(t)dt, B_{24}=\int_a^b \frac{-2e^{i\xi x}}{(k+1)}\phi(t)dt \qquad (18)$$

while the four homogeneous boundary conditions (3)-(6) are expressed in terms of $A_{n1}, A_{n2}, A_{n3}, A_{n4}$ ($n=1,2$) and $\phi(t)$:

$$e^{-\eta h_1}[2\eta A_{11}+(k-1-2\eta h_1)A_{12}]+e^{\eta h_1}[-2\eta A_{13}+(k-1+2\eta h_1)A_{14}] = P\sin(\eta L)/(G_1\eta) \qquad (19)$$

$$e^{-\eta h_1}[-2\eta A_{11}-(k+1-2\eta h_1)A_{12}]+e^{\eta h_1}[-2\eta A_{13}+(k+1+2\eta h_1)A_{14}] = 0 \qquad (20)$$

$$e^{\eta h_2}[-2\eta A_{21}-(k+1+2\eta h_2)A_{22}]+e^{-\eta h_2}[-2\eta A_{23}+(k+1-2\eta h_2)A_{24}] = \int_a^b F_1(\eta,t)\phi(t)dt \qquad (21)$$

$$e^{\eta h_2}\{(2\eta\beta-1)A_{21}+[\beta(k-1+2\eta h_2)-h_2]A_{22}\}$$
$$+e^{-\eta h_2}\{(-2\eta\beta-1)A_{23}+[\beta(k-1-2\eta h_2)-h_2]A_{24}\} = \int_a^b F_2(\eta,t)\phi(t)dt \qquad (22)$$

$$G_1[2\eta A_{11}+(k-1)A_{12}-2\eta A_{13}+(k-1)A_{14}]$$
$$-G_2[2\eta A_{21}+(k-1)A_{22}-2\eta A_{23}+(k-1)A_{24}] = G_2\int_a^b F_3(\eta,t)\phi(t)dt \qquad (23)$$

$$G_1[-2\eta A_{11}-(k+1)A_{12}-2\eta A_{13}+(k+1)A_{14}]$$
$$-G_2[-2\eta A_{21}-(k+1)A_{22}-2\eta A_{23}+(k+1)A_{24}] = G_2\int_a^b F_4(\eta,t)\phi(t)dt \qquad (24)$$

$$A_{11}+A_{13}-A_{21}-A_{23} = \int_a^b F_5(\eta,t)\phi(t)dt \qquad (25)$$

$$(-\eta A_{11} - kA_{12} + \eta A_{13} - kA_{14})/\eta$$
$$-(-\eta A_{21} - kA_{22} + \eta A_{23} - kA_{24})/\eta = \int_a^b F_6(\eta,t)\phi(t)dt \tag{26}$$

where $\beta = G_2/\gamma$, G_1 is the shear modulus of the material in the first layer, G_2 is the shear modulus of the material in the second layer.

$$F_1(\eta,t) = 2e^{-\eta(t-h_2)}[\eta(h_2-t)-1](k+1), F_3(\eta,t) = 2e^{-t\eta}\eta t(k+1),$$
$$F_2(\eta,t) = e^{-\eta(t-h_2)}\{8\beta\eta^2(h_2-t)+[2(k-1)+4\eta(h_2-t)]\}/4\eta(k+1),$$
$$F_4(\eta,t) = 2e^{-t\eta}(\eta t-1)(k+1), F_5(\eta,t) = e^{-t\eta}(k-1-2\eta t)/2\eta(k+1),$$
$$F_6(\eta,t) = 1/2 - e^{-t\eta}/2\eta + te^{-t\eta}/(k+1).$$

Solving Eqs.(19)-(26) for $A_{11}, A_{12}, A_{13}, A_{14}, A_{21}, A_{22}, A_{23}$ and A_{24} in terms of $\phi(t)$ and substituting $A_{21}, A_{22}, A_{23}, A_{24}, B_{23}$ and B_{24} into the fist equation of Eq.(9), it yields:

$$\int_a^b K_1(x,t)\phi(t)dt + \int_a^b K_2(x,t)\phi(t)dt = \frac{(k+1)\pi}{4G_2}p(x) \tag{27}$$

where $K_1(x,t) = \int_{-\infty}^{\infty} \frac{1}{2}(i\xi/|\xi|)e^{i\xi(x-t)}d\xi$. With the assistance of the result

$i\int_{-\infty}^{\infty} \text{sgn}(\xi)e^{i\xi(t-x)}d\xi = -\frac{2}{t-x}$, one can obtain: $K_1(x,t) = \frac{1}{t-x}$ and

$$K_2(x,t) = \frac{G_2 t}{2(kG_1+G_2)(G_1+kG_2)}[\frac{-3G_1+k^2G_1-2kG_2}{(t+x)^2}$$
$$+\frac{2(2xG_1+2kxG_2)}{(t+x)^3}] + \int_0^{\infty}\varphi(x,t,\eta)d\eta$$

$\varphi(x,t,\eta)$ and $p(x)$ can be solved by Gauss—Laguerre numerical quadrature formula.

To solve the integral equation numerically by using a collocation technique, the interval (a, b) in Eq. (27) is normalized and expressed as:

$$\int_{-1}^{1}\frac{g(r)}{r-s}dr + \int_{-1}^{1}K(s,r)g(r)dr = \frac{(k+1)\pi}{4G_2}f(s) \tag{28}$$

The cases a>0 represent an embedded crack problem. For an embedded crack, the solution of the integral equation (28) can be expressed as: $g(r) = F(r)/\sqrt{1-r^2}$.

Lobatto—Chebyshev integration formula is then used to solve the singular integral equations. The singular integral equation is converted to a system of linear equations by means of this numerical method. The expression of Eq. (28) can be written as:

$$\sum_{j=1}^{n} \omega_j F(r_j)[\frac{1}{r_j - s_i} + K(r_j, s_i)] = \frac{(k+1)\pi}{4\mu} f(s_i) \qquad (29)$$

where ω_j is the weight function $\omega_j = \pi/(n-1), j = 2, \ldots, n-1$,

$r_j = \cos((j-1)\pi/(n-1)), j = 1, \ldots, n$; $s_i = \cos((2i-1)\pi/(2n-1)), i = 1, \ldots, n-1$

In Eq. (29), there are n-1 equations and the last equation comes from the property of dislocation density function which is expressed as $\int_{-1}^{1} g(r)dr = 0$ for the embedded crack. It is also written as: $\sum_{j=1}^{n} \omega_j F(r_j) = 0$

The stress intensity factors are defined and evaluated as[9]:

$$K_{Ia} = \lim_{x \to a} \sqrt{2(a-x)} \sigma_{yy2}(x,0), K_{Ib} = \lim_{x \to b} \sqrt{2(x-b)} \sigma_{yy2}(x,0) \qquad (30)$$

Substituting $A_{11}, A_{12}, A_{13}, A_{14}, A_{21}, A_{22}, A_{23}, A_{24}, B_{13}, B_{14}, B_{23}$ and B_{24} into Eqs. (13)-(17), stress at any positions shown in Figure 2 can be derived. The same method can also be used to solve the model shown in Figure 1.

4 Numerical Examples and Discussion

In order to verify the achieved formulation and compare the stress intensity factors of model 1 (as shown in Figure 1) and model 2 (as shown in Figure 2). The parameters for the model 1 are: P=700000N/m, L=0.15m, E=3.1×10^{10}N/m^2, G=1.1482×10^{10}Mpa, γ=1.3×10^{9}N/m^3, h=0.25m, v=0.35. In the case of plane strain problem, k=1.6, β=G/γ=8.832, length of crack l=b-a. The parameters for the model 2 are: P=700000N/m, L=0.15m, E_1=4.32×10^{9}N/m^2, G_1=1.6×10^9 N/m^2, h_1=0.1m, v_1=0.35, k_1=1.6, E_2=3.1×10^{10}N/m^2, G_2=1.1482×10^{10}N/m^2, γ=1.3×10^{9}N/m^3, h_2=0.25m, v_2=0.35, k_2=1.6, β=G_2/γ=8.832. For comparing the old cement concrete pavement with and without asphalt overlay, the stress intensity factors of the crack tips are calculated for model 1 and model 2, and the results are shown in Figures 3 and 4. As the thickness of the overlay and the shear modulus of the material are important factors which affect the stress intensity factors, the results calculated with different thickness and shear modulus are shown in Figures 3-8.

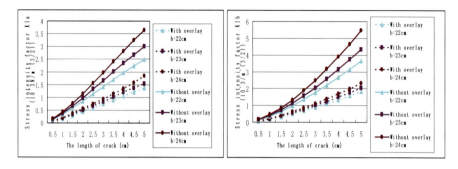

Figs. 3-4. Comparison of Stress intensity factors of the crack tip (a,b) with and without overlay on old cement concrete pavement

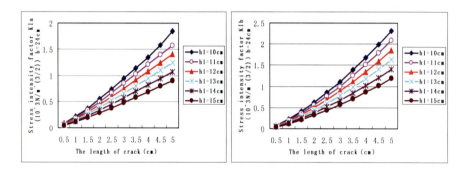

Figs. 5-6. Stress intensity factors of the crack tips (a,b) with different thickness (thickness from 10cm to 15cm) of overlay on old cement concrete pavement

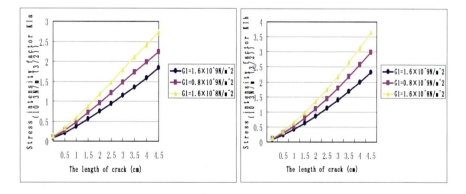

Figs. 7-8. Stress intensity factors of the crack tips (a,b) with different shear modulus of overlay material

In Figures 3 and 4, regardless of the length of the crack, the stress intensity factors of the crack tips in the pavement with 10 cm height overlay are about half of those in pavement without overlay. It indicates that using the overlay is an effective way for lowering the crack extension. From Figures 5- 6, with the thickness of the overlay increasing from 10cm to 15cm, the stress intensity factors decrease by around 50%. For shear modulus of the material, it is obvious that stress intensity factors of the crack tips increase as shear modulus of the material decreases. This can be found in Figures 7-8. Therefore, the shear modulus of the material is also considered as one significant factor that influences the crack propagation.

5 Conclusions

The method of Fourier transform and dislocation density function in association with solving the singular integral equations are introduced to calculate the stress and stress intensity factors in a cement concrete pavement which contains a crack perpendicular to the interface and with asphalt overlay on it. This method can be used to analyze effect of asphalt overlay for resisting crack propagation. Current numerical simulation indicates that the asphalt overlay on top of the old concrete pavement plays an important role for protecting the structure from the crack damage. The thickness of the overlay and the shear modulus of the material are concerned as two significant factors influencing the crack propagation.

References

[1] Ding, S.: Fracture mechanics. China Machine Press, Beijing (1997)
[2] Zak, A.R., Williams, M.L.: Crack point stress singularities at a bi-material interface. ASME 30, 142–143 (1963)
[3] Erdogan, F., Gupta, G.: The stress analysis of multi-layered composites with a flaw. International Journal of Solids Structures 7, 39–61 (1971)
[4] Ueda, S., Mukai, T.: The surface crack problem for layered elastic medium with a functionally graded non-homogeneous interface. JSME International Journal 45(3) (2002)
[5] Long, G., Wang, C.: Three-dimensional numerical analysis for reflective crack of asphalt pavement. Highway Engineering 33(2) (2008)
[6] Yang, D., Zhao, W., Li, L., Qi, J.: Analysis of reflection cracks of asphalt concrete overlay over used cement concrete pavements by finite element analysis. Huazhong Univ. of Sci.& Tech. (Natural Science Edition) 37(1) (2009)
[7] Li, Y.: Theory and application of fracture mechanics. Science Press, Beijing (2000)
[8] Zhao, H.: Fracture and Fatigue Analysis of Functionally Graded and Homogeneous Materials Using Singular Integral Equation Approach (1998)
[9] SERKAN DAG, Crack Problems in a Functionally Graded Layer under Thermal Stresses (1997)

Calibration of Asphalt Concrete Cracking Models for California Mechanistic-Empirical Design (CalME)

Rongzong Wu and John Harvey

University of California Pavement Research Center

Abstract. Cracking is one of the major distress mechanisms for pavements with asphalt concrete surfaces. Given the composite nature of asphalt concrete, simulation methods such as the Discrete Element Method (DEM) that can incorporate material microstructure are required for properly describing the formation and progression of cracking in flexible pavements. These methods are however typically too time-consuming for use in routine design. As a trade off, a simplified approach based on continuum damage mechanics is taken in the California Mechanistic-Empirical method, called CalME, for practical considerations. The effect of cracking (broken contacts) is described as decreases in overall stiffness, which is indicated by damage. The rate of damage increase is in turn empirically related to peak strain energy endured by the material. The format and constants of this relationship are determined from laboratory fatigue testing of the asphalt concrete. Except for the first few loads, where temperature effects may be pronounced, all of the stiffness versus number of load applications curve are used. Once damage history is calculated, visual surface cracking history can be derived as an empirical function of damage and asphalt layer thickness. This paper presents the CalME fatigue and reflective cracking model and its calibration process using deflection data collected from various Heavy Vehicle Simulator (HVS) tests and the WesTrack accelerated pavement testing experiment.

1 Introduction

Cracking is one of the major distress mechanisms for pavements with asphalt concrete surfaces. Given the composite nature of asphalt concrete, simulation methods such as Discrete Element Method (DEM) [1, 2] that can incorporate material microstructure are required for properly describing the formation and progression of cracking in flexible pavements. These methods are however typically too time-consuming for use in routine design. As a trade off, a simplified approach based on continuum damage mechanics is taken in the California Mechanistic-Empirical method, called CalME, for practical considerations.

In CalME, the effect of cracking (broken contacts) in asphalt concrete is described as decreases in overall stiffness, which is indicated by damage. The rate of damage increase (i.e., damage evolution) is in turn empirically related to peak strain energy endured by the material. Once damage history is calculated, visual surface crack density history can be derived as an empirical function of damage. The correlation between visual surface crack density and asphalt concrete are established through field observations.

This paper presents the CalME fatigue and reflective cracking model and its calibration process using deflection data collected from various Heavy Vehicle Simulator (HVS) tests and the WesTrack accelerated pavement testing experiment. The objective is to present a simple and reasonable mechanistic-empirical design framework.

2 Asphalt Concrete Cracking Models in *CalME*

The asphalt concrete (AC) cracking model in CalME can be divided into three components: (a). stiffness model; (b). damage evolution model; and (c). damage to surface crack density correlation. They are explained in detail below.

2.1 Asphalt Concrete Stiffness Model

Asphalt concrete (AC) is a rate and temperature dependent material. In CalME however, pavement responses such as stress, strain and deflection are calculated using multilayer elastic theory with every layer assumed to be linear elastic. Following conventional approach, the rate and temperature dependency of AC is described by the stiffness master curve:

$$\log E = \delta + \frac{\alpha}{1 + \exp(\beta + \gamma \log t_r)} \tag{1}$$

where δ, α, β, and γ are model parameters and t_r is the reduced loading time. The above equation is exactly the same as the one used in the mechanistic empirical pavement design guide (MEPDG) [3]. The reduced time is in turn a function of physical loading time and layer temperature:

$$t_r = t_l \cdot \left(\frac{\eta_{ref}}{\eta}\right)^{aT} \tag{2}$$

where t_l is the loading time, η is binder viscosity at given layer temperature, η_{ref} is binder viscosity at reference temperature and aT is a model parameter. Binder viscosity is in turn calculated as:

$$\log\log(\eta) = A + VTS \times \log(T_k) \tag{3}$$

where T_k is binder temperature in Kelvin (°K), η is binder viscosity in cpoise, and A and VTS are model parameters.

Loading time is a rather uncertain notion, as it will vary for different types of responses. For example, the loading time for transverse strain will be much longer than it is for longitudinal strain because the transverse strain is tangential to the load, whereas the longitudinal is radial and therefore has a sign change. The loading time is calculated as:

$$t_l = 2 \times \frac{r_{load} + h/3}{v} \qquad (4)$$

where r_{load} is the load radius of the tire, h is the AC layer thickness and v is wheel speed. The reference temperature is typically 20°C, which is selected arbitraryly and does not affect AC layer stiffness for any given temperature and loading rate.

Once damage is induced in the AC layer, its stiffness will change. The model for the master curve of the damaged AC layer is:

$$\log E = \delta + \frac{\alpha \cdot (1-\omega)}{1 + \exp\left(\beta + \gamma \log t_r\right)} \qquad (5)$$

where ω is a scalar representing damage.

2.2 Asphalt Concrete Damage Evolution Model

In CalME, the asphalt concrete damage ω as defined in Equation (5) is driven by strain energy experienced by the AC layer. Specifically, the ω is calculated using the following function:

$$\omega = \left(\frac{MN}{SF \cdot MN_p}\right)^\alpha \qquad (6)$$

where MN is the number of load repetitions applied in millions, SF is the shift factor for AC damage hereafter referred to as damage shift factor (DSF), while MN_p is the number of allowable load repetitions (i.e. fatigue life) for a given load and α is a model parameter. MN_p is in turn determined as:

$$MN_p = A \times \left(\frac{\epsilon}{\epsilon_{ref}}\right)^\beta \times \left(\frac{E}{E_{ref}}\right)^{\beta/2} \qquad (7)$$

where ϵ is bending strain at the bottom of AC layer, E is stiffness of AC layer after accounting for damage, aging and compaction etc. A and β are model parameters while ϵ_{ref} and E_{ref} are normalizing constants. The fact that the exponent for strain is twice the value of the exponent for stiffness implies that fatigue life of an AC layer is determined by strain energy input into the layer.

Equation (6) and (7) only describes how damage can be calculated when bending strain and AC layer stiffness remain constant, which does not happen in the field. For damage accumulation in the field, CalME adopts an incremental-recursive (IR) procedure. "Incremental" refers to the fact that CalME simulates pavement performance one increment at a time. The default duration of each increment is 30 days, but this may be as short as one day as determined by the user. "Recursive" refers to the fact that CalME uses the output from one increment, recursively, as input to the next increment.

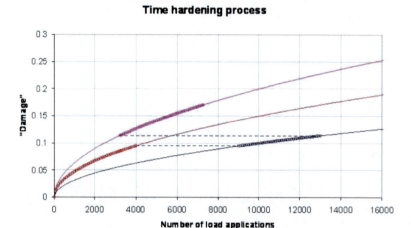

Fig. 1. Time hardening process for accumulating damage following multiple damage evolution curves, each corresponding to a different level of strain energy experience by the material

During a simulation of pavement performance over its design life, the parameters on the right hand side of the equation (7) may change from increment to increment. Essentially, there are multiple damage evolution functions in effect with one for each loading condition (wheel load level and loading temperature etc.). The first step in the process is, therefore, to calculate the "effective" number of load applications that would have been required, with the present parameters, to produce the condition at the beginning of the increment. In the second step the new condition, at the end of the increment, is calculated for the "effective" number of load applications plus the number of applications during the increment. This must be repeated for each load and load position within the increment. This process is illustrated in Figure 1, which shows how damage is accumulated over three time increments each with a different loading condition.

Note that in CalME, damage caused by fatigue cracking and reflective cracking are both calculated using equations (6) and (7) with the only difference in how the bending strain ϵ is calculated. For fatigue cracking, ϵ is the bending strain at the bottom of the AC layer. For reflective cracking, ϵ is the critical tensile strain at the crack tip. This is explained in more detail next.

CalME currently can only account for traffic-induced reflective cracking. For reflective cracking in AC overlay over old AC pavements, the overlay and the underlying layer are assumed to be fully bonded. The strain at the crack tip is singular in linear elasticity analysis. As a simple alternative to linear elastic fracture mechanics, the average first principle strain along centerline (see Figure 2) within 10-mm radius of the crack tip is calculated as the strain ϵ to be used in equation (7).

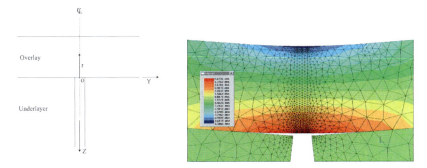

Fig. 2. Geometry at the crack tip for bonded AC overlay over old AC underlayer

Fig. 3. Local debonding between AC overlay and the underlying PCC layer

For reflective cracking in AC overlay over Portland cement concrete (PCC) pavements, the overlay and the underlying layer is assumed to be locally debonded near the underlying crack/joint. The debonding removes singularity in strain at the crack tip. An example of the bending strain field is shown in Figure 3. The maximum bending strain at the crack tip is used as ϵ in equation (7) for calculating reflective cracking damage. The exact equations for calculating reflective cracking strain were developed based on regression of thousands of finite element runs and can be found in [4].

2.3 Correlation between Damage and Surface Crack Density

The correlation between surface layer damage and surface crack density is described by an empirical equation:

$$C = \frac{C_{max} \times (\omega_{initiation}^{\alpha} - 1)}{\omega_{initiation}^{\alpha} - \frac{C_{max}}{C_i} + \left(\frac{C_{max}}{C_i} - 1\right) \times \omega^{\alpha}} \tag{8}$$

where C is the surface crack density and α is a model parameter. $\omega_{initiation}$ is the damage correspond to crack initiation. C_i is the surface crack density corresponding to crack initiation damage $\omega_{initiation}$, while C_{max} is the maximum surface crack density. C_i is assumed to be 5% wheelpath cracking while C_{max} is 100% wheelpath cracking. The damage to the surface layer at crack initiation is determined from:

$$\omega_{initiation} = \frac{1}{1 + \left(\frac{h_{AC}}{h_0}\right)^a} \tag{9}$$

where h_{AC} is the combined thickness of the surface AC layers, h_0 and a are empirical model parameters.

3 Determination of Mechanistic Model Parameters

The model parameters introduced in previous section can be divided into two groups: (1). mechanistic model parameters that can be determined using laboratory test data; and (2). empirical model parameters that can only be determined by incorporating field observations through model calibration.

Specifically, all of the model parameters introduced in previous section for AC layer stiffness i.e., Equations (1) to (3)) and AC layer damage evolution (i.e., Equations (6) and (7)) mechanistic parameters except the shift factor SF in Equation (6), and all the other model parameters in Equations (8) and (9) are empirical parameters. This section discusses determination of mechanistic model parameters, while the next section focuses on determination of empirical parameters (typically referred to as model calibration in M-E design).

Mechanistic model parameters are determined by fitting laboratory test data. In other words, these model parameters are determined by minimizing the difference between model predictions and laboratory measurements.

Table 1. Standard laboratory tests to obtain mechanical properties for AC Layer

Property	Test Type	Experiment Design
Stiffness Master Curve	Beam bending frequency sweep (AASHTO T321 with modifications on testing duration and frequency)	3 Temperatures (10, 20 and 30°C) x 2 Replicates = 6 tests
Fatigue Resistance	Beam bending fatigue (AASHTO T321)	2 Strains (200 and 400 microstrains) x 3 Replicates = 6 tests

AC stiffness model parameters can be determined by fitting any test data that fully describe the rate and temperature dependency of AC stiffness. AC damage evolution model parameters can be determined by fitting any test data that describe how AC stiffness decreases with number of load repetitions under various loading conditions. Note however, the empirical parameter SF needs to be set to 1.0 during this process. The standard laboratory test data used in California are listed in **Table 1**. When these test data are not available, many other alternatives can be used.

Error! Reference source not found. shows examples of a set of stiffness reduction curves from beam bending fatigue tests. A total of six beams are tested with air void contents within the range of 7.5±0.6%. Except for the first few loads, where temperature effects may be pronounced, all of the stiffness versus number of load applications curve are used. Thixotropic effects are not currently considered in the AC damage evolution. The influence of thixotropy is currently handled in a very simplistic manner through consideration of the effects of rest periods which is not discussed in this paper. The AC damage evolution model shown in Section 0 was used to fit the stiffness reduction curves, with shift factor

SF set to 1.0. The comparison of calculated and measured residual stiffness ratios is shown in Figure 4. As shown in Figure 4, the model predictions and lab measurements matched very well for residual stiffness ratios between 0.5 and 1.0, which typically correspond to the stage when microcracks in the beams are still coalescing and no major cracks have formed yet.

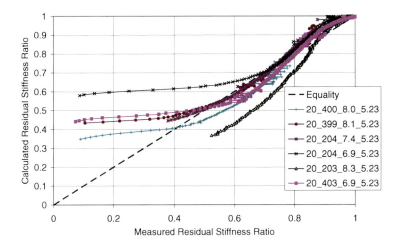

Fig. 4. Comparison of measured and calculated residual stiffness ratio after fitting beam bending fatigue test data, model parameters used are A = 17.9, α=0.768, $\beta = -4.49$, $\epsilon_{ref} = 200\mu\varepsilon$ and $E_{ref} = 3000$MPa.

4 Calibration of Empirical Model Parameters

Empirical model parameters can only be determined using field observation data through a process typically referred to as M-E model calibration. For CalME, model calibration was done using WesTrack [5] and heavy vehicle simulator (HVS) test data [6].

As mentioned in Section 0, the shift factor SF in Equation (6), and all of the model parameters introduced in Section 0 are empirical parameters. The calibration procedures and results for these empirical parameters are presented in this section.

4.1 Calibration of Shift Factor in AC Damage Evolution

The damage shift factor (DSF) SF in Equation (6) accounts for the difference in damage evolution rates between laboratory and field conditions. As the first step in calibrating damage the shift factor, CalME was used to predict the evolution of pavement deflections caused by trafficking. For WesTrack tests, trafficking was applied using four triple-trailers running at 64 km/hour while pavement deflections were measured at the surface at various intervals using falling weight deflectometer (FWD). For HVS tests, trafficking was typically applied using half-axle dual wheel truck tires running at around 9.0 km/hour while deflections were measured regularly using Multi Depth Deflectometers (MDDs) and Road Surface Deflectometer (RSD, similar to a Benkelman beam). The empirical model parameters were adjusted to allow best match between predicted and measured pavement deflections. One DSF was determined for each test section (either in WesTrack or HVS testing). An example of comparison between damage predicted by CalME and damage back-calculated from FWD deflection data is shown in Figure 5.

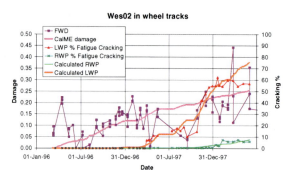

Fig. 5. Measured and calculated damage and cracking from section 02FLM, damage shift factor (DSF) = 0.5, LWP and RWP stands for left and right wheel path respectively

CalME has a series of models that account for factors such as AC stiffness change due to pavement temperature, traffic characteristics (wheel load, tire pressure, tire configuration, wheel wander, traffic volume, and wheel speed), AC stiffness change due to aging and traffic-induced densification and the effect of rest period. Nevertheless, there are differences between idealized CalME model predictions and measured pavement responses. These differences can be attributed to two sources: (a). inaccuracies in CalME models; and (b). factors neglected in CalME but affects pavement responses.

An example of source (a) is the multi-layer elastic theory used in CalME, which assumes pavement layers to be homogeneous, linear elastic, isotropic and uniform in layer thickness. These assumptions can all cause inaccuracy in the stress and strain calculated for driving damage evolution.

An example of source (b) is the effect of pavement temperature on AC fatigue performance. The mechanistic model parameters for equations (6) and (7) are temperature dependent. In CalME, this temperature dependency is simply neglected and its effect is incorporated into the damage shift factor SF in equation (6).

As expected, the combined effects of source (a) and source (b) factors can lead to different values of DSFs for individual test sections. The empirical cumulative distribution functions for DSF from HVS tests and Westrack sections are shown in Figure 6.

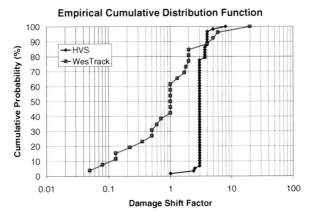

Fig. 6. Empirical cumulative distribution function for all of the damage shift factors used in CalME calibration using WesTrack and HVS test data

Figure 6 shows that DSF is not really a constant, indicating that the difference between CalME model predictions and field observations from WesTrack and HVS testing is not systematic. In theory, DSF should be a random variable with its statistics such as mean and standard deviations determined through model calibration. Since DSF is a random variable, the M-E performance prediction should be probabilistic by nature. In practice however, a constant DSF is desired to provide mean deterministic performance prediction. A median value should be chosen for this purpose to provide 50-th percentile performance prediction. The median value for DSF is 1.0 based on WesTrack data and 3.0 based on HVS data.

HVS testing was primarily used for the critical step of calibrating response models and initial damage shift factors, however the slow speed of the wheel, and the controlled temperature environment that eliminates the stresses and opening and closing of cracks from temperatures changes leads to the need to also consider the WesTrack DSF for design. Since WesTrack is believed to be closer to field condition compared to HVS testing, it is decided to use a value 1.0 for constant DSF based on WesTrack data. Additional calibration is a goal for future work using improved data from work being done on the Caltrans pavement management system, and from other test tracks.

Since Figure 6 shows large difference between individual DSF values obtained from WesTrack and HVS data, it is necessary to evaluate the effect of using a constant DSF of 1.0 on the quality of fit for surface deflection data measured in HVS. Figure 7 shows the comparison of empirical cumulative distribution functions for the ratio between calculated and measured surface deflections in HVS tests using different DSF values. As shown in Figure 7, the two empirical CDF functions are not significantly different, indicating that surface deflections are not very sensitive to the DSF values. This implies the need for using more pavement response data (in addition to surface deflection) when calibrating CalME AC damage models. At the same time, using a constant DSF value of 1.0 is still allows CalME to provide reasonable prediction of surface deflections.

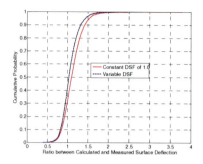 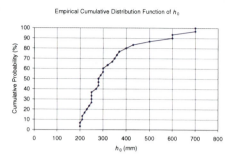

Fig. 7. Empirical cumulative distribution functions for the ratio between calculated and measured surface deflections in HVS tests using different DSF values

Fig. 8. Empirical cumulative distribution function for empirical model parameter h_0

Note that even though the same constant DSF value of 1.0 is used in CalME Monte Carlo simulation, the probabilistic nature of DSF is accounted for in Monte Carlo simulations by allowing mechanistic model parameter A in equation (7) to be a random variable.

4.2 Calibration of Damage to Surface Crack Density Model Parameters

Calibration of damage to surface crack density (D2SCD) model involves determination of model parameters α, a, and h_0 as defined in equations (8) and (9). This is done for each individual test section after DSF has been determined through the procedure shown in Section 0. Specifically, since both damage evolution history and surface damage crack density history are known, the D2SCD model parameters can be determined by minimizing the difference between calculated and measured surface crack density. An example comparison between calculated and measured surface crack density is shown in Figure 5 for both left and right wheel paths for WesTrack section 02.

Based on WesTrack test data, $\alpha = -8.0$, $a = -2.0$, while h_0 varies between 200 and 700mm. The empirical cumulative distribution function for h_0 is shown in Figure 8. According to equation (9), damage corresponding to crack initiation decreases as h_0 increases, which in turn means more surface crack for the same damage. A h_0 value of 250mm was selected as a round number close to the median found from calibration studies.

5 Summary and Conclusions

This paper presents a simplified mechanistic-empirical (M-E) fatigue and reflective cracking model based on continuum damage mechanics. Details about how damage evolves and how surface crack density is calculated are provided. The model parameters are divided into two groups: (a) mechanistic parameters that are determined using laboratory test data; and (b) empirical parameters that are determined using data measured in the field. The procedures for determination of both mechanistic and empirical parameters are presented along with examples and actual values for both groups of model parameters.

As shown in the paper, empirical parameters tend to have a high variability even though CalME is able to account for many of the important factors that affect pavement performance. This means that M-E model predictions are probabilistic in nature and one should account for this uncertainty properly while at the same time working to reduce variability with improved models, while maintaining practicality of the analysis method. The M-E design approach allows one to fully evaluate the reliability of pavement performance of a given structural design. Further more, with factors such as material properties, pavement temperatures and traffic spectra accounted for properly, the uncertainty in M-E performance predictions are far less than empirical estimates.

Acknowledgement. This paper describes research activities requested and sponsored by the California Department of Transportation (Caltrans), Division of Research and Innovation. Caltrans sponsorship is gratefully acknowledged. The contents of this paper reflect the views of the authors and do not reflect the official views or policies of the State of California or the Federal Highway Administration. The authors would also like to thank Per Ullidtz for his work in developing the models used in CalME.

References

[1] Zelelew, H.M., Papagiannakis, A.T.: Micromechanical Modeling of Asphalt Concrete Uniaxial Creep Using the Discrete Element Method. Road Materials and Pavement Design 11(3), 613–632 (2010)
[2] You, Z.P., Adhikari, S., Kutay, M.E.: Dynamic modulus simulation of the asphalt concrete using the X-ray computed tomography images. Materials and Structures 42(5), 617–630 (2009)
[3] ARA Inc. Guide for Mechanistic-Empirical Design of New and Rehabilitated Pavement Structures, ERES Consultants Division, ARA Inc, National Cooperative Highway Research Program, Transportation Research Board, National Research Council (2004)
[4] Wu, R.: Finite Element Analyses of Reflective Cracking in Asphalt Concrete Overlays. University of California, Berkeley (2005)
[5] Ullidtz, P., Harvey, J., Tsai, B.-W., Monismith, C.: Calibration of CalME models using WesTrack Performance Data, UCPRC-RR-2006-14. D. Spinner, Editor, California Department of Transportation Division of Research and Innovation Office of Roadway Research (2007)
[6] Ullidtz, P., Harvey, J.T., Tsai, B.-W., Monismith, C.L.: Calibration of Incremental-Recursive Flexible Damage Models in CalME Using HVS Experiments. in Report prepared for the California Department of Transportation (Caltrans) Division of Research and Innovation, UCPRC-RR-2005-06, University of California Pavement Research Center, Davis and Berkeley (2006)

Shear Failure in Plain Concrete as Applied to Concrete Pavement Overlays

Yi Xu and John N. Karadelis

Department of Civil Engineering, Architecture and Building, Coventry University, CV1 5FB, UK

Abstract. This study applied the modified Iosipescu loading configuration on beams and direct shear loads on cylinders to investigate the concrete behaviour under minimal flexural and prominent shear stress conditions, particularly in plain concrete tests. It aims to make a contribution in understanding the behaviour of a concrete pavement under shear loading and failure and ultimately design an adequate overlay system. A finite element model corresponding to the modified Iosipescu beam test was set up to assist with the recognition and study of the complex stress patterns developing at high stress concentration regions and the evaluation of principal stresses. Both, normal and high strength concrete were used, to imitate the performance of the existing (old) pavement and the prospect overlay under shear. On the basis of the laboratory results and the preliminary finite element analysis, it was found that all specimens failed abruptly without warning in a typical brittle-material / shear-failure manner. Considerably higher shear strengths were achieved compared to those specified in the standards. Unexpectedly, the high strength concrete developed slightly lower shear strength than the normal concrete. This is probably attributed to insufficient compaction and relatively less effective aggregate interlock in high shear regions. The research is in progress.

1 Introduction and Literature Survey

Shear failure at cracks/joints is a major cause of degradation of concrete pavements. It leads to serviceability problems and introduces reflective cracks and becomes an issue of structural integrity, durability, riding quality and safety of the deteriorated pavement.

To address the problem, a Rapid Pavement Repair and Strengthening Management System (RPRSMS) is under development at the Department of Civil Engineering, Coventry University. It aims to bring to light the structural and functional deficiencies and extend the life of the pavement and at the same time introduce substantial benefits, such as saving construction materials, time and labour costs, by bonding a layer of special-quality concrete on top of the existing damaged pavement. Therefore, all the aforementioned contribute to a more sustainable pavement repair system.

In this study, the modified Iosipescu loading configuration [1] was adopted for testing the beams. Direct shear loads were applied on cylindrical specimens to investigate the concrete behaviour under high shear and low moment conditions. Cylinder torsion tests were carried out to assist with the investigation of the concrete shear performance. A finite element model simulated the behaviour of the specimens and assisted with the recognition and study of the loading configuration that otherwise would not be easily identifiable (complex stress patterns developing at high stress concentration regions, evaluation of stresses and stress intensity factors). Both, normal and high strength concrete were used, to imitate the performance of the existing/old pavement and the prospect overlay under shear. Glass fibre was added to the high strength concrete mix in an effort to enhance its shear resistance further. The influence of coarse aggregate, mixing procedure, compaction and curing method on the shear resistance was also assessed experimentally. Emphasis was given to the experience built up so far, so essential for outlining future similar type of work.

1.1 Shear Problems in Pavement Overlays

Three possible main failure modes were identified:

- Flexure failure: The crack propagates upwards into the new overlay due to exceeding its flexural strength.
- Shear failure: The crack propagates upwards into the new overlay mainly due to insufficient shear resistance.
- Delamination: The crack develops along the interface.

In a real pavement scenario, failure takes place under the combined effects of bending moment, shear force and possible involvement of axial force due to boundary conditions. This article isolates and focuses on the problem of shear failure. Parallel studies on flexural and delamination problems are currently under scrutiny by other members of our research group at Coventry University. The findings will be combined to formulate optimum design guidelines for a "Green" concrete overlay.

When the load is located either side of an existing, old crack, high shear stress becomes dominant and a relative vertical displacement is expected at crack edges, which may lead to *reflective cracking* into the overlay under repetitive traffic loads, as shown in Figure 1.

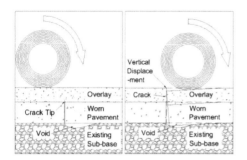 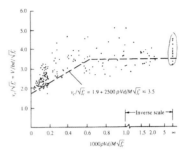

Fig. 1. Reflective shear cracking under repetitive traffic loading (Left)

Fig. 2. Correlation of ACI 318-02 with experimental data [2] (Right)

Depending on the nature of crack edge movement, it can be categorized into mode I *openning*, mode II *shearing/sliding* and mode III *tearing*. Mode II corresponds to the shearing problem but it usually coexists with mode I. The Paris' law [3], as shown below is usually deployed to describe the crack propagation under repetitive loading.

$$dc/dN = A(K_{eq})^n \qquad (1)$$

where, K_{eq} = equivalent stress intensity factor; A, n= fracture mechanics factors; c= crack length; N = number of loading cycles; dc/dN = increase of crack length per load cycle.

A fracture mechanics approach may be needed to highlight the crack initiation and propagation process. The stress intensity factor, K, due to bending and shear is one of the fundamental input parameters. Therefore, it was investigated in the following experimental study, assisted by numerical modelling.

1.2 Design Codes and Standards on Shear

A selection of design codes and published formulae from previous researchers are summarized in Table 1. The following assumptions are made to derive representative shear strengths using each code: A relatively high compressive strength of $f_c'=f_{ck}=50\text{MPa}$ is assumed. The shear span-to-depth ratio $\frac{a}{d} = \frac{46}{73} \approx 0.63$ is employed based on the geometry of the single notch shear beam test as shown in Figure 3. Since the steel ratio, ρ, is deemed necessary in some of the codes, to execute the calculation and return a non-zero result, a minimal value of ρ=0.00001 is assumed. It is understood that errors may be introduced by the latter and some empirical formulae may become invalid in the case of plain concrete.

Table 1. Summary of Design Codes and Published Shear Design Formulae [4-8] 1) Not all standards/codes take a/d into account, commonly acknowledged as one of the most influential parameters in evaluation of the shear strength. 2) steel ratio, ρ, not applicable in plain concrete. 3) Generally, $f'_c \leq 68.9$ MPa in ACI 318-02 [5]. 4) A wide scatter was presented in the correlation of ACI 318-02 with experimental data, as shown in Figure 2.

Design Codes and Published Formulae	Shear Strength (MPa)	Shear strength (MPa)	Comments
Eurocode 2 [4]	$\tau_{Rd} = (0.25 f_{ctk0.05})/\gamma_c$ $f_{ctk0.05} = 0.7 f_{ctm}$ $f_{ctm} = 0.3(f_{ck})^{0.67}$ $\gamma_c = 1.5$	0.48	
		0.72	If the safety factor $\gamma_c = 1.5$ is omitted.
ACI 318-02 [5]	$v_c = 0.16\sqrt{f'_c} + 17\frac{V_u d}{M_u}\rho_w$, but $\leq 0.16\sqrt{f'_c}$	≤1.13	for $\frac{a}{d} \geq 2.5$, i.e. relatively large shear span-to-depth ratio
	$v_c = \left(3.5 - \frac{2.5 M_u}{V_u}\right)\left(0.16\sqrt{f'_c} + 17\frac{V_u d}{M_u}\rho_w\right) \leq 0.29\sqrt{f'_c}$	≤2.05	for $\frac{a}{d} < 2.5$, i.e. relatively small shear span-to-depth ratio
	$v_c = 0.166\sqrt{f'_c}$	1.17	General
	$v_c \leq 0.83\sqrt{f'_c}$	≤5.87	Deep beams ($l_n \leq 4d$)
A23.3-94 [6]	$v_c = 0.2\sqrt{f'_c}$	1.414	No consideration for a/d
Zsutty Eqn [7, 8]	$v_c = 11.42 \left(f'_c \rho \frac{d}{a}\right)^{1/3}$	1.057	a/d > 2.5 or a/d < 2.5 under indirect load, as provided by side flanges
	$v_c = 28.55 (f'_c \rho)^{1/3} \left(\frac{d}{a}\right)^{4/3}$	3.204	1.5 < a/d < 2.5 with direct loading (top load & bottom supports), lower bound strength predictor

2 Experimental, Pilot Study

2.1 Materials

Stage 1. Three concrete mixes were tested in the preliminary experimental work. They were Ordinary Portland Cement Concrete (OPC1) as control specimens, plain Polymer Modified Cement Concrete (PMC1) and Polymer Modified Cement Concrete with 4kg/m³ glass fibre (PMC2). Following previous studies [9, 10], the PMC mix developed is of relatively high strength and high modulus of rapture and suitable for roller compaction. A specially modified Kango hummer was used to consolidate the specimens. 10mm granite was employed for coarse aggregate.

Stage II. In addition to the mixes in stage I, three more concrete mixes were produced and tested in cylinder direct shear mode. All six mixes are summarized in Table 2. Gritstone replaced granite in OPC2, PMC3 and PMC4 as its aggregate crushing value (ACV), tested according to BS 812-110:1990 [11], indicated that the former has a higher resistance to crashing than the latter (11.40% and 19.98%). Also, PMC4 was cast following a new mixing and curing method in accordance with ASTM 1439 [12] (4-day moist curing at $23\pm2^{\circ}C$ and relative humidity of approx. 95%, followed by 24-days air curing at room temperature environment). A new compaction device simulated the hammering and vibrating actions.

Table 2. Summary of Concrete Mixes

	Coarse Aggregate	Fibre Used	Mixing Method	Curing Method	Compaction Equipment
OPC1	Granite	N/A	Traditional	Traditional	Kango
OPC2	Gritstone	N/A	Traditional	Traditional	Kango
PMC1	Granite	N/A	Traditional	Traditional	Kango
PMC2	Granite	Glass	Traditional	Traditional	Kango
PMC3	Gritstone	N/A	Traditional	Traditional	Kango
PMC4	Gritstone	N/A	New	New	Vib. Comp/tor

2.2 Single Notch Shear Beam Test (SNSBT)

This paper adopts the concept of single edge notched beams employed by Iosipescu [1] and Arrea and Ingraffea [13]. This single notch short beam shear test created a concentrated shear zone in the (near) absence of bending moments at the notch and evaluated the shear strength level of OPC1, PMC1 and PMC2. Twelve beams were tested comprising four beams per mix. The geometry and loading arrangements are shown in Figure 3. All the beams failed abruptly with the crack

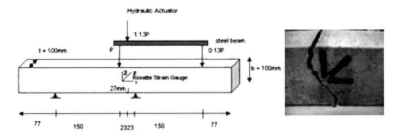

Fig. 3. Specimen Geometry and Loading and Crack Trajectory

propagating from the tip of the notch in an inclined 45° direction, reaching the surface of the specimen to the opposite side of the applied force P (Figure 3). The results were summarized in Table 3. Despite the high compressive strengths of PMC, the latter failed earlier under shear than ordinary concrete. This is in harmony with existing literature [14-16].

Figure 4 shows the variation of shear stress versus shear strain for a magnitude of shear stress up to 2.5MPa, before failure occurs. Four beams were tested for each mix. A wide range of scatter points (hatched areas) was present in the results. The same scatter was reported in Figure 2. Both PMC1 and PMC2 mixes show steeper slopes, i.e. higher shear modulus than the OPC1. Comparing plain PMC1 with glass-fibre reinforced PMC2 it is observed that the shear stress develops faster in PMC2. This indicates that the introduction of glass fibre did enhance the shear resistance of PMC but not the ductility.

Overall, PMC has high compressive strength but lower shear strength, possibly due to insufficient compaction. The addition of glass fibre enhanced the shear properties of the PMC. A good consistency exists between the average saturated densities and the shear strengths. In descending order, both rank as: OPC1 > PMC2 > PMC1, as listed in Table 3. As compaction during PMC casting is critical, insufficient compaction may induce voids (low density). When subject to shear, these voids become weak points and eventually lead to failure at a "lower-than-expected" load. However, in compression, the microstructures developed by superior ingredients still allow for good performance.

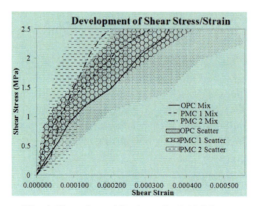

Fig. 4. Shear Stress/ Strain at the Initial Stage

For a single edge crack subject to in-plane shear, a relationship has been developed by Tada [17] for K_{II}, the stress intensity factor, as shown below.

$$K_{II} = \frac{2Q}{B\sqrt{\pi a}} \frac{1.30 - 0.65\left(a/b\right) + 0.37\left(a/b\right)^2 + 0.28\left(a/b\right)^3}{\sqrt{1 - \left(a/b\right)}} \qquad (2)$$

Where: Q = applied shear at the notch at the onset of slow crack growth; a = crack depth, 27mm in the testing geometry; b = height of the beam, 100mm; B = width of the beam, 100mm.

Substituting the above values into Eqn. (2), the latter is reduced to:

$$K_{II} \approx 0.002940621 \times Q \text{ in } N \cdot mm^{-3/2} \qquad (3)$$

All beams failed abruptly. Owing to the rapid crack development, the load at the onset of slow crack growth was not detected. If the failure loads were to be used in the assessment, the average K_{II} for OPC1, PMC1 and PMC2 would be 2435, 2136 and 2390 $kN \cdot m^{-3/2}$ respectively. As failure occurs at a higher load than that needed for initial crack growth, the actual K-values are likely to be smaller than those calculated. Although the definitive estimates of K_{II} are not available at the current stage, the PMC returned with smaller toughness than the OPC. Further investigation into the aggregate interlock and friction is underway. Additional tests conducted to increase confidence are cylinder torsion tests and direct shear tests as explained in the following paragraphs.

2.3 Cylinder Torsion Test

A total of eight concrete cylinders, four OPC1 and four plain PMC1 with 100mm diameter and 200mm length were cast. Table 3 provides all necessary properties and results. Studying failed specimens of OPC and PMC showed that the fractured plane in PMC was smoother than that of OPC. Hence, the aggregate failed to demonstrate good interlocking abilities.

2.4 Cylinder Direct Shear Test

The Iowa Testing Method 406-C [18] on four 150mm diameter by 300mm long OPC1 cylinders and four PMC2 (glass fibre) cylinders was adopted. To enhance the shear strength performance of PMC, the coarse aggregate as well as the mixing, compacting and curing methods were revised in stage II. Gritstone replaced granite in OPC2, PMC3 and PMC4. The new mixing, compaction and curing methods were deployed for PMC4, as described earlier. The test arrangement is pictured in Figure 5. All cylinders were sheared off in a vertical plane, as shown in Figure 6. PMC2 (with glass fibre) delivered higher shear strength than plain PMC1, similar to OPC1. The results were summarized in Figure 7. Unexpectedly, they failed to demonstrate a clear enhancement in shear strength, beside the use of gritstone. It is likely that the water absorption value for gritstone was used in the granite mix (PMC1), by mistake. As the water absorption of granite is higher than that of gritstone, the w/c ratio of PMC1 was found to be 0.33, while all other PMC mixes revealed 0.37. Had the correct value been used, PMC1 should have a lower shear strength than the current value, or even lower than that of PMC3 and PMC4. An investigation is currently under way. Nevertheless, the new mixing /compaction /curing methods did exhibit a great improvement to the strength and density, proving that effective compaction is essential in strength enhancement. This good practice shall be continued in the next phase of experiments.

2.5 Summary

The results in all three tests are summarized in the same Table 3. Results involving gritstone and the new mixing/compaction/curing methods have been discussed in the previous chapter and therefore are not repeated herein.

Fig. 5. Cylinder Dir Shear Test   **Fig. 6.** Failed Section (PMC2)

It is noted that the tests carried out have excellent consistency on the evaluation of shear performance. Despite high compressive strength, plain PMC failed to provide an equivalent shear resistance as the OPC. This is attributed to the high strength hydrated cement paste (HCP) in PMC and the relatively weak coarse aggregate, which led to less efficient interlock at the failure planes. The OPC failure face exposes a great amount of aggregates and most of the failure takes place at the HCP-aggregate interface. In contrast, the PMC shows a much smoother face through crushed aggregates. Hence, the replacement of granite with gritstone.

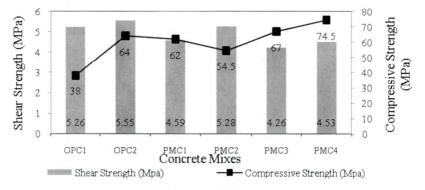

Fig. 7. Summary of Results

Compaction was also blamed for the "lower than expected" shear resistance of PMC. Different from high/normal-workability concrete, potent compaction in PMC is essential to attain the desired strength. After examining the cube densities, it was deduced that there were voids in the mix due to insufficient compaction. Hence, the investment in the new vibrating-compactor.

Table 3. Summary of Test Results

Mat. Prop.	Compr. Stren. f_c (MPa)	Density (kg/m^3)	Mean Shear Strength v_c (MPa)			Estimated Mean Shear Modulus G (GPa)	
Tests	Cube Compre. Test		SNSBT	Torsion	Direct Shear	SNSBT	Torsion
OPC1	38.0	2361.1	3.59	3.59	5.26	15.5	15.9
PMC1	60.5	2298.7	2.90	2.79	-	24.7	17.2
PMC2	54.5	2323.6	3.52	-	5.28	26.2	-

PMC2 with a glass fibre input did improve the shear properties of plain PMC to a similar level of OPC, but failed to provide the required ductility. Hence, glass fibres were replaced with steel fibres and later with synthetic fibres. A "cock-tail" of reinforcement has been introduced in the mix to enhance the shear performance and to provide the required resistance to reflective cracking. This is currently under investigation.

A quick comparison of Table 3 with Table 1 reveals that all experimentally obtained shear strengths are much higher than most of the values obtained by the codes. The following reasoning is stated:

- All tests had very small shear span-to-depth (a/d) ratio. The a/d ratio was 0.067 and 0.46 respectively in cylinder direct shear test and SNSBT. It is generally agreed that the influence of a/d is not insignificant [5]. A low a/d implies small bending moment, in which case, most of the cross section is available to resist shear. Hence, a high shear property is expected. Also, a low a/d ratio corresponds to a strong "strut" (arch) action. Hence, the high shear resistance could be attributed to the high compressive strength in concrete.
- It is noted that in some of the formulae in the codes, the shear strength, v_c, is derived from $\sqrt{f_c'}$. The correlation between $\sqrt{f_c'}$ and v_c is summarized from diagonal – tension failure, a form of indirect shear failure, whereas the tests conducted were direct shear. Additionally, even in the case of diagonal tension, a wide scatter is usually present in experimental investigations as shown in Figure 2. Therefore, in this case, the real concrete shear capacity is under-estimated by the empirical formulae.

3 Finite Element (FE) Analysis

A 2D FE-model corresponding to the single notch shear beam test was setup using ANSYS software [19] to study the stress distribution and the associated parameters. The material properties obtained in the experiment (Table 3) were used in the model. The 6-node triangular PLANE183 element, with two degrees of freedom (DOF) per node, supporting plasticity, large deflection and strain behaviour was used. This is defined by a quadratic shape function, considered to be well suited to model irregular meshes occurring at crack regions. All usual output is supported by the analysis, plus the displacement extrapolation method in calculating stress intensity factors. A "singularity" point was created at the crack tip with skewed nodes at quarter positions. The failure loads from the experimental tests were introduced. A non-linear, large displacement analysis was performed, converging after a few iterations. It is stressed here that this is a preliminary model designed to help with experimentation rather than the other way round.

Figure 8 portrays the high stress concentration at the crack tip. The maximum shear strengths and the mode II stress intensity factor derived at the crack tip are listed in Table 4 with the corresponding experimental results for comparison. It was found that shear stress at the crack tip is nearly twice the average shear strength of the reduced cross section above the crack in OPC1 (6.63 corresponding to 3.59). The predicted K_{II} results are lower than the measured values calculated using Tada's formula [17] in OPC1 (1216 compared to 2435). Swartz conducted a similar single notch shear test configuration both experimentally and numerically on concrete beams [20], noticing a similar discrepancy between K_{II} values. This leads to suggestions that Tada's formula may need modification to reflect the real geometry and loading conditions. Note that to calculate K using FEM, the crack width has to be zero. However, a 2mm width crack was created in the test beams. This should be allowed for. A 3D model is currently under development.

Table 4. Numerical Modelling Results

	Predicted Shear Stress at Crack Tip (MPa)	Measured Average Shear Strength (MPa)	Predicted K_{II} @ Crack Tip ($kN \cdot m^{-3/2}$)	Measured K_{II} @ Crack Tip ($kN \cdot m^{-3/2}$)
OPC1	6.63	3.59	1216	2435
PMC1	5.82	2.90	1047	2136
PMC2	6.51	3.52	1172	2390

Fig. 8. SNSBT Model Shear Stress, S_{xy} - Plot

4 Conclusions

All specimens failed abruptly in a typical shear failure and brittle material manner. The introduction of glass fibre did not provide acceptable results. Other forms of reinforcement are currently under scrutiny. Steel fibres seemed to be a successful alternative to conventional shear reinforcement.

On the basis of the laboratory results, it was found that considerably higher shear strengths were achieved than those specified in the standards. This is attributed to the nature of the tests with relatively low a/d ratio. For reasons explained earlier, it is considered inappropriate to apply the indirect and empirical shear design procedure on direct shear tests involving minimal bending, especially in the case of plain concrete.

The high strength concrete developed slightly lower shear strength than the normal concrete and this is attributed to the relatively less effective aggregate interlock in high shear regions and possibly insufficient compaction. A replacement coarse aggregate was introduced contributing to the shear capacity but requires further confirmation. The new mixing, compaction and curing method was also tested and proved to be effective in increasing the shear strength. This successful practice shall be continued and developed further to achieve better results. This is an on-going research programme funded by an EPSRC grant. The final results should be published next year.

Acknowledgements. The financial support of the Engineering and Physical Science Research Council (EPSRC), UK and Aggregate Industries (AI) is gratefully acknowledged. The authors would like to express their gratitude to Dr. Salah Zoorob and all the colleagues and technical staff at Coventry University for their valuable suggestions and comments. Special mention should also be made to Tarmac, Everbuild Products, AGS Mineraux and Power Minerals for providing the research materials.

References

[1] Iosipescu, N.: J. Mater. 2(3), 537–566 (1967)
[2] Leet, K.M., Bernal, D.: Reinforced Concrete Design: Conforms to 1995 ACI Codes. McGraw-Hill, New York (1997)
[3] Paris, P.C., Erdogan, F.A.: J. Basic Eng. 85(4), 528–534 (1963)
[4] Beeby, A.W., Narayanan, R.S.: Designers' Handbook to Eurocode 2 Part 1.1, Design of Concrete Structures, Telford, London (1995)
[5] ACI Committee 318. Building Code Requirements for Structural Concrete (ACI 318-02) and Commentary (ACI 318R-02). American Concrete Institute, USA (2002)
[6] CSA Technical Committee on Reinforced Concrete Design, Design of Concrete Structures A23.3-94, Canadian Standards Association, Canada (1994)
[7] Zsutty, T.C.: ACI Struct. J. 65(11), 943–951 (1968)
[8] Zsutty, T.C.: ACI J. Proceedings 68(2), 138–143 (1971)
[9] Karadelis, J.N., Koutselas, K.: In: Proceeding of 10th International Conference, Structural Faults and Repair., Engineering Technical Press, Edinburgh (2003)

[10] Koutselas, K.: Sustainable 'Green' Overlays for Strengthening and Rehabilitation of Concrete Pavements, PhD Thesis, Coventry University, Coventry (2010) (unpublished)
[11] British Standard, BS 812-110:1990 Testing Aggregates - Part 110: Methods for Determination of Aggregate Crushing Value (ACV), BSI, UK (1990)
[12] ASTM, ASTM Designation: C1439-99 Standard Test Methods for Polymer-Modified Mortar and Concrete, ASTM International, West Conshohocken (1999)
[13] Arrea, M., Ingraffea, A.R.: Mixed-mode Crack Propagation in Mortar and Concrete. Cornel University, New York (1982)
[14] Song, J., Kang, W., Kim, K.S., Jung, S.: Struct. Eng. Mech. 34(1), 15–38 (2010)
[15] Taylor, H.P.J.: ACI J. SP42, 43–78 (1974)
[16] Walraven, J.C.: J. Struct. Division 107, 2245–2270 (1981)
[17] Tada, H., Paris, P.C., Irwin, G.R.: The Stress Analysis of Cracks Handbook. The American Society of Mechanical Engineers, New York (2000)
[18] Iowa Department of Transportation, Method of Test for Determining the Shearing Strength of Bonded Concrete, Iowa Department of Transportation, Iowa (2000)
[19] ANSYS, ANSYS 12, SAS IP, Inc., Canonsburg, PA, USA (2009)
[20] Swartz, S.E., Lu, L.W., Tang, L.D., Refai, T.M.E.: Exp. Mech. 28(2), 146–153 (1988)

Influence of Residual Stress on PCC Pavement Potential Cracking

Xinkai Li, Decheng Feng, and Jian Chen

School of Transportation Science and Engineering, Harbin Institute of Technology, Harbin, P.R. China

Abstract. The load-induced stress and enviromental stress (curling and warping stress) are considered for PCC pavement slab cracking analysis in many design methods. However, the residual stress is not considered in design. Cement concrete slab volume changes are prevented by the structure surrounding the concrete or stable phases such as aggreagates will cause residual stress developing, and the residual stress may lead to early cracking in slab even before traffic load applying on it. Early cracking in cement concrete occurs frequently, however, currently residual stress are not considered in rigid pavement design method. Therefore, it is impossible to know the potential cracking of PCC pavement without knowing the residual stress level in a concrete pavement. In this paper, core-ring drilling is applied to release the residual stress in concrete pavement slab based on refering to measurements method in metal engineering. In the process of drilling strain gages are used to record the variation of stains on the surface of concrete slab. From test results, core-ring drilling can release stress and this method appears to be valid for measuring residual stress in concrete. Residual stress on the concrete slab surface can arrive at 7~16.8% of concrete flextual strength approximately based on finite element analysis of core-ring dirlling process.From the tests results and theoritical results if the residual stress is considered, PCC pavement top-down cracking risk will increase, which will influence the damage style of PCC pavement.

1 Introduction

Cement paste experiences volumetric shrinkage changes as a result of cement hydration, thermal variation or moisture losses in process of concrete strength development. Concrete structure surrounding and volumetrically stable phases inside the mixture prevents these volume changes and residual stress in concrete can develop. Especially in PCC pavement field, the concrete slab undergoes the complicated environmental factors, subbase conditions, boundary conditions and so on after paved, which could lead to residual stress existing on the top of slab. For instance, in 2005, the monitored curling behaviour of a single slab at the FAA's (Federal Aviation Administration) NAPTF (National Airport Pavement Testing Facility) indicates that residual stress exists and the effects are significant [1].

This phenomenon has been known well, but most design and testing approaches didn't consider the residual stress in concrete. In China, PCC pavement design assumes that the critical stress position is at the bottom of slab and bottom-up cracking is the dominating distress. However, in many field investigations the top-down cracking is also very common. Therefore, to properly evaluate the potential cracking in PCC pavement slab, the influences of residual stress should be investigated.

The critical stress of Concrete pavement is the combination of traffic load and environmental factors. Load-induced stresses can be well estimated from strains measurements and can be predicted closely to measured results by many computer models [2], but residual stress can't be well evaluated due to its forming complexity. Such as, with development of concrete strength, differential thermal expansion and differential drying shrinkage both can create residual stress. Creep further complicates matters as the history of loading and temperature/moisture effects all the stresses distribution in the slab [3]. Therefore, the direct measurement of residual stress in concrete pavement becomes a necessary step to quantify the critical stress, which influences the potential cracking of PCC pavement.

Since 1992 the first author uses the hole drilling technique for the deduction of the stress states in structural elements [4, 5]. This technique is used by means of the perforation procedure. In 2004 digital stereography method was used to evaluate residual stress of hole drilling technique in concrete structures [6]. In 2005 S. Pessiki and H. Turker provided the idea that using sensors to measure displacement changes before and after core drilling in concrete structure [7].Core-ring strain gage (CRSG) method was developed for concrete pavement and core-ring drilling tests in concrete beam were conducted at FAA's NAPTF in 2007. Based on FAA's concrete beam test, the feasibility of CRSG method in concrete structure was verified in 2009 [3, 8].

In this paper focus is on the CRSG method applied to measure the residual stress obtained from the strains registered by strain gages and the effects of residual stress level on PCC pavements potential cracking.

2 Brief Description of the Core-Ring Drilling Technique in PCC Pavement

The theory of CRSG method for measuring residual stress in concrete is similar to Hole-drilling strain gage method in metal engineering. The principle of full depth hole-drilling method which is first proposed by Mathar is the foundation of residual stress measurement [9]. *Figure 1* indicates that stress state of a thin plate which is subjected to a uniform residual stress, σ_x, before and after full depth hole drilling. The initial stresses, σ_r' and σ_θ', in radial and tangential direction at any point $P(R,\alpha)$ can be expressed in polar coordinates as following:

$$\sigma_r' = \frac{\sigma_x}{2}(1+\cos(2\alpha))$$
$$\sigma_\theta' = \frac{\sigma_x}{2}(1-\cos(2\alpha))$$
(1)

After a full depth hole is drilled through the plate, the stresses in the vicinity of the hole are significantly changed. The radial and tangential stress, $\sigma_r^{''}, \sigma_\theta^{''}$, at the point $P(R,\alpha)$ are:

$$\sigma_r^{''} = \frac{\sigma_x}{2}(1-\frac{1}{r^2}) + \frac{\sigma_x}{2}(1+\frac{3}{r^4}-\frac{4}{r^2})\cos(2\alpha)$$
$$\sigma_\theta^{''} = \frac{\sigma_x}{2}(1+\frac{1}{r^2}) - \frac{\sigma_x}{2}(1+\frac{3}{r^4})\cos(2\alpha) \qquad (2)$$

Where: $r = \frac{R}{R_0}(R \geq R_0)$; R_0 = hole radius; R = arbitrary radius from hole center

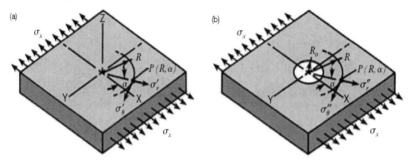

Fig. 1. Stress State before and after Hole is Drilled

Subtracting the initial stresses from the final stresses (after drilling) gives stress release at point $P(R, \alpha)$ due to drilling the hole as following.

$$\Delta\sigma_r = \sigma_r^{''} - \sigma_r^{'}$$
$$\Delta\sigma_\theta = \sigma_\theta^{''} - \sigma_\theta^{'} \qquad (3)$$

If the material of the plate is assumed to be homogeneous, isotropic, and linear-elastic behaviour, the relieved normal strains, ε_r, and tangential strains, ε_θ, at the point $P(R, \alpha)$ can be calculated by following equation:

$$\varepsilon_r = -\frac{\sigma_x(1+\upsilon)}{2E}\left[\frac{1}{r^2} - \frac{3}{r^4}\cos 2\alpha + \frac{4}{r^2(1+\upsilon)}\cos 2\alpha\right]$$
$$\varepsilon_\theta = -\frac{\sigma_x(1+\upsilon)}{2E}\left[-\frac{1}{r^2} + \frac{3}{r^4}\cos 2\alpha - \frac{4\upsilon}{r^2(1+\upsilon)}\cos 2\alpha\right] \qquad (4)$$

From the theoretical background for core-ring drilling method, substituting measured strain release into Eqn. (4) gives the stress release.

3 Parameters in CRSG Method

Effects of different parameters, including core-ring depth, strain gage locations, on CRSG method should be determined before slab test. The finite element model of square plate subjected to a uniform residual stress in one direction is selected for parameters analysis.

The size of core-ring should be larger than the max particle diameter of aggregate, so the diameter of core-ring 10cm is recommended and the gage length is usually 4.0cm. The size of square block is 20cm thick×2.0m square. Uniform stress field is represented by applying $\sigma_x=1.0$MPa on two opposite sides of plate. Poisson's ratio and Young's modulus of concrete plate is 0.18 and 30Gpa, respectively. *Figure 2* presents the finite element model and stress after a core-ring drilled.

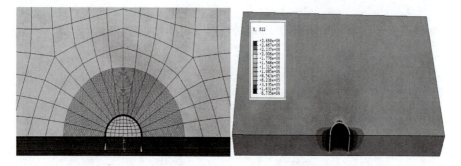

(a) FEM of Core-ring drilling (b) Stress contour plot after core-ring drilled

Fig. 2. Finite Element Model for Core-ring Method

Then results under different conditions are caculated by this finite element model. *Figure 3* a) and b) present the radial stresses variation in direction of gage 1 and gage 3 at different drilling depth, respectively, which are parallel and perpendicular to initial stress direction.

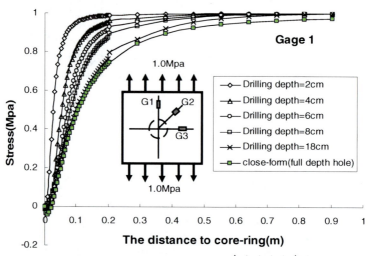

(a) The radial stress after different core-ring drilling depth ($\alpha = 0°$)

Fig. 3. The Radial Stress after Different Core-ring Drilling

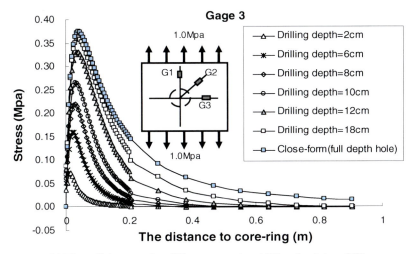

(b) The radial stress after different core-ring drilling depth ($\alpha = 90°$)

Fig. 3. (*continued*)

Figure 3 indicates that when the drilling depth is close to full depth the radial stresses at gage 1 and gage 3 are close to close-form solutions, which validate the right of FEM. From *Figure 3a)* the closer the distance to core-ring edge, the higher stress release is, as a result gage 1 should be place closer to core-ring edge. From *Figure 3b)* the distance of location of maximum stress to core-ring edge is from 1.0cm to 3.0cm, therefore, the distance of gage center to core-ring edge should be from 4.0cm to 6.0cm.

Figure 4 gives the radial stress variations in direction of gage 1 at different distance to core-ring edge.

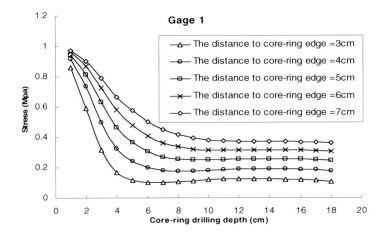

Fig. 4. The radial stress at different gage location ($\alpha = 0°$)

Figure 4 indicates that for gage 1 stress reduction changes quickly when the core-ring depth is small and when the depth is larger than 6cm~10cm (0.25~0.50h, where h is the thickness of slab) stress tends to be steady. Therefore, 0.50h is recommended for the core-ring drilling depth.

From above analysis, blind hole can not release the whole stress, and the strain recorded by strain gage is only a part of whole strain. Therefore, it is necessary to know the ratio of relieved stress for back caculation whole stress through recorded strain. When the distance of gage to core-ring edge is from 5.0cm to 7.0cm and the core-ring depth is from 0.25h to 0.50h, the variations of relieved strain is presented in *Figure 5*.

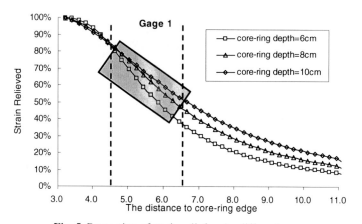

Fig. 5. Proportion of strain relief under different case

Figure 5 indicates that through core-ring drilling the ratio of relieved stress which is recorded by gage to total residual stress is about 50% to 70%.

4 Slab Test

In *Figure 6* six strain gages are pasted on the surface of concrete slab according to the parameters determined by finite element analysis.

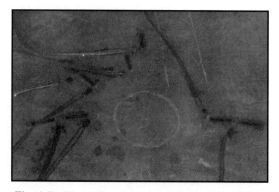

Fig. 6. Positions of strain gages and core-ring drilling

Influence of Residual Stress on PCC Pavement Potential Cracking 567

It should be noted that the friction in core-ring drilling process generates a massive heat, so that strain gages will take serious changes as a result of temperature increasing in slab. Strains will be stable with temperature decreasing after core-ring drilling stop. When the heat generated from core-ring drilling friction have dispersed completely, the core-ring drilling process continues for next depth.

Strain data from 1, 3, 5# gages during the process of core-ring drilling are plotted in *Figure 7*.

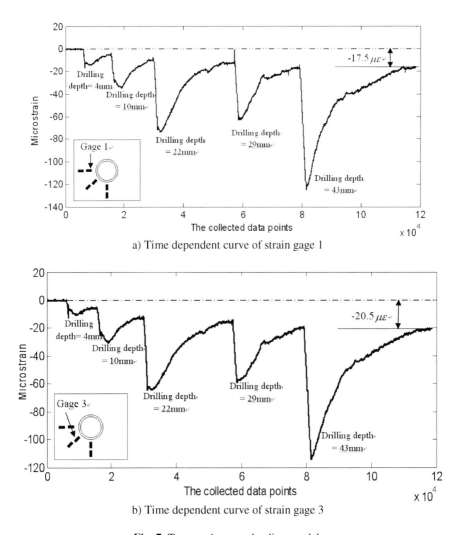

a) Time dependent curve of strain gage 1

b) Time dependent curve of strain gage 3

Fig. 7. Test results as no loading on slab

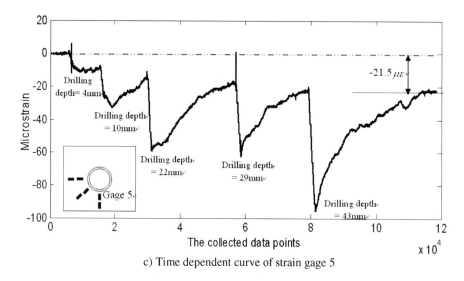

c) Time dependent curve of strain gage 5

Fig. 7. *(continued)*

Figure 7 indicates that core-ring drilling on slab can produce the release of residual stress without any load induced stress. For the residual stress on slab surface even shallow drilling depth can cause more stress release from test results. However, the stress release proportion decreases with drilling depth increasing, as core-ring drilling depth is more than 20mm stress release begin to slow down, as core-ring drilling depth exceeds 40mm stress release becomes stable, that is, as a result core-ring drilling depth is 0.25~0.5h (h is the slab thickness), stress release tends tobe steady which is close to FEM's results.

In this paper core-ring drilling test in slab is repeated for 6 times and the results at different drilling depth are listed in Table 1. The elastic modulus of concrete is 34.1 Gpa through elastic modulus test. According to Hooker's law the relieved stress can be caculated through the relieved strain which is recorded by strain gage.

Table 1. The strain recorded by strain gages in different CRSG tests

Drilling Step	Test 1				Test 2				Test 3			
	Drilling depth (mm)	Strian relief ($\mu\varepsilon$)			Drilling depth (mm)	Strian relief ($\mu\varepsilon$)			Drilling depth (mm)	Strian relief ($\mu\varepsilon$)		
		1#	3#	5#		1#	3#	5#		1#	3#	5#
1	4	5.0	6.0	9.0	3	7.0	4.0	bad	5	6.0	8.0	5.0
2	10	6.0	7.0	5.5	12	5.5	6.0		16	4.5	6.0	3.5
3	22	3.0	2.5	3.5	24	1.5	2.5		28	2.0	4.5	bad
4	29	2.5	4.0	4.0	36	bad	3.0		40	3.0	2.5	
5	43	1.0	0.5	-0.5					52	-0.5	0.5	
Total		17.5	20.0	21.5		14.0	15.5	/		15.0	21.5	8.5
Relied stress(MPa)		0.60	0.68	0.73		0.48	0.53			0.51	0.73	0.30

Table 1. *(continued)*

Drilling Step	Test 4				Test 5				Test 6			
	Drilling depth (mm)	Strian relief ($\mu\varepsilon$)			Drilling depth (mm)	Strian relief ($\mu\varepsilon$)			Drilling depth (mm)	Strian relief ($\mu\varepsilon$)		
		1#	3#	5#		1#	3#	5#		1#	3#	5#
1	5	7.0	6.0	7.0	4	4.0	6.0	6.0	5	7.0	5.0	8.0
2	11	6.0	4.5	5.0	12	6.5	4.5	3.5	12	5.5	4.5	6.0
3	20	4.0	3.5	3.5	21	3.5	2.5	bad	26	2.5	2.0	4.5
4	32	2.0	2.0	3.0	33	2.0	1.5		38	1.0	0	2.0
5	48	0.5	1.0	1.0	44	bad	0.5		50	-0.5		0.5
Total	43	19.0	16.0	18.5		16.0	14.5	9.5		16.0	11.5	20.5
Relied stress(MPa)		0.65	0.55	0.63		0.55	0.49	0.32		0.55	0.39	0.70

5 Influence of Residual Stress on PCC Pavement Potential Cracking

The maximum stress at top and bottom of slab under different axle loading are caculated through FEM software and listed in *Table 2*.

Table 2. Maximum stress at top and bottom of slab as different axle loading

Load (kN)	Stress under Single-axle load (MPa)		Stress under tandem axle load(MPa)		Stress under tridem axle load(MPa)	
	Top of slab	Bottom of slab	Top of slab	Bottom of slab	Top of slab	Bottom of slab
100	0.77	1.10	0.90	1.16	0.94	1.12
120	0.92	1.32	1.08	1.40	1.13	1.35
140	1.07	1.54	1.26	1.63	1.32	1.57
160	1.23	1.76	1.44	1.86	1.51	1.80
180	1.38	1.99	1.62	2.10	1.69	2.02
200	1.53	2.21	1.80	2.33	1.88	2.25
220	1.69	2.43	1.98	2.56	2.07	2.47
240	1.84	2.65	2.16	2.79	2.26	2.70
260	1.99	2.87	2.35	3.03	2.45	2.92
280	2.15	3.09	2.53	3.26	2.63	3.15
300	2.30	3.31	2.71	3.49	2.82	3.37

Table 2 indicates that the load induced maximum tensile stress is at the bottom of slab, so if only consider the load induced stress the bottom-up cracking will develop. However, from the previous test results residual stress exists on the top of slab and the residual stress accounts for 7%~16.8% of concrete flexural strength. If the flexural strength of concrete assumes to be 5.0MPa, the residual stress on the top of slab will be 0.35MPa~0.84MPa. Therefore, adding this part stress it will be found that the stress at the bottom of slab is close to the stress on the top of slab, which means the risk of top-down cracking is close to the risk of bottom-up cracking. As a result, the residual stress should be evaluated in PCC pavement design.

6 Conclusion

Residual stress induced by concrete volume changes is very complicated and influence the PCC pavement potential cracking. In this study, the core-ring drilling is applied to release the residual stress and strain gages are used to record the variations in the process of core-ring drilling. First, through 3D finite element analysis, core-ring depth, locations of strain gage and portion of stain relief in CRSG method are determined. Then, core-ring drilling tests are repeated for 6 times and the test results show that residual stress on the concrete slab surface can arrive at 7~16.8% of concrete flexural strength. Through finite element analysis, the residual stress increases the risk of top-down cracking of PCC pavement slab, which is close to the risk of bottom-up cracking.

Acknowledgements. The authors would show their appreciation to the FAA Airport Technology Research and Development Branch, Manager, Dr. Satish K. Agrawal and Yoh Foundation for its technical support through its representative Dr. Edward H Guo.

References

[1] Edward, G.H.O., DONG, M.Y., Daiutolo, H.: Curling under Different Environmental Variations as Monitored in a Single Concrete Slab. In: Proceeding of The 7th International Conference on Concrete Pavements, vol. I, pp. 1189–1203. Colorado Spring, Colorado (2005)
[2] Edward, G.U.O., Pecht, F.: Critical Gear Configuration and Positions for Rigid Airport Pavements – Observation and Analysis. In: Geotechnical Special Publication No. 154, Pavement Mechanics and Performance, pp. 4–14. ASCE (2006)
[3] Lange, D., Graham, M.: Development of Residual Stress Measurement for Concrete Pavements through Cantilevered Beam Testing. Master thesis, University of Illinois at Urbana-Champaign (2009)
[4] ASTM 837-95 Standard. Standard test method for determining residual stresses by the hole-drilling strain-gage method. American Society for Testing Materials
[5] ASTM 837-01 Standard. Standard test method for determining residual stresses by the hole-drilling strain-gage method. American Society for Testing Materials
[6] Hung, Y.Y., Long, K.W.: Evaluation of Residual Stresses in Concrete Structures by Digital Stereography. Interferometry VI: Applications (2004)
[7] Pessiki, S., Turker, H.: Theoretical formulation of the core drilling method to evaluate stresses in concrete structures. Experimental Mechanics 45(6), 359–367 (2005)
[8] Li, X., Feng, D.: Measurement of Residual stress in concrete by core-ring drilling method: influence of parameters. In: Proceeding of 89th Annual Meeting of Transportation Research Board, Washington, DC (2010)
[9] Mathar, J.: Determination of Initial Stresses by Measuring the Deformation around Drilled Holes. Trans., ASME 56(4), 249–254 (1934)

Plain Concrete Cyclic Crack Resistance Curves under Constant and Variable Amplitude Loading

Nicholas A. Brake[1] and Karim Chatti[2]

[1] Ph.D. Candidate, Michigan State University
[2] Professor, Michigan State University

Abstract. Concrete pavement structures are subjected to a complex combination of environmental and traffic loads which produce a unique distribution of stresses at the critical mid-slab edge. Moreover, the fracture propagation caused by this unique distribution of stresses is a complex process because it is both size and load history dependent. In this study, a series of quasi-static, constant and variable amplitude fatigue tests on simply supported single edge notched beam specimens were conducted. It is shown that variable amplitude testing can providea comprehensive assessment of fatigue life because the R-ratio, peak stress intensity, and load history effects can be assessed. The results of this study also suggest that the fatigue resistance curve under variable amplitude has a similar quality to that under constant amplitude loading; there is a positive decreasing slope that asymptotes to a zero slope condition beyond the critical crack extension. However, the results show the magnitude of the critical crack extension and the maximum fracture resistances are not the same under the two different loading conditions.

Keywords: fatigue fracture, load sequence, fatigue threshold, size effect, variable amplitude loading.

1 Introduction

Concrete pavements are subjected to a complex combination of environmental and traffic loads which produce a unique distribution of peak stress and stress ranges at the critical mid-slab edge. Moreover, the fracture propagation caused by thisunique distribution of stresses is complex because it is both size and history dependent due to the quasi-brittle nature of the material. Traditionally, high cycle fatigue damage has been quantified using Miner's Law that defines damage through the number of cycles to failure, $D = 1/N_f$, which is a function of the applied stress ratio (σ/MR) under constant amplitude loading [1; 2; 3]. One of the advantages of using this type of damage model is its computational efficiency. It is able to rapidly accountfor, process, and convertmillions of load repetitions to damageallowing multiple designs to be considered within minutes[4]. Some of the disadvantages however, are that it is insufficient in determining the in-situ state of damage because no information is given on the state of the material itself (no information on the stress-strain behavior and the reduction of the Elastic Modulus). In addition, it cannot account for size effect, load history effect, and variable amplitude loading without using some empirical calibration factors. Thus, there is a need to develop a concrete fatigue model that can account for all three of the aforementioned effects, and be able to maintain a comparable level of computational efficiency to the S-N approach.

In this paper, both constant and variable amplitude fatigue data are presented and discussed. A new model to predict the fatigue crack propagation under variable amplitude loading in both the transient and steady state is then presented. The model is substantiated by variable amplitude fatigue tests. In addition, differences/similarities between the fatigue response under both constant and variable amplitude loading are discussed.

1.1 Fatigue Fracture

In plain Portland Cement Concrete (PCC), the fatigue cracking process is similar to other quasi-brittle materials in that two distinct stages are observed: a transient stage where the crack growth rate is decreasing and a steady state stage where the rate is increasing [5]. This is due to an increase in residual stress behind the crack tip which causes the net stress intensity to decrease [6]. Under constant amplitude loading, this phenomenon has been investigated and modeled by simply dividing each cracking stage separately [7].

1.2 Constant Amplitude Loading

Numerous studies have been conducted showing that concrete fatigue fracture in the steady state range follows the well-known Paris Law shown in eqn. (1) [8; 9; 7]. The fatigue crack propagation rate (da/dN), where a is the crack length and N is the number of cycles, follows a power law in which stress intensity K_I is the argument. Eqn. (2) shows the expression for stress intensity as a function of the far field stress, σ_N, geometry, $k(\alpha)$, and size, D; where α is the crack (a) to depth (D) ratio. Eqn. (3) is the shape factor for a three point bending beam with a span to depth ratio (S/D) of 4 [13]. The coefficients, C and n, are considered to be material properties.

$$\frac{da}{dN} = C(\Delta K_I)^n \tag{1}$$

$$K_I = \sigma_N \sqrt{D} k(\alpha) \tag{2}$$

$$k(\alpha) = \frac{\sqrt{\alpha}\left[1.99 - \alpha(1-\alpha)(2.15 - 3.93\alpha + 2.7\alpha^2)\right]}{(1+2\alpha)(1-\alpha)^{3/2}} \tag{3}$$

However, eqn. (1) can only be used to describe the steady stage region. Subramaniam et al. [7] modeled the transient stage separately as a function of crack extension (Δa). The two regions were separated at a unique point called Δa_{bend}, which they found to be equal to the critical crack extension at failure, Δa_c, in the quasi-static crack resistance curve at peak load under quasi-static loading (see Figure 1).

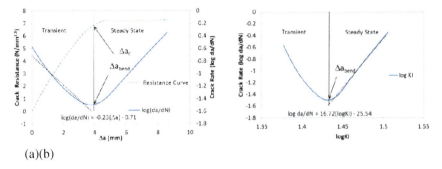

(a)(b)

Fig. 1. (a) Transient Stage Crack Propagation, (b) Steady-state Crack Propagation [10]

The crack propagation law in the transient stage is shown in eqn. (4)[7].

$$\frac{da}{dN} = C_1(\Delta a)^{n_1} \qquad (4)$$

Using this method, four fatigue parameters need to be defined: C and n in eqn. (1) and C_1, and n_1 in eqn. (4). The major limitation however is that it cannot predict crack growth under variable amplitude loading since stress intensity is not accounted for in the transient stage. This limitation can be overcome by inserting a crack resistance curve (K_{th}) into the Paris Law [6; 5; 10]. Note that in the literature, K_{th} is commonly referred to as the fatigue threshold. However, for other quasi-brittle materials, e.g. Alumina, it has been shown that before reaching the critical crack extension, the fatigue threshold will increase as a function of crack extension in similar fashion to the quasi-static resistance curve (R-curve) [5]. Thus, in this paper K_{th} will be referred to as the fatigue crack resistance curve.

Recently, Brake and Chatti [10] developed a method to determine the fatigue crack resistance curve under constant amplitude loading. The curve can be obtained if the crack propagation rate (da/dN), the crack extension (Δa), and the stress intensity (K_I) are known. They argue that there is a direct relationship between the decreasing crack propagation rate (transient stage) and the crack resistance curve which can be obtained by solving eqn. (5) for K_{th} and satisfying the following three conditions: 1) An intrinsic linear relationship between $\log(K_I-K_{th})$ and $\log(da/dN)$ exists [6], 2) The initial fatigue cracking resistance is zero, and 3) The fatigue fracture resistance in the post peak region (after Δa_c has been reached) should have zero-slope. This observation was made by Kruzic et al. for Alumina (another type of quasi-brittle material) under constant amplitude fatigue loading [5]. Figure 2 shows the behavior of a typical fatigue resistance curve under constant amplitude loading that satisfies the three conditions.

$$log\left(\frac{da}{dN}(\Delta a)\right) = logC + nlog(K_I(\Delta a) - K_{th}(\Delta a)) \qquad (5)$$

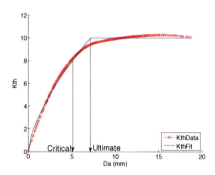

Fig. 2. Fatigue crack resistance curve (denoted here as K_{th})

The ultimate crack length corresponds to the crack length at which the resistance asymptotes to a constant value, $K_{th,ult}$. The crack resistance curve can then be described with eqn. (6) and used to predict fatigue cracking in eqn. (7). The mathematical form for K_{th} was chosen because it describes the general shape of a resistance curve; i.e., a monotonic increase pre-peak followed by a plateau region post-peak[11].

$$K_{th} = \begin{Bmatrix} K_{th,ult}\left[1 - \left(1 - \frac{\Delta a}{\Delta a_{ult}}\right)^m\right], \Delta a \leq \Delta a_{ult} \\ K_{th,ult}, \Delta a > \Delta a_{ult} \end{Bmatrix} \qquad (6)$$

$$\frac{da}{dN} = C(K_I - K_{th}(\Delta a))^n \qquad (7)$$

1.3 Variable Amplitude Loading

Slowik et al. [12] investigated the fatigue crack propagation (*da/dN*) under variable amplitude loading and developed the modified Paris Law shown in eqn. (8); where *ΔK₁* and *K₁* are the stress intensity range and the peak stress intensity, respectively; K_{IC} and K_{Imax} are the fracture toughness and the maximum K_I. The function *F* represents the jumps in the crack caused by abrupt changes in load. However, eqn. (8) is only applicable to the steady state cracking region.

$$\frac{da}{dN} = C \frac{\Delta K_I{}^n K_I{}^p}{(K_{IC}-K_{Imax})^q} + F(\sigma, \Delta a) \qquad (8)$$

2 Experimental Test Setup and Results

A total of 20 three-point single edge notched plain concrete beam specimens (TPB-SEN) were tested under quasi-static, constant, and variable amplitude fatigue loading. The concrete specimens had a span of 400 mm, a depth of 100 mm (S/D=4), and a width of 100 mm. The notch to depth ratio (α) was varied from 0.15 to 0.5 to assess the geometric effect. A Crack Opening Displacement (COD) gage was used to measure the crack mouth opening (CMOD) and was attached to a pair of knife edges which were mounted to the bottom face of the beam by a fast drying epoxy resin, as recommended by RILEM [4]. Six beams were tested under a cyclic quasi-static loading condition using a COD loading rate of 0.0005 mm/s. The remaining 14 specimens were tested in fatigue. Each specimen was subjected to a 2 Hz cyclical load. The constant amplitude specimens were subjected to a stress ratio (max load/peak load) of 0.85 (5 specimens) and 0.95 (5 specimens), and an R-ratio (min load/max load) of 0.05. The variable amplitude specimens were subjected to a uniform random distribution of stress ratios ranging from 0.5 to 0.9 and R ratios ranging from 0.5 to 0.05. Note the average peak load was obtained from the quasi-static tests.

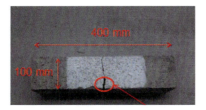

Fig. 3. (a)Notched concrete beam

2.1 Mix Characteristics

The concrete mix used in this research used an ASTMC-150 Type I cement, natural sand, and a limestone coarse aggregate (nominal maximum size of 25 mm). The water to cement ratio was 0.45 and the air content was 6.5%. The density was 2274.62 kg/m^3. The average 28 day modulus of rupture, MR, and the split tensile strength, f'_t, were 5.23 and 2.89 MPa, respectively. The 28 day compressive strength, f'_c, was 25 MPa. The specimens were cured for one year inside of a humidity room and then placed in ambient temperature for one more month to ensure minimal strength gain during fatigue testing.

2.2 Quasi-static Experiments

The average fracture toughness, $K_{IC} = K_{RC}$, and critical crack extension, Δa_c, for the six quasi-static specimens was 31 N/mm$^{3/2}$ and 6.22 mm, respectively. Figure 4 shows one quasi-static load displacement curve (a) and the resulting fracture resistance curve, K_R (b). The crack length at each cycle (for both quasi-static and fatigue loading) was calculated by the Jenq-Shah compliance technique [15].

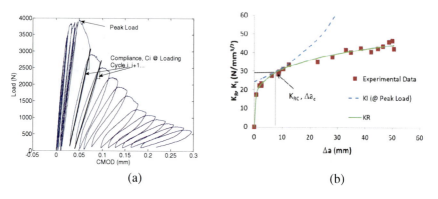

Fig. 4. Experimental Quasi-static Crack Resistance Curve [10]

2.3 Fatigue Experiments

2.3.1 Constant Amplitude Loading

Ten specimens were tested under constant amplitude fatigue loading. The crack lengths were calculated using the same Jenq-Shah compliance method used under quasi-static loading [7, 10, 15]. Throughout the fatigue test, the crack lengths and cycle number were recorded each time the crack grew by an amount of 0.25 mm. This data was then used to calculate log(da/dN), which was then filtered by a least squares cubic spline fit. Figure 5 shows the crack length, crack rate and the number of cycles. In addition, the cubic spline fit has been superimposed over the raw crack rate data.

The average Paris fatigue coefficients log C and n are -23.46 and 16.85, respectively (in eqn. 7). The average maximum fatigue resistance, $K_{th,ult}$, and the critical crack length, Δa_c, are 13.25 N/mm$^{3/2}$ and 5.80 mm, respectively (averaging both the 0.85 and 0.95 stress ratio results). Recall, the critical crack extension is defined as the crack extension at failure under quasi-static loading [7], which happens to correspond to a_{bend}, as shown in Figure 5a. $K_{th,ult}$ is defined as the constant post peak crack resistance that occurs after the ultimate crack extension has been reached. The average ultimate crack extension for these tests was 7.88 mm [10].

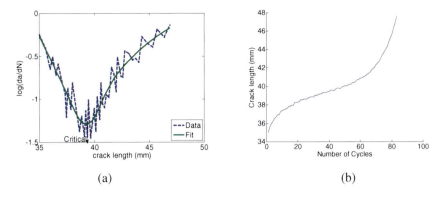

Fig. 5. (a) Crack rate & crack length, and (b) Crack length & Number of Cycles

2.3.2 Variable Amplitude Loading

Four variable amplitude fatigue tests were conducted. In this section a new fatigue model is presented which is founded on these four experimental tests. Eqn. (8) originally proposed by Slowik et al. [12] has been modified to account for the rising crack resistance observed in the transient region, as shown in eqn. (9). In addition, the stress intensity range was replaced with the term $(1-R)$; where K_{Ieff} is defined as the difference between K_I and K_{th}.

$$\frac{da}{dN} = C \frac{K_{Ieff}{}^n (1-R)^p}{(K_{IC} - K_{Imax})^q} \qquad (9)$$

In total, there are 4 fatigue and 3 crack resistance parameters governing the fatigue process. Each parameter was estimated using eqn. (10) and a least squares technique. Eqn. (10) describes the cumulative crack extension at each 0.25 mm crack growth interval; where Δa_f is the crack extension at fatigue failure and j is the total number of cycles in each 0.25 mm interval.

$$\Delta a_i = \sum_{i'=1}^{i} \sum_{1}^{j(i')} C \frac{K_{Ieff}{}^n (1-R)^p}{(K_{IC} - K_{Imax})^q} \Bigg|_{i=1}^{i=\frac{\Delta a_f}{0.25}} \qquad (10)$$

This method was used because in each sub interval i, there is a random distribution of R-ratios and stress ratios. This does not make it possible to use typical constant or variable amplitude methods to determine the coefficients because neither of the inputs is constant over a finite region. In addition, the crack rate cannot be monitored at each cycle because the instrumentation noise at this data acquisition frequency is larger than the actual readings. Thus, the fatigue parameters were evaluated using eqn. (10) at each crack interval i, and assessed by the actual cumulative crack extension, Δa_i.

The coefficients were determined by a two tier non-linear regression technique in MATLAB. The first tier optimized the 4 fatigue parameters using an initial guess for the crack resistance curve parameters (all greater than zero). The resistance curve parameters were then altered iteratively until the sum of the squares reached a minimum value. The average values of the fatigue parameters from all eight tests were the following: log C= -8.33, n=6.00, p=2.54, and q= -0.26. The ultimate crack extension, Δa_{ult}, was 13 mm, and the ultimate crack resistance, $K_{th,ult}$, was 17.25 N/mm$^{3/2}$.

Figure 6 shows the comparison between the experimental results and the prediction using eqn. (10) for one specimen. Figure 6a shows the crack extension as a function of the number of cycles, N. Figure 6b shows the crack rate versus crack extension. Reasonable predictions were observed.

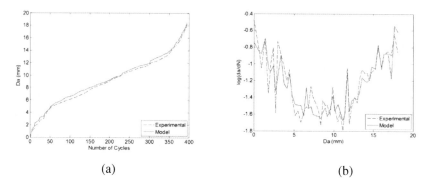

Fig. 6. Comparison between Experimental and Model: (a) Crack length v. Number of Cycles, and (b) Crack rate and Crack Extension

3 Discussion

Dominant fatigue crack growth in concrete is influenced by the stress intensity at the crack tip as well as crack bridging stresses distributed along the face of the crack (which are directly linked to crack resistance) [6,10,16]. Moreover, the accumulation of bridging stresses may be influenced by the applied peak stress and R-ratio. This may be why the post-peak behavior under quasi-static loading does not exhibit the same behavior under fatigue loading, i.e. the resistance curve has an increasing crack resistance. This observation was also reported for Alumina by Kruzic et al. [5]. The zero-slope post peak in fatigue may be caused by the steady degradation of the bridging stresses under fatigue loading. Under quasi-static loading, the degradationof the bridging stresses is not as severe and thus the resistance curve continues to increase (see Figure 4).

The fatigue resultsalso indicate load history dependence. Under constant amplitude loading, the average $K_{th,ult}$value was 13.25 N/mm$^{3/2}$ and the corresponding Δa_{ult} was 7.88 mm. Under variable amplitude loading, the average $K_{th,ult}$value was 17.25 N/mm$^{3/2}$ and the corresponding Δa_{ult} was 13 mm. The discrepancy may stem from the fact that on average, the applied stress ratio was lower under variable amplitude loading (0.7 compared to 0.85/0.95 under constant amplitude loading). This means that the bridging stresses should be higher under variable amplitude loading conditions, thus leading to a higher fatigue crack resistance. Table 1 shows the results of the constant and variable amplitude loading tests. Note that C and V represent constant and variable amplitude loading, respectively.

Table 1. Fatigue resistance parameters under constant and variable amplitude load

Constant Load	$K_{th,ult}$	Δa_{ult}	K_{IC}	Variable Load	$K_{th,ult}$	Δa_{ult}	K_{IC}
C-1	9.98	7.14	37.65	V-1	15	13	41.99
C-2	17.04	9.2	43.5	V-2	18	17	39.43
C-3	13.14	5.07	39.22	V-3	18	12	45.7
C-4	17.94	8.59	44.36	V-4	18	10	36.71
C-5	13.1	11.75	38.6	Mean V	17.25	13	40.96
C-6	7.19	4.04	32.04				
C-7	9.35	7.01	31.9				
C-8	17.59	11.36	38.9				
C-9	12.59	4.55	33.97				
C-10	14.55	10.12	38.22				
Mean C	13.25	7.88	37.84				

In addition, the transition between the transient and steady state stage is much shorter under constant amplitude loading. The transition under variable amplitude loading begins to occur at approximately 5 to 6 mm (refer to Figure 6b) and remains in this region until approximately 12 to 13 mm. The reason may be because the sudden overloads under variable amplitude loading cause an increase in crack tip plasticity which can lead to an increase in the critical crack extension.

4 Conclusion

In this paper, laboratory results from a series of quasi-static, constant and variable amplitude fatigue tests on simply supported single edge notched beam specimens were presented. It was shown that variable amplitude testing can provide a comprehensive assessment of fatigue life because the R-ratio, peak stress intensity, and load history effects can be assessed. The results of this study also suggest that although the fatigue resistance curve under variable amplitude shows a similar quality to that under constant amplitude loading (similar pre-peak shape), the magnitude of the parameters governing the curve, $K_{th,ult}$ and Δa_{ult} are significantly different. The following conclusions can be stated:

- Introducing the resistance term, K_{th}, into the modified Paris Law established by Slowik et al. enables one to predict fatigue cracking in both transient and steady-state regions.

- $K_{th,ult}$ and Δa_{ult} differ depending on loading history: Both of the parameters seem to be lower under constant amplitude loading as compared to variable amplitude loading.

The differences between the fatigue resistance curves under constant and variable amplitude loading may be caused by the stress range (difference between peak and valley stresses): Under constant amplitude loading, the stresses were higher and thereforethe crack opening should be larger, which will lead to lower bridging stresses and thus a lower crack resistance.

References

[1] Miner, M.A.: Journal of Applied Mechanics 12 (1945)
[2] Okamoto, P.A.: PCA R&D Serial No. 2213. Portland CementAssociation, Skokie, Ill (1999)
[3] Oh, B.H.: ACI Materials Journal 88, 41–48 (1991)
[4] Guide for Mechanistic-Empirical Design of New and Rehabilitated Pavement Structures, TRB, National Research Council, Washington, DC, NCHRP Report I-37A (2004)
[5] Kruzic, J.J., Cannon, R.M., Ager III, J.W., Ritchie, R.O.: Acta Materialia 53, 2595–2605 (2005)
[6] Li, V.C., Matsumoto, T.: Cement and Concrete Composites 20, 339–351 (1998)
[7] Subramaniam, K.V., Oneil, E.F., Popovics, J.S., Shah, S.P.: Journal of Engineering Mechanics 126, 891–898 (2000)
[8] Perdikaris, P.C., Calomino, A.M.: In: Proceeding of the SEM/RILEM International Conference on Fracture of Concrete and Rock, pp.64–69 (1987)
[9] Bazant, Z.P., Xu, K.: ACI Materials Journal 88, 390–399 (1991)
[10] Brake, N.A., Chatti, K.: Journal of Engineering Mechanics 138(4) (2012)
[11] Bazant, Z.P., Planas: Fracture and Size Effect in Concrete and Other Quasibrittle Materials. CRC Press, Boca Raton (1998)
[12] Slowik, V., Plizzari, G.A., Saouma, V.E.: ACI Materials Journal 93, 272–283 (1996)
[13] Tada, H., Paris, P.C., Irwin, G.R.: In: The Stress Analysis of Cracks Handbook. Del Research Corporation, Hellertown (1973)
[14] Shah, S.P., Swartz, S.E., Ouyang, C.: Fracture Mechanics of Concrete: Applications of Fracture Mechanics to Concrete, Rock and other Quasi-Brittle Materials, p. 552. Wiley-IEEE (1995) 0471303119
[15] Jenq, Y.S., Shah, S.P.: Journal of Engineering Mechanics 111, 1227–1241 (1985b)
[16] Fett, T., Munz, D., Geraghty, R.D., White, K.W.: Journal of the European Ceramic Society 20(12), 2143–2148 (2000)

Influence of External Alkali Supply on Cracking in Concrete Pavements

C. Sievering and R. Breitenbücher

Instiute for Building Materials, Department for Civil and Environmental Engineering, Ruhr-University Bochum, Universitätsstr. 150, 44801 Bochum, Germany
{christoph.sievering,rolf.breitenbuecher}@rub.de

Abstract. In the last few years cracking in concrete pavements was observed in several highway sections in some regions of Germany. Within the scope of extensive investigations no definite single cause of cracking could be observed in the majority of these cases. Besides incremental load-induced stresses by increasing heavy traffic, within this scope especially load-independent stresses due to hygral and thermal changes are of relevance. Several investigations additionally substantiate reaction products of an alkali silica reaction (ASR). Because of this multiplicity of potential causes for cracking it has not been definitely clarified up to now to which extent especially the ASR contributes to cracking in concrete pavements. Rather it seems that superposition and/or interactions of different mechanisms are responsible.

1 Introduction

In current discussions cracking in concrete pavements is often associated with an alkali silica reaction (ASR). However, even if relevant reaction products were detected in appropriate samples, cracks result in the rarest cases from an ASR solely. Rather it has to be assumed, that cracks were caused by a superposition of several stress impacts. Besides restraint stresses due to disabled thermal and hygral self-deformations, which mainly are raised by restraining of warping concrete pavements mainly are also subjected to traffic loads.

However, the influence of the alkali silica reaction in concrete pavements cannot be neglected. In such constructions especially the supply of alkalis from external sources has a substantial importance. Alkaline de-icing agents applied in the winter period penetrate more or less intensively into the concrete structure, which leads to a continuous increase of the alkali potential. In case of high traffic volume this penetration process is intensified by the following vehicles. The intrusion of the alkalis furthermore is particularly forwarded by already existing microcracks.

The durability of concrete pavements is strongly impaired by cracks. Therefore a comparatively high flexural strength of at least 5.5 N/mm² is generally required in plain concrete slabs to minimize the risk of cracking. Nevertheless cracking in concrete pavements cannot be avoided completely. However, not each crack results generally in an impairment of serviceability or leads to a hazard. In this context it should be considered that only 1 to 2 percent of the total 3,600 kilometres of concrete pavements are really damaged in Germany nationwide.

As long as the crack width remains sufficiently small (crack width < 0.3 - 0.4 mm) the serviceability of the pavement usually is not impaired. A critical stage has to be stated, if the edges of the cracks break out, particularly in the surroundings of joints, due to the over-rolling traffic. In this way the crack widths are expanding continuously by themselves which finally will result in a complete destruction of the material in this area. If larger particles are bursted out, the following traffic will be seriously endangered. By this the general conditions of the concrete pavement will be drastically worsened [1].

2 Influences Affecting Cracking in Concrete Pavements

2.1 Restraint and Residual Stresses

Due to the endless extension in longitudinal direction, deformations in pavements are practically completely restrained. In case of non-load-induced deformations, e.g. example by thermal or hygral changes, which extent constant over the complete cross-section, longitudinal restraint stresses are generated. Although in transverse direction a movement is enabled to some extent, restraint stresses also cannot be excluded completely in this case. If a thermal or a hygral gradient yield over the slab thickness, the slab would tend to warp and curl. These deformations are restrained by the dowels and tie bars resp. as well as by the dead load of the slab to a large extent. In addition the slabs are pressed down by traffic loads. In consequence appropriate flexural stresses with the tensile zone on the "cold" and/or „dry" side are generated. While due to longitudinal restraint stresses transversal cracks with nearly constant crack width are formed, flexural stresses result in wedge-shaped cracks with the opening on the dry or cold side. Furthermore residual stresses due to non-linear distributions of deformations are generated, which can result in map cracking [1].

2.2 Impact of Traffic on Microcracking

Within a case study of cracking in concrete pavements both load-independent influences (thermal and hygral) as well as load-dependent influences (i.e. traffic) have to be considered. Even if the permissible axle loads have not significantly increased in the past decades in most countries, the volume of the overall traffic and in particular of the heavy traffic has drastically grown in this period. Between 1970 and 2005 the transported cargo volume in Germany increased from 80 billion ton kilometers up to 400 billion ton kilometers, i.e. by five times. Besides the static loads especially the dynamic / cyclic loads can affect the microstructure of the concrete.

In appropriate tests the behavior of the concrete under cyclic loadings was investigated. It could be proved by a reduction in the dynamic E modulus, that already at upper stress levels between about 45 and 70 % of the concrete strength degradation processes take place in the microstructure far before fatigue failure is exceeded (Fig. 1) This degradation process is attended by the formation of microcracks in the concrete microstructure [2].

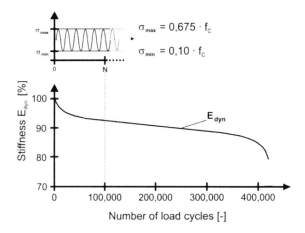

Fig. 1. Decreasing stiffness of concrete under cyclic load [2]

Such degradation processes can also be detected in concrete pavements by comparing the dynamic E-modulus between various lanes. In appropriate tests ultrasonic velocities of Rayleigh waves were measured at shoulder, near side lane and passing lane of various German highways. From these US-velocities then the dynamic E modulus of each lane was calculated. Figure 2 shows relative dynamic E-moduli of five highway sections, in which the dynamic E moduli of the near side lane and passing lane was related to the E modulus of the shoulder (same concrete in all lanes). It demonstrates that the relative dynamic E moduli of the traffic lanes are approximately 10 percent lower than the E modulus of the shoulder, which is assumed to be nearly unloaded within the investigated period and thus was used as reference-basis. This loss of stiffness can be related to cyclic loading due to traffic.

2.3 Expansion due to Alkali-Silica-Reaction

In the well-known alkali silica reaction amorphous silica (SiO_2) reacts with alkali hydroxide ions (NaOH, KOH), when simultaneously an appropriate amount of humidity is available. The reaction product is an alkali silica gel, which expands dramatically due to absorption of water. The expansion caused by interior swelling

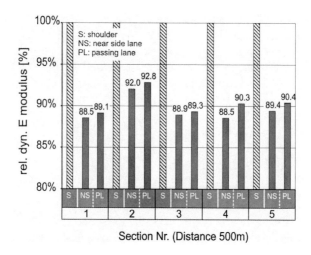

Fig. 2. Relative dynamic E moduli of different highway lanes / sections

is restrained by the surrounding concrete to a large extend. Thus an internal pressure occurs, which can accumulate up to 20 MPa [3]. Such stresses can easily exceed the tensile strength of the concrete with the consequence of internal cracking. Cracks due to an ASR are usually map- distributed and netlike. They are not limited only to the visible surface area but extend over the complete concrete structure. In the last few years cracking in concrete pavements often is associated with alkali silica reactions (ASR) [1]. However, even if relevant ASR-products are detected in appropriate samples, cracks result only in the rarest cases from that. In this context it should be considered, that ASR is the only cracking cause, which can be identified later by their reaction products. Other stresses, e.g. thermal and hygral influences or traffic loads do not leave their marks subsequently.

2.4 Alkalis from External Sources

The production of the swellable gel in the ASR is mainly influenced by solved alkalis in the pore solution of the concrete. In the case of concrete pavements not only internal sources but also external sources should be considered. As shown in Figure 5b below the external alkali supply / de-icing agents) plays a decisive role to the extent of an ASR in concrete pavements. In Figure 3 the amount of applied de-icing agents on German highways is illustrated.

It demonstrates that the amount of applied de-icing agents has successively increased in the last decade, which was caused by more and more prophylactic application in this period. However, due to the prophylactic application of sodium chloride nowadays this salt is pressed quite more intensive into the concrete by the over rolling traffic and hence forces the ASR subsequently. The intrusion of the alkalis is especially forwarded by the already existent cracks and microcracks respectively.

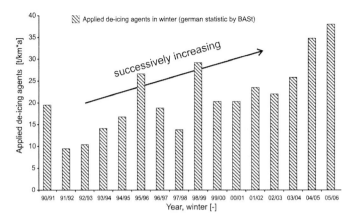

Fig. 3. Applied de-icing agents in winter in Germany

3 Investigations on Cracked Pavements by Analysis of Construction Documents and Additional Tests

3.1 Thermal Influences on Cracking in Concrete Pavements

Within the scope of extensive investigations highway sections (with and without cracking, map cracking as well as longitudinal cracks) were selected for studies referring to the causes for the cracking. In order to quantify these cracks, typical characteristics, e.g. the average quantity of cracks per slab, the average width of cracks per slab und the average cracking opening area, which is defined by the product of the total crack length and width per slab were determined.

In addition to the evaluation of the crack formation all available construction documents, i.e. daily construction records, reports on initial tests and quality control tests, weather records etc. were compiled and evaluated. In this context also the air temperatures during the concrete casting were compiled (Fig. 4).

In these analyses a significant influence of the temperature at concrete casting on the cracking could be proved in general (Fig. 4), whereat it can be assumed that the concrete temperature correlates more or less with the air temperature at casting. Except two subsections (10-6 and 10-8), which were additionally damaged by intensive hygral effects (rising ground–water), it could be verified, that nearly no cracking was observed in the concrete pavements after an average age of 14 years, when the temperature during casting was below 15°C to 20°C. Exceeding this temperature range (concreting in the summer season) cracking increases progressively.

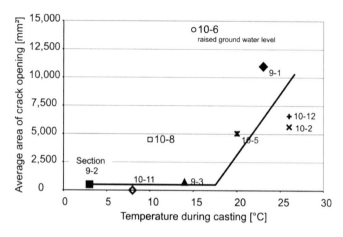

Fig. 4. Influence of the air temperature at concrete casting on crack opening area

3.2 Specific ASR- Tests on Concrete Cores at 60°C – with and without External Alkali Impact

To evaluate the ASR-potential within the inspected fields, specimens (cores) were sampled, on which the residual potential of ASR was determined by specific ASR-tests in the lab. To obtain information as soon as possible, the tests were performed in special climates at 60°C to accelerate the ASR. In this context also the influence of an external alkali supply was considered. So for this purpose the cores were divided into two parts. One subspecimen was stored constantly at 60°C above water according to the "Alkali-Guideline (2006)" [4], which is very similar to the "RILEM Recommended Test Method TC 191-ARP AAR-4" [5]. The other subspecimen was exposed to following cyclic procedure: 6 days above water at 60°C, 1 day at 20°C, 5 days at 60°C in a dry air and 2 days immersed in a sodium chloride solution with a concentration of 0.6 mol/l. With this procedure the test specimens were loaded for 16 cycles. During the complete test phase the deformations of the specimens were determined under both storing conditions. Even when no ASR takes place deformations to some extent will occur in the samples only due to thermal and hygral effects within the test procedure. In the tests without external alkali supply these can lead to an expansion up to about 0.3 to 0.4 mm/m, in case of an external alkali supply they can increase up to 0.5 mm/m. Thus, only deformations exceeding these values may be linked with an ASR and can be considered within the evaluation.

In Figure 5 the results of the deformation measurements without (Fig.5a) and with (Fig. 5b) external alkali supply are documented.

The comparison of Figure 5a and Figure5b demonstrates that the expansions increased significantly when alkalis are additionally supplied by external sources. This has been known also from other investigations [3]. Furthermore it could be proved, that the degree of expansion mainly is impacted by the degree of degradation / microcracking within the concrete microstructure.

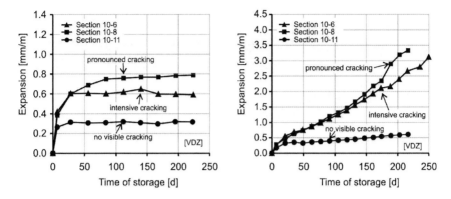

Fig. 5. Expansion of drilled cores due to storage at 60°C a) left: without external alkali supply b) right: with external alkali supply

4 Laboratory Investigations on Alkali-Penetration

In order to determine the impact of the degree of degradation in the concrete on the penetration of de-icing agents (sodium chloride) into the concrete structure, further laboratory investigations were carried out. Therefore beams (140 x 40 x 27 cm³) made of typical pavement concrete were cyclically loaded. The load applied simulated a superposition of thermal caused stresses and stresses due to traffic loads. This means that the maximum stress σ_{max} corresponded to thermal restraint stresses, which appear due to casting the concrete pavement in summer months. The amplitude of the cyclical load (σ_{max} - σ_{min}) characterizes the traffic load, which appears due to an over rolling truck with an axle load of 10 tons. Based on these configurations the concrete beams were cyclically loaded up to 5 million load cycles N with a frequency of 7 Hz in order to accelerate the test.

After various load cycles the dynamic E modulus was determined by measuring the ultrasonic speed of the Rayleigh wave at the tensile zone of the concrete beams. All dynamic E modulus were furthermore related to the E modulus of the beams before the first load was applied. On this basis a relative dynamic E modulus was determined in order to describe the deterioration of the concrete beam due to cyclic loading (Fig. 6).

Figure 6 demonstrates that the relative dynamic E modulus decreases continuously due to the applied cyclic loading. So the relative dynamic E Modulus decreases up to 90 percent after 5 million load cycles for instance, which can be related to an adequate degradation of the concrete microstructure.

In order to determine the influence of such a degradation on the penetration of a sodium chloride solution (3.0 w % sodium chloride) further experiments were conducted. For this purpose a sodium chloride solution was pressed into the tensile zone of variously pre-damaged beams (0, 2 million and 5 million load cycles N) by tire passes (TP), whereat the wheel was additionally loaded with a weight of 1 ton.

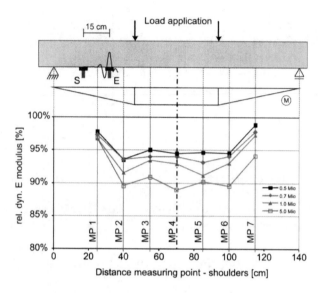

Fig. 6. Decrease of the relative dynamic E modulus due to cyclic loading

After 1 million tire passes concrete specimens (40 x 10 x 10 cm³) were cut transversally out of the beam. These specimens then were split in longitudinal direction (transversal axis of the beam) thereupon. Afterwards silver nitrate was sprayed on the fracture surface to detect the depth of chloride penetration by a color change ($c_d = 0.07$ mol / l).

Even if the silver nitrate test identifies only the presence of chlorides – the presence of sodium, which mainly influences the ASR, cannot be identified by such rapid tests – , this test is suitable for scanning large concrete specimens in a very short time as it is assumed, that the penetration of chloride and sodium are similar or even in line. At this point it should be mentioned either, that the effect of chromatography was disregarded so far. However, after measuring the penetration depth in steps of 1 cm a two-dimensional mapping was generated. In Figure 7 three penetration maps are shown, which differ in the applied load cycles (N) before the tire passes (TP) started.

If the concrete beams were not damaged before tire passes started, the sodium chloride penetrates 19 mm (mean value) into the concrete due to 1 million tire passes. Because of the increasing load cycles before the tire passes the penetration depth of the sodium chloride increases as well (22 mm after 2 million load cycles and 28 mm after 5 million load cycles). This means that the penetration depth of the sodium chloride increases noticeably by increasing degradation within the concrete microstructure.

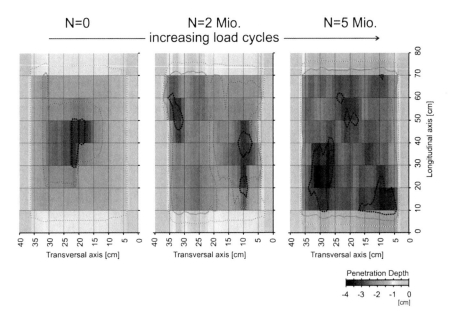

Fig. 7. Penetration depth of the sodium chloride solution after 1 Mio. tire passes

5 Final Discussion and Conclusion

Because of the multiplicity of potential causes for pavement cracking all possible influences have to be considered in order to reveal the relevant causes. In the minority of the investigated sections only one single reason could be found as responsible for the crack formation.

As presented in Figure 4, the temperature at concrete casting has significant influence on the initial cracking of concrete pavements. With increasing pouring temperature an increase of the average area of crack opening could be observed in the majority of the investigated highway sections. As a result of the interpretation of construction documents it can be assumed that the initial cracking in concrete pavements basically is influenced to a larger extent by thermal and hygral circumstances.

If microcracks due to cyclic traffic loads in combination with restraint stresses have already been formed in a concrete pavement, moisture as well as external supplied alkalis can easily penetrate deeper into the concrete microstructure. As demonstrated in Figure 5 in consequence such pre-existing degradations or damages play a decisive role for the further development of ASR-expansions, provided that simultaneously adequate portions of reactive aggregates are present in the concrete.

If the concrete has not been impaired in any way before, the impact of external alkalis plays only a minor role for the further ASR-expansion. Thus the extent of premature degradations, e.g. raised by thermal / hygral effects and traffic loads, has been proved to be the most important impact for the formation of an ASR in concrete pavements in cases of external alkali supply.

Acknowledgments. The authors are grateful to the Federal Ministry of Transport, Building and Urban Affairs (Germany) for financial support of this research project. The authors wish to acknowledge the cooperation with the Finger-Institute for Building Materials Science and the Research Institute of the German Cement Industry whose laboratory ASR tests were an important part of this project.

References

[1] Breitenbücher, R.: Potentielle Ursachen der Rissbildung in Betonfahrbahndecken. In: Proceedings of the 16, Ibausil, Weimar, Germany, September 20-23, vol. 1, pp. 1239–1254 (2006)
[2] Breitenbücher, R., Ibuk, H.: Experimentally based investigations on the degradation-process of concrete under cyclic load. Materials & Structures 39, 717–724 (2006)
[3] Stark, J., Freyburg, E., Seyfarth, K., Giebson, C.: AKR-Prüfverfahren zur Beurteilung von Gesteinskörnungen und projektspezifischen Betonen. Beton 12, 574–581 (2006)
[4] DAfStB, Vorbeugende Maßnahmen gegen schädigende Alkalireaktion im Beton (Alkali-Richtlinie). Deutscher Ausschuss für Stahlbeton (ed.) Beuth, Berlin (2006)
[5] Rilem, Detection of Potential Alkali-Reactivity– Accelerated method for testing aggregate combinations using concrete prisms. Recommended Test Method TC 191-ARP AAR-4. Materials and Structures 33, 290–293 (2000)

Plastic Shrinkage Cracking Risk of Concrete -- Evaluation of Test Methods

Patrick Fontana[1], Stephan Pirskawetz[1], and Pietro Lura[2,3]

[1] BAM Federal Institute for Materials Research and Testing, Berlin, Germany
[2] Empa Swiss Federal Laboratories for Materials Testing & Research, Dübendorf, Switzerland
[3] Institute for Building Materials, ETH Zurich, Switzerland

Abstract. Concrete pavements subjected to fast evaporation during the first hours after concrete placing are prone to plastic shrinkage cracking. In this paper the mechanisms by which evaporation of water leads to capillary pressure, and finally to shrinkage and cracking, are briefly discussed. In the main part, the paper presents results of a study on the evaluation of three different measurement setups. With two of these setups usually the efficiency of fibres in reducing plastic shrinkage cracking of concrete is assessed in Germany and Austria. The third one is an advanced setup that was developed in collaboration of BAM and Empa. It allows performing measurements of parameters that are important for the evaluation of the cracking risk.

1 Introduction

Plastic shrinkage of concrete can be defined as volume changes, which occur between the time of placement and the time of setting. It is increased under unfavourable climatic conditions due to accelerated evaporation from the fresh concrete surface and may result in substantial cracking, particularly in elements with large surfaces exposed to the environment (e.g. pavements and slabs). When the bleeding water on the surface of the concrete is consumed by evaporation, water menisci form. These menisci are accompanied by capillary pressure that compresses the concrete and cause it to shrink. In the literature, plastic shrinkage cracking of concrete is generally attributed to four main driving forces. Already in the plastic state of the concrete differential settlement, e.g. above reinforcing steel or at locations with a change in the thickness of the cross-section, and differential thermal strain due to a temperature gradient that is generated in the concrete by evaporative cooling, may occur [1]. The third driving force is the capillary pressure that may lead to initial cracking due to stress or strain concentration at locations at the concrete surface with existing flaws [2] or air entry [3]. Finally, plastic shrinkage is inducing tensile stresses in the case of restraint in the period of concrete setting, which may result in cracking when the stresses exceed the low strength of the concrete in the early age [4].

In many cases the cracks induced by plastic shrinkage are not limited to the superficial area but penetrate deeply the concrete element. This happens even in the case of concrete reinforced with rebars, since the plastic concrete is not able to sufficiently transfer forces to the reinforcements due to the weak bond. Therefore plastic shrinkage cracking may degrade significantly the durability of the concrete and requires often a complex and expensive rehabilitation [5]. Of course, plastic shrinkage can be avoided or at least reduced by proper curing, e.g. by keeping the surface of the fresh concrete wet. This is state of technology and limiting the evaporation of water immediately after placing the concrete is required by standards and guidelines. However, in some cases appropriate curing is difficult and in some cases it is applied too late or it is not applied at all. Therefore the use of shrinkage-reducing admixtures (SRA) might be reasonable. By reducing the surface tension of the water, SRA reduce capillary pressure, evaporation and settlement, and thus, the potential for plastic shrinkage cracking [1].

Another possibility for the reduction of plastic shrinkage cracking is given by a fibre-reinforcement. Results of a comprehensive experimental study are reported in [6]. They reveal that polymer as well as steel fibres may reduce plastic shrinkage cracking. The efficiency of the fibres was depending on their geometry and on the amount of fibre addition. Qi et al. [7] conclude that fibres increase the stiffness of the concrete, and thus reduce its settlement and plastic shrinkage cracking.

2 Test Methods

2.1 Test Method Recommended by DIBt

In Germany, a test method for plastic shrinkage cracking of concrete is recommended by Deutsches Institut für Bautechnik (DIBt) for the approval of fibres intended for the use as concrete additive [8]. The measurements are performed on a slab ($160 \times 60 \times 8$ cm^3) that is cast in a steel frame. Restraint of the concrete is provided by reinforcements fixed to the steel frame (Figure 1a).

Fig. 1. Measurement setup recommended by DIBt [8]

After compacting and finishing the fresh concrete, the slab is covered with a transparent wind tunnel. In the presented study, a wind speed of approx. 5 m/s inside the tunnel was generated by a ventilator (Figure 1b). The tests were performed in an environmental chamber at 30 °C and 50 % relative humidity. Evaporation and horizontal plastic shrinkage were measured on a separate concrete specimen (30 × 30 × 8 cm^3) that was also placed in the wind tunnel (Figure 2, left). In addition the settlement of the concrete surface was measured with non-contact laser sensors and measurements of the capillary pressure were performed with tensiometers [3], consisting of a pressure sensor connected to a metallic tube by a rubber hose (Figure 2, right).

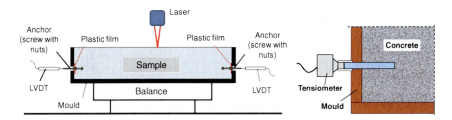

Fig. 2. Schematics of DIBt setup for measurement of plastic shrinkage and evaporation (left) and of setup for measurement of capillary pressure (right).

Cracking in the hardening concrete was measured at the end of the test after 7 hours. According to DIBt recommendations, the fibres are approved if the cracking and plastic shrinkage of the fibre-reinforced concrete is apparently less pronounced than in the control concrete mixture without addition of fibres. It is noticed that DIBt recommendations do not require the quantification of the cracks.

2.2 Test Method According to ÖVBB Guideline

Similar to German requirements, plastic shrinkage cracking of concrete has to be tested for the approval of fibres in Austria. Measurement setup and experimental procedure are determined in the guideline of Österreichische Vereinigung für Beton und Bautechnik (ÖVBB) for fibre-reinforced concrete [9]. The measurements are performed on concrete rings with a thickness of 150 mm and a width of 40 mm. The concrete specimens are cast in a mould consisting of two concentric steel rings fixed on a rigid panel (Figure 1). Inside the outer ring of the mould, 12 steel sheets are placed to constrain the concrete and to initiate cracking. The air on top of each specimen shall be removed by a fan connected to a circular air funnel that is placed directly above the concrete rings generating a wind speed of 4 m/s on the concrete surface. In the presented study, of each concrete mixture, three specimens were cast and placed in the wind tunnel as shown in Figure 1b, one of them on a balance to determine the water loss. Measurements of settlement, capillary pressure and evaporation were performed on separate samples as shown in Figure 2, which were also placed in the wind tunnel. The duration of the measurements was 5 hours at 20 °C and 50 % relative humidity. The wind speed was approx. 4 m/s. After the end of the test, the lengths of all observed cracks in each specimen were accumulated and the total crack lengths of fibre-reinforced concrete and control concrete were compared. According to ÖVBB guideline, the fibres are approved if the reduction of the total crack length is ≥ 80 %. A minimum threshold of crack width reduction is not defined in the ÖVBB guideline.

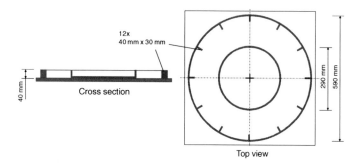

Fig. 3. Shrinkage ring according to Austrian guideline for fibre-reinforced concrete [9]

2.3 Proposed Advanced Test Method

A new measurement setup was developed that allows precise and repeatable testing. Homogeneous and rectified air flow on the surface of the concrete specimens is generated by a special blower unit. Measurements of evaporation, settlement and plastic shrinkage cracking were performed simultaneously on the same concrete samples. The moulds according to ASTM C 1579 [10] are provided with steel inserts (stress riser) to initiate cracking (Figure 4). The moulds were covered with a transparent wind tunnel made of acrylic glass that facilitated inspections of the concrete surface during the measurements (Figure 5). The capillary pressure was measured on separate samples as shown in Figure 2, which were also placed in the wind tunnel.

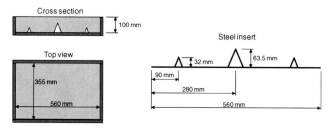

Fig. 4. Mould for measurement of plastic shrinkage cracking according to ASTM C 1579 [10]

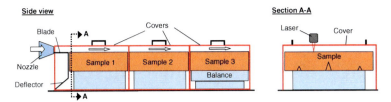

Fig. 5. Schematic of the new measurement setup using moulds according to ASTM C 1579 [10]

The complete measurement setup was placed in an environmental chamber at (37 ± 1) °C and (27 ± 1) % relative humidity. Temperature, velocity and relative humidity of the air flow on the surface of the concrete samples can be monitored with coupled temperature/RH sensors and anemometers being inserted into the wind tunnel through holes in the covers at different locations. The measurements revealed a homogeneous air flow with a velocity of (6 ± 0.5) m/s. The standard deviation of the air velocity in the cross section of the wind tunnel was ≤ 0.2 m/s. Typically the velocity and the temperature of the air flow gradually decreased along the wind tunnel while the relative humidity gradually increased from 27 % at the front to 32 % at the end of the wind tunnel.

According to ASTM C 1579 [10] the measurements were stopped when the end of setting of the concrete occurred and the samples were covered with plastic sheets to avoid further evaporation. The crack measurements were performed 24 hours after mixing and the crack reduction ratio (CRR) was calculated according to Eqn. (1).

$$\text{CRR} = \left[1 - \frac{\text{average crack width of fibre-reinforced concrete}}{\text{average crack width of control concrete}} \right] \times 100\% \qquad (1)$$

3 Materials

The investigated polypropylene (PP) fibres had a rectangular cross section $(1.4 \times 0.1 \text{ mm}^2)$ and a length of 40 mm. Their efficiency was evaluated using two basic concrete mixtures, which differed in cement type and w/c ratio (Table 1).

Table 1. Mixture compositions and fresh concrete properties

Constituent	Mixture C1		Mixture C2	
	Control	FRC	Control	FRC
Cement content (kg/m³)	360 [1]		360 [2]	
Water (kg/m³)	270		234	
Limestone powder (kg/m³)	161		148	

Table 1. *(continued)*

PP fibres (kg/m^3)	-	2.3	-	2.3
Aggregate 0/8 mm (kg/m^3)	1469	1461	1576	1569
w/c ratio	0.75		0.65	
Density (kg/m^3)	2243	2237	2256	2225
Flow (mm)				
5 min after end of mixing	695	640	500	470
30 min after end of mixing	635	610	430	430
60 min after end of mixing	605	570	430	420
Air content (%)	0.8	1.0	2.5	3.2

[1] CEM I 32.5 R, [2] CEM I 42.5 R

Limestone powder with a Blaine fineness of 5100 cm^2/g was added to increase the content of particles smaller than 250 µm to 680 kg/m^3. It is noticed that the flow of the fibre-reinforced concrete (FRC) mixtures directly after mixing was lower than that of the control mixtures without addition of PP fibres, but already after 30 minutes this was not the case anymore with mixture C2.

4 Results and Discussion

For the tests performed according to DIBt recommendations the basic mixture C1 was used. After finishing the samples they were covered with the wind tunnel and an "average" wind speed of 5 m/s was generated. The time between initial mixing and start of the measurements was approx. 1 hour. It has to be noticed that a precise determination of the wind speed was not possible with this measurement setup because the axial blower generated a very heterogeneously rotating air roll inside the wind tunnel. In the centre of the wind tunnel cross section, the measured air speed was approx. 3 m/s only, whereas it increased to approx. 6 m/s in the border areas.

In the control concrete slab the first cracks were observed 2 ½ hours after start of the measurements. In the slab with the FRC, the first cracks occurred approx. 30 minutes later. At the time of initial cracking no bleeding water was visible anymore on the surface of both concrete slabs. Figure 6 shows on the left that the addition of PP fibres resulted in a reduction of plastic shrinkage cracking. The measurement of the cracks after the end of the test revealed that not only the number of the cracks but also the crack lengths and in particular the crack widths were reduced.

The settlement of the concrete was slightly reduced by the addition of the PP fibres (Figure 6, right). This can be attributed to the higher stiffness of the FRC (smaller flow) compared to the control concrete. The faster decrease of the capillary pressure in the FRC indicates as well its minor settlement, which comes

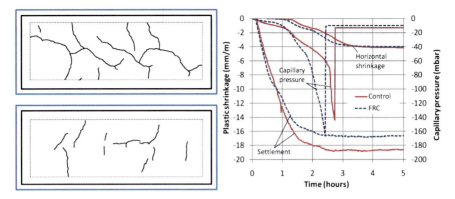

Fig. 6. Test results according to DIBt recommendations. Left: Crack patterns of control concrete (top) and FRC (bottom). Right: Plastic shrinkage and capillary pressure of the concrete mixtures.

along with reduced bleeding. With both mixtures the horizontal shrinkage started roughly at the time when the capillary pressure significantly decreased and the settlement slowed down. Due to addition of the PP fibres the horizontal shrinkage started slightly earlier, but there was no influence on the total amount of shrinkage. Settlement and horizontal shrinkage were completed in both mixtures after 4 hours. The rate of evaporation was almost equal in both tests.

For the tests performed according to the ÖBVV guideline the basic mixture C2 was used. After finishing the samples and covering with the wind tunnel an "average" wind speed of 4 m/s was generated. The time between initial mixing and start of the measurements was as well approx. 1 hour. Initial cracking was observed after 2 hours. As in the test with the large slabs according to DIBt recommendations, the addition of the PP fibres resulted in reduced plastic shrinkage cracking (Figure 7, left). The total crack length in the FRC was reduced by 56 % compared to the control mixture, which did not meet the requirements (CRR ≥ 80 %) of ÖBVV guideline. Since the PP fibres obviously reduce crack lengths as well crack widths, it seems more reasonable to evaluate the efficiency of the fibres with a CRR that is based on the total crack area. The total crack area can be calculated as the sum of single crack length multiplied with the corresponding average crack width. In this case the CRR would amount to 87 % (see also Table 2). Moreover, the approach of a CRR based on the total crack area is more adequate to the relevance of cracks concerning the durability of concrete structures, since transport of water and harmful ions is much more depending on crack width than on crack length.

Interestingly the shrinkage rings in the centre and at the end of the wind tunnel, i.e. the samples most distant from the ventilator, showed a more pronounced cracking than the shrinkage rings that were placed directly in front of the ventilator (with no cracking and only one crack, respectively). This was probably due to the non-uniform air flow inside the wind tunnel, which removed the humidity efficiently only in the rear part of the wind tunnel.

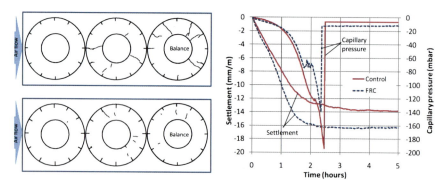

Fig. 7. Test results according to ÖVBB guideline. Left: Crack patterns of control concrete (top) and FRC (bottom). Right: Settlement and capillary pressure of the concrete mixtures.

In contrast to the mixture C1, with the mixture C2 larger settlement was measured in the case of the PP fibre reinforced concrete (Figure 7, right). This cannot be explained by the stiffness of the concrete mixtures, which was almost equal after 30 minutes (see Table 1). Also the capillary pressure developed equally in both mixtures during the first 90 minutes. At the end of the test after 5 hours, the plastic settlement of both mixtures was completed.

For the tests performed with the new measurement setup using moulds according to ASTM C 1579, the basic concrete mixture C1 was used. Cracking was observed after approx. 2½ hours for the control mixture without fibres and the FRC. At this time also the end of setting occurred for both mixtures. All cracks formed above the stress riser of the steel insert in the centre of the samples (Figure 8, left). In the control concrete slabs only a single large crack occurred, whereas several smaller cracks were observed in the FRC slabs. Figure 8, right shows that the addition of the PP fibres apparently did not affect the settlement. The decrease of the capillary pressure occurred only slightly earlier for the FRC compared to the control concrete, corresponding with the settlement, which was more similar for FRC and control concrete than in the tests with the same basic concrete mixture shown in Figure 6. The settlement of both mixtures was probably not fully completed after 2½ hours when the measurements were stopped at the end of concrete setting.

The analysis of the crack measurements of all tests is given in Table 2. At first it makes clear that the fibre-reinforcement reduces the crack lengths as well as the crack widths and that it has a minor influence on the crack length when the ASTM shrinkage cracking moulds are used. In this case the fibres predominantly reduce the width of the cracks. Table 2 shows also that the scatter of experimental results is lower in the case of the proposed advanced test setup, where a homogeneous air flow is applied to the concrete samples.

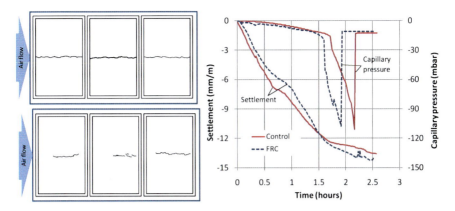

Fig. 8. Test results with new measurement setup using ASTM shrinkage cracking moulds. Left: Crack patterns of control concrete (top) and FRC (bottom). Right: Settlement and capillary pressure of the concrete mixtures.

Table 2. Analysis of the crack measurements

Type of test / concrete mixture / no. of samples		Total crack length / STDV [1] (mm)	Average crack width / STDV [1] (mm)	Crack area / STDV [1] (mm^2)
DIBt (mixture C1) 1 sample each	Control	3733	0.57 / -	2729
	FRC	1838	0.21 / -	421
	CRR (%)	*51*	*63*	*85*
ÖVBB (mixture C2) 3 samples each	Control	1108 / 349	0.31 / 0.18	477 / 164
	FRC	487 / 153	0.12 / 0.01	61 / 18
	CRR (%)	*56*	*61*	*87*
Advanced setup (mixture C1) 3 samples each	Control	980 / 8	0.65 / 0.22	642 / 73
	FRC	661 / 52	0.12 / 0.02	96 / 15
	CRR (%)	*33*	*81*	*85*

[1] The standard deviation (STDV) of the test results is calculated based on the number of samples used in each test.

5 Conclusions

The experimental results showed that the investigated test methods are basically appropriate for the evaluation of the efficiency of fibres in reducing the plastic shrinkage cracking of concrete. However, it was observed that a heterogeneous air flow, as it is generated for example with an axial ventilator, may affect negatively the reproducibility of the test results, since the wind velocity and generally the evaporation might be very different for samples of different sizes located at different places in a wind tunnel. In this case, the measurement of the evaporation rate is very important in order to detect potential anomalies during the test procedure.

The evaluation of the fibre efficiency by visual inspection only, as recommended by DIBt has to be regarded as inadequate. It is suggested to quantify cracking by means of a crack reduction ratio in order to evaluate the efficiency of fibres objectively. Since a fibre-reinforcement is reducing the crack widths as well as the crack lengths, it is more reasonable to calculate the crack reduction based on the crack areas. This approach would also consider the fact that the crack width is more important for the durability of concrete structures than the crack length. This is not taken into account by the requirements of ÖVBB guideline and would need to be improved.

Moreover it is recommended to establish a minimum crack reduction ratio for the proof of the fibre efficiency. To avoid the overestimation of their efficiency, it is also reasonable to define a minimum crack width in the control concrete and to adjust the rate of evaporation by variation of the experimental parameters temperature, relative humidity and air velocity.

The results of the presented study have indicated that the shrinkage, measured as horizontal deformation, is not necessarily completed at the time of the end of concrete setting. Subsequent shrinkage would not be denoted as plastic shrinkage anymore by definition. However, since further crack growth can be expected, it seems not reasonable to limit the duration of the shrinkage cracking test to the time of setting but to extend it until significant horizontal shrinkage or crack growths is completed.

The measurement setups are applied in current research that is focussed on the effect of different types of fibres and of shrinkage reducing admixtures. Future studies will comprise also the evaluation of curing compounds.

References

[1] Lura, P., Pease, B., Mazzotta, G.B., Rajabipour, F., Weiss, J.: ACI Mat. J. 104(2), 187–194 (2007)
[2] Scherer, G.W.: J. Non-Cryst. Solids 144, 210–216 (1992)
[3] Slowik, V., Schmidt, M., Fritzsch, R.: Cem. Concr. Comp. 30, 557–565 (2008)
[4] Wischers, G., Manns, W.: Beton 23(4), 167–171 (1973)
[5] Schmidt, D., Slowik, V., Schmidt, M., Fritzsch, R.: Beton- Stahlbetonbau 102(11), 789–796 (2007)
[6] Balaguru, P.: ACI Mat. J. 91(3), 280–288 (1994)
[7] Qi, C., Weiss, J., Olek, J.: Mater. Struct. 36(6), 386–395 (2003)
[8] Schriften des DIBt: Zulassungs- und Überwachungsgrundsätze Faserprodukte als Betonzusatzstoff. Reihe B, Heft 18 (2005)
[9] ÖVBB: Richtlinie Faserbeton (2008)
[10] ASTM C 1579, Standard Test Method for Evaluating Plastic Shrinkage Cracking of Restrained Fiber Reinforced Concrete (Using a Steel Form Insert) (2006)

Compatibility between Base Concrete Made with Different Chemical Admixtures and Surface Hardener

M.T. Pinheiro-Alves[1,2], A.R. Sequeira[1], M.J. Marques[1], and A. Bettencourt Ribeiro[3]

[1] University of Évora, School of Sciences and Technology, Portugal
[2] Centre of Territory, Environment and Construction (C-TAC) - Group of Sustainable Construction, Portugal
[3] Nacional Laboratory of Civil Engineering (LNEC), Portugal

Abstract. Many cases of cracking and detachment of the concrete surface have appeared in concrete pavements where surface hardeners were used in Portugal. The main causes for cracking and delamination of trowelled concrete pavements are several and it is essential to control bleeding and the time available to perform the finishing operations. Several base concretes were made with different chemical admixtures and one type of surface hardeners. The purpose of this study was evaluating the influence of the "open time" for each chemical admixture. Results show that the type of chemical admixture has a great importance in the control of the "open time" and consequently avoiding cracking and detachment.

1 Introduction

Concrete pavements can be used in airports, highways, city streets, parking lots and industrial zones. They have been a mainstay of our infrastructure and has evolved greatly in recent years, mostly due to the incorporation of chemical admixtures. In this paper will focus primarily on concrete pavements applied in parking lots and industrial zones.

A significant increase of problems in concrete pavements has been detected in Portugal, concerning pavements where surface hardeners were used, especially in car parks, industrial areas, warehouses, etc.. Many cases of cracking and detachment of the concrete surface have appeared. The negative effect of these abnormal situations has important economic implications, since the necessary work to correct them involves large areas of reconstruction and the use of expensive repairing materials, and also because it often involves the delay of the period of construction and the beginning of the infrastructure operation.

The main causes for cracking and delamination of trowelled concrete pavements referred to in the literature are two [1]: entrapped bleeding water and air beneath the top layer, closed by finishing operations (less permeable); and the spreading of soft mortar over already hardened adjacent areas of floor surface (over-layering). These two causes depend on several factors such as the finishing operations, curing and concrete mixture.

The increase of cracking and delamination problems in recent years may be related to the fact that the characteristics of both cement and concrete have been changed, particularly the increase in the fineness and quantity of tricalcium silicates (C_3S) in cements and the use of more effective water reducers in concrete.

It is thus essential to control bleeding [2] and the time available to perform the finishing operations due to the more rapid evolution between plastic and hardened phases of concrete (open time).

2 Materials and Methods

In laboratory, several base concretes (0.60 x 0.30 x 0.10 m^3) were made with four different chemical admixtures. Then, they were finished with one type of surface hardener (quartz). The cement used in the mixes was Portland limestone cement CEM II/A-L 42,5R, the chemical properties of which are presented in Table 1.

Table 1. Chemical properties of Portland limestone cement

SiO_2 %	16.36
Al_2O_3 %	4.40
Fe_2O_3 %	2.81
CaO %	60.84
MgO %	0.95
Na_2O %	
K_2O %	
SO_3 %	2.69
L.O.I. 950°C	7.64
R.I. %	1.18

L.O.I. – loss of ignition; R.I. – insoluble residue.

The chemical admixtures used were: two plasticiser/water reducer (Pozzolith 390NP - water and lignosulphonate, Pozzolit 540 - water and lignosulphonate), a superplasticiser/strong water reducer (Sikaplast 898 - combination of modified polycarboxylates) and a superplasticiser/strong water reducer with high efficiency (Rheobuild 561 - water and naphthalene sulphonate).

For all the compositions the consistency was determined by slump test and maintained at the class of S3 (100-150 mm) according EN 206-1:2007, as indicated in Table 2.

Table 2. Mixture proportions of concretes (kg/m³)

	390	**540**	**898**	**561**
Cement	340	341	340	341
Sand	730	730	730	730
Coarse aggregate 1	558	558	558	558
Coarse aggregate 2	556	556	556	556
Water	170	155	143	155
Superplasticiser	2.7	3.4	3.4	3.4
w/c	0.50	0.46	0.42	0.46
Slump (mm)	143	110	150	125

Major problems occur in aggressive situations. These aggressive situations are high temperatures and wind speed. To simulate them, two tunnels were built with a fan placed in each one of them. These allowed controlling temperature and humidity and also measuring the different evaporation rates inside the tunnels. Also, new equipment was developed to trowel the small slabs, Fig. 1.

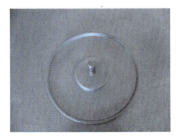

Fig. 1. Equipment used to trowel the slabs

The surface hardener was applied to each one of the base concretes at different times. The times considered were: the end of setting time (27,6 N/mm²) and 45 minutes after the setting time. These times were chosen because most of the problems normally occur when surface hardener is applied too late.

Due to the adverse situations created, there was not enough water at the surface of the base concrete and at the end of the setting time. Additional water had to be introduced. This replicates real application conditions whenever the times are exceeded.

Some of the tests carried out were slump test, setting times of base concretes and pull-off force. The pull-off test was used to determinate the resistance of the surfaces finishes and the homogeneity between the surface finish and the concrete slab. The dollies used had 50 mm of diameter.

3 Results and Discussion

The different setting times of base concretes are presented on Fig. 2.

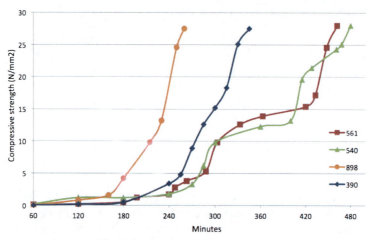

Fig. 2. Setting times of base concretes

The chemical admixtures influence the setting times in different ways. The results indicate that there is not a specific correlation between the use of plasticiser or superplasticiser, with regard to setting times.

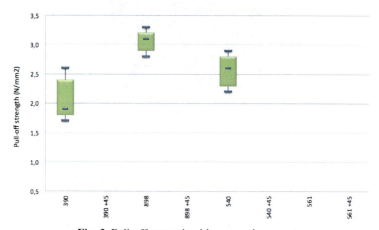

Fig. 3. Pull-off strength with rupture in concrete

Table 2 shows that despite the fact that 390 have more water in mixing, its behavior (Fig. 3-4) is not very different from the others, especially from the 898. Both of them break on the concrete zone at the end of setting time and break on the hardener zone after the 45 minutes. For these two mixtures the "open time" lasts until the end of setting time.

For the 540, some of the dollies at the end of setting time break on the concrete zone and the others on the hardener zone. This reveals that the end of "open time" was achieved before the end of setting time (8 hours). The 540 +45, broke all of them thru the hardener.

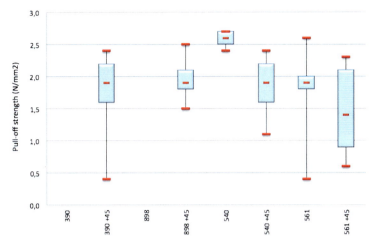

Fig. 4. Pull-off strength with rupture in hardener

For all the situations the 561 broke thru the hardener. This means that the "open time" is less than the end of setting time.

We can realize that, not always, the end of the setting time matches the "open time".

Even without available water in sight, and with the introduction of some additional water on the surface of base concrete, just before the placement of surface hardener, some compositions can interact (898 and 390). The end of setting time is already considered too late to place the surface hardener, but extreme conditions and late placement of surface hardener, for example at the end of setting time, is not necessarily a condition, to obtain bad results.

In a construction, work can easily be delays, which will affect the application of surface hardener over large areas of slabs.

4 Conclusions

The application of a surface hardener on four different base concretes at delay time, with high temperature and wind speed, was studied. The following conclusions can be drawn:

- The type of chemical admixture has a great importance in the control of the "open time" and can make all the difference on the application of surface hardener over large areas of slabs;
- Extreme situation of environmental conditions are needed to obtain bad results and the application of the surface hardener at delayed time also increases the bad results;
- It was observed that no direct relation exists between the use of plasticizers or superplasticizers;

- The superplasticiser/strong water reducer, 898, presented the best results at the end of setting time (4 hours), but also the 390, showed good results at the end of setting time (5:30 hours);
- Forty five minutes after the setting time all the compositions showed bad results;
- The composition that behaved worst was the 561;
- The end of setting time does not always matches the end of the "open time".

Acknowledgements. This work was funded by the Portuguese Foundation for Science and Technology (PTDC/ECM/105075/2008). The authors thanks to Britobetão, BASF and SIKA. The chemical admixtures indicated in this study should not be understood as a recommendation from the authors.

References

[1] The Concrete Society, Delamination of Concrete Floor Surfaces, Concrete Advice No. 18, The Concrete Society Bookshop (2003)
[2] Topçu, I.B., Elgun, V.B.: Cement and Concrete Research 34, 275–281 (2003)

Compatibility between a Quartz Surface Hardener and Different Base Concrete Mixtures

M.T. Pinheiro-Alves[1,2], A. Fernandes[1], M.J. Marques[1], and A. Bettencourt Ribeiro[3]

[1] University of Évora, School of Sciences and Technology, Portugal
[2] Center of Territory, Environment and Construction (C-TAC) – Group of Sustainable Construction, Portugal
[3] Nacional Laboratory of Civil Engineering (LNEC), Portugal

Abstract. Many cases of cracking and detachment of the concrete surface have appeared in concrete floors, namely pavements where surface hardeners were used, especially in car parks, industrial areas and warehouses. This paper studies the behaviour of ten different base concretes mixtures made with two chemical admixtures and four additions, when a quartz surface hardener is applied. The objective is to identify the relevant parameters of the concrete constituents that influence the open time. Results show that the type of chemical admixture used to fabricate the concrete is important, but also the kind of admixture due to the water available to make the interconnection between the base concrete and surface hardener.

1 Introduction

Previous works [1] indicate that the main causes for cracking and delamination of trowelled concrete pavements are: entrapped bleeding water and air beneath the top layer, closed by finishing operations (less permeable); and the spreading of soft mortar over already hardened adjacent areas of floor surface (over-layering). These two causes depend on several factors such as the finishing operations, curing and concrete mixture.

The increase of cracking and delamination problems in recent years may be related to the fact that the characteristics of both cement and concrete have changed, particularly the increase in the fineness and quantity of tricalcium silicates (C_3S) in cements and the use of more effective water reducers in concrete. These changes decrease the time available to perform the finishing operations due to the more rapid evolution between plastic and hardened phases of concrete. Indeed, the finishing operations may only begin when the fresh concrete supports the weight of the trowelling equipment and must end before the end of the setting time, which may be described as open time.

The objective of this study was to identify the relevant parameters of the concrete constituents that influence the open time, related not only with their characteristics but also with their content. The relationships between hydration and the type and dosage of binders and chemical admixtures, are the main objective of this paper.

2 Materials and Methods

In laboratory, several base concretes (60x30x10 cm^3) were made, combining two different chemical admixtures and partial replacement of the coarse cement with fly ash (FA), limestone filer (L), a special cement with less C_3S (C) and gypsum (G).

The chemical admixtures used were a plasticiser/water reducer (Pozzolith 390NP - Water and lignosulphonate) and a superplasticiser/strong water reducer (Sikaplast 898 - Combination of modified polycarboxylates). The base concrete was made with the CEM II/A-L 42,5R cement and the surface hardener was made with quartz sand.

The conditions of the experiments tried to replicate the extreme conditions found under real situations. Initially, high temperature and wind were considered. A tube cut in half with two fans allowed simulation of windy and hot environment, as shown in Fig. 1.

Fig. 1. Tunnel used in the simulation

Secondly, late implementation of surface hardeners was considered. The surface hardener was applied in each one of the base concretes at different setting times. The times considered for the application of the hardeners were the setting time (27,6 N/mm^2) and 45 minutes after the setting time. These times were chosen because most of the problems normally occur when surface hardener is applied too late.

New equipment was created to trowell the slabs of 60x30x10 cm^3, Fig. 2.

Some of the tests that were carried out were slump test, setting times of base concrete and pull-off strength.

Compatibility between a Quartz Surface Hardener and Different Base Concrete Mixtures   609

Fig. 2. Equipment used to trowel the slabs

The compositions of the concretes are indicated in Table 1. The consistency was determined by slump test and maintained at the class of S3 (100-150 mm) according EN 206-1:2007.

Table 1. Mixture proportions of concretes (kg/m^3)

	898	899G	898C	898FA	898F	390	390G	390C	390FA	390F
Cement	340	340	239	239	272	340	340	239	239	272
Sand	730	730	730	730	730	730	730	730	730	730
Coarse aggr 1	558	558	558	558	558	558	558	558	558	558
Coarse aggr 2	556	556	556	556	556	556	556	556	556	556
Water	143	143	142	134	139	170	169	169	160	165
Addition	0	7	102	102	68	0	7	102	102	72
Chemical add	3.4	3.4	3.4	3.4	3.4	2.7	2.7	2.7	2.7	2.7
w/c	0.4	0.4	0.6	0.6	0.5	0.5	0.5	0.7	0.7	0.6
Slump (mm)	150	150	150	137	148	143	100	132	123	124

To obtain the setting times were used different needles that allowed measuring the time of setting of concrete by penetration resistance, according ASTM C403/C403M:2008 (Fig. 3).

Fig. 3. Needle to measure the resistance to penetration

Was only used the material that passed on the 5 mm slieve, that is, only the mortar part of the concrete, Fig. 4.

Fig. 4. Sieving of the concrete

Additional water had to be introduced when applied the surface hardener, because enough water from the base concrete was not available, which is what happens under real application conditions.

To determinate the resistance of the surfaces finishes and the homogeneity between the surface finish and the concrete slab the pull-off test was realized according EN 1542. The pull-off test was done at the seventh day because problems begin to emerge in the early days. The dollies used in the study had 50 mm of diameter.

It is very important to control bleeding and the time available to perform the finishing operations due to the more rapid evolution from the plastic to the hardened phases of concrete. This time is known as "open time".

The purpose of the work was to determinate the resistance of the surfaces finishes and the homogeneity between the surface finish and the concrete slab under experimental conditions.

3 Results and Discussion

The end of setting times and end of setting times plus 45 minutes of base concretes are presented on Table 2. As can be observed, the chemical admixtures influence differently the setting times. The 898 has a shorter end of setting time when compared with the 390.

Table 2. Setting times of base concretes

	ST	ST +45min
898	4:08h	4:53h
898G	4:18h	5:03h
898C	4:40h	5:25h
898FA	4:35h	5:20h
898F	4:41h	5:26h
390	5:29h	6:14h
390G	5:43h	6:28h
390C	6:13h	6:58h
390FA	7:29h	8:14h
390F	5:39h	6:24h

Results of pull-off test show that most of the ruptures occur in hardener, as can be observed in Figs. 5 and 6.

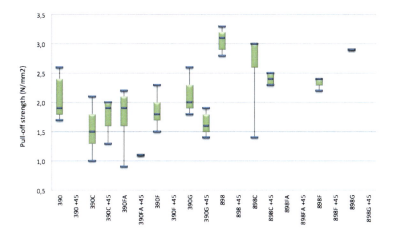

Fig. 5. Pull-off strength at 7th day with rupture in concrete

In Fig. 5, the average value of pull-off strength for the compositions with the chemical admixture 898 was 2.8 N/mm^2 and for the 390 was 1.7 N/mm^2.

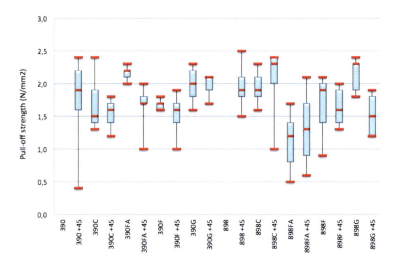

Fig. 6. Pull-off strength at 7th day with rupture in hardener

In regard to Fig. 6, the results are opposite. The 898 have an average value of pull-off strength of 1.8 N/mm^2 and the 390 have 1.9 N/mm^2.

The results indicated on Figs. 5 and 6, shows that the end of the setting time is not necessarily the end of the "open time". The compositions with only chemical admixtures (898 and 390) showed good results at the end of setting time with the incorporation of additional water. This means that some times, the "open time" is longer than the end of setting time.

The compositions with the chemical admixture 898 had a ratio w/c between 0.4-0.6, and the ones with the 390 had a ratio w/c ranging from 0.5-0.7. Although the compositions with the 390 had more water available, they were also exposed longer to the adverse situations.

The composition 898 presented the best result in the pull-off test, but the application of the surface hardener cannot exceed the 4:08h. The second best result was obtained with the composition 390, with an "open time" of 5:30h. Although the 898 has better results, when applying surface hardeners to pavements with large areas, the 390 could be a better solution, because the "open time" lasts later.

The Fig. 5,6 shows that the incorporation of additions reduces the "open time", when compared to the compositions without additions.

The addition that showed better results was the cement with less C_3S, probably because this cement is not so fineness. The other additions showed similar results, breaking mostly thru the hardener. The use of these new cements with more C_3S contents and lower fineness increases the problems.

Under conditions of high temperature and/or wind it is extremely important to choose the right composition and also to have a good team of workers specially when dealing with large areas.

4 Conclusions

The influence of additions and chemical admixtures in the application of a surface hardener at different setting times was studied. The following conclusions can be drawn:

- The compositions 898 and 390 presented good results even at the end of setting time;
- Composition with chemical admixture 390 could be a better solution for large areas;
- At the end of setting time, concretes made with only chemical admixtures showed better results than the ones with chemical admixtures and additions;
- The addition that showed better results was the cement with less C_3S, C;
- The additions FA, F and G showed similar results with both chemical admixtures;
- The end of the setting time does not necessarily coincide with the end of the "open time";
- Under conditions of high temperature and/or wind it is extremely important to choose the right composition and also to have a good team of workers specially when dealing with large areas.

Acknowledgements. This work was funded by the Portuguese Foundation for Science and Technology (PTDC/ECM/105075/2008). The authors would like to thank Britobetão, BASF and SIKA. The chemical admixtures and admixtures indicated in this study should not be understood as a recommendation from the authors.

Reference

[1] The Concrete Society, Delamination of Concrete Floor Surfaces, Concrete Advice No. 18, The Concrete Society Bookshop (2003)

Suitable Restrained Shrinkage Test for Fibre Reinforced Concrete: A Critical Discussion

Adriano Reggia, Fausto Minelli, and Giovanni A. Plizzari

DICATA - Department of Civil, Architectural, Environmental and Land Planning Engineering, University of Brescia, Italy

Abstract. Concrete performance traditionally refers to compressive strength and workability. Recently, high performance concrete evidenced the possibility of enhancing other material properties. Among these, resistance to shrinkage cracking is gaining more attention among practitioners, due to its strict relation to durability requirements. Shrinkage cracks occur in restrained structures: for this reason, material characterisation should be made on the basis of a restrained shrinkage test. The ring test is an easy-to-use tool since one can measure the time-to-cracking of a concrete mix. Focus of this paper is to critically discuss the actual standard test procedure and then to propose enhancements of the test set-up with the aim of reducing the time-to-cracking and making the test duration more suitable for practical uses. Furthermore, the effect of fiber reinforcement on shrinkage cracking is presented.

1 Introduction

Recent advances in construction methods, new materials, and admixtures have renewed the interest on early age cracking in cement based material due to restrained shrinkage. The latter has been recognized as the main cause of damage for thin structures such as highway pavements, industrial floors, bridge decks, but also for partial depth repairs, jacketing and overlays on existing structures. During the last few years, concrete technologists have been studying the mechanism governing early age cracking in concrete members, especially after the diffusion in the market of High Strength Concrete (HSC), where the autogenous shrinkage plays a major role. Initially, this occurs due to hydration reaction in the cement matrix. At a later stage, when the material is exposed to a low relative humidity environment, drying shrinkage takes place. In Normal Strength Concrete (NSC), where autogenous shrinkage plays a minor role, it can be assumed that the whole shrinkage strain is given by drying shrinkage.

When shrinkage is restrained by a structural element, or an internal restrain (rebars), tensile stresses occur and may give rise to crack formation. However, tensile stresses are relaxed by creep phenomena that may delay or prevent cracking [1]. Therefore, shrinkage crack formation and development depends on several factors such as air environmental humidity, creep, restrain conditions, tensile strength and stiffness as well as fracture toughness of concrete.

Currently, there is no general consensus on a standard method to investigate the shrinkage cracking behaviour of concrete. Many studies have been carried out, mainly using three types of test: the linear test [2], the plate test [3] and the ring test according to standard AASHTO PP 34-99 [4].

The linear restrained column type test gives a simple uniaxial stress development but suffers from the disadvantage of not providing a constant degree of restraint that makes this test complex. The plate test provides a biaxial restraint to evaluate both biaxial and plastic shrinkage. The ring test provides a nearly constant degree of reaction through an axi-symmetric specimen geometry. In this type of test, a concrete specimen is cast around an inner steel ring which provides a constant degree of restraint to the shrinkage deformation; the steel ring is also used to evaluate the induced tensile stresses in concrete through the measure of the steel compressive strains with strain gauges. However, this test method can be ineffective: for instance, for a concrete with a high tensile strength, the induced tensile stresses might not be able to generate cracks in the composite. Beside the material properties, the low crack-sensitivity is associated to the specimen geometry.

It is now commonly accepted that fiber reinforcement reduces cracking phenomena but it is not clear how to best quantify this contribution; in other words, the minimum material performance necessary to reduce the crack width below a design value is still an open question. This paper provides a critical discussion on the ring test method for assessing the performance of Fiber Reinforced Concrete (FRC) starting from AASHTO requirements and evaluating the influence of different set-up geometries and materials through numerical analyses based on Non Linear Fracture Mechanics (NLFM) [5]. The main phenomena governing the restrained shrinkage behaviour of early-age concrete, such as elastic and post-cracking material properties (with special emphasis on FRC toughness), shrinkage strains and creep are considered. Stress evolution and crack formation are determined through a model based on a discrete crack approach with an iterative time-step procedure.

2 Testing Procedure

As already mentioned, the AASHTO designation PP 34-99 is a testing procedure for the determination of the cracking tendency of ring-shaped concrete specimens. The time-to-cracking of the concrete ring is considered as the age when compressive strains in the steel ring suddenly decrease. This procedure represent a standard method and is not intended to determine the time-to-cracking of any particular structure cast with the same material.

The standard steel ring have a wall thickness of 12.7 mm (1/2 in), an outside diameter of 305 mm (12 in) and an height of 152 mm (6 in). The ring surface in contact with the concrete is coated with a form-release agent to minimize bond between the concrete and the steel. The form is made of a non-absorbent material and have an outside diameter of 457 mm (18 in).

Two or more strain gauges are applied on the inside surface of the steel ring to measure the strain. After a wet curing of 24 hours, specimens are stored in a controlled environment room with a constant air temperature of 21°C (73.4°F) and relative humidity of 50%. During the test, the strains in the steel ring are recorded every 30 minutes; a sudden decrease of the compressive strains in the steel ring higher than 30 µε usually indicates a crack formation.

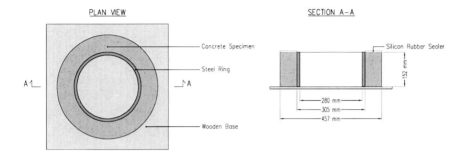

Fig. 1. AASHTO PP34-99 test apparatus

An essential parameter for this type of test is the degree of restraint that represents a measure of the effectiveness of the restraint provided by the steel ring. The degree of restraint R should be defined as the stiffness of the steel ring over the stiffness of both the concrete and steel ring, that is:

$$R = \frac{A_{st}E_{st}}{A_{st}E_{st} + A_c E_c} \tag{1}$$

where A_{st} and A_c are the cross-sectional areas of the steel and concrete rings, respectively, and E_{st} and E_c are the moduli of elasticity of the steel and concrete, respectively.

3 Numerical Model

A parametric study of the ring test was carried out through several numerical analyses performed with TNO DIANA 9.4 and a discrete crack approach [6]. The main phenomena governing early-age cracking in cement-based materials, such as elastic and post cracking properties, shrinkage strains and tensile creep are considered by a set of features provided by the program.

The inner steel ring and the outer concrete ring are modelled by four-point plane stress elements. Between the two rings, the steel-to-concrete interface is modelled using linear interface elements with a brittle Mode-I behaviour and a Mode-II behaviour given by a simple bond-slip relationship suitable for smooth rebars [7] and a Coulomb friction criterion. The discrete crack was simulated by zero-thickness interface elements with a bilinear tension softening behaviour (Mode-I). Due to the axi-symmetric geometry the discrete crack is arbitrary placed (Figure 2). Material non-linearity is localized in the discrete crack, while the concrete specimen and steel ring are defined as linear elastic.

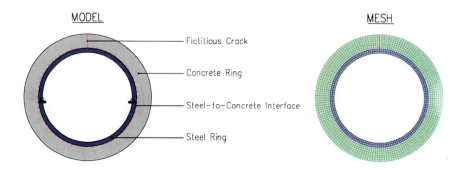

Fig. 2. Ring-test model and mesh

The numerical analyses were performed with a time step procedure considering the evolution in time of the tensile stresses induced by shrinkage to be compared with the rising tensile strength of concrete. When the maximum tensile stress exceeds concrete tensile strength, a crack starts to open in the discrete crack. The adopted time step was 1 day; analyses were performed up to 60 days. The time evolution of drying shrinkage $\varepsilon_{sh}(t)$ (Figure 3a) and compressive strength $f_c(t)$ (Figure 3b) were evaluated according to ACI 209R-92 [8]. The time evolution of tensile strength $f_t(t)$, elastic modulus $E_c(t)$ and fracture energy $G_f(t)$ were evaluated according to the following relationships:

$$f_t(t) = f_{t28}\sqrt{\frac{f_c(t)}{f_{c28}}} \quad (2)$$

$$E_c(t) = E_{c28}\sqrt{\frac{f_c(t)}{f_{c28}}} \quad (3)$$

$$G_f(t) = G_{f28}\sqrt{\frac{f_c(t)}{f_{c28}}} \quad (4)$$

where f_{t28}, E_{c28} and G_{f28} are, respectively, tensile strength, elastic modulus and fracture energy after 28 days of curing.

Concerning the viscoelasticity related to tensile creep at early age, it was chosen to develop creep functions in a Taylor series as DIANA does with a Power Law model for the stress calculation [9]. The compliance function for the Power Law model is given by the following relationship:

$$J(t,\tau) = \frac{1}{E(\tau)}(1 + \alpha \tau^{-d}(t-\tau)^p) \tag{5}$$

For these analyses the power of the creep function $p = 0.5$, the development point $t_d = 15.0$, the coefficient $\alpha = 0.16$ and the power of the time dependent part of the creep function $d = 0.1$ have been adopted. These values were chosen in order to best fit the compliance function J in Eqn. 5 with that included in ACI [8].

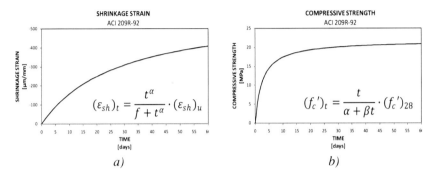

Fig. 3. Free shrinkage a) and compressive strength b) according to ACI209R-92

4 Parametric Study

4.1 Effect of Geometry

The parametric study on the specimen geometry is performed by considering as a reference the test setup given by AASHTO PP 34-99 and changing the two most significant parameters: the steel and the concrete thickness. These parameters are related to the cross-sectional areas of the steel and concrete rings and then to the degree of restraint provided by the test set-up. A variation on the degree of restraint produces a direct effect on the time required to form a crack: an increase of steel thickness or a decrease of concrete thickness reduces the time-to-cracking.

A first set of analyses has been performed with a constant steel ring thickness equal to 12.7 mm and a variable concrete thickness equal to 15, 30, 45, 60 76 and 90 mm, respectively. A second set of analyses has been carried out with a constant concrete ring thickness, equal to 76 mm, and a variable steel thickness equal to 6.3 mm (1/4 in), 12.7 (1/2 in), 25.4 (1 in), 50.8 (2 in) and 76.2 (3 in), respectively. The specimen dimensions have been chosen to favour crack formation in a reasonable time for the use in a laboratory. For each specimen geometry, three analyses were carried out by considering three different concrete grades: C12, C20 and C30. The mean values of elastic and post-cracking properties were determined according to *fib* Model Code 2010 [10], as reported in Table 1. Table 2 reports the degree of restraint R for the different specimen geometries considered for concrete C20. In the same table, the steel thickness (t_{st}), the concrete thickness (t_c) and elastic modulus (E_c) are reported. The elastic modulus of steel (E_{st}) was assumed as 210 GPa.

Table 1. Mean values of elastic and post-cracking properties for NSC

	f_{ck} [MPa]	f_{cm} [MPa]	f_{ctm} [MPa]	G_F [N/m]	E_c [GPa]	ν [-]
C12	12.0	20.0	1.6	125	22.9	0.2
C20	20.0	28.0	2.2	133	26.2	0.2
C30	30.0	38.0	2.9	141	29.7	0.2

Table 2. Degree of restraint R of ring-test setup for concrete C20

	t_{st} [mm]	A_{st} [mm^2]	t_c [mm]	A_c [mm^2]	E_c [GPa]	R [%]
Model 1	12,5	1900	15	2280	26,2	87%
Model 2	12,5	1900	30	4560	26,2	77%
Model 3	12,5	1900	45	6840	26,2	69%
Model 4	12,5	1900	60	9120	26,2	63%
Model 5	12,5	1900	76	11552	26,2	57%
Model 6	12,5	1900	90	13680	26,2	53%
Model 7	6,3	958	76	11552	26,2	40%
Model 8	25,4	3861	76	11552	26,2	73%
Model 9	50,8	7722	76	11552	26,2	84%
Model 10	76,2	11582	76	11552	26,2	89%

4.2 Effect of Material Toughness

The parametric study on material toughness focused on the evaluation of cracking behaviour of plain (NSC) and Fibre Reinforced Concrete (FRC) having a post-cracking softening behaviour. A second aim of this study is the evaluation of the effect of the enhanced concrete toughness on crack development. The tensile behaviour of both plain and FRC in tension was described with a linear relationship up to the cracking strength followed by a post-peak bilinear relationship. For plain concrete the latter was assumed according to *fib* MC2010 [10] (Figure 4) while, for FRC, the two branches of the bilinear law were varied as shown in Figure 5.

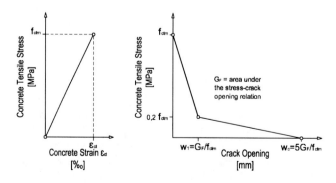

Fig. 4. Constitutive laws for concrete under uniaxial tension

In particular, a first set of analyses for FRC was performed on Model 3, which was considered the most efficient specimen geometry, with a steel thickness of 12.7 mm and a concrete thickness of 45 mm, by varying the first branch of the softening law (Figure 5a). A second set of analyses was carried out by considering a variation on the second branch of the post-cracking softening law (Figure 5b).

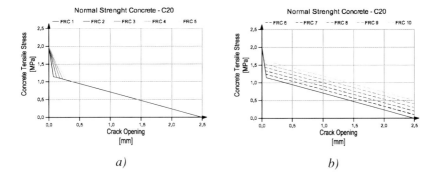

Fig. 5. Bilinear tension softening laws for FRC

5 Numerical Results

An overview of the results of the parametric study on the effect of different geometries with regard to the time-to-cracking is given in Figure 5. From a numerical point of view the time-to-cracking was identified as the age at which the crack interface is completely open and the tensile stress in all interface elements has reached the tensile strength; this assumption is reasonable by considering that a macro-crack has to form for releasing the steel ring, whereas a micro-cracking would probably not have any significant impact on the strain-gauge measurements.

As previously suggested, the time-to-cracking strongly depends on concrete thickness for two principal reasons: a decrease of concrete thickness implies, on one hand, an increase of the degree of restraint R and, on the other hand, an increase of the intensity of drying shrinkage due to a higher moisture exchange with the environment. Figure 6a shows the increase of the time-to-cracking with the concrete thickness for all the concrete grades considered. The results suggest to employ a concrete with a thickness lower than 60 mm (Models 1, 2 and 3) to maintain the time-to-cracking in the first two weeks after initiation of drying. Numerical results also show that the time-to-cracking depends on the steel thickness (Figure 6b): in fact, an initial strong decay is followed by a smoother decrease for a steel thickness larger than 25.4 mm. Therefore, a steel ring thicker than 25.4 mm (Models 9 and 10) does not provide an appreciable decrease of the time-to-cracking and, for this reason, it is not recommended for this test.

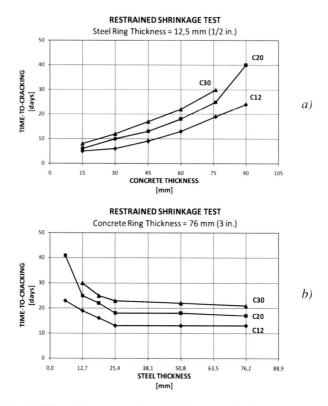

Fig. 6. Effects of concrete and steel thickness on the time-to-cracking

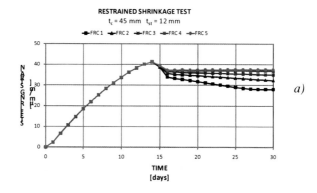

Fig. 7. Influence of different tension softening laws on ring-test response

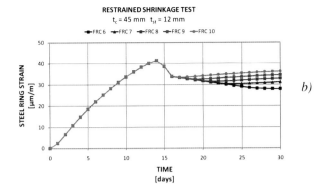

Fig. 7. *(continued)*

The influence of fiber reinforcement on shrinkage cracking is shown in Figure 7; both sets of tension softening models adopted were able to prevent a sudden strain drop in the steel ring and allowed to control cracking by transfer tensile stresses after the appearance of the crack. The first set of analyses shows how a different slope of the first branch influences the model response from the early crack formation (Figure 7a). The second set of analyses illustrates how a modification on the second branch changes the numerical response only for considerable crack widths (Figure 7b). The numerical results are consistent with experimental results available into the literature, stating the effectiveness of fiber reinforcement in controlling shrinkage cracking. However, FRC with a higher residual strength for micro-cracking is more suitable in controlling shrinkage cracking.

6 Concluding Remarks

Shrinkage cracking requires special attention from structural designers for many aspects: with regard to durability, early-age cracking advances the deterioration process of concrete and reduces the service life of a structure; from an aesthetical and psychological point of view, it is not acceptable because makes structure appearance worse and people feel unsafe. Thus, the evaluation of cracking sensitivity of a material is strongly recommended for an effective and durable design of a structure. Based on the numerical study presented herein, the following conclusions can be drawn:

1. Time-to-cracking strongly depends on the specimen geometry and on the degree of restraint: an increase of steel ring thickness or a decrease of concrete thickness can accelerate the occurrence of cracking. For the normal concrete considered, starting from the standard AASTHO test, results suggest to employ concrete thickness lower than 60 mm to maintain the time-to-cracking within the first two weeks after drying initiation. It is also recommended the use of a

steel ring having a thickness in the range of 12.7÷25.4 mm (0.5÷1 in.) to provide enough stiffness to ensure crack formation.
2. Fiber reinforcement has a considerable effect on behaviour of FRC under restrained shrinkage, by preventing sudden drops in the strain of the steel ring, controlling cracking and transmitting tensile stresses after the appearance of the crack. The crack control is mainly influenced by the residual strength for smaller crack opening (first branch of the tension softening law).

References

[1] See, H.T., Attiogbe, E.K., Miltenberger, M.A.: ACI Mat. Jour. 100(3), 239–245 (2003)
[2] Klover, K.: Mat. and Stru. 27(170), 324–330 (1994)
[3] Kraai, P.P.: Conc. Cons. 30(9), 775–778 (1985)
[4] AASHTO, Standard Practice for Estimating the Crack Tendency of Concrete, pp. 34–99 (2006)
[5] Hillerborg, A., Modèer, M., Petersson, P.E.: Cem. and Conc. Res. 6, 773–782 (1976)
[6] TNO Building and Construction Research, TNO DIANA User's Manual Release 9.4, TNO DIANA BV, Delft, The Netherlands (2009)
[7] CEB-FIP: Bull. d'Info. (195), 480 (1990)
[8] ACI, ACI 209R-92, Farmington Hills, Michigan, USA (1992)
[9] de Borst, R., van den Boogaard, A.H.: J. Eng. Mech. Div. ASCE 120(12), 2519–2534 (1994)
[10] Walraven, et al.: Fib Bull. 1(55), 318 (2010)

Influence of Chemical Admixtures and Environmental Conditions on Initial Hydration of Concrete

A. Bettencourt Ribeiro[1], V. Aguiar Medina[2], and A. Martins Gomes[2]

[1] LNEC-Portuguese National Laboratory of Civil Engineering
[2] Instituto Superior Técnico, Technical University of Lisbon

Abstract. There has been an increasing number of problems in concrete pavements, such as cracks and delamination, in which surface hardeners have been used, especially in car parks, industrial areas, warehouses etc. Factors that can potentiate the occurrence of these problems are the increasing use of more effective water reducers in concrete and also the use of binders with higher fineness and higher tricalcium silicate content than the former ones. Concrete and surface hardeners are both cementitious materials, which however are applied at different instants. Distinct stages of hydration may lead to lack of homogeneity and, eventually, to cracking and delamination. This study evaluates early hydration and bleeding behaviour of concrete, which are relevant factors for finishing with surface hardeners. The influence of chemical admixtures and environmental conditions were tested by penetration resistance and bleed water, which highlights the shortening of the period in which surface hardener application is possible.

1 Introduction

A significant increase in problems occurring on concrete floors has been detected, concerning pavements in which surface hardeners were used, especially in car parks, industrial areas, warehouses, etc. The negative effect of these abnormal situations has important economic implications, since the necessary work to correct them involves large areas of reconstruction and the use of expensive repairing materials, and also because it often delays the completion of the construction work and, hence the beginning of the infrastructure operation. There is no systematic survey on the number of works in which this the problem occurs, but LNEC (National Lab. of Civil Eng.) has been asked to intervene in many cases of cracking and detachment of the concrete surface, in various regions of Portugal and for different entities, which shows that is a widespread problem, rather than just local or circumscribed [1-2]. The application of surface hardeners to pavements is not a novelty. However, the occurrence of systematic problems is a current issue and it seems to be caused by changes in the constituents used, whether in relation to the concrete base or to the hardener used. One factor that potentiates the occurrence of cracks and delamination is the increasing use of more effective water reducers in concrete. The consequent reduction in the water content of concrete and the pressure to increase productivity in the construction industry leads to poorer conditions for achieving a homogenous concrete pavement. In order to obtain an appropriate embedment of the hardener, the concrete mixture must have such proportions that excessive bleed water does not appear on the surface after application of the hardener [3]. However, lack of bleeding decreases the hardener hydration. This lack of hydration is even more significant when the hardener is added later than it should be, as often happens. The control of concrete bleeding behaviour is a crucial aspect for obtaining good finishing in concrete pavements. This work analyses the influence of several chemical admixtures on concrete bleeding. The bleeding was measured after mixing and during the "open time" (period considered suitable for application of the hardener).

2 Materials and Methods

2.1 Materials

Mortar tests were performed to evaluate bleeding. A reference mortar, formulated without chemical admixtures was obtained by screening of a concrete mixture suitable for pavement construction. Table 1 shows the mixture proportion of the reference concrete. After screening, the workability of the mortar was measured for reference, (150 mm spread) [4]. Mortar mixtures with chemical admixtures were formulated keeping the cement/sand constant (c/s=0.375), but changing the W/C to achieve the reference flow (140-154%). Potable tap water, Lisbon-Portugal, was used. Mortars with chemical admixtures were prepared with cement CEM II/A-L 42.5 R, currently used in Portugal, Table 2. Five chemical admixtures, plasticizers or superplasticizers, were tested. Table 3 shows the characteristics of the products. Mortar mixtures are presented in Table 4.

2.2 Methods

The following tests were performed on mortars: flow [4]; compressive strength [5]; penetration resistance [6]; bleeding without compaction; bleeding with compaction. Bleeding without compaction was performed according to the following procedure: *i)* fill 100 mm cube moulds with mortar, up to a height of 90±3 mm; *ii)* with a round steel tamping rod (16 mm in diameter and approximately 600 mm long) compact the mortar with 25 strokes; *iii)* after consolidation, finish smoothly the top surface and measure the net weight of the mortar; *iv)* depending on the cure method, either cover the specimens with a plastic sheet to avoid moisture exchange or expose them to dry in an aggressive environment (forced air circulation with or without heating); *v)* collect every hour the bleeding liquid on the mortar surface with a syringe, beginning the measurements 2 hours after mixing and weigh the specimens exposed to an aggressive environment to evaluate the evaporation rate. Bleeding with compaction has been performed according to the following procedure: *i)* to *iv)* as described above; *v)* when the penetration resistance reaches approximately 1, 2, 5, 15, and 25 MPa (measured in a parallel cube), the bleeding liquid on mortar the surface (if any) is collected with a syringe, then tare the weight of the specimen and then consolidate it in a Vebe Vibrating Table [7], using a 10.8 kg surcharge (see Figure 1), to simulate the pressure of the consolidating and finishing tools used in pavements construction practice, and, simultaneously collect the bleeding liquid on the mortar surface; *vi)* after vibrating the specimen, remove the surcharge and measure the weight loss.

Table 1. Reference concrete

Material	Content (kg/m^3)
Cement CEM I 32.5 R	295
Natural sand	785
Coarse aggregate (5-15 mm)	556
Coarse aggregate (15-25 mm)	524
Water	196

Table 2. Cement properties (II/A-L 42.5 R, information supplied by the producer)

Chemical property	Result	Physical property	Result
Loss on ignition (%)	5.93	Water for standard consistence (%)	29
SiO_2 (%)	17.65	Initial setting time (min.)	125
Al_2O_3 (%)	5.18	Final setting time (min.)	180
Fe_2O_3 (%)	2.92	Soundness (mm)	0.5
Total CaO (%)	62.81	Compressive strength 2 days (MPa)	30.8
MgO (%)	1.61	Compressive strength 7 days (MPa)	43.0
SO_3 (%)	2.67	Compressive strength 28 days (MPa)	53.0
K_2O (%)	0.99		
Free CaO (%)	1.57		
Filler content (%)	12.9		

As referred to above, and with a view to simulate distinct possible environments, three different curing conditions have been performed, with different rates of evaporation imposed by wind and temperature [8]: controlled environment, without drying and at a temperature of 21±2 °C; forced air circulation without heating; forced air circulation with heating. The 2 environments with forced air circulation have been simulated using a tunnel (0.345 m radius, 2.10 m length, 1 fan and 2 heaters, as presented in Figure 2). Since the fan and the fan heaters were put on one end of the tunnel, the rate of evaporation was not constant throughout its length. The gradients in temperature, humidity, wind speed and rate of evaporation between the entrance and the exit are presented in Table 6. With each mortar mixture subjected to aggressive environments, 11 numbered cubic moulds were used, positioned in the tunnel according Figure 2. Specimens with numbers 8, 9 and 11 were used for penetration resistance tests, specimens 7 and 10 were used for bleeding without compaction and rate of evaporation measurements, and the remaining ones served to bleeding with compaction determinations.

With the mortars exposed to the non aggressive environment (protected from evaporation), 11 cubic moulds were used, 4 for penetration resistance, 1 for bleeding without compaction, and the remaining 6 for bleeding with compaction. The penetration resistance in this case was performed for 2 curing conditions: protected from evaporation with a plastic sheet (2 specimens); exposed to drying (2 specimens).

Table 3. Chemical admixtures (information supplied by the producers, except for the solid content of the first four products which were determined)

Property	Rheobuild 561	Glenium C 313	Melment L10	Pozzolith 390 NP	Sikaplast 898
Main function	Superplast.	Superplast.	Superplast.	Plasticizer	Superplast
Chemical base	Naphthal. sulphonate	Policarbox ether	sulfonated *melamine*	Modified lignosulph.	Policarbox ether(Mod.)
Colour	Dark brown	Dark brown	Colourless	Dark brown	Light brown
pH	7±1	6±1	9±1	8.5±1	5.0±1
Density	1.18±0.03	1.03±0.03	1.12±0.03	1.17±0.03	1.07±0.02
Viscos. (cps)	≤ 100	≤ 50	≤ 30	≤ 75	
Rec.dos.(%)	0.9-1.4	0.6-2.6	0.8-3.9	0.5-1.2	0.5-1.5
Chlorides (%)	≤ 0.1	≤ 0.1	≤ 0.1	≤ 0.1	≤ 0.1
Alkalis (%)	≤ 9	≤ 1.5	≤ 6.5	≤ 6.8	-
Sol. cont. (%)	50.3	11.5	19.2	40.0	32±2

Table 4. Mortars: mixture proportions

Mortar	Cement (kg)	Sand (kg)	Water (kg)	Chemical admixture (ml)	W/C
Reference	7.97	21.26	4.35	-	0.55
Rheobuild 561			3.52	85	0.44
Glenium C 313			3.80	70	0.48
Melment L10			3.95	100	0.50
Pozzolith 390 NP			4.00	58.5	0.50
Sikaplast 898			3.70	50	0.46

Table 5. Aggressive environments inside the tunnel

Property	Measurement location	Windy condition	Windy and heating condition
Temperature (°C)	Entrance	22±2	68±2
	Exit		48±2
Relative Humidity (%)	Entrance	50±5	10 ± 5
	Exit	55±5	20 ± 5
Wind speed (km/h)	Entrance	15-16	4.5-5.5
	Exit	7-8	3
Rate of water evaporation kg/(m^2.h)	Entrance	0.55±0.05	0.90±0.05
	Exit	0.37±0.02	0.45±0.05

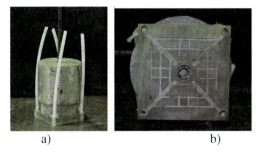

Fig. 1. Surcharge used in the bleeding with compaction test: a) upright position, as placed on the mortar surface; b) bottom of the plate, showing the flow channels

3 Results and Discussion

3.1 Flow

Table 6 shows the flow obtained on the six mortars tested. The values given are within the established range, and evidence similar workability in all the mortars.

Table 6. Mortars: flow (%)

Reference	Rheobuild 561	Glenium C 313	Melment L10	Pozzolith 390 NP	Sikaplast 898
150	142	146	152	142	153

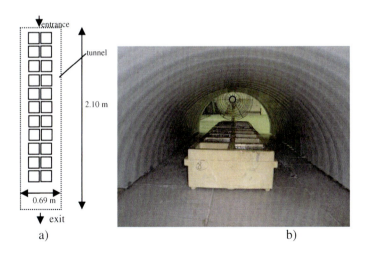

Fig. 2. a) Schematic presentation of the cubes in the tunnel; b) picture from inside the tunnel

3.2 Penetration Resistance

Penetration resistance was measured on cubes 8, 9 and 11, at different points in the tunnel. Without heating (only forced air circulation), the differences between the 3 specimens are small, as shown in Figure 3a). However, a small delay is observed on cube 11, probably due to the proximity to the fan (cooling due to evaporation).

Conversely, with heating there is a distinct behaviour between the 3 specimens, due to the temperature gradient. A hardening delay occurs, as shown in Figure 3b), which is higher in the farthest cube.

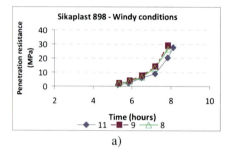
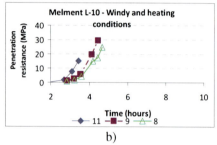

a)　　　　　　　　　　　　　　　　　b)

Fig. 3. Penetration resistance of mortar on cubes 8, 9, and 11: a) Sikaplast 898 air circulation without heating; b) Melment L-10 air circulation without heating

Figure 4a) shows the penetration resistance of cube 11, for specimens kept under sealed conditions. The set retarder effect of the chemical admixtures is clearly observed, being the most efficient retarder the product with naphthalene (Rheobuild 561). Under windy conditions, without heating, Figure 4b), there is a hardening delay, which is dependent on the type of product used, but within the range of about 1-2 hours. This delay should be related with cooling due to evaporation. The relative positions of different curves remain constant, except for the mortar with lignosulphonate (Pozzolith 390 NP).

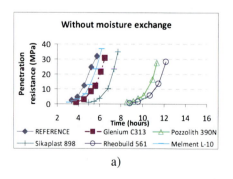
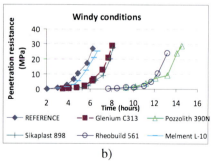

a)　　　　　　　　　　　　　　　　　b)

Fig. 4. Penetration resistance of cube 11 of mortars: a) specimens covered with a plastic sheet; b) specimens subjected to forced air circulation without heating

For the most aggressive environment, in windy and heating conditions (Figure 5), the hardening process begins much earlier and lasts for a shorter period. The naphthalene and lignosulphonate products remain the ones with a retarding effect, when compared with the reference mortar.

These results show a significant influence of chemical admixtures and environmental conditions on the onset of the hardening of mortars, which is a relevant parameter for pavement construction.

3.3 Bleeding without Compaction

The bleeding without compaction was measured on one cube for sealed conditions and on two cubes for aggressive environments (cubes 7 and 10). Figure 6a) shows the bleeding without compaction obtained on the six mixtures for sealed conditions, expressed by grams of collected liquid (cumulate) per kilogram of mortar specimen. As can be seen, the mortar with the naphthalene product has the highest bleeding value, followed by the mortar with the lignosulphonate product. This is in agreement with the results obtained in the penetration resistance test, which show that these two products have a significant retarding effect. This effect gives more time for the liquid rise to the surface. The results obtained on other mortars with chemical admixtures show that the retarding effect provided is insufficient to compensate for the lower W/C, when compared to the reference mixture. Another significant advantage of the reference mortar is the early supply of water, which is due to the higher water content.

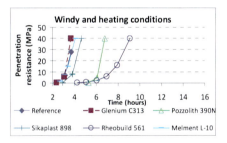

Fig. 5. Penetration resistance of cube 11 of mortars (specimens subjected to forced air circulation with heating)

In aggressive environments, the bleeding values are obtained by the sum of the liquid collected on the surface, using a syringe, with the weight difference of the specimen due to evaporation. Figure 6b) shows an example of the bleeding without compaction obtained on cubes 7 and 10. As can be seen, the highest bleeding is measured on cube 10, the nearest to the fan. A similar relative behaviour was observed on the other 5 mortars. So, for easier interpretation, Figure 7a) only shows the results obtained on cube 10. This figure indicates that the occurrence of wind is a crucial factor in the bleeding behaviour, since the scatter of the curves is very small. Compared to sealed conditions, the bleeding is higher for windy conditions but the differences between mortars are almost insignificant.

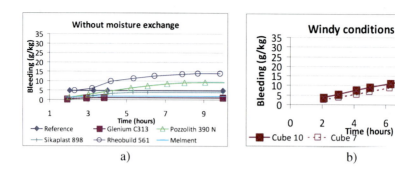

Fig. 6. Bleeding without compaction: a) specimens covered with a plastic sheet; b) Glenium C313 (specimens subjected to air circulation without heating)

Figure 7b) shows the results obtained in the most aggressive condition. With the increase in temperature the bleeding also grows. This growth can not be solely attributed to an increase in temperature but also to the deep relative humidity decrease. Comparing Figure 6a) with Figure 7b), is possible to observe a similar pattern of relative behaviours, being the highest bleeding obtained on the naphthalene mortar and the lowest on the mortar with policarboxylic ether.

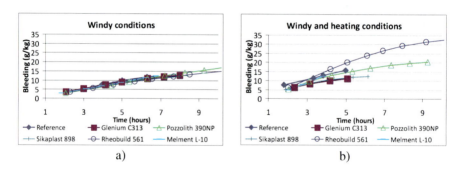

Fig. 7. Bleeding without compaction: a) specimens subjected to air circulation without heating; b) specimens subjected to air circulation with heating

3.4 Bleeding with Compaction

Figure 8a) shows the bleeding with compaction. The results presented are the values obtained in each measurement and not the cumulative bleeding, as shown in the previous figures. In this case, the evaporation occurring between mixing and measurements is not taken in consideration. Reference must be made to the fact that, for aggressive conditioning the cubes were not subjected to the same conditions, since the aggressiveness decreases from the entrance to the exit positions. However, the gradient in temperature and humidity is limited and all the mortars were tested in the same conditions. So, the comparative evaluation would not be affected by the distinct values in temperature and humidity at the different points in the tunnel.

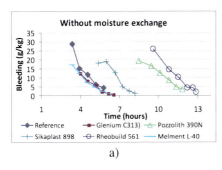 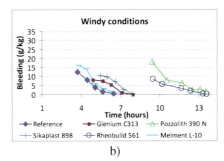

Fig. 8. Bleeding with compaction: a) specimens covered with a plastic sheet; b) specimens subjected to forced air circulation without heating

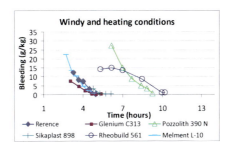

Fig. 9. Bleeding with compaction
(specimens subjected to forced air circulation with heating)

The results presented in Figure 8a) show that, with compaction, the water is available for longer periods. The energy provided by vibration also leads to an increase in the water rising to the surface. Comparing Figure 7a) with Figure 6a), the relative behaviour of different mortars is similar. Mortars with the naphthalene and lignosulphonate products have water available for a longer period of time. The curves of reference mortar and mortars with melamine (Melment L-10) and policarboxylic ether (Glenium C313) almost overlap, which did not occur without vibration, Figure 6a). This may be due to the lower W/C of mortars with admixtures, and consequently to a higher solid content, which hinders the rise of the water. Figure 8b) presents the results for windy conditions. The availability of water decreases, due to evaporation, but the relative behaviour of mortars is not significantly affected. Figure 9, reveals a similar pattern, for windy and heating conditions, but at earlier ages, due to the increase in hydration rate by heating. Nevertheless, the bleeding values are higher than the ones obtained with only windy conditions, due to a briefer evaporation time.

4 Final Remarks

This study evaluates the early hydration and bleeding behaviour of concrete, which are relevant factors for finishing with surface hardeners. The influence of chemical admixtures and environmental conditions were tested by penetration resistance and bleed water. The results show a significant influence of the type of chemical admixture on the initial hydration, being the delay, in the initial setting time, of 0.5-8 hours, when using recommended common dosages of the products. For heating conditions, the setting times decreases substantially, in particular by shortening the period of time between the initial and final set.

The differences in initial hydration influence the workable period and water availability, which are important factors for surface finishing. For aggressive environments, mainly windy conditions, longer setting times lead to greater water evaporation volumes, leaving less water available for surface hardeners hydration, hence increasing the proneness to surface defects. With heating, there is a shortening in the setting time, therefore decreasing the workable period, but also decreasing the evaporation time, with advantages in water availability.

Acknowledgements. This work has been funded by the Portuguese Foundation for Science and Technology (PTDC/ECM/105075/2008).

References

[1] Ribeiro, A.B., Monteiro, A.: Análise das causas do destacamento do endurecedor de superfície aplicado no armazém industrial Keramic, LNEC, Mat. Dep., Concrete Division (2009)
[2] Monteiro, A., Gonçalves, A.: Delamination of troweled floors – Shopping center Dolce Vita Tejo, LNEC, Mat. Dep. Concrete Division (2009)
[3] ACI 302.1R – 96, Guide for concrete floor and slab construction, ACI Committee 302 (1996)
[4] ASTM C1437 – 07, Standard Test Method for Flow of Hydraulic Cement Mortar, ASTM Volume 04.01 Cement, Lime, Gypsum (2007)
[5] CEN EN 196-1, Methods of testing cement - Part 1: Determination of strength, European Committee for Standardization (2005)
[6] ASTM C403/C403M – 08, Standard Test Method for Time of Setting of Concrete Mixtures by Penetration Resistance, ASTM Volume 04.02 Concrete and Aggregates (2008)
[7] ASTM C1170/C1170M – 08, Standard Test Method for Determining Consistency and Density of Roller-Compacted Concrete Using a Vibrating Table, ASTM Volume 04.02 Concrete and Aggregates (2008)
[8] Concrete Society, Standard Non-structural cracks in concrete, Technical report, 22 (1992)

Application of Different Fibers to Reduce Plastic Shrinkage Cracking of Concrete

Tara Rahmani[1], Behnam Kiani[1], Mehdi Bakhshi[2], and Mohammad Shekarchizadeh[1]

[1] School of Civil Engineering, College of Engineering, University of Tehran, Tehran, Iran
[2] Department of Civil, Environmental, and Sustainable Engineering,
   School of Sustainable Engineering and the Built Environment,
   Arizona State University, Tempe, USA

Abstract. Cracking generated by shrinkage is a major concern, particularly in structures with a high surface area to volume ratio. It has been well established that the inclusion of fibers in concrete contribute to the shrinkage crack reduction. In the present study, the efficiency of different fibers in arresting the cracks in cementitious composites due to restrained plastic shrinkage was investigated. This paper focused on the effects of using steel, glass and polypropylene fibers at volume fraction of 0.1%. Crack characteristics including the maximum crack width, average crack width, and total crack area were measureed using the image analysis. The test results indicate that steel fibers were more effective in reducing restrained plastic shrinkage cracking compared to others whereas glass fibers had better performance than polypropylene fibers.

1 Introduction

Plastic shrinkage cracks may often appear during the first few hours after casting while concrete is still in a plastic state and has not attained any significant strength. When the rate of water loss due to evaporation exceeds the rate of bleeding, negative capillary pressures generate which pull the solid particles together and consequently result in shrinkage [1-3]. At this time, the restraints provided by a rough substrate or steel reinforcement cause tensile stresses within the concrete and may lead to cracking [4, 5]. Plastic shrinkage cracks facilitate the ingress of aggressive agents into the concrete and affect the long term durability, serviceability, and aesthetic aspects of the structure [6, 7]. These cracks are more commonly observed in thin concrete elements with a high surface area to volume ratio like slabs on grade, industrial floors, bridge decks, tunnel linings, etc [8, 9].

Precautionary measures minimizing the loss of water from the concrete surface including prolonged curing, erecting wind breaks and sunshades, fogging, and reduced use of admixtures that prevent bleeding can help reduce the plastic shrinkage cracking [10]. Fiber reinforcement has proven to be one of the most effective methods to reduce plastic shrinkage cracking [5, 8]. Fibers inhibit further crack propagation by providing the bridging forces across the cracks [4].

In this paper, the behavior of steel, polypropylene, and glass fibers at low volume fractions in controlling plastic shrinkage cracks were evaluated by percent reduction of crack width (PRCW) and percent reduction of crack area (PRCA) relative to plain concrete. To achieve this purpose, an image analysis technique was used to estimate the crack measurements.

2 Mechanism of Plastic Shrinkage

Capillary pressure is the source of shrinkage in the plastic stage of cementitious materials and may eventually lead to cracking. Figure 1 shows the process of capillary pressure build up and plastic shrinkage. After casting, bleeding may occur, i.e. the solid particles settle due to gravitational forces and on the surface a plane water film is formed (Figure 1A). Evaporation at the upper surface continuously reduces the thickness of the water film and, eventually, the near surface particles are no longer covered by a plane water surface. As a result of adhesive forces and surface tension, water menisci are formed between the solid particles. The curvature of the water surface causes a negative pressure in the capillary water. This pressure acts on the solid particles resulting in the contraction of the still plastic material (Figure 1B). The ongoing evaporation at the surface causes a continuing reduction of the main radii of the menisci resulting in an increase of the absolute capillary pressure value as well as of the shrinkage strain (Figure 1C). If a certain pressure is reached, the largest gaps between the particles at the surface can no longer be bridged by the menisci and air penetrates locally into the pore system accompanied by a local pressure break down (Figure 1D). When air starts to penetrate the pores, the plastic cracking risk is assumed to reach its maximum because the drained pores are weak points in the system. The contracting forces between the particles in the air penetrated regions are considerably smaller than those in the water filled regions. Hence, a localization of strains is taking place leading to visible cracks [1, 11].

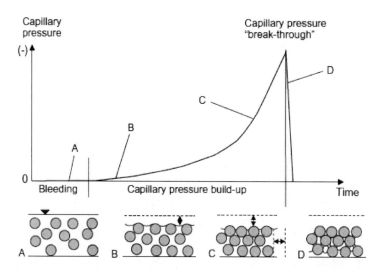

Fig. 1. Capillary pressure build up during plastic shrinkage

It has to be pointed out that air entry does not necessarily result in cracking. Cohesive stresses between the particles resulting from hindered shrinkage strain are also required. The combination of the inherent shrinkage and restraint develops tensile stresses. If the restrained tensile stresses exceed the tensile strength of concrete, the cracks can develop throughout the surface of concrete [12]. However, cracking is impossible without air entry into the drying suspension [1, 11].

3 Experimental Program

3.1 Material

ASTM type II Portland cement was used in all concrete mixtures. The coarse aggregate was 19-mm maximum size crushed natural stone with a bulk specific gravity of 2.66 while natural river sand with a specific gravity of 2.62 and fineness modulus of 3.31 was used as the fine aggregate. The water to cement ratio by weight was 0.47. The mix proportions are given in Table 1. Three types of fibers namely hooked steel, glass, and polypropylene fibers were used at 0.1 volume fraction, and their properties are presented in Table 2. It should be noted that the typical dosage of fibers to reduce plastic shrinkage cracking is smaller than 0.3% by volume [8].

The average compressive strengths of the plain and fiber reinforced concretes at 28 days and the workability of mixtures are also given in Table 3.

Table 1. Mix proportions used for mixtures

Cement (kg/m^3)	Fine aggregate (kg/m^3)	Coarse aggregate (kg/m^3)		Water (kg/m^3)
		12.5 mm	19 mm	
350	957	174	609	164.5

Table 2. Properties of the fibers

Fiber	Length (mm)	Diameter (mm)	Aspect ratio (l/d)	Specific gravity (gr/cm^3)	Tensile strength (MPa)
Steel	35	0.55	64	8	1100
Glass	15	0.012	1250	2.74	2450
Polypropylene	12	0.022	545	0.91	300-400

Table 3. Fiber dosage and compressive strength results of concrete mixtures

Mixture	Type of fiber	Fiber dosage (kg/m^3)	Slump (cm)	Compressive strength (MPa)
Plain	-	-	14-16	38.4
SFRC	Steel	7.8	7-9	39.2
GFRC	Glass	2.6	5-7	35.8
PFRC	Polypropylene	0.91	5-7	35.9

3.2 Mixing and Sample Preparation

The mixing procedure used for concrete mixtures was as follows. The coarse aggregate, sand, and cement were first mixed dry for a period of 1 min. Then, the water was added, and mixing was continued for another 2 min. This was followed by 2 min of rest. With fiber reinforced concrete mixtures, the fibers were dispersed by hand and mixed for another 2 min to ensure the dispersion of fibers throughout the concrete.

The plastic shrinkage test method used in this study was similar to that prposed in ASTM C 1579-06 [13]. The plywood mold with a depth of 85 mm and rectangular dimensions of 360 by 560 mm has been used. The mold was provided with a stress riser of 63.5 mm height at the center and two base restraints of 32 mm height at 90 mm from both ends, along the transverse direction, as shown in Figure 2. After casting, the concrete specimens were kept for 24 h in a 2 meter by 1 meter and 1.5 meter height room. Inasmuch as environmental conditions play a significant role on early age properties of fiber reinforced concrete, the required temperature and humidity were generated by electric heaters and temperature controller. Fans were used to achieve a wind speed of more than 4.7 m/s over the entire surface area of the specimens (see Figure 3). The fresh concrete specimens were exposed to an average temperature of 36, a relative humidity of 25%, and a wind velocity of 6 m/s. The average evaporation rate, measured by detecting the loss of water from a plastic bowl placed, was 2 $kg/m^2/h$. The specimens were monitored visually for any signs of cracking at approximately 30 min intervals.

According to ASTM C 1579-06, a test unit is comprised of at least two plain specimens and two fiber reinforced specimens. So for each type of fiber reinforced concrete, plastic shrinkage test was carried two times, and in each test, a plain specimen tested as well. To summarize, two series of data for each type of fiber and six series for plain concrete were detected and the average of measurements were reported.

Fig. 2. Plywood mold with the stress risers

Fig. 3. Environmental chamber for keeping specimens

4 Results and Discussion

From the cracking data, several values including maximum crack width, average crack width, total crack area, and time of first crack appearance were acquired and presented in Table 4. For each specimen, the cracks were represented by 6 digital images. These images were analyzed by the algorithms developed in the MATLAB programming environment. Then, the analysis results were superposed to obtain the crack width (average and maximum) and total crack area. The width of cracks was measured every 10 pixels along the crack. It is assumed that the lens was parallel to the surface of the concrete, and that it produced no optical errors.

Table 4. Plastic shrinkage test results

Mixture	Time of first visible crack (min)	Average crack width (mm)	Maximum crack width (mm)	Total crack area (mm^2)	Percent reduction of crack width (%)	Percent reduction of crack area (%)
Plain	90	0.763	2.623	253.90	-	-
SFRC	125	0.359	0.997	120.34	52	53
GFRC	120	0.379	1.123	109.24	43	59
PFRC	110	0.517	1.845	144.86	30	43

4.1 Time of First Crack Appearance

The age of first visible crack, which could be identified by eyes, was detected for each specimen. The specimens were checked for cracks every 30 min. For plain concrete, 90 min after casting a fine crack was observed whereas in the case of fiber reinforced concretes, this time was detected more than 110 min. This is because the fibers in concrete act as bleeding channels, which supply water to replenish the drying surface and reduce the magnitude of capillary stresses developed [14]. After the first visible crack, for plain specimens the cracks will widen while in the case of fiber reinforced concrete, owing to the fiber bridging effect, the widening of the cracks are prevented. The microscopic images revealing the fiber bridging effect across the cracks are shown in Figure 5.

4.2 Crack Characteristics

Addition of steel, glass, and polypropylene fibers at the same volume fraction were effective in controlling the restrained plastic shrinkage cracking. The effect of fiber reinforcement on crack pattern is shown in Figure 4. As shown, plane concrete exhibited a denser crack pattern while in the case of fiber reinforced concretes, many subparallel cracks occured. The reduction of maximum crack width relative to plain concrete for different fiber reinforced concretes varied from 30% to 50%. The percent reduction of average crcak width for steel, glass, and polypropylene fibers were obtained 52%, 43%, and 30% respectively. In all cases, total crack area decreased with adding fibers in the range of 40% to 55%, and fibers of all types and lengths were extremely effective.

Steel fibers have the best performance among the other fibers, reducing the total crack area by 53% and maximum crack width by 48% to less than 0.1 mm. This may be due to the hooked end shape of steel fibers, which improves the bonding between fibers and matrix and results in higher tensile strain capacity. Glass fibers had better performance compared to polypropylene fibers. It can be attributed to the fact that polypropylene fibers are shorter than glass fibers thus they could have got pulled out of the matrix (see Figure 5).

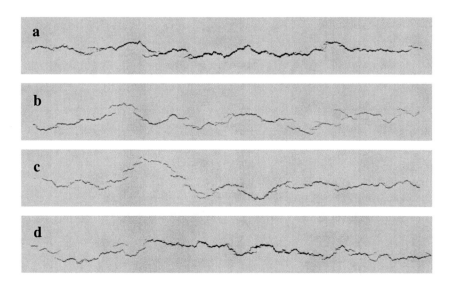

Fig. 4. Crack pattern obtained using image analysis. (a) Plain, (b) SFRC, (c) GFRC, and (d) PFRC

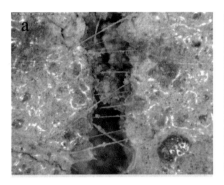 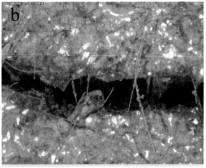

Fig. 5. Microscopic images of crack bridging effect induced by fibers. (a) glass fiber, and (b) polypropylene fiber

5 Conclusions

The addition of fibers significantly reduced the restrained plastic shrinkage cracking and delayed the first crack appearance. The reduction of maximum crack width and total crack area were ranged 30-50% and 40-60%, respectively. Among the various fibers investigated at a volume fraction of 0.1%, steel fiber was found to be most effective in reducing the crack width and area, because the steel fibers improves the tensile strain capacity of concrete and restricts the growth of cracks. Glass fibers performed significantly better than polypropylene fibers.

References

[1] Slowik, V., Schmidt, M., Fritzsch, R.: Cement and Concrete Research 30(7), 557–565 (2008)
[2] Turcry, P., Loukili, A.: ACI Materials Journal 103(4), 272–279 (2006)
[3] Brown, M.D., Sellers, G., Folliard, K., Fowler, D.: Restrained shrinkage cracking of concrete bridge decks: State-of-the-Art Review, Center for Transportation Research, University of Texas at Austin, FHWA/TX-0-4098-1 (2001)
[4] Kim, J.H.J., Park, C.G., Lee, S.W., Lee, S.W., Won, J.P.: Composites Part B: Engineering 39(3), 442–450 (2008)
[5] Banthia, N., Gupta, R.: Cement and Concrete Research 36(7), 1263–1267 (2006)
[6] Pelisser, F., Santos Neto, A.B.S., Rovere, H.L.L., Pinto, R.C.A.: Construction and Building Materials 24(11), 2171–2176 (2010)
[7] Naaman, A.E., Wongtanakitcharoen, T., Hauser, G.: ACI Materials Journal 102(1), 49–58 (2005)
[8] Toledo Filho, R.D., Ghavami, K., Sanjuán, M.A., England, G.L.: Cement and Concrete Composites 27(5), 537–546 (2005)
[9] Sivakumar, A., Santhanam, M.: Cement and Concrete Research 29(7), 575–581 (2007)
[10] Uno, P.J.: ACI Materials Journal 95(4), 365–375 (1998)

[11] Slowik, K., Hübner, T., Schmidt, M., Villmann, B.: Cement and Concrete Composites 31(7), 461–469 (2009)
[12] Wongtanakitcharoen, T.: Effect of randomly distributed fibers on plastic shrinkage cracking of cement composites, PhD Thesis, University of Michigan, Ann Arbor, USA (2005)
[13] ASTM C 1579-06, Standard test method for evaluating plastic shrinkage cracking of restrained fiber reinforced concrete (using a steel form insert), Annual Book of ASTM Standards, vol. 04(02), American Society for Testing and Materials, West Conshohocken, PA (2006)
[14] Zollo, R.F., Alter, J., Bouchacourt, B.: Developments in fiber reinforced cement and concrete. In: Proceedings RILEM Symposium, England (1986)

Printed by Publishers' Graphics LLC